Communications and Control Engineering

Series Editors
E.D. Sontag • M. Thoma • A. Isidori • J.H. van Schuppen

Published titles include:

Stability and Stabilization of Infinite Dimensional Systems with Applications
Zheng-Hua Luo, Bao-Zhu Guo and Omer Morgul

Nonsmooth Mechanics (Second edition)
Bernard Brogliato

Nonlinear Control Systems II
Alberto Isidori

L_2-Gain and Passivity Techniques in Nonlinear Control
Arjan van der Schaft

Control of Linear Systems with Regulation and Input Constraints
Ali Saberi, Anton A. Stoorvogel and Peddapullaiah Sannuti

Robust and H_∞ Control
Ben M. Chen

Computer Controlled Systems
Efim N. Rosenwasser and Bernhard P. Lampe

Control of Complex and Uncertain Systems
Stanislav V. Emelyanov and Sergey K. Korovin

Robust Control Design Using H_∞ Methods
Ian R. Petersen, Valery A. Ugrinovski and Andrey V. Savkin

Model Reduction for Control System Design
Goro Obinata and Brian D.O. Anderson

Control Theory for Linear Systems
Harry L. Trentelman, Anton Stoorvogel and Malo Hautus

Functional Adaptive Control
Simon G. Fabri and Visakan Kadirkamanathan

Positive 1D and 2D Systems
Tadeusz Kaczorek

Identification and Control Using Volterra Models
Francis J. Doyle III, Ronald K. Pearson and Babatunde A. Ogunnaike

Non-linear Control for Underactuated Mechanical Systems
Isabelle Fantoni and Rogelio Lozano

Robust Control (Second edition)
Jürgen Ackermann

Flow Control by Feedback
Ole Morten Aamo and Miroslav Krstić

Learning and Generalization (Second edition)
Mathukumalli Vidyasagar

Constrained Control and Estimation
Graham C. Goodwin, María M. Seron and José A. De Doná

Randomized Algorithms for Analysis and Control of Uncertain Systems
Roberto Tempo, Giuseppe Calafiore and Fabrizio Dabbene

Switched Linear Systems
Zhendong Sun and Shuzhi S. Ge

Subspace Methods for System Identification
Tohru Katayama

Digital Control Systems
Ioan D. Landau and Gianluca Zito

Multivariable Computer-controlled Systems
Efim N. Rosenwasser and Bernhard P. Lampe

Dissipative Systems Analysis and Control (Second edition)
Bernard Brogliato, Rogelio Lozano, Bernhard Maschke and Olav Egeland

Algebraic Methods for Nonlinear Control Systems
Giuseppe Conte, Claude H. Moog and Anna M. Perdon

Polynomial and Rational Matrices
Tadeusz Kaczorek

Simulation-based Algorithms for Markov Decision Processes
Hyeong Soo Chang, Michael C. Fu, Jiaqiao Hu and Steven I. Marcus

Iterative Learning Control
Hyo-Sung Ahn, Kevin L. Moore and YangQuan Chen

Distributed Consensus in Multi-vehicle Cooperative Control
Wei Ren and Randal W. Beard

Control of Singular Systems with Random Abrupt Changes
El-Kébir Boukas

Nonlinear and Adaptive Control with Applications
Alessandro Astolfi, Dimitrios Karagiannis and Romeo Ortega

Stabilization, Optimal and Robust Control
Aziz Belmiloudi

Control of Nonlinear Dynamical Systems
Felix L. Chernous'ko, Igor M. Ananievski and Sergey A. Reshmin

Periodic Systems
Sergio Bittanti and Patrizio Colaneri

Discontinuous Systems
Yury V. Orlov

Michael Malisoff • Frédéric Mazenc

Constructions of Strict Lyapunov Functions

Springer

Michael Malisoff, PhD
Department of Mathematics
Louisiana State University
Baton Rouge
LA 70803-4918
USA
malisoff@lsu.edu

Frédéric Mazenc, PhD
INRIA Sophia-Antipolis
UMR Analyse des Systèmes et Biométrie
2 place Pierre Viala
34060 Montpellier Cedex 1
France
frederic.mazenc@supagro.inra.fr

ISSN 0178-5354
ISBN 978-1-4471-5782-3 ISBN 978-1-84882-535-2 (eBook)
DOI 10.1007/978-1-84882-535-2
Springer Dordrecht Heidelberg London New York

British Library Cataloguing in Publication Data
A catalogue record for this book is available from the British Library

Softcover re-print of the Hardcover 1st edition 2009

Cover design: le-tex publishing services GmbH, Leipzig, Germany

Printed on acid-free paper

Springer is part of Springer Science+Business Media (www.springer.com)

To Laurent Praly and Eduardo Sontag, with thanks

Preface

During the past fifteen years, there have been many exciting developments at the interface of mathematical control theory and control engineering. Many of these developments were based on Lyapunov methods for analyzing and controlling nonlinear systems. Constructing strict Lyapunov functions is a challenging, central problem. By contrast, *non*-strict Lyapunov functions are often readily constructed from passivity, backstepping, or forwarding (especially in the time-varying context), or by using the Hamiltonian in Euler-Lagrange systems. Roughly speaking, strict Lyapunov functions are characterized by having negative *definite* time derivatives along all trajectories of the system, while non-strict Lyapunov functions have negative *semi-definite* derivatives along the trajectories. Even when a system is known to be globally asymptotically stable, one often still needs an explicit strict Lyapunov function, e.g., to build feedbacks that provide input-to-state stability to actuator errors.

One important research direction involves finding necessary and sufficient conditions for various kinds of stability, in terms of the existence of Lyapunov functions, such as Lyapunov characterizations for hybrid systems, or for systems with outputs and measurement uncertainty. Converse Lyapunov function theory guarantees the existence of strict Lyapunov functions for many globally asymptotically stable nonlinear systems. However, the Lyapunov functions provided by converse theory are often abstract and nonexplicit, because they involve suprema or infima over infinite sets of trajectories, so they may not always lend themselves to feedback design. Explicit strict Lyapunov functions are also useful for quantifying the effects of uncertainty, since for example they can be used to construct the comparison functions in the input-to-state stability estimate, or to guarantee that a model reduction based on singular perturbation analysis can be done. In fact, once an appropriate global strict Lyapunov function has been constructed, several important robustness and stabilization problems can be solved almost immediately, through standard arguments.

In some cases, non-strict Lyapunov functions are sufficient, because they can be used in conjunction with Barbalat's Lemma or the LaSalle Invariance Principle to prove global stability. In other situations, it is enough to analyze the system near an equilibrium point, or around a reference trajectory, so linearizations and simple local quadratic Lyapunov functions suffice. However, it has become clear in the past two decades that non-strict Lyapunov functions and linearizations are insufficient to analyze general nonlinear time-varying systems. Non-strict Lyapunov functions are not well suited to robustness analysis, since their negative semi-definite derivatives along trajectories could become positive under arbitrarily small perturbations of the dynamics. Moreover, there are important nonlinear systems (e.g., chemostat models) that naturally evolve far from their equilibria. This has motivated a great deal of significant research on methods to explicitly construct global strict Lyapunov functions.

One approach to building explicit strict Lyapunov functions, which has received a considerable amount of attention in recent years, is the so-called *strictification method*. This involves transforming given non-strict Lyapunov functions into strict Lyapunov functions. Strictification reduces strict Lyapunov function construction problems to oftentimes much easier non-strict Lyapunov function construction problems. This book brings together a broad but unifying repertoire of strictification based methods. Much of this work appears here for the first time. We cover many important classes of nonlinear dynamics, including Jurdjevic-Quinn systems, time-varying systems satisfying LaSalle or Matrosov Conditions, adaptively controlled dynamics, slowly and rapidly time-varying systems, and hybrid time-varying systems. In fact, under a very mild extra assumption, we show how strict Lyapunov functions can be constructed for systems satisfying the conditions of the LaSalle Invariance Principle. The simplicity of our constructions makes them suitable for quantifying the effects of uncertainty, and for feedback design, including cases where only an output is available for measurement. We illustrate this in several applications that are of compelling engineering interest.

This work complements several books on nonlinear control theory, such as [149] by Sepulchre, Janković, and Kokotović. While many texts include Lyapunov function constructions, our work provides a systematic, design-oriented approach to building global strict Lyapunov functions, including simplified constructions that are more amenable to feedback design and robustness analysis. In fact, many of the systems covered by our approaches are beyond the scope of the well-known explicit strict Lyapunov constructions. Our book will be easily understood by readers who are familiar with the nonlinear control theory in the textbooks of Khalil [70] and Sontag [161]. We review much of the prerequisite material in the first two chapters. The remaining chapters can be used as supplemental reading in a first graduate control systems theory, or for a second course on Lyapunov based methods. Engineers and applied mathematicians interested in nonlinear control will also find our book useful.

Much of this book is based on the authors' research, which was supported by the NSF Division of Mathematical Sciences through Mathematical Sciences Priority Area Grants 0424011 and 0708084. Some of the work was done while the second author visited Louisiana State University. His visits were partly funded by the Louisiana Board of Regents Support Fund. The authors appreciate the support, and the helpful comments, corrections, and suggestions they received from Rick Barnard, Aleksandra Gruszka, Zongli Lin, Eduardo Sontag, and others. The authors also appreciate the good research environments at INRIA and LSU, and the editorial assistance they received from le-tex in Leipzig and from Springer-London (especially the quick answers from Aislinn Bunning, Oliver Jackson, Nadja Kroke, and Sorina Moosdorf). Finally, they thank Jeff Sheldon and Nik Svoboda from the LSU Department of Mathematics for helping with several computer related issues.

Baton Rouge, LA — *Michael Malisoff* (LSU)
Montpellier, France — *Frédéric Mazenc* (INRIA)
May 11, 2009

Contents

Part I
Background

Chapter 1
Background on Nonlinear Systems

Abstract We review some basic concepts from the theory of ordinary differential equations and nonlinear control systems, as well as several notions of stability, including the input-to-state stability paradigm. An important feature is the distinction between uniform and non-uniform stability for time-varying systems. We also include an overview of the problem of stabilization of nonlinear systems, including the "virtual" obstacles to stabilization imposed by Brockett's Necessary Condition. Brockett's Criterion motivates our use of time-varying feedbacks to stabilize both autonomous and time-varying systems. We illustrate these notions in several examples. In later chapters, we revisit these notions using strict Lyapunov functions.

1.1 Preliminaries

Throughout this book, we use the following standard notation and classical results. We let $\mathbb{N}$ denote the set of natural numbers $\{1, 2, \ldots\}$, $\mathbb{Z}$ the set of all integers, and $\mathbb{Z}_{\geq 0} = \mathbb{N} \cup \{0\}$. Also, $\mathbb{R}$ (resp., $\mathbb{R}^n$) denotes the set of all real numbers (resp., real n-tuples for any $n \in \mathbb{N}$). We use the following norms for vectors $x = (x_1, ..., x_n) \in \mathbb{R}^n$:

$$|x|_\infty = \max_{1 \leq i \leq n} |x_i|, \quad |x|_1 = \sum_{i=1}^{n} |x_i|, \quad \text{and} \quad |x|_2 = \left(\sum_{i=1}^{n} |x_i|^2 \right)^{1/2} .$$

Unless we indicate otherwise, the norm on $\mathbb{R}^n$ is $|\cdot|_2$ which we often denote by $|\cdot|$. For a measurable essentially bounded[1] function $u : \mathcal{I} \to \mathbb{R}^p$ on an interval $\mathcal{I} \subseteq \mathbb{R}$, we let $|u|_\mathcal{I}$ denote its essential supremum, which we

[1] Readers who are not familiar with Lebesgue measure theory can replace "measurable essentially bounded" with "bounded and piecewise continuous" throughout our work, in which case the essential supremum is just the sup norm.

indicate by $|u|_\infty$ when $\mathcal{I} = \mathbb{R}$. For real matrices A, we use the matrix norm $||A|| = \sup\{|Ax| : |x| = 1\}$, and $I_n \in \mathbb{R}^{n\times n}$ is the identity matrix. Given an interval $\mathcal{I} \subseteq \mathbb{R}$ and a function $x : \mathcal{I} \to \mathbb{R}^n$ that is differentiable (Lebesgue) almost everywhere, we use $\dot{x}$ or $\dot{x}(t)$ to denote its derivative $\frac{dx}{dt}(t)$.

For each $k \in \mathbb{Z}_{\geq 0} \cup \{\infty\}$, a real-valued function defined on an open subset of Euclidean space is called C^k provided its partial derivatives exist and are continuous up to order k. A C^0 function is one that is continuous, and a C^∞ function is one that is a smooth function, that is, it has continuous partial derivatives of any finite order. We use the same C^k notation for vector fields on $\mathbb{R}^n$. We present all of our results under those differentiability assumptions that lead to the shortest and clearest proofs. Throughout the book, increasing means strictly increasing and similarly for decreasing.

If $f : \mathbb{R}^n \to \mathbb{R}^n$ is a smooth vector field and $h : \mathbb{R}^n \to \mathbb{R}$ is smooth, the *Lie derivatives* of h in the direction of f are defined recursively by

$$L_f h(x) \doteq \frac{\partial h}{\partial x}(x) f(x) \quad \text{and} \quad L_f^k h(x) = L_f(L_f^{k-1} h)(x) \quad \forall k \geq 2.$$

Recall the following classes of comparison functions. We say that a C^0 function $\gamma : [0,\infty) \to [0,\infty)$ belongs to class $\mathcal{K}$ and write $\gamma \in \mathcal{K}$ provided it is increasing and $\gamma(0) = 0$. We say that it belongs to class $\mathcal{K}_\infty$ if, in addition, $\gamma(r) \to \infty$ as $r \to \infty$. We say that a C^0 function $\beta : [0,\infty) \times [0,\infty) \to [0,\infty)$ is of class $\mathcal{KL}$ provided for each fixed $s \geq 0$, the function $\beta(\cdot, s)$ belongs to class $\mathcal{K}$, and for each fixed $r \geq 0$, the function $\beta(r, \cdot)$ is non-increasing and $\beta(r,s) \to 0$ as $s \to \infty$. The following lemma is well-known:

Lemma 1.1. *(Barbalat's Lemma) If $\phi : \mathbb{R} \to \mathbb{R}$ is uniformly continuous on $[0,\infty)$ and*

$$\lim_{t\to\infty} \int_0^t \phi(m)\, dm$$

exists and is finite, then $\lim_{t\to\infty} \phi(t) = 0$.

We also use Young's Inequality, which says that

$$ab \leq \frac{1}{p}|a|^p + \frac{p-1}{p}|b|^q$$

for all $a \in \mathbb{R}$ and $b \in \mathbb{R}$, and all $p > 1$ and $q > 1$ satisfying $\frac{1}{p} + \frac{1}{q} = 1$.

1.2 Families of Nonlinear Systems

The basic families of dynamics are autonomous systems, nonautonomous systems, and systems with inputs. We review these basic families next for the case where the dynamics are given by families of ordinary differential equations. We then discuss their analogs in discrete time. In later chapters, we consider more general systems with multiple time scales, such as hybrid time-varying systems. In general, we allow nonlinear systems, meaning the dynamics are nonlinear in the state variable.

1.2.1 Nonautonomous Systems

A general nonautonomous ordinary differential equation consists of a finite number of first-order one-dimensional differential equations:

$$\begin{cases} \dot{x}_1 = f_1(t, x_1, x_2, ..., x_n) \\ \dot{x}_2 = f_2(t, x_1, x_2, ..., x_n) \\ \quad \vdots \\ \dot{x}_n = f_n(t, x_1, x_2, ..., x_n) \end{cases} \tag{1.1}$$

where t is the time, and each function f_i is in general nonlinear in all of its arguments. The variables x_i are called *states*, and n is called the *dimension* of the system. The differential equations characterize the evolution of the states with respect to time. Frequently, we write (1.1) more compactly as

$$\dot{x} = f(t, x). \tag{1.2}$$

The state vector $x = (x_1, x_2, ..., x_n)$ is valued in a given open set $\mathcal{X} \subseteq \mathbb{R}^n$. Given a constant $t_0 \geq 0$, $x_0 \in \mathcal{X}$, and a constant $t_{\max} > t_0$, the corresponding initial value problem IVP$(t_{\max}, t_0, x_0)$ for (1.2) is that of determining an absolutely continuous function $y : [t_0, t_{\max}) \to \mathcal{X}$ such that $\dot{y}(t) = f(t, y(t))$ for almost all $t \in [t_0, t_{\max})$ and $y(t_0) = x_0$. We assume that the vector field $f : [0, \infty) \times \mathcal{X} \to \mathbb{R}^n$ is measurable in t and of class C^1 in x, meaning the function $x \mapsto f(t, x)$ is C^1 for each $t \geq 0$. We further assume that for each compact set $K \subseteq \mathcal{X}$, there is a locally integrable function α_K so that

$$\left| \frac{\partial f}{\partial x}(t, x) \right| \leq \alpha_K(t) \text{ for all } x \in K \text{ and } t \geq 0.$$

By classical results (reviewed, e.g., in [161]), these properties ensure that for each $t_0 \geq 0$ and $x_0 \in \mathcal{X}$, there exists a $t_{\max} > t_0$ so that IVP$(t_{\max}, t_0, x_0)$ has a solution $t \mapsto x(t, t_0, x_0)$ with the following uniqueness and maximality property: If $\tilde{t} > t_0$ and IVP$(\tilde{t}, t_0, x_0)$ admits a solution $z(t)$, then $\tilde{t} \leq t_{\max}$ and $z(t) = x(t, t_0, x_0)$ for all $t \in [t_0, \tilde{t})$. If $x(t, t_0, x_0)$ can be uniquely defined for all $t \geq t_0$ for all initial conditions $x(t_0, t_0, x_0) = x_0$, then we call (1.1) *forward complete*. Since f depends on time, the systems (1.2) are also called *time-varying systems*.

An equilibrium point $x^* = (x_1^*, ..., x_n^*)$ of (1.2) is defined to be a vector in $\mathbb{R}^n$ for which $f(t, x^*) = 0$ for all $t \geq 0$. Frequently, the equilibrium point is the origin $x^* = 0$. If a system $\dot{X} = g(t, X)$ admits a solution $X_s(t)$, then, through the time-varying change of variable $x = X - X_s(t)$, we can transform the system $\dot{X} = g(t, X)$ into a new time-varying x dynamics

$$\dot{x} = f(t, x), \quad \text{where} \quad f(t, x) = g(t, x + X_s(t)) - \dot{X}_s(t)$$

which admits $x^* = 0$ as an equilibrium point. This transformation is used to analyze the asymptotic behavior of a system with respect to a specific solution $X_s(t)$, i.e., tracking. Frequently, the time-varying systems in engineering applications are *periodic with respect to* t, meaning there is a constant $w > 0$ (called a *period*) such that f satisfies

$$f(t + w, x) = f(t, x)$$

for all (t, x) in its domain.

1.2.2 Autonomous Systems

If the right side of (1.1) or (1.2) is independent of the time variable t, then the systems are called *autonomous* or *time-invariant* systems. Naturally, they are written in compact form as

$$\dot{x} = f(x) \tag{1.3}$$

and their flow maps are denoted by $x(t, x_0)$. In this case, we view $t \mapsto x(t, x_0)$ as being defined on some maximal interval $\mathcal{I} \subseteq \mathbb{R}$, possibly depending on the initial state x_0. If $t \mapsto x(t, x_0)$ is uniquely defined on $\mathbb{R}$ for all $x_0 \in \mathcal{X}$, then we call (1.3) *complete*. The family of systems (1.3) is the simplest we consider in this book. However, the behavior of the solutions of (1.3) is a very general subject and by no means simple. No general prediction of the asymptotic behavior of the solutions exists as soon as the dimension n of the system is larger than 2. Rather, such a classification exists only for systems of dimension 1 and 2, by the celebrated Poincaré-Bendixson Theorem [23, 153].

1.2.3 Systems with Inputs

The general *time-varying continuous time control system* is

$$\dot{x} = f(t, x, u) \tag{1.4}$$

or, equivalently,

$$\begin{cases} \dot{x}_1 = f_1(t, x_1, x_2, ..., x_n, u_1, ...u_p) \\ \dot{x}_2 = f_2(t, x_1, x_2, ..., x_n, u_1, ...u_p) \\ \quad \vdots \\ \dot{x}_n = f_n(t, x_1, x_2, ..., x_n, u_1, ...u_p). \end{cases} \tag{1.5}$$

The variables $u_1, ..., u_p$ are called *inputs.* The state and input vectors are valued in a given open set $\mathcal{X} \subseteq \mathbb{R}^n$ and a given set $U \subseteq \mathbb{R}^p$, respectively. When discussing systems with inputs, we assume that $[0, \infty) \times \mathcal{X} \times U \ni (t, x, u) \mapsto f(t, x, u) \in \mathbb{R}^n$ is piecewise continuous in t and of class C^1 in (x, u). We refer to the preceding conditions as our usual (or standing) assumptions on (1.5). We also let $\mathcal{M}(U)$ denote the set of all measurable essentially bounded functions $u : [0, \infty) \to U$; i.e., inputs that are bounded in $|\cdot|_\infty$. Solutions of (1.5) are obtained by replacing $(u_1, u_2, \ldots, u_p)$ with an element $u \in \mathcal{M}(U)$. For all $u \in \mathcal{M}(U)$, $x_0 \in \mathcal{X}$, and $t_0 \geq 0$, we let $x(t, t_0, x_0, u)$ denote the solution of (1.4) with u as input that satisfies $x(t_0, t_0, x_0, u) = x_0$, defined on its maximal interval $[t_0, b)$. If $t \mapsto x(t, t_0, x_0, u)$ is uniquely defined on $[t_0, \infty)$ for all $t_0 \geq 0$, $x_0 \in \mathcal{X}$, and $u \in \mathcal{M}(U)$, then we call (1.4) *forward complete.* By an equilibrium state of (1.4), we mean a vector $x^* \in \mathbb{R}^n$ that admits a vector $u^* \in U$ such that $f(t, x^*, u^*) = 0$ for all $t \geq 0$. If the system (1.5) can be written in the form

$$\dot{x} = \mathcal{F}(t, x) + \mathcal{G}(t, x)u$$

for some vector fields $\mathcal{F}$ and $\mathcal{G}$, then we say that (1.5) is *affine in controls* or *control affine.*

Inputs are essential in nonlinear control theory. One of the principal aims of control theory is to provide functions $u(t, x)$ such that all or some of the solutions of the system $\dot{x} = f(t, x, u(t, x))$ possess a desired property. In this situation, we refer to $u(t, x)$ as a *controller* or a *feedback,* and the feedback controlled system $\dot{x} = f(t, x, u(t, x))$ as a *closed-loop system.* Inputs can also represent disturbances, which are uncertainties that may modify the behavior of the solutions (often in an undesirable way). Then, the problem of quantifying the effect of disturbances $u(t)$ on the solutions of (1.4) arises.

1.2.4 Discrete Time Dynamics

The general family of time-varying discrete time systems with inputs admits the representation

$$x_{k+1} = f(k, x_k, u_k) \tag{1.6}$$

or, equivalently,

$$\begin{cases} x_{k+1,1} = f_1(k, x_{k,1}, x_{k,2}, \ldots, x_{k,n}, u_{k,1}, \ldots u_{k,p}) \\ x_{k+1,2} = f_2(k, x_{k,1}, x_{k,2}, \ldots, x_{k,n}, u_{k,1}, \ldots u_{k,p}) \\ \quad \vdots \\ x_{k+1,n} = f_n(k, x_{k,1}, x_{k,2}, \ldots, x_{k,n}, u_{k,1}, \ldots u_{k,p}). \end{cases} \tag{1.7}$$

The variables $u = (u_1, ..., u_p)$ are again called *inputs*, which are now *sequences* $u_k = (u_{k,1}, ...u_{k,p})$ that take their values in some subset $U \subseteq \mathbb{R}^p$ for each time $k \in \{0, 1, 2, \ldots\}$. We use k for the time indices to emphasize that they are discrete instants rather than being on a continuum. We let $\mathcal{D}(U)$ denote the set of all such input sequences. The state vector $x_k = (x_{k,1}, x_{k,2}, \ldots, x_{k,n})$ at each instant k is assumed to be valued in a given open set $\mathcal{X} \subseteq \mathbb{R}^n$.

Since the solutions of (1.7) are given recursively, there is no need to impose the regularity on f that we assumed in the continuous time case. However, when discussing discrete time systems, we always assume that the recursion defining the solutions is *forward complete*, meaning that solutions of (1.7) exist for all integers $k \geq 0$, all initial conditions $x(k_0) = x_0 \in \mathcal{X}$, and all $u \in \mathcal{D}(U)$. As in the continuous time case, $x(k, k_0, x_0, u)$ then denotes the unique solution of (1.4) that satisfies $x(k_0, k_0, x_0, u) = x_0$ for all $u \in \mathcal{D}(U)$, $x_0 \in \mathcal{X}$, and $k \geq k_0 \geq 0$. We define equilibrium states for (1.6) and time-invariant discrete time systems analogously to the definitions for continuous time systems.

Discrete time dynamics are of significant interest in engineering applications. In fact, when time-varying continuous time systems with inputs are implemented in labs, this is often done using sampling, which leads to dynamics of the form (1.6). Discrete time systems are also important from the theoretical point of view, including cases where (1.6) is a sub-dynamics of a larger *hybrid* time-varying system that has mixtures of continuous and discrete parts and prescribed mechanisms for switching between the parts.

It is possible to define time-varying systems in a unifying, behavioral way that includes both continuous and discrete time systems. This was done in [161, Chap. 2]. However, strict Lyapunov function constructions for continuous and discrete time systems are often very different, so we treat continuous time and discrete time systems separately in most of the sequel.

1.3 Notions of Stability

Stability, instability, asymptotic stability, exponential stability and input-to-state stability are of utmost importance for nonlinear control systems. Stability formalizes the following intuition: an equilibrium point of a system is stable if any solution with any initial state close to the equilibrium point stays close to the equilibrium point forever. *Asymptotic* stability formalizes the following: an equilibrium point is asymptotically stable if it is stable and all solutions starting near the equilibrium point converge to the equilibrium point as time goes to infinity.

An equilibrium point of a system is *exponentially* stable if it is asymptotically stable and if the solutions are smaller in norm than a positive function of time that exponentially decays to zero. Finally, *input-to-state* stability roughly says that an equilibrium point of a system with inputs is asymptotically stable for the zero input and, in the presence of a bounded input, the solutions are bounded and asymptotically smaller in norm than a function of the sup norm of the input. We use the following abbreviations and acronyms:

LAS	local asymptotic stability
UGAS	uniform global asymptotic stability
GAS	global asymptotic stability
LES	local exponential stability
GES	global exponential stability
UGES	uniform global exponential stability
ISS	input-to-state stability

We also use ISS to mean input-to-state stable, and similarly for the other stability notions. We now make the various stability notions mathematically precise. We focus on continuous time systems but one can define these notions for discrete time systems in an analogous way. For any constants $\rho > 0, r \in \mathbb{N}$, and $q \in \mathbb{R}^r$, we use the notation $\rho\mathcal{B}_r(q) \doteq \{x \in \mathbb{R}^r : |x - q| \leq \rho\}$, which we denote simply by $\rho\mathcal{B}_r$ when $q = 0$.

1.3.1 Stability

Assume that the system (1.2) admits the origin 0 as an equilibrium point. This equilibrium point is *stable* provided for each constant $\varepsilon > 0$, there exists a constant $\delta(\varepsilon) > 0$ such that for each initial state $x_0 \in \mathcal{X} \cap \delta(\varepsilon)\mathcal{B}_n$ and each initial time $t_0 \geq 0$, the unique solution $x(t, t_0, x_0)$ satisfies $|x(t, t_0, x_0)| \leq \varepsilon$ for all $t \geq t_0$. Otherwise we call the equilibrium *unstable*.

1.3.2 Asymptotic and Exponential Stability

Assume that the system (1.2) admits the origin 0 as an equilibrium point. *Uniform globally asymptotic stability (UGAS)* of the equilibrium 0 means that there exists a function $\beta \in \mathcal{KL}$ such that for each initial state $x_0 \in \mathcal{X}$ and each initial time $t_0 \geq 0$, the solution $x(t, t_0, x_0)$ for (1.2) satisfies

$$\big|x(t, t_0, x_0)\big| \leq \beta\big(|x_0|, t - t_0\big) \;\; \forall t \geq t_0 \geq 0. \tag{1.8}$$

In this case, we also say that the system is UGAS to 0, or simply UGAS, and similarly for the other stability notions. When the system is autonomous, this property is called *global asymptotic stability (GAS)*. If there exists a function $\beta \in \mathcal{KL}$ and a constant $\bar{c} > 0$ independent of t_0 such that (1.8) holds for all initial conditions $x_0 \in \bar{c}\mathcal{B}_n \cap \mathcal{X}$, then we call the system *uniformly asymptotically stable.* Hence, uniform asymptotic stability of the equilibrium implies that it is stable and that there exists a constant $\bar{c} > 0$ such that for each initial state $x_0 \in \mathcal{X} \cap \bar{c}\mathcal{B}_n$ and each initial time $t_0 \geq 0$, the solution $x(t, t_0, x_0)$ satisfies $\lim_{t \to +\infty} x(t, t_0, x_0) = 0$. When the system is time-invariant, we call the preceding property *(local) asymptotic stability (LAS).*

When (1.2) admits the origin 0 as an equilibrium point, we call the equilibrium point *uniformly exponentially stable* provided there exist positive constants K_1, K_2, and r such that for each initial state $x_0 \in \mathcal{X} \cap r\mathcal{B}_n$ and each $t_0 \geq 0$, the corresponding solution $x(t, t_0, x_0)$ satisfies $|x(t, t_0, x_0)| \leq K_1 e^{-K_2(t-t_0)}$ for all $t \geq t_0$. When the system is autonomous, we call this property *local exponential stability (LES)* or, if r can be taken to be $+\infty$, *global exponential stability (GES).* The special case of uniformly exponentially stability where we can take $r = +\infty$ is called *uniform global exponential stability (UGES).* More generally, we say that an equilibrium point x^* of (1.2) (which may or may not be zero) is GES provided the dynamics of $x(t) - x^*$ is GES, and similarly for the other stability notions.

1.3.3 Input-to-State Stability

The input-to-state stability (ISS) condition for (1.4) is the requirement that there exist functions $\beta \in \mathcal{KL}$ and $\gamma \in \mathcal{K}$ such that for each $u \in \mathcal{M}(U)$ and each initial condition $x(t_0) = x_0 \in \mathcal{X}$, the solution $x(t, t_0, x_0, u)$ of (1.4) with input vector u satisfies

$$\big|x(t, t_0, x_0, u)\big| \leq \beta\big(|x_0|, t - t_0\big) + \gamma\big(|u|_{[t_0,t]}\big) \;\; \forall t \geq t_0 \, . \tag{1.9}$$

The ISS paradigm plays a fundamental role in nonlinear control, as do its extensions to systems with outputs; see [165] for an extensive discussion.

One immediate consequence of (1.9) is that if (1.4) admits an input $u \in \mathcal{M}(U)$ and an initial condition for which the corresponding trajectory is unbounded, then the system cannot be ISS. This gives a method for testing whether a system is ISS. In Chap. 2, we use this alternative method:

Lemma 1.2. *Assume that (1.4) has state space $\mathcal{X} = \mathbb{R}^n$. Let $\delta \in \mathcal{M}(U)$ be any non-zero input, let $L \in \mathbb{R}^{n \times n}$ be invertible, and set $z(t, t_0, z_0) = Lx(t, t_0, L^{-1}z_0, \delta)$ for each $t \geq t_0 \geq 0$ and $z_0 \in \mathbb{R}^n$. If there is an index $k \in \{1, 2, \ldots, n\}$ such that the kth component z_k of $z(t, t_0, z_0)$ satisfies $\frac{\partial}{\partial t} z_k(t, t_0, z_0) = 0$ for all $t \geq t_0 \geq 0$ and all $z_0 \in \mathbb{R}^n$, then (1.4) is not ISS.*

Proof. Suppose the contrary. Then the dynamics

$$\dot{z} = Lf(t, L^{-1}z, u) \tag{1.10}$$

is easily shown to be ISS as well.[2] Pick $\beta \in \mathcal{KL}$ and $\gamma \in \mathcal{K}_\infty$ such that

$$\big|z(t, t_0, z_0)\big| \;\leq\; \beta\big(|z_0|, t - t_0\big) + \gamma\big(|\delta|_\infty\big) \tag{1.11}$$

along all trajectories of (1.10). By assumption,

$$\big|z(t, t_0, z_0)\big| \;\geq\; \big|z_k(t, t_0, z_0)\big| \;=\; \big|z_{0,k}\big|$$

for all $t \geq t_0 \geq 0$ and $z_0 \in \mathbb{R}^n$, so we get a contradiction by picking

$$z_{0,k} = 2\gamma(|\delta|_\infty)$$

and letting $t \to +\infty$ in (1.11). □

If a system (1.4) is ISS, then necessarily the system

$$\dot{x} = f(t, x, 0) \tag{1.12}$$

is UGAS. However, if (1.12) is UGAS, then it does not follow that (1.4) is ISS. The one-dimensional system

$$\dot{x} = -\arctan(x) + u \tag{1.13}$$

illustrates this. When $u = 0$, the system (1.13) becomes $\dot{x} = -\arctan(x)$ which is GAS. However, (1.13) is not ISS because the bounded input $u = 2$ results in the system

$$\dot{x} = 2 - \arctan(x)$$

which has unbounded solutions.

On the other hand, the system (1.13) is *integral input-to-state stable (iISS)* [160]. For a general nonlinear system (1.4), the iISS condition says that there exist functions $\underline{\gamma}, \bar{\gamma} \in \mathcal{K}_\infty$ and $\beta \in \mathcal{KL}$ such that for each $u \in \mathcal{M}(U)$ and each initial condition $x(t_0) = x_0$, the unique solution $x(t, t_0, x_0, u)$ of (1.4) with input vector u satisfies

$$\underline{\gamma}(|x(t, t_0, x_0, u)|) \leq \beta(|x_0|, t - t_0) + \int_{t_0}^{t} \bar{\gamma}(|u(m)|)\mathrm{d}m \tag{1.14}$$

for all $t \geq t_0$. The fact that (1.13) is iISS will follow from the Lyapunov characterizations for ISS and iISS that we discuss in Chap. 2.[3] The ISS property is essentially global. Indeed, any system (1.4) such that the corresponding

---

[2] For example, if (1.4) has the ISS Lyapunov function V, then (1.10) has the ISS Lyapunov function $\tilde{V}(t, z) \doteq V(t, L^{-1}z)$; see Chap. 2 for the relevant definitions.

[3] In fact, (1.13) admits the iISS Lyapunov function $V(x) = x \arctan(x)$ and therefore is iISS.

system (1.12) admits the origin as a locally uniformly asymptotically stable equilibrium point is locally ISS, meaning there exists a neighborhood $\mathcal{G}$ of the equilibrium so that an ISS estimate holds along trajectories remaining in $\mathcal{G}$; this follows from the local Lipschitzness of the dynamics in the state.[4]

1.3.4 Linear Systems and Linearizations

Stability analysis is considerably simpler for linear systems than for nonlinear systems. For example, for a linear system

$$\dot{x} = Ax \tag{1.15}$$

with a constant matrix A, the properties GAS, LAS, GES, and LES are all equivalent, and they are satisfied if and only if all eigenvalues of A have negative real parts, in which case A is called *Hurwitz*. The solutions of (1.15) have the form $x(t) = e^{At}x_0$. Frequently, the local behavior of a nonlinear system $\dot{x} = f(x)$ can be analyzed using the fact that an equilibrium point x^* of a time-invariant nonlinear system is LES if and only if its variational matrix $A \doteq Df(x^*)$ is Hurwitz [161]. This can be equivalently formulated by saying that the equilibrium point of a nonlinear system is LES if and only if its linear approximation at the equilibrium point is LES.

Even when the variational matrix is not Hurwitz, the linearization can still provide important information. One important result in that direction is the following one from [131, p.120]:

Theorem 1.1. *(Hartman-Grobman Theorem) Let $\mathcal{X} \subseteq \mathbb{R}^n$ be a neighborhood of the origin, and let $f : \mathcal{X} \to \mathbb{R}^n$ be C^1 with equilibrium point* 0*. Assume that $A \doteq Df(0)$ has no eigenvalue with zero real part. Then we can find a homeomorphism H of an open neighborhood $\mathcal{V}_1$ of the origin into an open neighborhood $\mathcal{V}_2$ of* 0 *such that for each $x_0 \in \mathcal{V}_1$, there is an interval $\mathcal{I}$ containing* 0 *for which $H(x(t, x_0)) = e^{At}H(x_0)$ for all $t \in \mathcal{I}$.*

Here $x(t, x_0)$ is the flow of $\dot{x} = f(x)$ in the usual ODE sense.

1.3.5 Uniformity vs. Non-uniformity

For time-varying systems, asymptotic stability and uniform asymptotic stability are different. The one-dimensional linear time-varying system

[4] Given $n, p \in \mathbb{N}$, an interval $\mathcal{I} \subseteq \mathbb{R}$, and a subset $\mathcal{X} \subseteq \mathbb{R}^n$, we say that a function $g : \mathcal{I} \times \mathcal{X} \to \mathbb{R}^p$ is *locally Lipschitz in $x \in \mathcal{X}$* provided for each compact subset $K \subseteq \mathcal{X}$, there is a constant L_K so that $|g(t, x) - g(t, x')| \le L_K|x - x'|$ for all $t \in \mathcal{I}$ and $x, x' \in K$. If L_K can be chosen independently of K, then we say that g is Lipschitz in $x \in \mathcal{X}$.

$$\dot{x} = -\tfrac{x}{1+t} \tag{1.16}$$

is GAS in the sense that its solutions are

$$x(t, t_0, x_0) = x_0 \frac{1+t_0}{1+t}$$

and therefore go to zero when t goes to infinity. However, it is not UGAS. To prove this, we proceed by contradiction. Suppose that there exists a function $\beta \in \mathcal{KL}$ such that for all $t \geq t_0 \geq 0$, the inequality

$$\left|x(t, t_0, x_0)\right| \leq \beta\big(|x_0|, t - t_0\big) \tag{1.17}$$

is satisfied. Choosing $x_0 = 1$ and $t = 2t_0 + 1$, we have

$$\frac{1}{2} = \frac{1+t_0}{2+2t_0} \leq \beta(1, t_0 + 1). \tag{1.18}$$

Since $\beta(1, t_0 + 1)$ goes to zero when t_0 goes to infinity, the inequality (1.18) leads to a contradiction.

1.3.6 Basin of Attraction

The *region of attraction* (also called the *basin of attraction*) of a LAS equilibrium point of a system is the set of all initial states that generate solutions of the system that converge to the equilibrium point. Often, it is not sufficient to determine that a given system has an asymptotically stable equilibrium point. Rather, it is important to find the region of attraction or an approximation of this region. Such approximations can be found using Lyapunov functions. We revisit the problem of estimating the basin of attraction in Sect. 2.5.

1.4 Stabilization

Consider the classical problem of constructing a control law $u_s(t, x)$ such that the origin of (1.4) is asymptotically stable. Later, we will see how this problem can often be handled by Lyapunov function constructions. When the problem is restricted to local stabilization, techniques based on the stabilization of the linear approximation of (1.4) at the origin are frequently used.

However, when UGAS is desirable, linear techniques usually cannot be used. Then, nonlinear design techniques called *backstepping* and *forwarding* apply, provided the system admits a special structure. Backstepping applies to lower triangular systems

$$\begin{cases} \dot{x}_1 = f_1(t, x_1, x_2) \\ \dot{x}_2 = f_2(t, x_1, x_2, x_3) \\ \quad\vdots \\ \dot{x}_n = f_n(t, x_1, ..., x_n, u). \end{cases} \tag{1.19}$$

These systems are called *feedback* systems. Forwarding applies to systems having the upper triangular form

$$\begin{cases} \dot{x}_1 = f_1(t, x_1, ..., x_n, u) \\ \dot{x}_2 = f_2(t, x_2, ..., x_n, u) \\ \quad\vdots \\ \dot{x}_n = f_n(t, x_n, u) \end{cases} \tag{1.20}$$

which are called *feedforward* systems. We discuss backstepping in detail in Chap. 7.

When a nonlinear system admits a linear approximation around an equilibrium point that is not exponentially stabilizable, it may not be easy to tell whether the equilibrium point is locally asymptotically stabilizable. Besides, in some cases, an equilibrium point is asymptotically stabilizable by a C^1 time-varying feedback but not stabilizable by a C^1 *time-invariant* state feedback. For example, this phenomenon occurs for the origin of

$$\begin{cases} \dot{x}_1 = u_1 \\ \dot{x}_2 = u_2 u_1. \end{cases} \tag{1.21}$$

The fact that the origin of this system is not asymptotically stabilizable by a C^1 time-invariant feedback can be proven using the following necessary condition from [18]:

Theorem 1.2. *(Brockett's Stabilization Theorem) Consider a system*

$$\dot{x} = f(x, u) \tag{1.22}$$

with $f \in C^1$. Assume that there exist an equilibrium point x_* and a C^1 feedback $u_s(x)$ such that the system

$$\dot{x} = f\big(x, u_s(x)\big)$$

admits x_* as a LAS equilibrium point. Then the image of the map f contains some neighborhood of x_*.

The system (1.21) does not satisfy the necessary condition of Brockett's Theorem at the origin, because for any $\varepsilon \neq 0$, there is no pair (x, u) such that

$$(u_1, u_2 u_1) = (0, \varepsilon)$$

and for any open neighborhood of the origin $\mathcal{V} \subseteq \mathbb{R}^2$, there exists $\varepsilon \neq 0$ such that $(0, \varepsilon) \in \mathcal{V}$. On the other hand, it can be globally stabilized by a *time-varying* C^1 feedback; see p.19. These considerations show one reason why time-varying systems are important.

1.5 Examples

In many cases, one can use Lyapunov functions to establish the various stability properties. However, in the following examples, we establish the stability properties using other techniques. In later chapters, we primarily use Lyapunov function methods to establish stability. As we will see later, strict Lyapunov functions have the advantage that they can also be used to quantify the effects of uncertainty, especially when they are given in explicit closed form.

1.5.1 Stable System

An example of a nonlinear system that is stable but not asymptotically stable is given by the two-dimensional pendulum dynamics

$$\begin{cases} \dot{\theta} = \omega \\ \dot{\omega} = -\frac{g}{l}\sin(\theta) \end{cases} \tag{1.23}$$

where g and l are positive real numbers. To simplify, we assume

$$\frac{g}{l} = 1.$$

The local stability of the origin can be proved as follows. Let $\varepsilon \in \left(0, \frac{1}{4}\right]$. Consider the non-negative function

$$H(\theta, \omega) \;=\; 1 - \cos(\theta) + \frac{1}{2}\omega^2.$$

Let $\delta(\varepsilon) = \frac{1}{8}\varepsilon^2$. Take any solution $(\theta(t), \omega(t))$ of (1.23) with any initial condition satisfying $|(\theta(0), \omega(0))|_\infty \le \delta(\varepsilon)$. Since $\delta(\varepsilon) \le 1/128$, we get

$$H\big(\theta(0), \omega(0)\big) \;\le\; |\theta(0)| + \frac{1}{2}\omega^2(0) \;\le\; 2\delta(\varepsilon).$$

Simple calculations yield

$$\frac{d}{dt}H\big(\theta(t), \omega(t)\big) = 0 \;\; \forall t \ge 0.$$

Hence, (1.23) cannot be asymptotically stable. On the other hand, since H is constant along the trajectories of (1.23),

$$1 - \cos(\theta(t)) + \frac{1}{2}\omega^2(t) = H(\theta(t), \omega(t)) \le \frac{1}{4}\varepsilon^2 \le \frac{1}{64}$$

for all $t \ge 0$. Therefore, $|\omega(t)| \le \varepsilon$ for all $t \ge 0$. Also, since $1 - \cos(\theta(t)) \le \frac{1}{64}$ for all $t \ge 0$ and $|\theta(0)| \le \pi/4$, we deduce that $|\theta(t)| \le \pi/4$ for all $t \ge 0$, which implies that

$$\frac{1}{4}\theta^2(t) \le 1 - \cos(\theta(t)) \le \frac{1}{4}\varepsilon^2$$

for all $t \ge 0$. This gives $|(\theta(t), \omega(t))|_\infty \le \varepsilon$ for all $t \ge 0$, which is the desired stability estimate.

Remark 1.1. One can also construct unstable autonomous systems all of whose trajectories converge to the origin. An example of this phenomenon is

$$\dot{x}_1 = \frac{x_1^2(x_2 - x_1) + x_2^5}{(x_1^2 + x_2^2)(1 + (x_1^2 + x_2^2)^2)}, \quad \dot{x}_2 = \frac{x_2^2(x_2 - 2x_1)}{(x_1^2 + x_2^2)(1 + (x_1^2 + x_2^2)^2)}. \tag{1.24}$$

For the proof that (1.24) satisfies the requirements, see [54, pp. 191-194].

1.5.2 Locally Asymptotically Stable System

When a friction term is added to (1.23), the system becomes

$$\begin{cases} \dot{\theta} = \omega \\ \dot{\omega} = -\frac{g}{l}\sin(\theta) - \frac{k}{m}\omega \end{cases} \tag{1.25}$$

where k and m are positive real numbers. The origin of (1.25) is a LES equilibrium point that is not GAS because the system admits multiple equilibrium points.

The proof that the origin of (1.25) is LES is a consequence of the fact that its linear approximation at the origin is

$$\begin{cases} \dot{\theta}_e = \omega_e \\ \dot{\omega}_e = -\frac{g}{l}\theta_e - \frac{k}{m}\omega_e\,, \end{cases} \tag{1.26}$$

which is an exponentially stable linear system because the eigenvalues of the matrix

$$\begin{bmatrix} 0 & 1 \\ -\frac{g}{l} & -\frac{k}{m} \end{bmatrix} \tag{1.27}$$

have negative real parts.

1.5.3 Globally Asymptotically Stable System

The two-dimensional system

$$\begin{cases} \dot{S} = D(S_e - S) - \dfrac{KS}{L+S}x \\ \dot{x} = \left(\dfrac{KS}{L+S} - D\right)x \end{cases} \tag{1.28}$$

with positive constant parameters D, K, L, and S_e has the invariant domain $\mathcal{X} = (0, \infty) \times (0, \infty)$. It is a simplified model of a bio-reactor with dilution rate D, input nutrient concentration S_e, and Monod growth rate

$$\mu(S) = \frac{KS}{L+S};$$

see [153] for generalizations. Assume that

$$K > D \quad \text{and} \quad S_e > \frac{DL}{K-D}. \tag{1.29}$$

We show that the equilibrium point

$$(S_*, x_*) = \left(\frac{DL}{K-D}, S_e - \frac{DL}{K-D}\right) \tag{1.30}$$

for (1.28) is GAS and LES.

The variable $Z = S + x - S_e$ satisfies

$$\dot{Z} = -DZ. \tag{1.31}$$

We easily deduce that all of the trajectories of (1.28) enter

$$B = (0, 2S_e) \times (0, 2S_e).$$

One can readily check that (S_*, x_*) and $(S_e, 0)$ are the unique equilibrium points of (1.28) in the closure $\overline{B}$ of B. Also, (S_*, x_*) is the unique LES equilibrium point in $\overline{B}$ (by considering the linearization of (1.28) around (S_*, x_*), and using (1.29) to show that $(S_e, 0)$ is not an asymptotically stable equilibrium, because $\dot{x} > 0$ when $S > S_e$ and S is near S_e).

We next consider any trajectory $(S(t), x(t))$ of (1.28) with any initial condition in B and prove that it converges asymptotically to (S_*, x_*). This will show that (S_*, x_*) is a GAS equilibrium of (1.28) with state space $\mathcal{X} = B$. Our analysis uses basic results from dynamic systems theory; see, e.g., [53, 153].

Let Ω denote the ω-limit set of this trajectory. One can easily prove that $\Omega \neq \{(S_e, 0)\}$ because

$$\frac{KS_e}{L + S_e} - D > 0.$$

We claim that $(S_e, 0) \notin \Omega$. To prove this claim, we proceed by contradiction. Suppose that $(S_e, 0) \in \Omega$. Since $\Omega \neq \{(S_e, 0)\}$, the well-known Butler-McGehee Theorem (e.g., from [153, p.12]), applied to the hyperbolic rest point $(S_e, 0)$, provides a value $S_c \neq S_e$ such that $(S_c, 0)$ belongs to Ω. This is impossible because

$$Z(t) = S(t) + x(t) - S_e \; \rightarrow \; 0.$$

Therefore $(S_e, 0) \notin \Omega$. Similarly, one can prove that there is no point of the form $(S_p, 0) \in \overline{B}$ in Ω.

Therefore, we deduce from the Poincaré-Bendixson Trichotomy [153, p.9] that either $\Omega = \{(S_*, x_*)\}$ or it is a periodic orbit which does not contain any point of the form $(S_p, 0)$. Suppose Ω is a periodic orbit, and set

$$f_1(S, \xi) = D(S_e - S) - \frac{KS}{L + S} e^{\xi} \text{ and } f_2(S, \xi) = \frac{KS}{L + S} - D.$$

Then, the system

$$\begin{cases} \dot{S} = f_1(S, \xi) \\ \dot{\xi} = f_2(S, \xi) \,, \end{cases} \tag{1.32}$$

which is deduced from (1.28) through the change of coordinate $\xi = \ln x$, also admits a periodic trajectory. On the other hand,

$$\frac{\partial f_1}{\partial S}(S, \xi) + \frac{\partial f_2}{\partial \xi}(S, \xi) < 0,$$

so Dulac's Criterion [53] implies that (1.32) admits no periodic orbit. This contradiction shows that Ω is reduced to (S_*, x_*), as claimed.

1.5.4 UGAS Time-Varying System

The one-dimensional linear time-varying system

$$\dot{x} = -\sin^2(t)x \tag{1.33}$$

admits the origin as a UGAS equilibrium point. For all $t \geq t_0$ and initial states x_0, its solutions are

$$x(t, t_0, x_0) = \exp\left(-\int_{t_0}^{t} \sin^2(m)\mathrm{d}m\right) x_0 \ ,$$

which satisfy

$$\begin{aligned} |x(t, t_0, x_0)| &= \exp\left(-\frac{1}{2}(t - t_0) + \frac{1}{4}(\sin(2t) - \sin(2t_0))\right) |x_0| \\ &\leq \beta(|x_0|, t - t_0) \end{aligned} \tag{1.34}$$

with $\beta(r, s) = e^{-\frac{1}{2}s + \frac{1}{2}}r$. The function β is of class $\mathcal{KL}$.

1.5.5 Systems in Chained Form

We have seen that the origin of the system (1.21) is not asymptotically stabilizable by any feedback of class C^1 that is independent of t. However, the origin of this system can be globally uniformly asymptotically stabilized by time-varying control laws of class C^1. To prove this, let us choose

$$\begin{aligned} u_1 &= -x_1 + \sin(t)[\cos(t)x_1 + x_2] \\ u_2 &= -\sin(t) - \cos(t). \end{aligned} \tag{1.35}$$

This choice yields the chained form system

$$\begin{cases} \dot{x}_1 &= -x_1 + \sin(t)\big[\cos(t)x_1 + x_2\big] \\ \dot{x}_2 &= \big[-\sin(t) - \cos(t)\big]\big[-x_1 + \sin(t)(\cos(t)x_1 + x_2)\big]. \end{cases} \tag{1.36}$$

It follows that the time derivative of $\zeta \doteq \cos(t)x_1 + x_2$ satisfies

$$\begin{aligned} \dot{\zeta} &= -\sin(t)x_1 + \cos(t)\big[-x_1 + \sin(t)(\cos(t)x_1 + x_2)\big] \\ &\quad + \big[-\sin(t) - \cos(t)\big]\big[-x_1 + \sin(t)(\cos(t)x_1 + x_2)\big] \\ &= -\sin^2(t)\zeta. \end{aligned} \tag{1.37}$$

We showed in Sect. 1.5.4 that for all $t \geq t_0$ and any initial state (x_{10}, x_{20}),

$$|\zeta(t, t_0, \zeta_0)| \ \leq \ \exp\left(-\frac{1}{2}(t - t_0) + \frac{1}{2}\right) |\zeta_0| \ , \tag{1.38}$$

where $\zeta_0 = \cos(t_0)x_{10} + x_{20}$. On the other hand, we have

$$\dot{x}_1 = -x_1 + \sin(t)\zeta. \tag{1.39}$$

We deduce that for all $t \geq t_0$ and any initial condition (x_{10}, x_{20}),

$$\begin{aligned}|\zeta(t,t_0,\zeta_0)| &\le e^{-\frac{1}{2}(t-t_0)+\frac{1}{2}}|\zeta_0| \text{ and}\\ |x_1(t,t_0,x_{10},x_{20})| &\le e^{-(t-t_0)}|x_{10}|\\ &\quad +2e^{\frac{1}{2}}\left(e^{-\frac{1}{2}(t-t_0)}-e^{-(t-t_0)}\right)|\zeta_0|.\end{aligned} \tag{1.40}$$

Therefore,

$$|x_2(t,t_0,x_{10},x_{20})| \le e^{-(t-t_0)}|x_{10}| + 3e^{\frac{1}{2}}e^{-\frac{1}{2}(t-t_0)}(|x_{10}|+|x_{20}|), \tag{1.41}$$

because $|x_2| \le |x_1| + |\zeta|$ everywhere. These inequalities give the announced result.

1.6 Comments

The ISS paradigm was first announced by Sontag in [156]. This was a significant development, because it merged the state space framework of Lyapunov with the input-output operator approach of Zames. ISS enjoys invariance under coordinate changes, and can be stated in various equivalent forms including energy-like estimates that generalize the standard Lyapunov decay condition. Sontag and Wang characterized ISS by proving that a system is ISS if and only if it admits an ISS Lyapunov function [169]; see our discussion on ISS Lyapunov functions in the next chapter. This characterization simplifies the task of checking that a system is ISS. Another important property of ISS is the following ISS superposition principle [168]:

Theorem 1.3. *A time-invariant system*

$$\dot{x} = f(x,u), \quad x \in \mathbb{R}^n, \ u \in \mathbb{R}^m \tag{1.42}$$

is ISS if and only if the following are true: its zero-system $\dot{x} = f(x,0)$ is stable and (1.42) satisfies the asymptotic gain property.

The asymptotic gain property is the requirement that there exists a function $\gamma \in \mathcal{K}_\infty$ such the flow map $x(t,x_0,u)$ of (1.42) satisfies

$$\limsup_{t\to+\infty} |x(t,x_0,u)| \le \gamma(|u|_\infty)$$

for all $x_0 \in \mathbb{R}^n$ and $u \in \mathcal{M}(\mathbb{R}^m)$. It is tempting to surmise that the GAS property of $\dot{x} = f(x,0)$ (i.e., 0-GAS of (1.42)) guarantees boundedness of all trajectories of (1.42) under disturbances that converge to 0. This is true if (1.42) is a linear time-invariant system $\dot{x} = Ax + Bu$. In fact, 0-GAS linear time-invariant systems satisfy the *converging-input converging state (CICS) property* which says that trajectories converge to zero when the inputs do [165]. However, this does not carry over to nonlinear systems because as noted in [165], the system $\dot{x} = -x + (x^2 + 1)u$ has divergent solutions when $u(t) = (2t + 2)^{-1/2}$.

One can also give a superposition principle for iISS, using the following *bounded energy frequently bounded state (BEFBS) property* :

$$\begin{array}{l} \exists \sigma \in \mathcal{K}_\infty \text{ such that :} \\ \displaystyle\int_0^{+\infty} \sigma(|u(s)|)\mathrm{d}s < \infty \;\Rightarrow\; \liminf_{t \to +\infty} |x(t, x_0, u)| < \infty. \end{array} \tag{BEFBS}$$

In fact, (1.42) is iISS if and only if it satisfies the BEFBS property and is 0-GAS [6].

During the past ten years, ISS has been generalized in several different directions. There are now notions of ISS for hybrid systems, which involve discrete and continuous subsystems and rules for switching between the subsystems [47, 48]. There are also analogs of ISS for systems with outputs

$$\dot{x} = f(x, u), \quad y = H(x), \tag{1.43}$$

such as input-to-output stability (IOS), which is the requirement that there exist functions $\beta \in \mathcal{KL}$ and $\gamma \in \mathcal{K}_\infty$ such that

$$|y(t)| \;\le\; \beta\big(|x(0)|, t\big) + \gamma\big(|u|_{[0,t]}\big)$$

along all trajectories of the system [171]. One then shows that a system is IOS if and only if it admits an IOS Lyapunov function; see Sect. 6.7 for the relevant definitions and results on constructing explicit IOS Lyapunov functions for time-varying systems.

Some other output stability concepts for (1.43) include *input/output-to-state stability (IOSS)* and *output-to-state stability (OSS)* which are the requirements that there are functions $\gamma_i \in \mathcal{K}_\infty$ and $\beta \in \mathcal{KL}$ such that

$$|x(t)| \;\le\; \beta\big(|x(0)|, t\big) + \gamma_1\big(|u|_{[0,t]}\big) + \gamma_2\big(|y|_{[0,t]}\big) \tag{IOSS}$$

and

$$|x(t)| \;\le\; \beta\big(|x(0)|, t\big) + \gamma_3\big(|y|_{[0,t]}\big) \tag{OSS}$$

along all trajectories of (1.43), respectively. The IOSS and OSS properties can be characterized in terms of the existence of Lyapunov functions as well [73]. Input-measurement-to-error stability (IMES) is a significant generalization of ISS for systems

$$\dot{x} = f(x, u), \quad y = h(x), \quad w = g(x) \tag{1.44}$$

with error outputs $y = h(x)$ and measurement outputs $w = g(x)$ [165]. The IMES property says that there exist $\beta \in \mathcal{KL}$ and functions $\sigma, \gamma \in \mathcal{K}$ such that

$$|y(t)| \;\le\; \beta\big(|x(0)|, t\big) + \sigma\big(|w|_{[0,t]}\big) + \gamma\big(|u|_{[0,t]}\big) \tag{IMES}$$

along all trajectories of (1.44). However, to our knowledge, there is no smooth Lyapunov characterization for IMES available.

Backstepping is discussed in detail in [149]. See also Chap. 7. Some pioneering results on backstepping include [19, 31, 179].

The proof of Brockett's Stabilization Theorem uses basic facts from degree theory, combined with a homotopy argument. Here is a sketch of the proof; see [161, Sect. 5.9] for details. Using degree theory results from [15], one first proves the following:

Lemma 1.3. *Let $\rho > 0$ be a given constant and $H : [0,1] \times \rho\mathcal{B}_n \to \mathbb{R}^n$ be a continuous function such that the following hold:*

1. $H(1,x) = -x$ for all x; and
2. $H(t,x) \neq 0$ for all $x \in \mathrm{boundary}(\rho\mathcal{B}_n)$.

Then there is a constant $\varepsilon > 0$ such that the image of $\rho\mathcal{B}_n \ni x \mapsto H(0,x)$ contains $\varepsilon\mathcal{B}_n$.

Brockett's Theorem follows by applying Lemma 1.3 to

$$H(t,x) \doteq \begin{cases} f(x, u_s(x)), & \text{if } t = 0 \\ -x, & \text{if } t = 1 \\ \frac{1}{t}\left[\phi\left(\frac{t}{1-t}, x\right) - x\right], & \text{if } 0 < t < 1 \end{cases}$$

where ϕ is the flow map for the closed-loop system $\dot{x} = f(x, u_s(x))$ and $\rho > 0$ is chosen so that $\rho\mathcal{B}_n$ is in the domain of attraction of the closed-loop system. Brockett's Criterion is a far reaching result because it implies that no system of the form

$$\dot{x} = u_1 g_1(x) + \ldots + u_m g_m(x) = G(x)u$$

with $m < n$ and

$$\mathrm{rank}[g_1(0), \ldots, g_m(0)] \;=\; m$$

admits a C^1 pure state stabilizing feedback $u_s(x)$; see [163] for the simple proof. Hence, no totally nonholonomic mechanical system is C^1 stabilizable by a pure state feedback.

In [146], Samson provided important general results that use time-varying feedback to help overcome the obstructions imposed by Brockett's Criterion. See also [65], which uses backstepping to build a time-varying feedback stabilizer for a two degrees-of-freedom mobile robot. By [27], the system $\dot{x} = f(x,u)$ is stabilizable by a time-varying continuous feedback $u = k(t,x)$ when it is drift free (meaning $f(x,0) \equiv 0$) and completely controllable.

Another approach to stabilizing the system is to look for a *dynamic* stabilizer, meaning a locally Lipschitz dynamics

$$\dot{z} = A(z,x)$$

and a locally Lipschitz function $k(z,x)$ such that the combined system

$$\begin{cases} \dot{x} = f(x, k(z, x)) \\ \dot{z} = A(z, x) \end{cases}$$

is GAS. See [160] for a detailed discussion on dynamic stabilizers for linear systems. However, a dynamic feedback for $\dot{x} = f(x, u)$ may fail to exist, even if the system is completely controllable. An example from [173] where this happens is

$$\dot{x} = f(x, u) = \begin{bmatrix} (4 - x_2^2)u_2^2 \\ e^{-x_1} + x_2 - 2e^{-x_1}\sin^2(u_1) \end{bmatrix}, \quad x \in \mathbb{R}^2, \ u \in \mathbb{R}^2. \tag{1.45}$$

The fact that (1.45) is completely controllable (and therefore GAC to $\mathcal{A} = \{0\}$) was shown in [173], which also shows that it is impossible to pick paths converging to the origin in such a way that this selection is continuous as a function of the initial states. Since the flow map of any dynamic stabilizer would give a continuous choice of paths converging to 0, no dynamic stabilizer for (1.45) can exist, even if we drop the requirement that the state of the regulator converges to zero. As a special case, (1.45) cannot admit a continuous time-varying feedback $u = k(t, x)$. This does not contradict the existence theory [27] for time-varying feedbacks because (1.45) has drift.

Yet another approach to circumventing the "virtual obstacles" to feedback stabilization imposed by Brockett's Condition involves nonsmooth analysis and discontinuous feedbacks. See for example [94] where a nonsmooth (but time-invariant) feedback was constructed for Brockett's Nonholonomic Integrator using a generalized Lie derivative, which involves a proximal subgradient [22] and a semi-concave control-Lyapunov function (CLF). Discontinuous feedbacks complicate the analysis because they give differential equations with discontinuous right hand sides. Discontinuous dynamics can sometimes be handled using Filippov solutions, sample-and-hold solutions, or Euler solutions [94, 162].

In addition to "virtual" obstacles, there are also "topological" obstacles to time-invariant feedback stabilization. If a time-invariant system $\dot{x} = f(x, u)$ evolving on some manifold $\mathcal{M}$ is globally asymptotically controllable to a singleton equilibrium and has a continuous stabilizing feedback $k(x)$, then Milnor's Theorem [115] implies that $\mathcal{M}$ is diffeomorphic to Euclidean space. This follows because $k(x)$ would guarantee the existence of a smooth CLF that could be taken as a Morse function with a unique critical point, and manifolds admitting such Morse functions are known to be diffeomorphic to Euclidean space [163].

Chapter 2
Review of Lyapunov Functions

Abstract We turn next to some of the basic notions of Lyapunov functions. Roughly speaking, a Lyapunov function for a given nonlinear system is a positive definite function whose decay along the trajectories of the system can be used to establish a stability property of the system. In general, one also requires Lyapunov functions to be proper, but one can prove stability using non-proper Lyapunov-like functions as well. Even when a system is known to be stable, one often still needs explicit strict Lyapunov functions, e.g., to design stabilizing feedbacks, or to find closed form expressions for the comparison functions in the ISS condition.

Non-strict Lyapunov functions cannot in general be used for these purposes. As we will see, strict Lyapunov functions are also important when estimating domains of attraction and $\mathcal{L}_2$ gains. We also address the issue of whether a given time-invariant system admits a Lyapunov function that has a globally bounded gradient. This is important, because the existence of such a Lyapunov function guarantees robustness with respect to additive uncertainty in the dynamics. We illustrate these ideas in several examples.

2.1 Strict Lyapunov Function

2.1.1 Definition

A strict Lyapunov function is a CLF for a system with no controls. Strict Lyapunov functions are also called strong Lyapunov functions. We therefore begin by defining CLFs. In the rest of this section, we consider only *continuous* time nonlinear systems

$$\dot{x} = f(t, x, u) \tag{2.1}$$

under the assumptions of the previous chapter, where the state x and input vector u are valued in an open set $\mathcal{X} \subseteq \mathbb{R}^n$ and a set $U \subseteq \mathbb{R}^p$, respectively. We discuss analogs for discrete time systems in Sect. 2.3.

We assume that the system (2.1) has equilibrium state 0. Let $V : [0, \infty) \times \mathcal{X} \to [0, \infty)$. We say that V is *proper* provided the set $\{x \in \mathcal{X} : \sup_t V(t,x) \leq L\}$ is compact for each constant $L > 0$; we call it *positive definite* provided $\inf_t V(t,x) = 0$ if and only if $x = 0$. When V is C^1, we use the notation

$$\dot{V}(t,x,u) \doteq \frac{\partial V}{\partial t}(t,x) + \frac{\partial V}{\partial x}(t,x)f(t,x,u).$$

A C^1, proper, positive definite function $V : [0,\infty) \times \mathcal{X} \to [0,\infty)$ is then called a CLF for the system (2.1) provided for each $x \in \mathcal{X} \setminus \{0\}$, there exists a value $u \in U$ such that

$$\dot{V}(t,x,u) < 0$$

for all $t \geq 0$. When the system has no controls, we indicate this decay condition by $\dot{V}(t,x) < 0$. Also, when $\mathcal{X} = \mathbb{R}^n$, we use the term *radially bounded* to mean properness, which in this case gives the condition that

$$\lim_{|x| \to +\infty} \inf_t V(t,x) = +\infty.$$

For the special case of time-invariant control affine systems

$$\dot{x} = \varphi_1(x) + \varphi_2(x)u \tag{2.2}$$

with $\mathcal{X} = \mathbb{R}^n$, a positive definite time-invariant function $V(x)$ is a CLF for (2.2) provided the following hold:

1. V is radially unbounded; and
2. $L_{\varphi_1} V(x) \geq 0 \Rightarrow [x = 0 \text{ or } L_{\varphi_2} V(x) \neq 0]$.

We say that a CLF for (2.2) has the *small control property* provided: For each $\varepsilon > 0$, there is a $\delta > 0$ such that if $0 \neq |x| < \delta$, then there is a $u \in U$ such that $|u| < \varepsilon$ and $\nabla V(x)\varphi_1(x) + \nabla V(x)\varphi_2(x)u < 0$. A special case of Artstein's Theorem [10] says the following: *Let $V(x)$ be a positive definite radially unbounded function. There exists a continuous feedback $K(x)$ so that V is a strict Lyapunov function for (2.2) in closed-loop with $u = K(x)$ if and only V is a CLF for (2.2) that satisfies the small control property.*

Specializing to systems with no controls and $\mathcal{X} = \mathbb{R}^n$, the strict Lyapunov function decay condition $\dot{V}(t,x) < 0$ for all $x \neq 0$ and all $t \geq 0$ means that

$$\frac{d}{dt}V(t, x(t,t_0,x_0)) < 0 \tag{2.3}$$

for all $t \geq t_0 \geq 0$ as long as the trajectory $x(t,t_0,x_0)$ is not at 0. The decay condition (2.3) is equivalent to the existence of a positive definite function α such that

$$\dot{V}(t,x) \leq -\alpha(|x|) \quad \forall x \in \mathbb{R}^n \text{ and } \forall t \geq 0;$$

this is shown in [157] for time-invariant systems but the generalization to time-varying systems is straightforward. Using suitable transformations $\Gamma(V)$ of the Lyapunov function gives different possible functions α. In fact, a slight variant of an argument from [141, Sect. 4] shows that a suitable transformation $V_1 = \Gamma(V)$ that is C^1 on $\mathbb{R}^n \setminus \{0\}$, proper, and positive definite satisfies $\dot{V}_1(t,x) \leq -V_1(t,x)$ for all x and t. We then call V_1 an *exponential decay Lyapunov function*, although the norm of the trajectories will not in general decay exponentially.

It is sometimes useful to relax the properness requirement on Lyapunov functions. A positive definite function that satisfies all of the requirements for being a strict Lyapunov function except properness is called a *strict Lyapunov-like function.* Strict Lyapunov-like functions were constructed in [106], under Matrosov Conditions; see Chap. 3. Throughout the chapter, we use the convention that all (in)equalities should be understood to hold globally unless otherwise indicated, and we leave out the arguments of our functions when they are clear from the context.

2.1.2 Lemmas

The existence of a strict Lyapunov function for our system

$$\dot{x} = f(t,x), \quad x \in \mathcal{X} \tag{2.4}$$

is sufficient for the system to be UGAS. Strict Lyapunov-like functions can be used to prove asymptotic stability as well. The following result from [70, Sect. 4.5] illustrates these points:

Lemma 2.1. *Let* 0 *be an equilibrium for (2.4), and* $V : [0,\infty) \times \mathcal{X} \to [0,\infty)$ *be a* C^1 *function that admits continuous positive definite functions* W_i *so that the following conditions hold:*

1. $W_1(x) \leq V(t,x) \leq W_2(x)$; *and*
2. $\dot{V}(t,x) \leq -W_3(x)$ *for all* $t \geq 0$ *and* $x \in \mathcal{X}$.

Then 0 *is a uniformly asymptotically stable equilibrium for (2.4). If the preceding conditions hold with* $\mathcal{X} = \mathbb{R}^n$ *and* W_1 *is radially unbounded, then* 0 *is a UGAS equilibrium for (2.4). In the special case where there exist positive constants* c_i *and* p *so that the preceding assumptions hold with* $W_i(x) = c_i|x|^p$ *and* $\mathcal{X} = \mathbb{R}^n$*, then the equilibrium is GES.*

The preceding theorem reduces the stability analysis to a search for an appropriate Lyapunov function. On the other hand, even if a system (2.4) is known to be UGAS, it is often important to be able to go in the converse direction, by constructing a strict Lyapunov function for the system. As a simple time-invariant example, assume that we know that a control affine system (2.2) is rendered GAS to the origin by a given feedback $u_s(x)$. In general, there is no reason to expect the closed-loop system

$$\dot{x} = \phi_1(x) + \phi_2(x)[K(x) + d] \tag{2.5}$$

with the disturbance d to be ISS when we pick $K(x) = u_s(x)$.[1] On the other hand, if we know a strict Lyapunov function V for the closed-loop system

$$\dot{x} = f(x) \doteq \phi_1(x) + \phi_2(x)u_s(x) \tag{2.6}$$

for which $-L_f V$ is radially unbounded, then (2.5) is ISS if we choose

$$K(x) = u_s(x) - \left(L_{\phi_2} V(x)\right)^\top . \tag{2.7}$$

Standard converse Lyapunov function theory guarantees the *existence* of a strict Lyapunov function for the GAS system (2.6).

However, to have an implementable stabilizer (2.7), we need an explicit expression for the Lie derivative $L_{\phi_2} V(x)$, hence an explicit strict Lyapunov function V. The strict Lyapunov functions provided by converse Lyapunov theory are usually not explicit, even if the system is UGES. The following result from [70, Sect. 4.7] illustrates this point:

Lemma 2.2. *Assume that there exist constants $D > 1$ and $\lambda > 0$ such that all trajectories of (2.4) satisfy the UGES condition*

$$|x(t, t_0, x_0)| \leq D|x_0|e^{-\lambda(t-t_0)} \quad \forall x_0 \in \mathcal{X} \text{ and } \forall t \geq t_0 \geq 0 \tag{2.8}$$

and that there exists a constant $K > \lambda$ such that

$$\left|\frac{\partial f}{\partial x}(t, x)\right| \leq K \quad \forall x \in \mathbb{R}^n \text{ and } \forall t \in [0, \infty). \tag{2.9}$$

Then the function

$$V(t, \xi) = 2\int_t^{t+\delta} |x(\tau, t, \xi)|^2 \mathrm{d}\tau, \text{ where } \delta = \frac{\ln(2D^2)}{2\lambda} \tag{2.10}$$

admits constants $c_1, c_2, c_3 > 0$ such that

$$\begin{aligned} &c_1|\xi|^2 \leq V(t, \xi) \leq c_2|\xi|^2 , \ |V_\xi(t, \xi)| \leq c_3|\xi| , \text{ and} \\ &V_t(t, \xi) + V_\xi(t, \xi) f(t, \xi) \leq -|\xi|^2 \end{aligned} \tag{2.11}$$

hold for all $t \in [0, \infty)$ and $\xi \in \mathbb{R}^n$, and therefore is a strict Lyapunov function for the system.

---

[1] For example, $\dot{x} = -\arctan(x) + u$ is GAS when we choose $u \equiv 0$, but $\dot{x} = -\arctan(x) + d$ is not ISS, because the bounded disturbance $d \equiv 2$ produces unbounded trajectories.

Formula (2.10) is non-explicit, because the flow map in the integrand cannot ordinarily be obtained in closed form, except in basic cases where (2.4) is linear and time-invariant. In Chap. 10, we explicitly construct strict Lyapunov functions for a class of *nonlinear* time-varying systems that satisfy the conclusions of Lemma 2.2. Lemma 2.2 can be extended to time-varying systems that are not necessarily exponentially stable. For example, we have the following from [70, Chap. 4, p.167]:

Theorem 2.1. *Assume that (2.4) is UGAS to the origin, and that* $f : [0,\infty)\times\mathcal{X}\to\mathbb{R}^n$ *is* C^1. *Let* $r>0$ *be any constant such that* $r\mathcal{B}_n\subseteq\mathcal{X}$, *and assume that* $\frac{\partial f}{\partial x}$ *is bounded on* $[0,\infty)\times r\mathcal{B}_n$. *Let* $\beta\in\mathcal{KL}$ *and the constant* $r_0>0$ *be such that* $r_0\le r$, $\beta(r_0,0)<r$ *and*

$$|x(t,t_0,x_0)|\le\beta(|x_0|,t-t_0)\quad\forall t\ge t_0\ge 0\ \text{ and }\ x_0\in\mathcal{X}. \tag{2.12}$$

Then the following conclusions hold: (a) There exist a C^1 *function* $V : [0,\infty)\times(r_0\mathcal{B}_n)\to\mathbb{R}$ *and continuous positive definite increasing functions* $\alpha_i : [0,r_0]\to[0,\infty)$ *such that the following hold on* $[0,\infty)\times r_0\mathcal{B}_n$:

$$\begin{aligned}\alpha_1(|x|)\ &\le\ V(t,x)\ \le\ \alpha_2(|x|);\\ \frac{\partial V}{\partial t}+\frac{\partial V}{\partial x}f(t,x)\ &\le\ -\alpha_3(|x|);\ \ \text{and}\\ \left|\frac{\partial V}{\partial x}(t,x)\right|\ &\le\ \alpha_4(|x|).\end{aligned} \tag{2.13}$$

(b) If $\mathcal{X}=\mathbb{R}^n$ *and* $\frac{\partial f}{\partial x}$ *is bounded, then we can find a* C^1 *function* V *and functions* $\alpha_1,\ldots\alpha_4\in\mathcal{K}_\infty$ *such that (2.13) hold for all* $t\ge 0$ *and* $x\in\mathbb{R}^n$. *If, in addition, the system (2.4) is time-invariant, then* V *can be taken to be time-invariant; while if (2.4) is periodic in* t, *then* V *can be taken to be periodic in* t *as well.*

The strict Lyapunov function in the proof of Theorem 2.1 is also expressed in terms of the flow map and so is non-explicit; see Appendix B.1 for the main ideas from the proof. The challenge is to obtain explicit formulas for global strict Lyapunov functions that do not involve the flow map.

2.1.3 ISS Lyapunov Function

Consider the system with inputs

$$\dot{x}=f(t,x,u),\quad x\in\mathcal{X},\ u\in U \tag{2.14}$$

satisfying our standing assumptions from the previous chapter. For simplicity, we assume in the rest of this subsection that the state space $\mathcal{X}$ for (2.14) is all of $\mathbb{R}^n$. When $\mathcal{X} = \mathbb{R}^n$, we call a function $V : [0,\infty) \times \mathcal{X} \to [0,\infty)$ a *storage function* provided that there exist $\alpha_1, \alpha_2 \in \mathcal{K}_\infty$ so that $\alpha_1(|x|) \leq V(t,x) \leq \alpha_2(|x|)$ everywhere; in this case, we also say that V is *uniformly proper and positive definite*, or of class UPPD. For systems with inputs, we typically make the following more stringent assumption on V: A C^1 function $V : [0,\infty) \times \mathbb{R}^n \to [0,\infty)$ is said to be of class UBPPD (written $V \in$ UBPPD) provided (i) it is a storage function and (ii) its gradient is uniformly bounded in t, meaning there exists a function $\alpha_3 \in \mathcal{K}_\infty$ such that for all $t \geq 0$ and $x \in \mathbb{R}^n$, we have

$$|\nabla V(t,x)| \leq \alpha_3(|x|). \tag{2.15}$$

Notice that (2.15) is redundant when $V \in C^1$ is periodic in t. The corresponding notions of iISS and ISS Lyapunov functions are as follows:

Definition 2.1. Assume that $V : [0,\infty) \times \mathbb{R}^n \to [0,\infty)$ is a C^1 storage function. We say that V is an iISS Lyapunov function for (2.14) provided there exist a positive definite function α_3 and a function $\gamma \in \mathcal{K}_\infty$ such that

$$\dot{V}(t,x,u) \;\leq\; -\alpha_3(|x|) + \gamma(|u|) \tag{2.16}$$

for all $x \in \mathbb{R}^n$, $t \geq 0$, and $u \in U$. If, in addition, $\alpha \in \mathcal{K}_\infty$, then we call V an ISS Lyapunov function.

The following was established by Sontag and Wang in [169] for time-invariant ISS systems but the time-varying systems version can be shown by similar arguments [39]. The iISS statement was shown in [8].

Lemma 2.3. *Let (2.14) be periodic in t. The system (2.14) is iISS (resp., ISS) if and only if it admits a C^1 iISS (resp., ISS) Lyapunov function.*

As in the case where there are no controls, the converse parts of this lemma do not in general lead to explicit Lyapunov functions. On the other hand, if we know an explicit ISS Lyapunov function for (2.14), then we can use standard arguments to derive explicit formulas for the functions $\beta \in \mathcal{KL}$ and $\gamma \in \mathcal{K}$ in the ISS estimate. Let us sketch the derivation.

Let V be an ISS Lyapunov function for (2.14). Choose $\alpha_1, \alpha_2 \in \mathcal{K}_\infty$ so that $\alpha_1(|x|) \leq V(t,x) \leq \alpha_2(|x|)$ everywhere, and let the functions $\alpha_3, \gamma \in \mathcal{K}_\infty$ satisfy the requirements of Definition 2.1. Setting $\alpha(s) = \min\{s, \alpha_3 \circ \alpha_2^{-1}(s)\}$ gives

$$\dot{V} \leq -\alpha(V) + \gamma(|u|_\infty)$$

along all trajectories of (2.14). Hence, along any trajectory of (2.14),

$$V \geq \alpha^{-1}(2\gamma(|u|_\infty)) \quad \Rightarrow \quad \dot{V} \leq -\tfrac{1}{2}\alpha(V).$$

Let $u \in \mathcal{M}(U)$, $x_0 \in \mathcal{X}$, and $t_0 \geq 0$ be given, and let $x(t)$ denote the corresponding trajectory of (2.14) satisfying $x(t_0) = x_0$.

Arguing as in [82, Lemma 4.4, p.135] with the function $y(r) = V(t_0 + r, x(t_0 + r))$ shows that if $\dot{V} \leq -0.5\alpha(V)$ on any interval $[t_0, \bar{t}]$, then

$$V(t, x(t)) \le \beta_\alpha(V(t_0, x(t_0)), t - t_0)$$

on that interval, where $\beta_\alpha \in \mathcal{KL}$ is

$$\beta_\alpha(s,t) = \begin{cases} 0, & \text{if } s = 0 \\ s + \Phi^{-1}(\Phi(s) + t), & \text{if } s > 0 \end{cases} \tag{2.17}$$

and

$$\Phi(s) = \begin{cases} +\infty, & \text{if } s = 0 \\ -2\int_1^s \frac{\mathrm{d}r}{\alpha(r)}, & \text{if } s > 0 \end{cases} . \tag{2.18}$$

The fact that $\Phi(s) \to \infty$ as $s \to 0^+$ is used to show that $\beta(s,t) \to 0$ as $s \to 0^+$ for each $t \geq 0$ [82]. A standard invariance argument that is analogous to the one used in [157] shows that if

$$V(t, x(t)) \le \alpha^{-1}(2\gamma(|u|_\infty))$$

for a given trajectory $x(t))$ of (2.14) at a given time $t = \tilde{t}$, then this inequality remains true for all $t \geq \tilde{t}$. Hence, we can take

$$\beta(s,t) = \alpha_1^{-1} \circ \beta_\alpha(\alpha_2(s), t) \text{ and } \gamma(r) = \alpha_1^{-1} \circ \alpha^{-1}(2\gamma(r))$$

to satisfy our requirements.

This makes it possible to explicitly quantify the effects of the disturbance, while at the same time obtaining the decay rate on the norm of the state, which is valuable in applications. See Sect. 2.4.2 for a specific example where β and γ are computed. Analogous arguments can be carried out for the iISS case; see [8]. This motivates our search for explicit iISS Lyapunov functions as well.

2.2 Non-strict Lyapunov Function

Our main building blocks for strict Lyapunov functions will be non-strict Lyapunov functions (which are also called weak Lyapunov functions). Non-strict Lyapunov functions V are defined in exactly the same way as strict Lyapunov functions except instead of the decay condition $\dot{V} < 0$ outside the equilibrium state, we have $\dot{V} \le 0$. A positive definite function V that satisfies all requirements for being a (non-)strict Lyapunov function except for properness is called a *(non-)strict Lyapunov-like function.* In this subsection, we discuss three contexts in which non-strict Lyapunov functions naturally arise. In Chapters 3-5, we provide systematic mechanisms for building strict Lyapunov functions in each of these contexts.

2.2.1 Matrosov Theorems

Matrosov's Theorem [97] provides a Lyapunov approach to proving stability without having to construct a strict Lyapunov function. In its original formulation, it concludes uniform asymptotic stability of time-varying systems by using a non-strict Lyapunov function and an auxiliary function whose time derivative along trajectories is non-zero at all points $x \in \mathbb{R}^n \setminus \{0\}$ where the derivative of the non-strict Lyapunov function is zero. There are various generalizations of the original Matrosov result, involving an arbitrary number of auxiliary functions [86]. These generalizations are referred to as Matrosov Theorems, and they prove uniform asymptotic stability as well.

While the original motivation for Matrosov's Theorem was to eliminate the need for a strict Lyapunov function, it is still important to be able to construct explicit strict Lyapunov functions for systems satisfying Matrosov's Conditions [111]. However, the proofs in [86, 97] do not construct strict Lyapunov functions. Instead, they conclude uniform asymptotic stability by directly analyzing the trajectories of the system. One standard formulation is the following result from [145], where we maintain our standing assumptions on (2.4) from the previous chapter:

Theorem 2.2. *Assume that there are constants $\bar{R} > \underline{R} > 0$ and $L > 0$, functions $\underline{\alpha}, \bar{\alpha} \in \mathcal{K}$, and continuous functions*

$$\begin{aligned} &V_1 : [0, \infty) \times \mathrm{int}(\bar{R}\mathcal{B}_n) \to \mathbb{R}, \\ &V_2 : [0, \infty) \times \mathrm{int}(\bar{R}\mathcal{B}_n) \to \mathbb{R}, \text{ and} \\ &W : \bar{R}\mathcal{B}_n \to \mathbb{R} \end{aligned}$$

for which $\dot{V}_1(t,x)$ and $\dot{V}_2(t,x)$ are continuous and the following hold:

1. $V_1(t,0) = \dot{V}_1(t,0) = 0$ *for all* $t \geq 0$;
2. $\max\{|V_2(t,x)|, |f(t,x)|\} \leq L$ *for all* $(t,x) \in [0,\infty) \times \underline{R}\mathcal{B}_n$;
3. $\underline{\alpha}(|x|) \leq V_1(t,x) \leq \bar{\alpha}(|x|)$ *for all* $(t,x) \in [0,\infty) \times \underline{R}\mathcal{B}_n$;
4. $\dot{V}_1(t,x) \leq W(x) \leq 0$ *for all* $(t,x) \in [0,\infty) \times \underline{R}\mathcal{B}_n$; *and*
5. $\dot{V}_2(t,x)$ *is non-zero definite on* $\{x \in \underline{R}\mathcal{B}_n : W(x) = 0\}$.

Then $\lim_{t\to+\infty} x(t,t_0,x_0) = 0$ for each solution $x(\cdot,t_0,x_0)$ of (2.4) that remains in $\underline{R}\mathcal{B}_n$ for all $t \geq t_0$.

For the proof, see [145]. By non-zero definiteness of a function $G : [0,\infty) \times \bar{R}\mathcal{B}_n \to \mathbb{R}$ on a closed set $M \subseteq \bar{R}\mathcal{B}_n$, we mean that for each pair of constants (ν, ε) for which $0 < \nu < \varepsilon \leq \bar{R}$, there are values $\gamma, \delta > 0$ such that:

$$\Big[\{\nu \leq |x| \leq \varepsilon\} \text{ and } \{|x|_M < \gamma\} \text{ and } \{t \in [0,\infty)\}\Big] \;\Rightarrow\; |G(t,x)| > \delta,$$

where $|x|_M \doteq \inf\{|x - q| : q \in M\}$ is the distance of x from M. When G is independent of t, non-zero definiteness simply says that G is bounded away from zero on the part of any annulus around 0 that is close enough to M.

2.2.2 LaSalle Invariance Principle

Recall that a set $M \subseteq \mathbb{R}^n$ is called *positively invariant* for a forward complete time-invariant dynamics

$$\dot{x} = f(x) \tag{2.19}$$

evolving on an open set $\mathcal{X}$ provided for each $x_0 \in M$, the corresponding solution $t \mapsto x(t, x_0)$ remains in M for all times $t \geq 0$. The set M is called *invariant* for (2.19) if the system is complete and each such trajectory is in M for all $t \in \mathbb{R}$. LaSalle's Invariance Theorem is the following result, which is shown, e.g., in [70, Sect. 4.2]:

Lemma 2.4. *Let the compact set $\Omega \subseteq \mathcal{X}$ be positively invariant for (2.19), and $V : \mathcal{X} \to \mathbb{R}$ be a C^1 function for which $L_f V(x) \leq 0$ on Ω. Let M be the largest invariant subset of*

$$E \doteq \{x \in \Omega : L_f V(x) = 0\}$$

for (2.19). Then every solution of (2.19) converges to M as $t \to +\infty$. If the preceding assumptions hold except with $\Omega = \mathbb{R}^n$ and $f(0) = 0$, and if no solution of (2.19) can stay in E for all times $t \geq 0$ except for the trivial solution $x(t) \equiv 0$, then the origin is GAS.

The preceding result can be extended to time-varying systems. For example, we have the following from [148, Sect. 5.4]:

Lemma 2.5. *Consider the system (2.4) with state space $\mathcal{X} = \mathbb{R}^n$. Assume that $f(t, x)$ and $V(t, x)$ have the same period $T > 0$ in t, where V is a C^1 storage function. If $\dot{V}(t, x) \leq 0$ for all $t \geq 0$ and all $x \in \mathbb{R}^n$, and if the largest invariant set for (2.4) in*

$$S \doteq \left\{x \in \mathbb{R}^n : \dot{V}(t, x) = 0 \ \forall t \geq 0\right\}$$

is $\{0\}$, then 0 is a UGAS equilibrium for (2.4).

See also [148, Sect. 5.5] for generalized LaSalle Theorems for time-varying systems that are not necessarily periodic in time. LaSalle Invariance provides another method for proving stability without having to find a strict Lyapunov function, but it is of limited use when the system is subject to disturbances. This is because small perturbations of the dynamics can cause $\dot{V}(t, x)$ to become positive at some pairs (t, x).

2.2.3 Jurdjevic-Quinn Theorem

Consider a general control affine system

$$\dot{x} \;=\; f(x) + g(x)u \;\;,\;\; \text{where } g(x) = (g_1(x), \ldots, g_p(x)) \qquad (2.20)$$

evolving on $\mathbb{R}^n$ in which the vector fields $f : \mathbb{R}^n \to \mathbb{R}^n$ and $g_i : \mathbb{R}^n \to \mathbb{R}$ are smooth and $f(0) = 0$. We use the standard Lie bracket notation

$$\begin{aligned} &\mathrm{ad}_f^0(g) = g, \quad \mathrm{ad}_f(g) = [f, g] = g_* f - f_* g, \\ &\text{and } \; \mathrm{ad}_f^k(g) = \mathrm{ad}_f\left(\mathrm{ad}_f^{k-1}(g)\right) \end{aligned}$$

for all $k > 1$ and all smooth vector fields $f, g : \mathbb{R}^n \to \mathbb{R}^n$, where the star subscript indicates a Jacobian. In [68], Jurdjevic and Quinn proved the following single input result:

Theorem 2.3. *(Jurdjevic-Quinn Theorem) Consider the system (2.20) with state space $\mathcal{X} = \mathbb{R}^n$ and $p = 1$. Assume the following:*

1. $f(x) = Ax$ for some skew symmetric matrix A; and [2]

2. for all $x \in \mathbb{R}^n \setminus \{0\}$, we have $\mathrm{span}\{(ad_f^k(g))(x) : k = 0, 1, 2, \ldots\} = \mathbb{R}^n$.

Then the feedback $u(x) = -x^\top g(x)$ renders (2.20) GAS to zero.

Proof. By a simple calculation,

$$\frac{d}{dt}|x(t, x_0)|^2 \;=\; -2u^2(x(t, x_0)) \leq 0$$

for all $t \geq 0$ and $x_0 \in \mathbb{R}^n$ along all trajectories $t \mapsto x(t, x_0)$ of the closed-loop system, because $x^\top Ax = 0$ for all x. We show that no solution can stay in $E \doteq \{x : u(x) = 0\}$ except for the trivial solution. The GAS property will then follow from the LaSalle Invariance Principle.

Fix $x_0 \in E$ and consider the function $\Gamma(t) \doteq \langle e^{tA}x_0, g(e^{tA}x_0)\rangle$. Since $x(t, x_0) = e^{tA}x_0 \in E$ for all $t \geq 0$, we get $\Gamma(t) = 0$ for all $t \geq 0$. A simple inductive argument gives

$$0 = \frac{d^k}{dt^k}\Gamma(t) = \left\langle x_0, e^{-tA}\mathrm{ad}_f^k(g)(e^{tA}x_0)\right\rangle$$

for all $t \geq 0$. Evaluating these higher time derivatives at $t = 0$ shows that $\langle x_0, (ad_f^k(g))(x_0)\rangle = 0$ for all $k \geq 0$. Hence, $x_0 = 0$, by Assumption 2. □

---

[2] A slight variant of the argument that we are about to give applies if the skew symmetry assumption on A is replaced by the assumption that there is an invertible matrix M such that $MAM^{-1} = MAM^\top = J$ is skew symmetric, by showing that the dynamics for $y = Mx$ is GAS to the origin with the feedback $u(y) = -\langle y, Mg(M^\top y)\rangle$. In fact, the dynamics are $\dot{y} = F(y) + G(y)u$, where $F(y) = Jy$ and $G(y) = Mg(M^\top y)$, so the proof follows from the relations $(\mathrm{ad}_F^k(G))(y) = M(\mathrm{ad}_f^k(g))(M^\top y)$.

The Jurdjevic-Quinn Theorem has been generalized in several works. In general, conditions that provide a smooth asymptotically stabilizing control law using a first integral of the drift vector field, under some controllability conditions, are now called Jurdjevic-Quinn Conditions. A general set of Jurdjevic-Quinn Conditions for cases where the vector field f in (2.20) is not required to be linear is as follows.

Definition 2.2. We say that (2.20) satisfies the *(Weak) Jurdjevic-Quinn Conditions* provided there exists a smooth function $V : \mathbb{R}^n \to \mathbb{R}$ satisfying:

1. V is positive definite and radially unbounded;
2. for all $x \in \mathbb{R}^n$, $L_f V(x) \le 0$; and
3. there exists an integer l such that the set

$$W(V) = \left\{ \begin{array}{l} x \in \mathbb{R}^n : \forall k \in \{1, \ldots, p\} \text{ and } i \in \{0, \ldots, l\}, \\ L_f V(x) = L_{ad_f^i(g_k)} V(x) = 0 \end{array} \right\}$$

equals $\{0\}$.

If (2.20) satisfies the Weak Jurdjevic-Quinn Conditions, then it is globally asymptotically stabilized by any feedback

$$u = -\xi(x) L_g V(x)^\top,$$

where ξ is any everywhere positive function of class C^1 [41]. The proof of this result also follows from the LaSalle Invariance Principle. However, it is far from clear how to construct CLFs for systems satisfying the Weak Jurdjevic Quinn Conditions. We address this CLF construction problem in Chap. 4.

2.3 Discrete Time Lyapunov Function

The preceding definitions have analogs for discrete time systems

$$x_{k+1} = f(k, x_k, u_k) \tag{2.21}$$

with equilibrium state 0 by replacing the continuous time $t \ge 0$ with the discrete time $k \in \{0, 1, 2, \ldots\}$ and replacing the time derivative $\dot{V}(t, x, u)$ of the Lyapunov function along trajectories with the first difference

$$\Delta V(k, x, u) = V\big(k + 1, f(k, x, u)\big) - V(k, x) \tag{2.22}$$

in the conditions defining Lyapunov functions.

The definition of discrete time Lyapunov functions does not require V to be C^1 because there are no derivatives of V in (2.22). Also, discrete time strict Lyapunov functions have the property that along each trajectory sequence

$\{x_k\}$ of (2.21), the sequence $V(k, x_k)$ is decreasing in k, as long as $x_k \neq 0$. If instead V is only a *non-strict* Lyapunov function, then $V(k, x_k)$ is non-increasing. We focus on continuous time systems in much of the sequel, but most of the results to follow have discrete time analogs.

2.4 Illustrations

2.4.1 Strict Lyapunov Function

Let us again consider the system

$$\begin{cases} \dot{x}_1 = u_1 \\ \dot{x}_2 = u_2 u_1 \end{cases} \tag{2.23}$$

from Sect. 1.4. In closed-loop with the stabilizing feedbacks

$$\begin{aligned} u_1 &= -x_1 + \sin(t)(\cos(t)x_1 + x_2) \\ u_2 &= -\sin(t) - \cos(t), \end{aligned} \tag{2.24}$$

the system becomes

$$\begin{cases} \dot{x}_1 = -x_1 + \sin(t)\big[\cos(t)x_1 + x_2\big] \\ \dot{x}_2 = \big[-\sin(t) - \cos(t)\big]\big[-x_1 + \sin(t)\big(\cos(t)x_1 + x_2\big)\big]. \end{cases} \tag{2.25}$$

We now show that (2.25) admits the global strict Lyapunov function

$$V_s(t, x) = \frac{1}{2}x_1^2 + \left(4 + \frac{\pi}{2} - 2\sin(t)\cos(t)\right)[\cos(t)x_1 + x_2]^2. \tag{2.26}$$

In later chapters, we provide general methods for constructing global strict Lyapunov functions.

Since

$$\begin{aligned} [\cos(t)x_1 + x_2]^2 &\geq \cos^2(t)x_1^2 + x_2^2 - \left(\frac{3}{4}x_2^2 + \frac{4}{3}\cos^2(t)x_1^2\right) \\ &\geq -\frac{1}{3}x_1^2 + \frac{1}{4}x_2^2, \end{aligned}$$

one easily checks that the inequalities

$$\frac{1}{6}[x_1^2 + x_2^2] \leq V_s(t, x) \leq 17[x_1^2 + x_2^2] \tag{2.27}$$

are satisfied everywhere. Also, the time derivative of V_s along trajectories of (2.25) is

$$\begin{aligned}\dot{V}_s(t,x) = & -x_1^2 + \sin(t)x_1[\cos(t)x_1 + x_2] \\ & +2\left(\sin^2(t) - \cos^2(t)\right)[\cos(t)x_1 + x_2]^2 \\ & +2\left(4 + \tfrac{\pi}{2} - 2\sin(t)\cos(t)\right)[\cos(t)x_1 + x_2] \\ & \times[-\sin(t)x_1 + \cos(t)\dot{x}_1 + \dot{x}_2].\end{aligned} \tag{2.28}$$

Setting $\zeta = \cos(t)x_1 + x_2$, we obtain

$$\begin{aligned}\dot{V}_s(t,x) &= -x_1^2 + \sin(t)x_1\zeta + 2\left(\sin^2(t) - \cos^2(t)\right)\zeta^2 \\ &\quad +2\left(4 + \tfrac{\pi}{2} - 2\sin(t)\cos(t)\right)\zeta \\ &\quad \times\left[-\sin(t)x_1 + \cos(t)(-x_1 + \sin(t)\zeta)\right. \\ &\quad \left. +\{-\sin(t) - \cos(t)\}(-x_1 + \sin(t)\zeta)\right] \\ &= -x_1^2 + \sin(t)x_1\zeta + 2\left(\sin^2(t) - \cos^2(t)\right)\zeta^2 \\ &\quad -2\left(4 + \tfrac{\pi}{2} - 2\sin(t)\cos(t)\right)\sin^2(t)\zeta^2 \\ &= -x_1^2 + \sin(t)x_1\zeta \\ &\quad +\left[-2 - (4+\pi)\sin^2(t) + 4\sin^3(t)\cos(t)\right]\zeta^2 \\ &\le -\tfrac{1}{2}x_1^2 - \zeta^2 \\ &= -\tfrac{1}{2}x_1^2 - (\cos(t)x_1 + x_2)^2 \le -\tfrac{1}{6}[x_1^2 + x_2^2],\end{aligned} \tag{2.29}$$

where the first inequality used the relation

$$\sin(t)x_1\zeta \le \frac{1}{2}x_1^2 + \frac{1}{2}\zeta^2$$

and the second inequality used

$$-(\cos(t)x_1 + x_2)^2 \le -\cos^2(t)x_1^2 - x_2^2 + \frac{4}{3}\cos^2(t)x_1^2 + \frac{3}{4}x_2^2.$$

Since $\dot{V}_s(t,x)$ has a negative definite upper bound, it follows from (2.27) that V_s is a strict Lyapunov function for the system (2.25).

2.4.2 ISS Lyapunov Function

We next consider the case where there is additive noise in the u_1 input in (2.23). We show that the resulting closed-loop system

$$\begin{cases} \dot{x}_1 = -x_1 + \sin(t)[\cos(t)x_1 + x_2] + \delta_1(t) \\ \dot{x}_2 = [-\sin(t) - \cos(t)][-x_1 + \sin(t)(\cos(t)x_1 + x_2) + \delta_1(t)] \end{cases} \tag{2.30}$$

with the controllers (2.24) is ISS with respect to the disturbance δ_1. Our strategy is to show that (2.26) is an ISS Lyapunov function for (2.30), which will lead to explicit functions β and γ in the ISS estimate.

To this end, first note that (2.26) satisfies

$$\max_{i=1,2} \left| \frac{\partial V_s}{\partial x_i}(t,x) \right| \leq 17|x|_1 \;\; \forall x = (x_1\; x_2) \in \mathbb{R}^2 \text{ and } t \geq 0. \tag{2.31}$$

Hence, the last inequality of (2.29) implies that the time derivative of V_s along the trajectories of (2.30) satisfies

$$\begin{aligned} \dot{V}_s(t,x) &\leq -\tfrac{1}{6}|x|^2 + 51|x|_1|\delta_1(t)| \\ &\leq -\tfrac{1}{6}|x|^2 + \{|x|\}\{102|\delta_1(t)|\} \\ &\leq -\tfrac{1}{12}|x|^2 + 3 \times 102^2\delta_1^2(t), \end{aligned} \tag{2.32}$$

by the triangle inequality

$$pq \leq \tfrac{1}{2\varepsilon}p^2 + \tfrac{\varepsilon}{2}q^2$$

applied to the terms in braces with $p = |x|$, $q = 102|\delta_1(t)|$, and $\varepsilon = 6$. This and the proper positive definiteness condition (2.27) imply that V_s is an ISS Lyapunov function for (2.30). As we saw in Sect. 2.1.3, explicit ISS Lyapunov functions lead to explicit expressions for the functions $\beta \in \mathcal{KL}$ and $\gamma \in \mathcal{K}_\infty$ in the ISS estimate. We derive these expressions next for the dynamics (2.30).

Combining (2.32) with the inequalities (2.27) gives

$$\dot{V}_s(t,x) \leq -\tfrac{1}{204}V_s(t,x) + 3 \times 102^2\delta_1^2(t). \tag{2.33}$$

By integrating this inequality, we deduce that for any initial condition $x(t_0) = x_0$, the corresponding trajectories satisfy

$$V_s(x(t),t) \leq e^{-\frac{t-t_0}{204}}V_s(x(t_0),t_0) + 612 \times 102^2|\delta_1|^2_{[t_0,t]}. \tag{2.34}$$

Combining (2.34) with (2.27), and then using the relation

$$\sqrt{p+q} \;\leq\; \sqrt{p} + \sqrt{q}$$

for $p, q \geq 0$ gives the desired ISS estimate

$$|x(t,t_0,x_0,u)| \leq \beta(|x_0|, t-t_0) + \gamma(|\delta_1|_\infty)$$

with the choices $\beta(r,s) = 11re^{-s/408}$ and $\gamma(r) = 102\sqrt{3672}r$.

2.4.3 iISS Lyapunov Function

If we allow additive disturbances in both the u_1 and u_2 channels in the dynamics (2.23) and use the feedbacks (2.24) as before, then the corresponding closed-loop system

$$\begin{cases} \dot{x}_1 = -x_1 + \sin(t)\zeta + \delta_1(t) \\ \dot{x}_2 = [-\sin(t) - \cos(t) + \delta_2(t)][-x_1 + \sin(t)\zeta + \delta_1(t)] \end{cases} \tag{2.35}$$

where

$$\zeta = \cos(t)x_1 + x_2$$

is not ISS with respect to the disturbance $\delta = (\delta_1, \delta_2)$. This follows by applying Lemma 1.2 with the disturbance

$$\delta = (0, \sin(t) + \cos(t) + 1),$$

$Lx = (x_2 - x_1, x_2)$, and $k = 1$. On the other hand, (2.35) is iISS with respect to δ. We show this next using the strict Lyapunov function (2.26). Our arguments are a time-varying analog of those of [8, pp.1091-2].

Using (2.31) and the first inequality in (2.32) gives

$$\begin{aligned} \dot{V}_s(t,x) &\le -\tfrac{1}{6}|x|^2 + 51|x|_1|\delta| + 17|x|_1(2|x|_1 + |\delta|)|\delta| \\ &\le -\alpha_3(|x|) + \lambda(|x|)\Delta(|\delta|), \end{aligned} \tag{2.36}$$

and

$$\alpha_1(|x|) \;\le\; V_s(t,x) \;\le\; \alpha_2(|x|)$$

everywhere, where

$$\begin{aligned} &\alpha_1(r) = \alpha_3(r) = \tfrac{1}{6}r^2, \;\; \alpha_2(r) = 17r^2, \\ &\lambda(r) = 136(r + r^2), \;\; \text{and} \;\; \Delta(r) = r + r^2. \end{aligned}$$

Taking

$$\begin{aligned} &W(t,x) = \Pi(V_s), \;\; \text{where } \Pi(r) = \int_0^r \frac{\mathrm{d}s}{1 + \chi(s)} \;\; \text{and} \\ &\chi(s) = \lambda \circ \alpha_1^{-1}(s) = 136\sqrt{6s} + 816s, \end{aligned}$$

it follows that

$$\dot{W}(t,x) \;\le -\rho(|x|) + \Delta(|\delta|), \;\; \text{where } \rho(r) = \frac{\alpha_3(r)}{1 + \lambda(\alpha_1^{-1}(\alpha_2(r))}.$$

Moreover, W is also proper and positive definite, and ρ is positive definite. Therefore, W is an iISS Lyapunov function for (2.35). It follows from Lemma 2.3 that (2.35) is iISS.

2.4.4 LaSalle Invariance Principle

Consider the system

$$\begin{cases} \dot{x}_1 = x_2 \\ \dot{x}_2 = -x_1 - x_2^3 \end{cases} \tag{2.37}$$

with $x \in \mathbb{R}^2$. The function

$$V(x) = \frac{1}{2}[x_1^2 + x_2^2] \tag{2.38}$$

satisfies

$$\dot{V}(x) = -W(x), \tag{2.39}$$

where $W(x) = x_2^4$. However, this is not enough to conclude that the system is GAS to 0 because W is non-negative definite but not positive definite. Instead, we use W in conjunction with the LaSalle Invariance Principle to show the GAS property for (2.37).

Let $x(t)$ be any solution of (2.37) with any initial state x_0. By (2.39),

$$x(t) \in \Omega_{x_0} = \{x \in \mathbb{R}^2 : |x| \le |x_0|\} \ \forall t \ge 0.$$

We deduce from Lemma 2.4 that $x(t)$ converges to the largest invariant set $\mathcal{S}$ contained in

$$E_{x_0} = \{x \in \Omega_{x_0} : x_2 = 0\}.$$

We show that $\mathcal{S} = \{0\}$. Let $z(t) = (z_1(t)\ z_2(t))$ be any solution of the system (2.37) with initial condition $z_0 \in \mathcal{S}$. Since $\mathcal{S}$ is positively invariant, we have $z_2(t) = 0$ for all $t \ge 0$. Therefore $\dot{z}_2(t) = 0$ for all $t \ge 0$, so

$$-z_1(t) - z_2^3(t) = 0$$

for all $t \ge 0$. It follows that $z_1(t) = 0$ for all $t \ge 0$. We deduce that $\mathcal{S} = \{0\}$ which implies that all the solutions of (2.37) converge to the origin, by LaSalle Invariance.

2.4.5 Matrosov Theorems

The GAS of the origin of (2.37) can be established through the version of the Matrosov Theorem we gave in Theorem 2.2, as follows. We show that for all positive constants $\underline{R}$, all solutions of the system with initial states in $\underline{R}\mathcal{B}_2$ remain in $\underline{R}\mathcal{B}_2$ and converge to 0 as $t \to +\infty$. We apply the theorem with

$$\begin{array}{l} V_1(x) = \frac{1}{2}[x_1^2 + x_2^2]\,, \quad V_2(x) = x_1 x_2\,, \quad \text{and} \\ W(x) = -x_2^4. \end{array} \tag{2.40}$$

The conditions are verified as follows:

1. $V_1(0) = \dot{V}_1(0) = 0$ along the trajectories of (2.37);
2. for all $x \in \underline{R}\mathcal{B}_2$, we have $\max\{|V_2(x)|, |f(x)|\} \leq |x|^2 + \sqrt{x_2^2 + (x_1 + x_2^3)^2} \leq \underline{R}^2 + (\underline{R}^2 + (\underline{R} + \underline{R}^3)^2)^{1/2}$;
3. $\underline{\alpha}(|x|) \leq V_1(x) \leq \bar{\alpha}(|x|)$ for all x when we choose $\bar{\alpha}(r) = \underline{\alpha}(r) = \frac{1}{2}r^2$;
4. $\dot{V}_1(x) = W(x) = -x_2^4 \leq 0$ for all $x \in \underline{R}\mathcal{B}_2$; and
5. $\dot{V}_2(x) = \dot{x}_1 x_2 + x_1 \dot{x}_2 = x_2^2 + x_1(-x_1 - x_2^3)$ so for any constants $\nu, \varepsilon > 0$, we can find a constant $\gamma > 0$ so that $|\dot{V}_2(x)|$ is bounded away from zero on $\{x \in \mathbb{R}^2 : \nu \leq |x| \leq \varepsilon, |x_2| < \gamma\}$. Hence, $\dot{V}_2$ is non-zero definite on $\{x \in \underline{R}\mathcal{B}_2 : x_2^4 = 0\}$.

We conclude from Matrosov's Theorem and the fact that $\dot{V}_1 \leq 0$ that every solution of (2.37) with initial condition in $\underline{R}\mathcal{B}_2$ remains in $\underline{R}\mathcal{B}_2$ and converges to 0 as $t \to +\infty$. We deduce that the origin of (2.37) is GAS.

2.4.6 Non-strict Lyapunov-Like Function

Consider an experimental anaerobic digester used to treat waste water [16, 89, 172]. This process degrades a polluting organic substrate s with the anaerobic bacteria x and produces a methane flow rate y_1. The methane and substrate can generally be measured, so the system with output y is

$$\begin{cases} \dot{s} = u(s_{in} - s) - kr(s,x) \\ \dot{x} = r(s,x) - \alpha u x \\ y = (\lambda r(s,x), s) \end{cases} \tag{2.41}$$

where the biomass growth rate r is any non-negative C^1 function that admits everywhere positive functions $\underline{\Delta}$ and $\bar{\Delta}$ such that

$$s\bar{\Delta}(s,x) \;\geq\; r(s,x) \;\geq\; xs\underline{\Delta}(s,x) \tag{2.42}$$

for all $s \geq 0$ and $x \geq 0$; u is the non-negative input (i.e., the dilution rate); α is a known positive real number representing the fraction of the biomass in the liquid phase; and λ, k, and s_{in} are positive constants representing methane production and substrate consumption yields and the influent substrate concentration, respectively. Hence, the methane flow rate is $y_1 = \lambda r(s,x)$. This includes the single species undisturbed chemostat model from Sect. 1.5.3, in which case $r(s,x)$ is the product of the species concentration and the Monod growth rate function, namely,

$$r(s,x) = \frac{Asx}{B+s}$$

for appropriate positive constants A and B. However, (2.42) is far more general because it allows other growth laws such as those of Haldane and Cantois; see [14] for details.

Assume now that $s_* \in (0, s_{in})$ is a given constant. We wish to regulate s to s_*. We assume that there are known constants $\gamma_M > \gamma_m > 0$ such that

$$\gamma_* \doteq \frac{k}{\lambda(s_{in} - s_*)} \in (\gamma_m, \gamma_M) \quad \text{and} \quad \frac{k}{\lambda s_{in}} < \gamma_m \tag{2.43}$$

and we use the notation

$$v_* = s_{in} - s_* \quad \text{and} \quad x_* = \frac{v_*}{k\alpha}$$

in the sequel.

The work [89] leads to a non-strict Lyapunov-like function and an adaptive controller for an error dynamics associated with (2.41). We next review these earlier results. We treat adaptive control in detail in Chap. 9. Later, we will see how the constructions from [89] lead to a strict Lyapunov-like function for the error dynamics of $(\tilde{s}, \tilde{x}) = (s - s_*, x - x_*)$.

We introduce the dynamics

$$\dot{\gamma} = y_1(\gamma - \gamma_m)(\gamma_M - \gamma)\nu$$

evolving on (γ_m, γ_M), where ν is a function to be selected that is independent of x. With $u = \gamma y_1$, the system (2.41) with its dynamic extension becomes

$$\begin{cases} \dot{s} = y_1 \left[\gamma(s_{in} - s) - \frac{k}{\lambda}\right] \\ \dot{x} = y_1 \alpha \left[\frac{1}{\alpha\lambda} - \gamma x\right] \\ \dot{\gamma} = y_1(\gamma - \gamma_m)(\gamma_M - \gamma)\nu \end{cases} \tag{2.44}$$

by the definition of y_1, with the same output y as before. The dynamics (2.44) evolves on the invariant domain $E = (0, \infty) \times (0, \infty) \times (\gamma_m, \gamma_M)$. The following is easily checked:

Lemma 2.6. *For each initial value $(s(t_0), x(t_0), \gamma(t_0)) \in E$, there is a compact set $K_0 \subseteq (0, \infty)^2$ so that the corresponding solution of (2.44) is such that $(s(t), x(t)) \in K_0$ for all $t \geq t_0$.*

It follows from Lemma 2.6 and (2.42) that we can re-parameterize (2.44) in terms of the Erdmann Transformation

$$\tau = \int_{t_0}^{t} y_1(l)\,\mathrm{d}l\ .$$

Doing so and setting

$$\tilde{x} = x - x_*, \quad \tilde{s} = s - s_*, \quad \text{and} \quad \tilde{\gamma} = \gamma - \gamma_*$$

yields the error dynamics

$$\begin{cases} \dot{\tilde{s}} = -\gamma\tilde{s} + \tilde{\gamma}v_* \\ \dot{\tilde{x}} = \alpha\left[-\gamma\tilde{x} - \tilde{\gamma}x_*\right] \\ \dot{\tilde{\gamma}} = (\gamma - \gamma_m)(\gamma_M - \gamma)\nu \end{cases} \tag{2.45}$$

for $t \mapsto (\tilde{s}, \tilde{x}, \tilde{\gamma})(\tau^{-1}(t))$. The state space of (2.45) is

$$D = (-s_*, \infty) \times (-x_*, \infty) \times (\gamma_m - \gamma_*, \gamma_M - \gamma_*).$$

The system (2.45) has an uncoupled triangular structure; i.e., its $(\tilde{s}, \tilde{\gamma})$-subsystem does not depend on $\tilde{x}$ (because ν is independent of x), and the $\tilde{x}$-subsystem is globally input-to-state stable with respect to $\tilde{\gamma}$ with the ISS Lyapunov function $\tilde{x}^2$. Therefore, (2.45) is GAS to 0 if the system

$$\begin{cases} \dot{\tilde{s}} = -\gamma\tilde{s} + \tilde{\gamma}v_* \\ \dot{\tilde{\gamma}} = (\gamma - \gamma_m)(\gamma_M - \gamma)\nu \end{cases} \tag{2.46}$$

with state space

$$\mathcal{X} = (-s_*, \infty) \times (\gamma_m - \gamma_*, \gamma_M - \gamma_*)$$

is GAS to 0. Hence, we may limit our analysis to (2.46) in the following analysis.

For a given tuning parameter $K > 0$, the non-strict Lyapunov-like function for (2.46) provided by [89] is

$$V_1(\tilde{s}, \tilde{\gamma}) = \frac{1}{2\gamma_m}\tilde{s}^2 + \frac{v_*}{K\gamma_m}\int_0^{\tilde{\gamma}} \frac{l}{(l + \gamma_* - \gamma_m)(\gamma_M - \gamma_* - l)}dl, \tag{2.47}$$

which is positive definite on $\mathcal{X}$. In fact,

$$\dot{V}_1 = \frac{1}{\gamma_m}\left[-\gamma\tilde{s}^2 + \tilde{s}\tilde{\gamma}v_*\right] + \frac{v_*}{K\gamma_m}\tilde{\gamma}\nu$$

along the trajectories of (2.46), so choosing

$$\nu(\tilde{s}) = -K\tilde{s} \tag{2.48}$$

gives

$$\dot{V}_1 = -\frac{\gamma}{\gamma_m}\tilde{s}^2 \le -\tilde{s}^2.$$

This follows because $\gamma(t) \in (\gamma_m, \gamma_M)$ for all t. Using the LaSalle Invariance Principle, it follows [89] that (2.46) is globally asymptotically stable to 0 when we make the choice (2.48).

2.5 Basin of Attraction Revisited

Strict Lyapunov functions can be used to estimate the basins of attractions for locally asymptotically stable time-varying nonlinear systems

$$\dot{x} = f(t, x). \tag{2.49}$$

Let us review how this can be done. Assume that we know a C^1 storage function $V : [0, \infty) \times \mathbb{R}^n \to [0, \infty)$, a continuous positive definite function $W : \mathbb{R}^n \to [0, \infty)$, and an open set $D \subseteq \mathbb{R}^n$ containing $x = 0$ such that

$$\frac{\partial V}{\partial t}(t, x) + \frac{\partial V}{\partial x}(t, x) f(t, x) \leq -W(x) \quad \forall t \geq 0 \text{ and } x \in D\,. \tag{2.50}$$

Choose functions $\alpha_1, \alpha_2 \in \mathcal{K}_\infty$ such that

$$\alpha_1(|x|) \leq V(t, x) \leq \alpha_2(|x|) \quad \forall t \geq 0 \text{ and } x \in \mathbb{R}^n\,. \tag{2.51}$$

For convenience, we continue to write

$$\dot{V} := \frac{\partial V}{\partial t}(t, x) + \frac{\partial V}{\partial x}(t, x) f(t, x)\,.$$

Since D is an open set that contains the origin, there exists a constant $r > 0$ such that $r\mathcal{B}_n \subseteq D$. The non-decreasing function $\gamma : [0, r] \to [0, \infty)$ defined by

$$\gamma(s) = \inf_{\{\xi \in r\mathcal{B}_n : |\xi| \geq s\}} W(\xi)$$

then satisfies $\gamma(0) = 0$ and $\gamma(s) > 0$ for all $s \in (0, r)$, because W is positive definite. Moreover, for all $t \geq 0$ and $x \in r\mathcal{B}_n$, we have

$$\dot{V} \leq -\gamma(|x|)\,. \tag{2.52}$$

Also, (2.51) in combination with the facts that $\alpha_2 \in \mathcal{K}_\infty$ and γ is non-decreasing give

$$0 \leq \gamma\big(\alpha_2^{-1}\big(V(t, x)\big)\big) \leq \gamma(|x|) \quad \forall t \geq 0 \text{ and } x \in r\mathcal{B}_n \tag{2.53}$$

Combining (2.52) and (2.53) yields

$$\dot{V} \leq -\overline{\gamma}\big(V(t, x)\big) \quad \forall t \geq 0 \text{ and } x \in r\mathcal{B}_n \tag{2.54}$$

where $\overline{\gamma}(s) = \gamma(\alpha_2^{-1}(s))$.

Let us now consider the set $E_r = \{x \in r\mathcal{B}_n : V(t, x) < \alpha_1(r)\ \forall t \geq 0\}$. One readily checks that E_r is a positively invariant set for (2.49). To see why, let $x_0 \in E_r$ and $t_0 \geq 0$, and let $x(t)$ denote the solution of (2.49) such that $x(t_0) = x_0$. Suppose that there exists a time $t_1 > t_0$ such that $V(t_1, x(t_1)) = \alpha_1(r)$ and $V(t, x(t)) < \alpha_1(r)$ for all $t \in [t_0, t_1)$. Then (2.51) implies that

$\alpha_1(|x(t)|) < \alpha_1(r)$ for all $t \in [t_0, t_1)$, i.e., $x \in r\mathcal{B}_n$ for all $t \in [t_0, t_1)$. This allows us to conclude from (2.54) that

$$V(t_1, x(t_1)) \leq V(t_0, x(t_0)) < \alpha_1(r),$$

which is a contradiction. We deduce that for all $t \geq t_0$, $V(t, x(t)) < \alpha_1(r)$ and $|x(t)| < r$, which gives the positive invariance of E_r. Moreover, (2.54) implies that all solutions of the system starting in E_r converge to the origin. Also, $0.5\alpha_2^{-1}(\alpha_1(r))\mathcal{B}_n \subseteq E_r$ and $\alpha_2^{-1}(\alpha_1(r)) > 0$. Therefore E_r is a non-empty subset of the basin of attraction of (2.49). Hence, knowing a strict Lyapunov function for the system leads to an approximation of the basin of attraction.

2.6 $\mathcal{L}_2$ Gains

2.6.1 Basic Theorem

Strict Lyapunov functions can also help estimate the effect of a disturbance on a specific output. For instance, consider a locally Lipschitz forward complete nonlinear control affine system with output

$$\dot{x} = f(x) + g(x)u, \quad y = h(x) \in \mathbb{R}^q \tag{2.55}$$

with $x \in \mathbb{R}^n$ and $u \in \mathbb{R}^p$. We wish to determine a constant $\gamma > 0$ and a function $\Gamma \in \mathcal{K}_\infty$ such that for any continuous input $u(t)$ and any initial state x_0, the corresponding solution $x(t)$ of (2.55) satisfies

$$\sqrt{\int_0^T |y(s)|^2 \, \mathrm{d}s} \leq \gamma \sqrt{\int_0^T |u(s)|^2 \, \mathrm{d}s} + \Gamma(|x_0|) \tag{2.56}$$

for all constants $T > 0$. Here is a useful result in that direction from [70]:

Theorem 2.4. *Consider the system with output (2.55). Assume that f, g, h are Lipschitz continuous and $f(0) = 0$ and $h(0) = 0$. Assume that there is a C^1 function $V : \mathbb{R}^n \to [0, \infty)$ such that*

$$L_f V(x) + \frac{1}{2\gamma^2} \frac{\partial V}{\partial x}(x) g(x) \left(\frac{\partial V}{\partial x}(x) g(x) \right)^\top + \frac{1}{2} h(x)^\top h(x) \leq 0 \tag{2.57}$$

for all $x \in \mathbb{R}^n$. Then for all $x_0 \in \mathbb{R}^n$, the inequality

$$\sqrt{\int_0^T |y(s)|^2 \, \mathrm{d}s} \leq \gamma \sqrt{\int_0^T |u(s)|^2 \, \mathrm{d}s} + \sqrt{2V(x_0)} \tag{2.58}$$

is satisfied for all $T \geq 0$.

Proof. The time derivative of V along the trajectories of (2.55) satisfies

$$\begin{aligned}\dot{V} &= L_f V(x) + \frac{\partial V}{\partial x}(x) g(x) u \\ &= L_f V(x) + \frac{1}{2\gamma^2}\frac{\partial V}{\partial x}(x) g(x) \left(\frac{\partial V}{\partial x}(x) g(x)\right)^\top + \frac{1}{2} h(x)^\top h(x) \\ &\quad -\frac{\gamma^2}{2} u^\top u + \frac{\partial V}{\partial x}(x) g(x) u - \frac{1}{2\gamma^2}\frac{\partial V}{\partial x}(x) g(x) \left(\frac{\partial V}{\partial x}(x) g(x)\right)^\top \\ &\quad +\frac{\gamma^2}{2} u^\top u - \frac{1}{2} h(x)^\top h(x)\end{aligned} \tag{2.59}$$

From (2.57) and the fact that

$$\begin{aligned}&-\frac{\gamma^2}{2} u^\top u + \frac{\partial V}{\partial x}(x) g(x) u - \frac{1}{2\gamma^2}\frac{\partial V}{\partial x}(x) g(x) \left(\frac{\partial V}{\partial x}(x) g(x)\right)^\top \\ &= -\frac{1}{2}\left|\gamma u^\top - \frac{1}{\gamma}\frac{\partial V}{\partial x}(x) g(x)\right|^2 \leq 0,\end{aligned}$$

it follows that

$$\dot{V} \leq \frac{\gamma^2}{2}|u|^2 - \frac{1}{2}|y|^2. \tag{2.60}$$

By integrating (2.60) over $[0, T]$, we obtain

$$V(x(t)) - V(x_0) \leq \tfrac{\gamma^2}{2}\int_0^T |u(s)|^2 \, \mathrm{d}s - \frac{1}{2}\int_0^T |y(s)|^2 \, \mathrm{d}s \tag{2.61}$$

Using the non-negativity of $V(x(t))$, we get

$$\tfrac{1}{2}\int_0^T |y(s)|^2 \, \mathrm{d}s \leq V(x_0) + \tfrac{\gamma^2}{2}\int_0^T |u(s)|^2 \, \mathrm{d}s \tag{2.62}$$

and therefore (2.58) follows from the relation $\sqrt{a+b} \leq \sqrt{a} + \sqrt{b}$ for nonnegative values a and b. □

2.6.2 Illustration

The $\mathcal{L}_2$ gain of the system (2.55) is defined to be the infimum of the set of all constants $\gamma > 0$ such that the inequality (2.58) holds for all x_0, u, and T. Theorem 2.4 provides constants γ that are larger than the $\mathcal{L}_2$ gain of the system (2.55). The constants depend on the function V selected. Moreover, there are cases where the $\mathcal{L}_2$ gain is a finite number, but where an inadequate

Lyapunov function V does not allow us to determine an approximate upper bound for this gain.

We illustrate this using the simple one-dimensional nonlinear system

$$\begin{cases} \dot{x} = -x - x^3 + u \\ y = x \, . \end{cases} \tag{2.63}$$

If we choose $V_a(x) = \frac{1}{2}x^2$, then the inequality (2.57) becomes

$$-x^4 + \frac{1-\gamma^2}{2\gamma^2}x^2 \le 0 \tag{2.64}$$

and 1 is the smallest value of γ such that (2.64) holds for all x. If we instead choose $V_b(x) = x^2$, then the inequality (2.57) becomes

$$\frac{4-3\gamma^2}{2\gamma^2}x^2 - 2x^4 \le 0 \tag{2.65}$$

and $2/\sqrt{3}$ is the smallest γ such that (2.64) holds for all x. Finally, if we take $V_c(x) = \frac{1}{4}x^4$, then the inequality (2.57) reads

$$-x^4 - x^6 + \frac{1}{2\gamma^2}x^6 + \frac{1}{2}x^2 \le 0 \, , \tag{2.66}$$

which cannot be satisfied for all values of x. Hence, we cannot use V_c to get an upper bound for the $\mathcal{L}_2$ gain of the system (2.55).

This provides yet another reason for wanting explicit Lyapunov functions. Indeed, no estimate for the $\mathcal{L}_2$ gain of a system (2.55) can be deduced from the mere existence of a strict Lyapunov function, as provided by the converse Lyapunov theorem.

2.7 Lyapunov Functions with Bounded Gradients

The converse Lyapunov theorem ensures that UGAS systems admit global strict Lyapunov functions. However, it does not in general give information on the type of Lyapunov function that is associated with a given UGAS system, e.g., whether or not the gradient of the Lyapunov function is uniformly bounded in norm. This is a shortcoming of converse theory.

In fact, if $V(x)$ is a strict Lyapunov function for a system $\dot{x} = f(x)$, and if $\nabla V(x)$ is known to be globally bounded, then V is also an iISS Lyapunov function for $\dot{x} = f(x) + d$ with disturbance d, and then the iISS estimate bounds the trajectories if d is exponentially decaying to zero. This is yet another motivation for constructing strict Lyapunov functions. Many UGAS systems admit strict Lyapunov functions with globally bounded gradients; see Sect. 4.6 for related results.

2.7.1 Effect of Exponentially Decaying Disturbances

It is possible to construct a UGAS system that has unbounded trajectories in the presence of an additive exponentially decaying disturbance, and therefore cannot admit a strict Lyapunov function with a globally bounded gradient. For example, consider the following system from [177]:

$$\begin{cases} \dot{x}_1 = g(x_1x_2)x_1 \\ \dot{x}_2 = -2x_2 + d \,, \end{cases} \tag{2.67}$$

where $x = (x_1, x_2) \in \mathbb{R}^2$, $d \in \mathbb{R}$ is a disturbance, and the function g is such that:

1. g is Lipschitz continuous;
2. $|g(s)| \leq 1$ for all s;
3. $g(s) = -1$ for all $s \in \left(-\infty, \frac{1}{2}\right] \cup \left[\frac{3}{2}, \infty\right)$; and
4. $g(1) = 1$.

The system (2.67) has the additional property of being globally *exponentially* stable when $d = 0$. In fact, we can prove:

Proposition 2.1. *When $d \equiv 0$, the solutions of (2.67) satisfy*

$$|x(t)| \leq e^4 e^{-t} |x(0)| \tag{2.68}$$

for all $t \geq 0$ and all initial states $x(0) \in \mathbb{R}^2$.

Proof. Let

$$P(r) = \int_0^r H(m)\,\mathrm{d}m \tag{2.69}$$

where

$$H(m) = \begin{cases} \dfrac{1 + g(m)}{m[2 - g(m)]}, & \text{if } m \neq 0 \\ 0, & \text{if } m = 0 \,. \end{cases}$$

Then

$$0 \leq P(r) \leq 4 \tag{2.70}$$

for all $r \in \mathbb{R}$. We introduce the variable

$$\xi = x_1 x_2 \,. \tag{2.71}$$

Then

$$\begin{cases} \dot{x}_1 = g(\xi)x_1 \\ \dot{x}_2 = -2x_2 \\ \dot{\xi} = [-2 + g(\xi)]\xi \ . \end{cases} \tag{2.72}$$

Let

$$Z_1 = e^{P(\xi)} x_1 \ . \tag{2.73}$$

Then

$$\begin{aligned} \dot{Z}_1 &= \Big(g(\xi) + H(\xi)\big[- 2 + g(\xi)\big]\xi \Big) Z_1 \\ &= -Z_1 \ . \end{aligned} \tag{2.74}$$

Hence,

$$|Z_1(t)| \ \leq \ e^{-t}|Z_1(0)| \quad \text{and} \quad |x_2(t)| \ \leq \ e^{-2t}|x_2(0)| \tag{2.75}$$

for all $x(0) \in \mathbb{R}^2$ and all $t \geq 0$. We deduce that

$$|x_1(t)| \leq e^{-P(\xi(t))} e^{-t} e^{P(x_1(0)x_2(0))} |x_1(0)| \ . \tag{2.76}$$

Finally, from (2.70), we conclude that (2.68) holds, as claimed. □

Remark 2.1. In [177], it is proved that

$$|x(t)| \ \leq \ 9e^{-t}|x(0)|$$

along all trajectories of (2.67). We establish the inequality (2.68) because its proof relies on a slightly different technique from the one given in [177] that will be useful later to establish complementary results.

The next result establishes that the behavior of the solutions of the system (2.67) may change drastically in the presence of a decaying disturbance.

Proposition 2.2. *When*

$$x_1(0) \neq 0, \ \ x_2(0) = x_1(0)^{-1}, \ \text{and} \ \ d(t) = x_2(0)e^{-t}, \tag{2.77}$$

the solutions of (2.67) satisfy

$$x_1(t) = e^t x_1(0) \tag{2.78}$$

for all $t \geq 0$.

Proof. The functions

$$x_1(t) = e^t x_1(0) \ \text{and} \ \ x_2(t) = x_2(0)e^{-t}$$

are such that

$$\begin{cases} \dot{x}_1 = x_1 \\ \dot{x}_2 = -2x_2 + x_2(0)e^{-t} \ . \end{cases} \tag{2.79}$$

Since $g(1) = 1$ and $d(t) = x_2(0)e^{-t}$, we have

$$\begin{cases} \dot{x}_1 = g(1)x_1 \\ \dot{x}_2 = -2x_2 + d(t) \ . \end{cases} \tag{2.80}$$

In addition, for all t, we get $x_1(t)x_2(t) = e^t x_1(0)x_2(0)e^{-t} = x_1(0)x_2(0) = 1$. Therefore,

$$\begin{cases} \dot{x}_1 = g(x_1x_2)x_1 \\ \dot{x}_2 = -2x_2 + d(t) \ . \end{cases} \tag{2.81}$$

This proves the proposition. □

An immediate consequence of Propositions 2.1 and 2.2 is:

Proposition 2.3. *If $V(x)$ is a strict Lyapunov function for the system (2.67) with the disturbance $d \equiv 0$, then there does not exist a constant $C > 0$ such that*

$$\left| \frac{\partial V}{\partial x}(x) \right| \leq C \tag{2.82}$$

for all $x \in \mathbb{R}^2$.

Proof. Suppose the contrary. Since V satisfies (2.82) for some constant C, the time derivative of V along the trajectories of (2.67) satisfies

$$\dot{V} \leq C|d(t)| \ . \tag{2.83}$$

In particular, the choices (2.77) give

$$\dot{V} \leq C|x_2(0)e^{-t}| \ . \tag{2.84}$$

By integrating this inequality we deduce that

$$V(x(t)) \leq V(x(0)) + C|x_2(0)| \tag{2.85}$$

for all $t \geq 0$. According to Proposition 2.2, $x_1(t) = e^t x_1(0)$ for all $t \geq 0$, so

$$\lim_{t \to +\infty} V(x(t)) = +\infty.$$

This and (2.85) yield a contradiction. □

2.7.2 Dependence on Coordinates

We next show that the property of having no strict Lyapunov function with a bounded gradient is coordinate dependent.

Proposition 2.4. *Let P be as defined in (2.69). Then the variables*

$$Z_1 = e^{P(x_1x_2)}x_1 \quad and \quad Z_2 = x_2 \tag{2.86}$$

define a global change of coordinates that transforms (2.67) with $d \equiv 0$ into

$$\begin{cases} \dot{Z}_1 = -Z_1 \\ \dot{Z}_2 = -2Z_2 \ . \end{cases} \tag{2.87}$$

Hence, (2.67) is globally linearizable.

Proof. Routine calculations yield

$$\begin{aligned} \tfrac{\partial Z_1}{\partial x_1}(x) &= e^{P(x_1x_2)}\left[1 + x_1x_2P'(x_1x_2)\right] \\ &= e^{P(x_1x_2)}\left[1 + \frac{1+g(x_1x_2)}{2-g(x_1x_2)}\right] = e^{P(x_1x_2)}\frac{3}{2-g(x_1x_2)} > 0 \ . \end{aligned} \tag{2.88}$$

We deduce that (2.86) defines a global change of coordinates that transforms the system (2.67) with $d \equiv 0$ into the linear system (2.87). □

Remark 2.2. Notice that $\mathcal{V}(Z) = \ln(1+Z_1^2+Z_2^2)$ is a strict Lyapunov function for (2.87) that has a bounded gradient.

2.7.3 Strictification

It is tempting to surmise that non-strict Lyapunov functions with globally bounded gradients can be transformed into strict Lyapunov functions with globally bounded gradients. The following result shows how such a strictification transformation can sometimes be carried out:

Proposition 2.5. *Let*

$$\dot{x} = f(t,x) \tag{2.89}$$

be a UGAS system that is periodic in t and admits a non-strict Lyapunov function V such that

$$\left|\frac{\partial V}{\partial t}(t,x)\right| \leq 1 \quad and \quad \left|\frac{\partial V}{\partial x}(t,x)\right| \leq 1 \tag{2.90}$$

for all $t \in \mathbb{R}$ and $x \in \mathbb{R}^n$. Assume that $\frac{\partial f}{\partial x}$ is bounded. Then the system (2.89) admits a strict Lyapunov function U such that

$$\left|\frac{\partial U}{\partial t}(t,x)\right| \leq 1 \quad and \quad \left|\frac{\partial U}{\partial x}(t,x)\right| \leq 1 \tag{2.91}$$

for all $t \in \mathbb{R}$ and $x \in \mathbb{R}^n$.

Proof. By Theorem 2.1, there exists a C^1 function $\nu(t,x)$, a positive definite function $w(x)$, and functions $\gamma_1, \gamma_2, \gamma_3 \in \mathcal{K}_\infty$ such that

$$\gamma_1(|x|) \;\leq\; \nu(t,x) \;\leq\; \gamma_2(|x|) \text{ and} \tag{2.92}$$

$$\max\left\{\left|\frac{\partial \nu}{\partial t}(t,x)\right|, \left|\frac{\partial \nu}{\partial x}(t,x)\right|\right\} \;\leq\; \gamma_3(|x|) \tag{2.93}$$

hold for all $t \in \mathbb{R}$ and $x \in \mathbb{R}^n$, and such that the time derivative of $\nu(t,x)$ along the trajectories of (2.89) satisfies

$$\dot{\nu}(t,x) \leq -w(x)\ . \tag{2.94}$$

Let $\Gamma : [0,\infty) \to [1,\infty)$ be a continuous increasing function such that

$$1 + \gamma_3(\gamma_1^{-1}(s)) \leq \Gamma(s) \tag{2.95}$$

for all $s \geq 0$. By (2.92), we have

$$|x| \;\leq\; \gamma_1^{-1}(\nu(t,x)),$$

so (2.95) with the choice $s = \nu(t,x)$ gives

$$1 + \gamma_3(|x|) \;\leq\; 1 + \gamma_3(\gamma_1^{-1}(\nu(t,x))) \;\leq\; \Gamma(\nu(t,x))$$

everywhere. This inequality and (2.93) imply that

$$\left|\frac{\partial \nu}{\partial t}(t,x)\right| \leq \Gamma(\nu(t,x)) \text{ and } \left|\frac{\partial \nu}{\partial x}(t,x)\right| \leq \Gamma(\nu(t,x)) \tag{2.96}$$

everywhere.

Next, consider the function

$$U(t,x) = \frac{1}{2}\left[V(t,x) + \int_0^{\nu(t,x)} \frac{1}{\Gamma(m)}\mathrm{d}m\right]\ . \tag{2.97}$$

Then

$$\left|\frac{\partial U}{\partial t}(t,x)\right| \;\leq\; \frac{1}{2}\left[\left|\frac{\partial V}{\partial t}(t,x)\right| + \frac{1}{\Gamma(\nu(t,x))}\left|\frac{\partial \nu}{\partial t}(t,x)\right|\right] \;\leq\; 1 \tag{2.98}$$

everywhere, by (2.90) and (2.96). Similarly, one can show that

$$\left|\frac{\partial U}{\partial x}(t,x)\right| \leq 1\ . \tag{2.99}$$

Finally, observe that

$$\frac{1}{2}V(t,x) \leq U(t,x) \leq \frac{1}{2}[V(t,x) + \nu(t,x)] \tag{2.100}$$

implies that

$$\frac{1}{2}\alpha_1(|x|) \leq U(t,x) \leq \frac{1}{2}[\alpha_2(|x|) + \gamma_2(|x|)] , \tag{2.101}$$

where $\alpha_1, \alpha_2 \in \mathcal{K}_\infty$ are from the proper and positive definite conditions on V. Therefore, U is a global strict Lyapunov for (2.89) with a bounded gradient. This proves the proposition. □

2.8 Comments

Throughout our work, we assume that our (control) Lyapunov functions are at least C^1. One can generalize the definitions to allow Lyapunov functions that are continuous but not necessarily differentiable; see, e.g., the books [17, 184] for some early results. Nonsmooth analysis provides a unifying method for the analysis of nondifferentiable functions [22]. One pioneering result [155] by Sontag showed that a system $\dot{x} = f(x,u)$ evolving on $\mathbb{R}^n$ is asymptotically controllable to the origin if and only if it admits a continuous positive definite proper function V that satisfies: *For each $x \in \mathbb{R}^n$, there exists a relaxed control w such that $\dot{V}_w(x) < 0$. Moreover, there exist positive constants k and η such that w can be chosen to satisfy $||w|| < k$ whenever $|x| < \eta$.* Here

$$\dot{V}_w(x) = \liminf_{t \to 0^+} \frac{1}{t}\{V(\phi(t,x,w)) - V(x)\}$$

is the lower Dini Derivative, ϕ is the flow map of the system, and relaxed controls are measurable mappings of $[0,\infty)$ into the set of all probability measures on the control set U. Also, $||w||$ is the infimum of the set of all r's so that w is supported in $r\mathcal{B}_m$. One motivation for using nonsmooth analysis in control theory is that a system $\dot{x} = f(x,u)$ admits a C^1 control Lyapunov function $V(x)$ if and only if it is C^1 stabilizable by a time-invariant feedback $u_s(x)$, so C^1 control Lyapunov functions cannot exist unless Brockett's Condition is satisfied. This motivation becomes less important when we allow time-varying feedback stabilizers.

The system (2.67) is important because it shows how a globally exponentially stable system can be destabilized by an exponentially decaying disturbance that is arbitrarily small in the $\mathcal{L}_1$ norm. See [166] for an earlier example of a time-invariant GAS system $\dot{x} = f(x)$ that admits an integrable disturbance $\mathbf{d}$ for which $\dot{x} = f(x) + \mathbf{d}$ has unbounded solutions.

One can also develop Lyapunov function theory for *retarded functional differential equations*, which have the form

$$\dot{x}(t) = f(t, x_t) \tag{2.102}$$

where $x(t) \in \mathbb{R}^n$, $f : \mathbb{R} \times \mathcal{C}_n([-r, 0]) \to \mathbb{R}^n$ for a given choice of the constant $r > 0$, the function x_t is defined by $x_t(\theta) = x(t + \theta)$ for all $\theta \in [-r, 0]$, and the initial value is $\phi \in \mathcal{C}_n([-r, 0])$. In this context, $\mathcal{C}_n(\mathcal{I})$ is the set of all continuous functions $\phi : \mathcal{I} \to \mathbb{R}^n$ on any interval $\mathcal{I} \subseteq \mathbb{R}$. We assume that f has enough regularity to guarantee the existence and uniqueness of a maximal solution for each initial condition; see [109] for sufficient conditions for the existence of maximal solutions. Equations of this type are also called *(time) delayed differential equations.* Their study is an important subject that is best considered in books devoted only to systems with delays [51, 114, 124]. Here we only summarize some Lyapunov results for delay systems.

Lyapunov related functions are key for the stability analysis and control design for systems with delay. Two important theorems for delayed systems are the Razumikhin Theorem and the Lyapunov-Krasovski Theorem. Both rely on delayed Lyapunov functions or functionals, which are often constructed by first building Lyapunov functions for the corresponding undelayed systems (obtained by setting the delays equal to zero). For a given constant $r > 0$, the Razumikhin Theorem is the following [52]:

Theorem 2.5. *(Razumikhin Theorem) Let $f : \mathbb{R} \times \mathcal{C}_n([-r, 0]) \to \mathbb{R}^n$ map $\mathbb{R}\times$(bounded subsets of $\mathcal{C}_n([-r, 0])$) into bounded subsets of $\mathbb{R}^n$. Let $u, v, w : [0, \infty) \to [0, \infty)$ be continuous non-decreasing functions for which u and v are positive definite and v is increasing. Assume the following:*

1. *There exists a continuously differentiable function $V : \mathbb{R} \times \mathbb{R}^n \to \mathbb{R}$ whose time derivatives along the solutions $x(t)$ of (2.102) satisfy*

$$\dot{V}(t, x(t)) \le -w(|x(t)|) \tag{2.103}$$

whenever

$$V(t + \theta, x(t + \theta)) \le V(t, x(t)) \tag{2.104}$$

for all $\theta \in [-r, 0]$. Also, $u(|x|) \le V(t, x) \le v(|x|)$ everywhere.

Then the system (2.102) is uniformly stable. If, in addition,

2. *$w(s) > 0$ for all $s > 0$ and there exists a continuous non-decreasing function p such that $p(s) > s$ for all $s > 0$, and such that*

$$\dot{V}(t, x(t)) \le -w(|x(t)|) \tag{2.105}$$

whenever

$$V(t + \theta, x(t + \theta)) \le p(V(t, x(t))) \tag{2.106}$$

for all $\theta \in [-r, 0]$.

then the system (2.102) is uniformly asymptotically stable. If 1. and 2. hold and

$$\lim_{s \to +\infty} u(s) = +\infty,$$

then the system (2.102) is UGAS.

The stability concepts for (2.102) are defined as in the undelayed case, except that the initial state $x_0 \in \mathbb{R}^n$ is replaced by an initial *function* $\phi \in \mathcal{C}_n([-r,0])$. The Lyapunov-Krasovski Theorem is the following, where we use the notation

$$\dot{V}(t,\phi) = \frac{d}{dt}V(t,x_t)|_{t=\tau,x_t=\phi}$$

for any C^1 function $V : \mathbb{R} \times \mathcal{C}_n([-r,0]) \to \mathbb{R}$ [52]:

Theorem 2.6. *(Lyapunov-Krasovski Theorem) Let $f : \mathbb{R} \times \mathcal{C}_n([-r,0]) \to \mathbb{R}^n$ map $\mathbb{R}\times$(bounded subsets of $\mathcal{C}_n([-r,0])$) into bounded subsets of $\mathbb{R}^n$. Let $u,v,w : [0,\infty) \to [0,\infty)$ be continuous non-decreasing functions for which u and v are positive definite and v is increasing. Assume the following:*

1. *There exists a continuously differentiable function $V : \mathbb{R}\times\mathcal{C}_n([-r,0]) \to \mathbb{R}$ such that*
$$u(|\phi(0)|) \leq V(t,\phi) \leq v(|\phi|_{[-r,0]}) \tag{2.107}$$
and
$$\dot{V}(t,\phi) \leq -w(|\phi(0)|) \tag{2.108}$$
for all $\phi \in \mathcal{C}_n([-r,0])$ and $t \in \mathbb{R}$.

Then the trivial solution of (2.102) is uniformly stable. If, in addition,

2. *$w(s) > 0$ for all $s > 0$,*

then (2.102) is uniformly asymptotically stable. Finally, if 1. and 2. hold and if we also have

$$\lim_{s\to+\infty} u(s) = +\infty,$$

then the system (2.102) is UGAS.

Over the last two decades, Lyapunov-Krasovski functionals have been used extensively for the analysis of linear systems. For linear systems, Lyapunov-Krasovski functionals give stability criteria in terms of linear matrix inequalities, which can be analyzed through numerical methods; see for instance [125].

The ISS paradigm can be extended to delayed systems, using either the Razumikhin Theorem (as was done, e.g., in [62, 175]) or Lyapunov-Krasovski functionals (as in [109, 130]). For example, if we consider

$$\dot{x}(t) = f\big(x(t),t\big) + g(x(t),t)\big[u_s\big(\xi_\tau(t),t\big) + d(t)\big], \quad x(t) \in \mathbb{R}^n \tag{2.109}$$

with

$$\xi_\tau(t) = \big(x_1(t-\tau_1), x_2(t-\tau_2), \cdots, x_n(t-\tau_n)\big)$$

and constant delays $\tau = (\tau_1, \tau_2, \ldots, \tau_n)$ satisfying $0 \leq \tau_i \leq \bar{\tau}$ for all i for any bound $\bar{\tau} \geq 0$, and with unknown but bounded disturbances d and a known feedback u_s, an appropriate notion of Lyapunov-Krasovski functionals is as follows:

Definition 2.3. A continuous functional $U : [0,\infty)\times\mathcal{C}_n(\mathbb{R}) \to [0,\infty)$ is called an *ISS Lyapunov-Krasovski functional (ISS-LKF)* for (2.109) provided that for all $\tau \in (0,\bar{\tau}]^n$ and all trajectories $x(t) \doteq x(t;t_0,x_0,d,\tau)$ of (2.109) (corresponding to all possible initial conditions $x(t_0) = x_0$ and measurable essentially bounded disturbances d), the function $t \mapsto U(t,x_t)$ is locally absolutely continuous and there exist functions $\alpha_i \in \mathcal{K}_\infty$ for $i = 1,2,3,4$ and $\kappa \in \mathbb{N}$ such that for all $\phi \in \mathcal{C}_n([-\kappa\bar{\tau},0])$, all trajectories $x(t)$ of (2.109), and all $t \geq t_0+\kappa\bar{\tau}$, we have (a) $\alpha_1(|\phi(0)|) \leq U(t,\phi) \leq \alpha_2(|\phi|_{[-\kappa\bar{\tau},0]})$ and (b) the time derivative $D_tU(t,x_t)$ of $U(t,x_t)$ satisfies $D_tU(t,x_t) \leq -\alpha_3(U(t,x_t))+\alpha_4(|d|_{[t_0,t]})$ almost everywhere.

In this context, $\kappa \in \mathbb{N}$ represents the length of the time lag. The key ingredient in the ISS-LKF definition is that instead of being defined for points in the state space, $U(t,\phi)$ is evaluated at continuous $\mathbb{R}^n$-valued functions $\phi \in \mathcal{C}_n(\mathbb{R})$ defined on the real line and times $t \geq 0$, hence the term function*al* instead of function in the definition.

The explicit construction of ISS-LKFs is a challenging problem. One such construction in [109] shows that (2.109) admits an ISS-LKF of the form

$$U(t,x_t) \;=\; V\big(t,x(t)\big) + \frac{1}{4\bar{\tau}}\int_{t-2\bar{\tau}}^{t}\left(\int_r^t \sigma^2(\sqrt{n}|x(l)|)\mathrm{d}l\right)\mathrm{d}r \qquad (2.110)$$

where the proper positive definite function V, the undelayed dynamics

$$\dot{x} = F(t,x,u_s) \doteq f(t,x) + g(t,x)u_s(t,x),$$

and the function σ satisfy appropriate conditions, including the decay condition

$$V_t(l,x) + V_x(l,x)F(l,x,u_s) \leq -\sigma^2\left(\sqrt{n}|x|\right).$$

The function σ is of class $\mathcal{K}_\infty$, making V a strict Lyapunov function for the undelayed dynamics. In particular, u_s is assumed to stabilize the undelayed dynamics. The Lyapunov-Krasovski functional is valid when the delay bound $\bar{\tau}$ is small enough, but one can take any desired positive constant bound $\bar{\tau}$ when the drift term f in (2.109) is identically zero; see [109] for the explicit computation of the admissible delay bound $\bar{\tau}$ for (2.109). Using (2.110), we can get an explicit ISS decay estimate for (2.109) which is analogous to the usual ISS estimate, except with an initial function replacing the initial state. Therefore, we can quantify the combined effects of feedback delays τ_i and uncertainty d on the stability performance of the feedback u_s. Lyapunov functions are also useful for partial differential equations [28, 29, 76] and stochastic systems [72]. This book will focus on deterministic ODE systems without delays, although it would be of interest to extend many of the results to time-varying delayed systems or stochastic PDEs.

Stabilization problems for biological dynamics such as (2.41) have been studied by numerous authors. Many of the results are based on the (local) linearization approach [57] or linearizing state controllers that assume perfect knowledge of the model. For an alternative approach to uncertain biosystems based on interval observers, see [49]. Our biosystems analysis assumes that the reactor is well mixed, meaning all organisms have equal access to the nutrient; dropping this assumption leads to PDE models. Our approach takes the non-negativity of the state components into account and leads to an explicit strict Lyapunov function; see Sect. 3.1 for details.

Part II
Time-Invariant Case

Chapter 3
Matrosov Conditions: Simple Case

Abstract In the preceding chapter, we saw two ways to use non-strict Lyapunov functions to prove asymptotic stability. The first was the LaSalle Invariance Principle. A second involved Matrosov Theorems, which require a non-strict Lyapunov function and auxiliary functions that satisfy appropriate decay conditions. In general, the decay conditions in Matrosov type theorems are less restrictive than those in the strict Lyapunov function definition. Hence, the Matrosov method can be regarded as a way to prove stability without having to find strict Lyapunov functions.

On the other hand, it is very desirable to have explicit strict Lyapunov functions, even when the Matrosov Conditions are satisfied, because, e.g., strict Lyapunov functions make it possible to quantify the effects of uncertainty using the ISS paradigm. In this chapter, we discuss several methods for constructing strict Lyapunov functions for time-invariant systems that satisfy appropriate Matrosov Conditions. In Chapters 8 and 12, we generalize to much more complex time-varying systems, including Matrosov type theorems for hybrid systems.

3.1 Motivation

To motivate our constructions, let us return to the experimental anaerobic digester model

$$\begin{cases} \dot{s} = u(s_{in} - s) - kr(s, x) \\ \dot{x} = r(s, x) - \alpha ux \\ y = (\lambda r(s, x), s) \end{cases} \tag{3.1}$$

we considered in the preceding chapter, where the biomass growth rate r is any non-negative C^1 function that admits everywhere positive functions $\underline{\Delta}$ and $\bar{\Delta}$ such that

$$s\bar{\Delta}(s, x) \geq r(s, x) \geq xs\underline{\Delta}(s, x) \tag{3.2}$$

for all $s \geq 0$ and $x \geq 0$; u is the non-negative input (i.e., dilution rate); and the positive constants α, λ, k, and s_{in} are as defined in Sect. 2.4.6. This includes the one species chemostat model with a Monod growth rate, as a special case [107]. This time our objective is to construct a strict Lyapunov-like function for an appropriate adaptively controlled error dynamics for (3.1).

Arguing as in the previous chapter, we introduce the dynamics $\dot{\gamma} = y_1(\gamma - \gamma_m)(\gamma_M - \gamma)\nu$ evolving on (γ_m, γ_M), where ν is a function to be selected that is independent of x and the γ_i's are prescribed positive constants. With $u = \gamma y_1$, the system (3.1) with its dynamic extension becomes

$$\begin{cases} \dot{\tilde{s}} = -\gamma\tilde{s} + \tilde{\gamma}v_* \\ \dot{\tilde{x}} = \alpha\left[-\gamma\tilde{x} - \tilde{\gamma}x_*\right] \\ \dot{\tilde{\gamma}} = (\gamma - \gamma_m)(\gamma_M - \gamma)\nu. \end{cases} \tag{3.3}$$

Here $\tilde{s} = s - s_*$, $\tilde{x} = x - x_*$, $\tilde{\gamma} = \gamma - \gamma_*$, $s_* \in (0, s_{in})$ is the desired equilibrium substrate level, and

$$\gamma_* \doteq \frac{k}{\lambda(s_{in} - s_*)} \in (\gamma_m, \gamma_M) \quad \text{and} \quad \frac{k}{\lambda s_{in}} < \gamma_m, \tag{3.4}$$

where $v_* = s_{in} - s_*$ and $x_* = \frac{v_*}{k\alpha}$. The dynamics (3.3) follow by applying the Erdmann transformation

$$\tau = \int_{t_0}^{t} y_1(l)\mathrm{d}l,$$

and the state space for (3.3) is $D = (-s_*, \infty) \times (-x_*, \infty) \times (\gamma_m - \gamma_*, \gamma_M - \gamma_*)$. We first consider the subsystem

$$\begin{cases} \dot{\tilde{s}} = -\gamma\tilde{s} + \tilde{\gamma}v_* \\ \dot{\tilde{\gamma}} = (\gamma - \gamma_m)(\gamma_M - \gamma)\nu \end{cases} \tag{3.5}$$

with state space $\mathcal{X} = (-s_*, \infty) \times (\gamma_m - \gamma_*, \gamma_M - \gamma_*)$.

Let us transform the non-strict Lyapunov-like function

$$V_1(\tilde{s}, \tilde{\gamma}) = \frac{1}{2\gamma_m}\tilde{s}^2 + \frac{v_*}{K\gamma_m}\int_0^{\tilde{\gamma}} \frac{l}{(l + \gamma_* - \gamma_m)(\gamma_M - \gamma_* - l)}\mathrm{d}l \tag{3.6}$$

from [89] into a *strict* Lyapunov-like function for (3.5), where $K > 1$ is a tuning parameter. Later in this chapter, we will see how this transformation process is a special case of a general method for constructing strict Lyapunov-like functions.

Choosing

$$\nu(\tilde{s}) = -K\tilde{s} \tag{3.7}$$

as before gives

$$\dot{V}_1 = -\frac{\gamma}{\gamma_m}\tilde{s}^2 \leq -\mathcal{N}_1(\tilde{s}, \tilde{\gamma}), \text{ where } \mathcal{N}_1(\tilde{s}, \tilde{\gamma}) = \tilde{s}^2. \tag{3.8}$$

Set

$$V_2(\tilde{s},\tilde{\gamma}) \;=\; -\tilde{s}\tilde{\gamma}. \tag{3.9}$$

Along the trajectories of (3.5), in closed-loop with (3.7), simple calculations yield

$$\dot{V}_2 = \gamma\tilde{s}\tilde{\gamma} - \tilde{\gamma}^2 v_* + (\gamma-\gamma_m)(\gamma_M-\gamma)K\tilde{s}^2.$$

From the relation

$$\gamma\tilde{s}\tilde{\gamma} \le \frac{v_\star\tilde{\gamma}^2}{2} + \frac{\gamma^2\tilde{s}^2}{2v_\star}$$

and the fact that the maximum value of $(\gamma-\gamma_m)(\gamma_M-\gamma)$ over $\gamma\in[\gamma_m,\gamma_M]$ is $(\gamma_M-\gamma_m)^2/4$, we get

$$\dot{V}_2 \;\le\; -\mathcal{N}_2(\tilde{s},\tilde{\gamma}) + \left[\frac{\gamma_M^2}{2v_\star} + \frac{K(\gamma_M-\gamma_m)^2}{4}\right]\mathcal{N}_1(\tilde{s},\tilde{\gamma}), \tag{3.10}$$

where $\mathcal{N}_2(\tilde{s},\tilde{\gamma}) = \frac{v_*}{2}\tilde{\gamma}^2$.

Setting

$$\Upsilon_1 \;=\; 1 + \left[\min\left\{\frac{1}{\gamma_m}, \frac{4v_*}{K\gamma_m(\gamma_M-\gamma_m)^2}\right\}\right]^{-1}, \tag{3.11}$$

we can use the decay conditions (3.8)-(3.10) to check that

$$S(\tilde{s},\tilde{\gamma}) \;=\; V_2(\tilde{s},\tilde{\gamma}) + \left[\Upsilon_1 + \frac{2\gamma_M^2}{v_*} + K(\gamma_M-\gamma_m)^2\right]V_1(\tilde{s},\tilde{\gamma}) \tag{3.12}$$

is a strict Lyapunov-like function for (3.5) in closed-loop with (3.7). In fact,

$$\begin{aligned}&\dot{S} \;\le\; -W(\tilde{s},\tilde{\gamma}), \quad \text{where}\\ &W(\tilde{s},\tilde{\gamma}) = \mathcal{N}_2(\tilde{s},\tilde{\gamma}) + \Upsilon_1\mathcal{N}_1(\tilde{s},\tilde{\gamma}) = \tfrac{v_\star}{2}\tilde{\gamma}^2 + \Upsilon_1\tilde{s}^2\end{aligned} \tag{3.13}$$

along the closed-loop trajectories of (3.5), and S is also positive definite; see Sect. 3.6.1 for details and our reasoning behind the choice (3.12) of S.

The fact that the *full system* (3.3) in closed-loop with (3.7) is GAS to the origin now follows because (a) its $\tilde{x}$ sub-dynamics is ISS with respect to $(\tilde{s},\tilde{\gamma})$ and (b) the asymptotically stable $(\tilde{s},\tilde{\gamma})$ sub-dynamics does not depend on $\tilde{x}$. Let us now construct a strict Lyapunov-like function for the full closed-loop system. We claim that

$$M(\tilde{x},\tilde{s},\tilde{\gamma}) = \bar{\omega}S(\tilde{s},\tilde{\gamma}) + \tilde{x}^2, \;\text{ where }\; \bar{\omega} = \frac{4\alpha x_\star^2}{\gamma_m v_\star} \tag{3.14}$$

is a strict Lyapunov-like function for the system (3.3), in closed-loop with (3.7), for which

$$\dot{M} \;\le\; -\alpha\gamma_m\tilde{x}^2 - \frac{\alpha x_\star^2}{\gamma_m}\tilde{\gamma}^2 - \bar{\omega}\Upsilon_1\tilde{s}^2 \tag{3.15}$$

along the trajectories of (3.3). To see why, first notice that the relation

$$2\alpha x_* |\tilde{x}\tilde{\gamma}| \;\le\; \alpha\gamma_m \tilde{x}^2 + \alpha x_\star^2 \frac{\tilde{\gamma}^2}{\gamma_m}$$

implies that

$$\begin{aligned} \tfrac{d}{dt}\tilde{x}^2 &= -2\alpha\gamma\tilde{x}^2 - 2\tilde{x}\tilde{\gamma}\alpha x_* \;\le\; -2\alpha\gamma_m\tilde{x}^2 + 2\alpha x_*|\tilde{x}\tilde{\gamma}| \\ &\le -\alpha\gamma_m\tilde{x}^2 + \frac{\alpha x_\star^2}{\gamma_m}\tilde{\gamma}^2 \end{aligned} \tag{3.16}$$

along the trajectories of (3.3). Then (3.15) follows by adding the inequality

$$\bar{\omega}\dot{S} \;\le\; -\bar{\omega}\frac{v_\star}{2}\tilde{\gamma}^2 - \bar{\omega}\Upsilon_1\tilde{s}^2$$

to (3.16). We turn next to a general theory that leads to the preceding analysis as a special case.

3.2 Continuous Time Theorem

For simplicity, we first state our main result for time-invariant systems

$$\dot{x} = f(x) \tag{3.17}$$

evolving on an open set $\mathcal{X} \subseteq \mathbb{R}^n$. Later we generalize to time-varying systems. In the rest of this section, we assume that the relevant functions are sufficiently smooth. We also assume:

Assumption 3.1 *There exist an integer $j \ge 2$; known functions*

$$\begin{aligned} &V_i : \mathcal{X} \to \mathbb{R}, \\ &\mathcal{N}_i : \mathcal{X} \to [0,\infty), \quad \text{and} \\ &\phi_i : [0,\infty) \to (0,\infty); \end{aligned}$$

and real numbers $a_i \in (0,1]$ such that $V_i(0) = 0$ and $\mathcal{N}_i(0) = 0$ for all i;

$$\nabla V_1(x) f(x) \;\le\; -\mathcal{N}_1(x) \quad \forall x \in \mathcal{X} \text{ ; and} \tag{3.18}$$

$$\nabla V_i(x) f(x) \;\le\; -\mathcal{N}_i(x) + \phi_i\big(V_1(x)\big) \sum_{l=1}^{i-1} \mathcal{N}_l^{a_i}(x) V_1^{1-a_i}(x) \tag{3.19}$$

for $i = 2, \ldots, j$ and all $x \in \mathcal{X}$. The function V_1 is also assumed to be positive definite on $\mathcal{X}$.

Assumption 3.2 *The following conditions hold:*

1. *there exists a function* $\rho : [0,\infty) \to (0,\infty)$ *such that*

$$\sum_{l=1}^{j} \mathcal{N}_l(x) \;\geq\; \rho\big(V_1(x)\big)V_1(x) \quad \forall x \in \mathcal{X}; \;\; \textit{and} \tag{3.20}$$

2. *there exist functions* $p_2, \ldots, p_j : [0,\infty) \to [0,\infty)$ *and a positive definite function* $\bar{p}$ *such that for each* $i \in \{2, \ldots, j\}$*, the following hold: (a) If* V_i *is positive definite, then*

$$p_i(r) = 0 \;\; \textit{and} \;\; |V_i(x)| \leq \bar{p}\big(V_1(x)\big) \tag{3.21}$$

for all $r \geq 0$ *and* $x \in \mathcal{X}$*. (b) If* V_i *is not positive definite, then*

$$|V_i(x)| \;\leq\; p_i\big(V_1(x)\big)V_1(x) \tag{3.22}$$

holds for all $x \in \mathcal{X}$*.*

Assumptions 3.1 and 3.2 agree with the ones in [106], except that [106] requires the functions p_i to satisfy (3.22) for all i and all $x \in \mathcal{X}$ (instead of Condition 2. from Assumption 3.2). We refer to Assumptions 3.1 and 3.2 as our Matrosov Conditions, owing to their use of multiple functions V_i. However, there are several different sets of conditions that are referred to as Matrosov Conditions in the control literature. We prove:

Theorem 3.1. *Let Assumptions 3.1 and 3.2 be satisfied. Then one can explicitly determine* C^1 *functions* $k_l, \Omega_l \in \mathcal{K}_\infty$ *such that the function*

$$S(x) \;=\; \sum_{l=1}^{j} \Omega_l\Big(k_l\big(V_1(x)\big) + V_l(x)\Big) \tag{3.23}$$

satisfies

$$S(x) \;\geq\; V_1(x) \tag{3.24}$$

and

$$\nabla S(x) f(x) \;\leq\; -\frac{1}{4}\rho\big(V_1(x)\big)V_1(x) \tag{3.25}$$

for all $x \in \mathcal{X}$*.*

Remarks on Assumptions

Remark 3.1. If $\mathcal{X} = \mathbb{R}^n$ and V_1 is radially unbounded, then (3.24) implies that S is a strict Lyapunov function for (3.17). If V_1 is not radially unbounded, then S is not necessarily radially unbounded and therefore one cannot conclude from standard Lyapunov theory that the origin is GAS. However, in many cases, GAS can be proved through a Lyapunov-like function and extra arguments, e.g., by proving that any trajectory belongs to a compact set included in $\mathcal{X}$. This is often true in biological models that are based on mass conservation properties, such as the one we discussed in Sect. 3.1.

Remark 3.2. If V_1 is also lower bounded by a positive definite quadratic form in a neighborhood of 0, then (3.25) implies that the time derivative of S along the trajectories of (3.17) is upper bounded in a neighborhood of 0 by a negative definite quadratic function. Also, (3.24) gives a positive definite quadratic lower bound on S near the origin.

3.3 Proof of Continuous Time Theorem

Throughout the sequel, all inequalities should be understood to hold globally unless otherwise indicated, and we omit the arguments of our functions when they are clear from the context.

Construction of the k_i's and Ω_i's

Fix $j \geq 2$ and functions satisfying Assumptions 3.1 and 3.2. Fix $k_2, \ldots, k_j \in C^1 \cap \mathcal{K}_\infty$ such that

$$k_i(s) \geq s + p_i(s)s \quad \text{and} \quad k_i'(s) \geq 1 \tag{3.26}$$

for all $s \geq 0$ for $i = 2, 3, \ldots, j$. The following simple lemma is key:

Lemma 3.1. *The functions $\{U_i\}$ defined by*

$$U_1(x) = V_1(x) \quad \textit{and} \quad U_i(x) = k_i\big(V_1(x)\big) + V_i(x) \ \text{ for } \ i \geq 2$$

satisfy $2k_i\big(V_1(x)\big) + \bar{p}(V_1(x)) \geq U_i(x) \geq V_1(x)$ for $i = 2, \ldots, j$ and all $x \in \mathcal{X}$.

Proof. Assumption 3.2 and our choices of the k_i's give

$$\begin{aligned} &U_i(x) \geq V_1(x) + p_i\big(V_1(x)\big)V_1(x) - p_i\big(V_1(x)\big)V_1(x) = V_1(x) \ \text{ and} \\ &U_i(x) \leq k_i\big(V_1(x)\big) + p_i\big(V_1(x)\big)V_1(x) \leq 2k_i\big(V_1(x)\big) \end{aligned}$$

for all indices $i \geq 2$ for which V_i is not positive definite. For the other indices, the desired inequalities follow from (3.21) and the non-negativity of the corresponding functions V_i. This proves the lemma. □

Returning to the proof of the theorem, set $k_1(s) \equiv s$, and define the functions U_i according to Lemma 3.1. We can recursively define continuous non-decreasing functions $\mu_i : [0, \infty) \to [1, \infty)$ such that

$$\mu_i(V_1) \geq 2\Phi(V_1) \sum_{l=1+i}^{j} \mu_l^{\frac{1}{a_l}} (2k_l(V_1) + \bar{p}(V_1)) \tag{3.27}$$

everywhere, where

$$\Phi(V_1) = \max_{i=2,..,j} \left\{ \phi_i^{\frac{1}{a_i}}(V_1) \left[\frac{4(j-1)(i-1)}{\rho(V_1)} \right]^{(1-a_i)/a_i} \right\} \tag{3.28}$$

for $i = 1, 2, \ldots, j$. For convenience, we set $\mu_j(v) \equiv 1$, and we introduce the functions

$$\Omega_i(p) = \int_0^p \mu_i(r)dr.$$

Then $\Omega_i'(s) \geq 1$ for all $s \geq 0$ and i, and Lemma 3.1 gives

$$\Omega_i'(U_i) \;\geq\; 2\Phi(V_1) \sum_{l=1+i}^{j} \Omega_l'(U_l)^{\frac{1}{a_l}} \tag{3.29}$$

for all i and $x \in \mathcal{X}$. In particular, we have $\Omega_j(p) \equiv p$.

Stability Analysis

Since $\Omega_1'(s) \geq 1$ everywhere, we get $\Omega_1(U_1(x)) \geq U_1(x) = V_1(x)$ everywhere. Hence,

$$S(x) \;=\; \Omega_1(2V_1(x)) + \sum_{i=2}^{j} \Omega_i(U_i(x)) \tag{3.30}$$

satisfies (3.24). To check the decay estimate (3.25), first note that Assumption 3.1 and our choices of the k_i's give

$$\begin{aligned} \nabla S(x)f(x) &= 2\Omega_1'(2U_1)\dot{V}_1 + \sum_{i=2}^{j} \Omega_i'(U_i)\left[k_i'(V_1)\dot{V}_1 + \dot{V}_i\right] \\ &\leq \sum_{i=1}^{j} \Omega_i'(U_i)\dot{V}_i \\ &\leq -\sum_{i=1}^{j} \Omega_i'(U_i)\mathcal{N}_i \\ &\quad + \sum_{i=2}^{j} \Omega_i'(U_i)\left(\phi_i(V_1)\sum_{l=1}^{i-1} \mathcal{N}_l^{a_i} V_1^{1-a_i}\right) \end{aligned} \tag{3.31}$$

along the trajectories of $\dot{x} = f(x)$. Define the everywhere positive functions $\Gamma_2, \ldots, \Gamma_j$ by

$$\Gamma_i(x) = \frac{4(j-1)(i-1)\Omega_i'(U_i(x))\phi_i(V_1(x))}{\rho(V_1(x))}.$$

For any $i \geq 2$ for which $0 < a_i < 1$, we can apply Young's Inequality

$$v_1 v_2 \leq v_1^p + v_2^q, \text{ with } p = \frac{1}{a_i}, \quad q = \frac{1}{1-a_i},$$

$$v_1 = \Gamma_i^{1-a_i}(x)\mathcal{N}_l^{a_i}(x), \text{ and } v_2 = \left\{\frac{V_1(x)}{\Gamma_i(x)}\right\}^{1-a_i}$$

to get

$$\mathcal{N}_l^{a_i}(x)V_1^{1-a_i}(x) \; \leq \; \Gamma_i^{(1-a_i)/a_i}(x)\mathcal{N}_l(x) + \frac{V_1(x)}{\Gamma_i(x)}$$

for all $x \in \mathcal{X}$. The preceding inequality also holds when $a_i = 1$, so we can substitute it into (3.31) to get

$$\begin{aligned}
\nabla S(x)f(x) \leq & -\sum_{i=1}^{j} \Omega_i'(U_i)\mathcal{N}_i \\
& + \sum_{i=2}^{j}\left(\Omega_i'(U_i)\phi_i(V_1)\Gamma_i^{\frac{1-a_i}{a_i}}\sum_{l=1}^{i-1}\mathcal{N}_l\right) \\
& + \left(\sum_{i=2}^{j}\Omega_i'(U_i)\frac{\phi_i(V_1)(i-1)}{\Gamma_i}\right)V_1 \\
\leq & -\sum_{i=1}^{j}\Omega_i'(U_i)\mathcal{N}_i + \frac{1}{4}\rho(V_1)V_1 \\
& + \sum_{i=2}^{j}\left(\Omega_i'(U_i)\phi_i(V_1)\Gamma_i^{\frac{1-a_i}{a_i}}\sum_{l=1}^{i-1}\mathcal{N}_l\right) \\
\leq & -\sum_{i=1}^{j}\Omega_i'(U_i)\mathcal{N}_i + \frac{1}{4}\rho(V_1)V_1 \\
& + \Phi(V_1)\sum_{i=2}^{j}\left(\Omega_i'(U_i)^{\frac{1}{a_i}}\sum_{l=1}^{i-1}\mathcal{N}_l\right),
\end{aligned} \tag{3.32}$$

by our choices of the Γ_i's and the formula for Φ in (3.29).

Since $\Omega_i' \geq 1$ for all i, Assumption 3.2 gives

$$\sum_{i=1}^{j}\Omega_i'(U_i)\mathcal{N}_i \; \geq \; \rho(V_1)V_1.$$

Hence, (3.32) gives

$$\nabla S(x)f(x) \leq -\tfrac{1}{4}\rho(V_1)V_1 - \tfrac{1}{2}\sum_{i=1}^{j} \Omega_i'(U_i)\mathcal{N}_i \\ +\Phi(V_1)\sum_{i=2}^{j}\left(\Omega_i'(U_i)^{\frac{1}{a_i}}\sum_{l=1}^{i-1}\mathcal{N}_l\right).$$

By reorganizing terms, one can prove that

$$\sum_{i=2}^{j}\left(\Omega_i'(U_i)^{\frac{1}{a_i}}\sum_{l=1}^{i-1}\mathcal{N}_l\right) = \sum_{i=1}^{j-1}\left(\sum_{l=1+i}^{j}\Omega_l'(U_l)^{\frac{1}{a_l}}\right)\mathcal{N}_i\,. \tag{3.33}$$

It follows that

$$\nabla S(x)f(x) \leq -\frac{1}{4}\rho(V_1)V_1 \\ +\sum_{i=1}^{j-1}\left[-\frac{1}{2}\Omega_i'(U_i) + \Phi(V_1)\sum_{l=1+i}^{j}\Omega_l'(U_l)^{\frac{1}{a_l}}\right]\mathcal{N}_i\,.$$

Since the $\mathcal{N}_i$'s are non-negative, (3.25) now readily follows from (3.29). □

Remark 3.3. When $a_2 = \ldots = a_j = 1$, Assumption 3.2 can be relaxed by replacing (3.20) by the assumption that

$$x \mapsto \sum_{l=1}^{j}\mathcal{N}_l(x) \tag{3.34}$$

is positive definite, in which case we instead conclude that $\nabla S(x)f(x)$ is negative definite. The proof proceeds as in the proof of Theorem 3.1 through (3.31). Then we can directly apply (3.29) and (3.33) to get

$$\nabla S(x)f(x) \leq -\frac{1}{2}\sum_{i=1}^{j}\Omega_i'(U_i(x))\mathcal{N}_i(x)$$

everywhere. The result follows because $\Omega_i' \geq 1$ everywhere for all i.

3.4 Discrete Time Theorem

We turn next to an analog of Theorem 3.1 for the discrete time system

$$x_{k+1} = f(x_k), \quad x_k \in \mathcal{X} \tag{3.35}$$

where $\mathcal{X} \subseteq \mathbb{R}^n$ is open and contains the origin. Throughout this subsection, we make these two assumptions:

Assumption 3.3 *There exist a constant $a \in (0,1]$; an integer $j \geq 2$; continuous functions $\nu_1, \mathcal{M}_i, \phi_i : \mathcal{X} \to [0,\infty)$ for $i = 1,2,\ldots,j$; and continuous functions $\nu_i : \mathcal{X} \to \mathbb{R}$ for $i = 2,\ldots,j$ such that*

$$\nu_1(f(x)) - \nu_1(x) \leq -\mathcal{M}_1(x) \tag{3.36}$$

for all $x \in \mathcal{X}$ and

$$\nu_i(f(x)) - \nu_i(x) \leq -\mathcal{M}_i(x) + \phi_i(\nu_1(x))\nu_1(x)^{1-a}\sum_{l=1}^{i-1}\mathcal{M}_l^a(x) \tag{3.37}$$

for all $x \in \mathcal{X}$ and $i = 2,3,\ldots,j$.

Assumption 3.4 *There are continuous functions $C_k : [0,\infty) \to (0,\infty)$ for $k = 1,2,3,4$ such that the functions from Assumption 3.3 satisfy*

$$\sum_{l=1}^{j}\mathcal{M}_l(x) \;\geq\; C_1(\nu_1(x))|x|^2 \tag{3.38}$$

and

$$C_2(\nu_1(x))|x|^2 \;\leq\; \nu_1(x) \leq C_3(\nu_1(x))|x|^2 \tag{3.39}$$

for all $x \in \mathcal{X}$ and

$$|\nu_i(x)| \;\leq\; C_4(\nu_1(x))|x|^2 \tag{3.40}$$

for all $x \in \mathcal{X}$ and $i = 2,3,\ldots,j$.

Assumption 3.3 is the discrete time analog of the continuous time Matrosov Condition in Assumption 3.1 except for simplicity, we took all of the a_i's to be equal. Notice that we are not requiring the auxiliary functions ν_i to be non-negative for $i \geq 2$, although ν_1 is non-negative.

Theorem 3.2. *Assume that the system (3.35) satisfies Assumptions 3.3 and 3.4. Then we can explicitly determine non-decreasing everywhere positive C^1 functions κ_l such that the function*

$$S(x) \;=\; \sum_{l=1}^{j}\kappa_l\big(\nu_1(x)\big)\nu_l(x) \tag{3.41}$$

satisfies

$$S(x) \;\geq\; |x|^2 \tag{3.42}$$

and

$$S(f(x)) - S(x) \;\leq\; -\nu_1(x) \tag{3.43}$$

for all $x \in \mathcal{X}$. Therefore, S is a strict Lyapunov function when $\mathcal{X} = \mathbb{R}^n$.

Remark 3.4. We have chosen to study the case where the auxiliary functions $\nu_2, \nu_3, \ldots, \nu_j$ are bounded from above by a positive definite quadratic function in a neighborhood of the origin. We made this choice because it leads to reasonably simple calculations. We strongly conjecture that a strict Lyapunov function construction can also be carried out without making this local quadratic upper bound assumption.

3.5 Proof of Discrete Time Theorem

Throughout our proof, all inequalities should be understood to hold for all $x \in \mathcal{X}$ unless otherwise indicated. We can easily find a C^1 non-decreasing function $\Gamma : [0, \infty) \to [1, \infty)$ such that

$$\frac{C_4(r)}{C_2(r)} + 1 \leq \Gamma(r) \ \forall r \geq 0. \tag{3.44}$$

Hence, (3.36) and the non-negativity of ν_1 give

$$\Gamma(\nu_1(f(x)))\nu_1(f(x)) - \Gamma(\nu_1(x))\nu_1(x) \ \leq \ -\Gamma(\nu_1(x))\mathcal{M}_1(x) \tag{3.45}$$

for all $x \in \mathcal{X}$. Also, (3.39) and (3.40) give

$$\frac{C_4(\nu_1(x))}{C_2(\nu_1(x))}\nu_1(x) + \nu_i(x) \geq 0. \tag{3.46}$$

We introduce the functions

$$\bar{\nu}_1 \doteq \nu_1, \ \text{ and } \overline{\nu}_i(x) \doteq \Gamma(\nu_1(x))\nu_1(x) + \nu_i(x) \ \text{ for } i = 2, 3, \ldots, j. \tag{3.47}$$

Then

$$\overline{\nu}_i(x) \geq \nu_1(x) \ \forall i. \tag{3.48}$$

Also, (3.37) and (3.45) give

$$\overline{\nu}_i(f(x)) - \overline{\nu}_i(x) \leq -\mathcal{M}_i(x) + \phi_i(\nu_1(x)) \sum_{l=1}^{i-1} \mathcal{M}_l^a(x)\nu_1^{1-a}(x) \tag{3.49}$$

for $i \geq 2$. We define the functions $V_1, V_2, \ldots, V_j$ by

$$V_1(x) = \nu_1(x) \ \text{ and } \ V_l(x) = \sum_{r=1}^{l} \overline{\nu}_r(x) \ \text{ for } \ l \geq 2\,. \tag{3.50}$$

Each function V_l is positive definite, because ν_1 is positive definite. Moreover, a simple calculation yields

$$V_i(f(x)) - V_i(x) \leq -\mathcal{N}_i(x) + \psi_i(V_1(x))V_1^{1-a}(x)\mathcal{N}_{i-1}^a(x) \tag{3.51}$$

for all $i \geq 2$, where

$$\mathcal{N}_i(x) = \sum_{r=1}^{i} \mathcal{M}_r(x) \tag{3.52}$$

and

$$\psi_i(V_1(x)) = i\sum_{r=2}^{i} \phi_r(V_1(x)) \tag{3.53}$$

everywhere for $i = 1, \ldots, j$. We also set

$$\psi_1(m) = 0 \ \forall m\,. \tag{3.54}$$

We can recursively define everywhere positive non-decreasing C^1 functions $\alpha_j, \alpha_{j-1}, \ldots, \alpha_1$ that satisfy

$$\frac{\alpha_j(r)C_1(r)}{2C_3(r)} \geq 1 \tag{3.55}$$

and

$$(2j)^{\frac{1-a}{a}} \frac{C_3^{\frac{1-a}{a}}(s)}{C_1^{\frac{1-a}{a}}(s)} \frac{\alpha_{i+1}^{\frac{1}{a}}(s)}{\alpha_j^{\frac{1-a}{a}}(s)} \psi_{i+1}^{\frac{1}{a}}(s) \leq \frac{1}{2}\alpha_i(s) \tag{3.56}$$

for $i = 1, 2, \ldots, j-1$.

Consider the functions

$$\begin{aligned} U_i(x) &= \alpha_i\big(V_1(x)\big)V_i(x) \ \text{ for } i = 1, 2, ..., j \ \text{ and} \\ \mathcal{U}(x) &= \sum_{r=1}^{j} U_r(x). \end{aligned} \tag{3.57}$$

Notice that for all $i \in \{1, \ldots, j\}$, we have

$$U_i(f(x)) - U_i(x) = \alpha_i\Big(V_1\big(f(x)\big)\Big)V_i\big(f(x)\big) - \alpha_i\big(V_1(x)\big)V_i(x).$$

Since $V_1\big(f(x)\big) \leq V_1(x)$ and each α_i is non-decreasing, and since each V_i is positive definite, we deduce that

$$U_i(f(x)) - U_i(x) \leq \alpha_i\big(V_1(x)\big)\big[V_i\big(f(x)\big) - V_i(x)\big]. \tag{3.58}$$

It follows from (3.51) that

$$\begin{aligned} &U_i(f(x)) - U_i(x) \\ &\leq \alpha_i\big(V_1(x)\big)\Big[-\mathcal{N}_i(x) + \psi_i\big(V_1(x)\big)V_1^{1-a}(x)\mathcal{N}_{i-1}^a(x)\Big]. \end{aligned} \tag{3.59}$$

Therefore,

$$\begin{aligned}\mathcal{U}(f(x)) - \mathcal{U}(x) \leq & \sum_{r=1}^{j}\Big[-\alpha_r\big(V_1(x)\big)\mathcal{N}_r(x)\\ & +\alpha_r\big(V_1(x)\big)\psi_r\big(V_1(x)\big)V_1^{1-a}(x)\mathcal{N}_{r-1}^a(x)\Big]\\ \leq & -\sum_{r=1}^{j}\alpha_r\big(V_1(x)\big)\mathcal{N}_r(x)\\ & +\sum_{r=2}^{j}\Big[\alpha_r(V_1(x))\psi_r(V_1(x))V_1^{1-a}(x)\mathcal{N}_{r-1}^a(x)\Big]\,,\end{aligned} \tag{3.60}$$

where the last inequality is from (3.54). We deduce that

$$\begin{aligned}\mathcal{U}(f(x)) - \mathcal{U}(x) \leq & -\sum_{r=1}^{j}\alpha_r\big(V_1(x)\big)\mathcal{N}_r(x)\\ & +\sum_{r=1}^{j-1}\alpha_{r+1}\big(V_1(x)\big)\psi_{r+1}\big(V_1(x)\big)V_1^{1-a}(x)\mathcal{N}_r^a(x).\end{aligned} \tag{3.61}$$

Using the fact that

$$\begin{aligned}\sum_{r=1}^{j}\alpha_r(V_1(x))\mathcal{N}_r(x) &= \alpha_j\big(V_1(x)\big)\mathcal{N}_j(x) + \sum_{r=1}^{j-1}\alpha_r\big(V_1(x)\big)\mathcal{N}_r(x)\\ &\geq \alpha_j\big(V_1(x)\big)C_1\big(\nu_1(x)\big)|x|^2 + \sum_{r=1}^{j-1}\alpha_r\big(V_1(x)\big)\mathcal{N}_r(x)\end{aligned}$$

and therefore also

$$\begin{aligned}\sum_{r=1}^{j}\alpha_r(V_1(x))\mathcal{N}_r(x) \geq & \frac{\alpha_j(V_1(x))C_1(V_1(x))}{C_3(V_1(x))}V_1(x)\\ & +\sum_{r=1}^{j-1}\alpha_r\big(V_1(x)\big)\mathcal{N}_r(x),\end{aligned} \tag{3.62}$$

we deduce that

$$\begin{aligned}\mathcal{U}(f(x)) - \mathcal{U}(x) \leq & -\frac{\alpha_j\big(V_1(x)\big)C_1\big(V_1(x)\big)}{C_3\big(V_1(x)\big)}V_1(x)\\ & +\sum_{r=1}^{j-1}\Big[-\alpha_r(V_1(x))\mathcal{N}_r(x)\\ & +\alpha_{r+1}\big(V_1(x)\big)\psi_{r+1}\big(V_1(x)\big)V_1^{1-a}(x)\mathcal{N}_r^a(x)\Big].\end{aligned} \tag{3.63}$$

Setting

$$\Gamma_r(s) = \frac{C_3^{\frac{1-a}{a}}(s)\,\alpha_{r+1}^{\frac{1}{a}}(s)}{C_1^{\frac{1-a}{a}}(s)\,\alpha_j^{\frac{1-a}{a}}(s)}$$

for $r = 1, 2, \ldots, j-1$, Young's Inequality $pq \leq p^{1/(1-a)} + q^{1/a}$ applied with

$$p = \frac{\alpha_j^{1-a}(V_1(x))C_1^{1-a}(V_1(x))V_1^{1-a}(x)}{(2j)^{1-a}C_3^{1-a}(V_1(x))} \quad \text{and}$$
$$q = (2j)^{1-a}\psi_{r+1}(V_1(x))\Gamma_r^a(V_1(x))\mathcal{N}_r^a(x)$$

for $a \neq 1$ gives

$$\begin{aligned} &\alpha_{r+1}(V_1(x))\psi_{r+1}(V_1(x))V_1^{1-a}(x)\mathcal{N}_r^a(x) \\ &\leq \frac{\alpha_j(V_1(x))C_1(V_1(x))}{2jC_3(V_1(x))}V_1(x) \\ &\quad +(2j)^{\frac{1-a}{a}}\Gamma_r(V_1(x))\psi_{r+1}^{\frac{1}{a}}(V_1(x))\mathcal{N}_r(x) \end{aligned} \tag{3.64}$$

for all possible $a \in (0,1]$.

Combined with (3.63), this gives

$$\begin{aligned} &\mathcal{U}(f(x)) - \mathcal{U}(x) \\ &\leq -\frac{\alpha_j(V_1(x))C_1(V_1(x))}{2C_3(V_1(x))}V_1(x) \\ &+\sum_{r=1}^{j-1}\left[-\alpha_r(V_1(x)) + (2j)^{\frac{1-a}{a}}\Gamma_r(V_1(x))\psi_{r+1}^{\frac{1}{a}}(V_1(x))\right]\mathcal{N}_r(x), \end{aligned} \tag{3.65}$$

for all possible $a \in (0,1]$. Since our functions α_i satisfy (3.56), we get

$$\mathcal{U}(f(x)) - \mathcal{U}(x) \leq -\frac{\alpha_j(V_1(x))C_1(V_1(x))}{2C_3(V_1(x))}V_1(x). \tag{3.66}$$

Using (3.40), we can determine an increasing C^1 function $\Lambda : [0,\infty) \to [1,\infty)$ such that

$$\begin{aligned} &|\mathcal{U}(x)| \leq \Lambda(V_1(x))V_1(x) \ \ \forall x \in \mathcal{X} \quad \text{and} \\ &\Lambda(r) \geq \tfrac{1}{C_2(r)} \ \ \forall r \geq 0. \end{aligned} \tag{3.67}$$

We easily deduce that

$$S(x) \;=\; \mathcal{U}(x) + 2\Lambda\big(V_1(x)\big)V_1(x) \tag{3.68}$$

satisfies

$$S(x) \geq \Lambda(V_1(x))V_1(x) \geq \frac{V_1(x)}{C_2(V_1(x))} \geq |x|^2 \tag{3.69}$$

and

$$S(f(x)) - S(x) \leq -\frac{\alpha_j(V_1(x))C_1(V_1(x))}{2C_3(V_1(x))}V_1(x). \tag{3.70}$$

Combined with our condition (3.55) on α_j, this proves the theorem. □

3.6 Illustrations

3.6.1 Continuous Time: One Auxiliary Function

Let us show how the strict Lyapunov-like function (3.12) we constructed in Sect. 3.1 follows as a special case of Theorem 3.1. Choose V_1 and V_2 as defined in (3.6) and (3.9), respectively. Then our decay conditions (3.8) and (3.10) imply that Assumption 3.1 is satisfied with $j = 2$, $\mathcal{N}_1(\tilde{s}) = \tilde{s}^2$, $\mathcal{N}_2(\tilde{\gamma}) = \frac{v_*}{2}\tilde{\gamma}^2$, $a_2 = 1$, and the constant function

$$\phi_2(s) \equiv \frac{\gamma_M^2}{2v_\star} + \frac{K(\gamma_M - \gamma_m)^2}{4}.$$

Moreover, since V_1 is bounded from above by a positive definite quadratic function near 0, we can find an everywhere positive function ρ so that

$$\rho(V_1)V_1 \leq \min\left\{1, \frac{v_*}{2}\right\}(\tilde{s}^2 + \tilde{\gamma}^2) \leq \sum_{i=1}^{2} \mathcal{N}_i(\tilde{s}, \tilde{\gamma})$$

on $\mathcal{X}$. (In fact, we can choose ρ so that outside a neighborhood of zero,

$$\rho(v) = \frac{c}{1+v}$$

for a suitable constant c.) Thus, the first part of Assumption 3.2 is also satisfied.

Next note that because $\max\{(\gamma_M - \gamma)(\gamma - \gamma_m) : \gamma \in [\gamma_m, \gamma_M]\} = \frac{1}{4}(\gamma_M - \gamma_m)^2$, we know that

$$V_1(\tilde{s}, \tilde{\gamma}) \geq \frac{1}{2\gamma_m}\tilde{s}^2 + \frac{2v_*}{K\gamma_m(\gamma_M - \gamma_m)^2}\tilde{\gamma}^2 \geq \frac{1}{2}\underline{v}(\tilde{s}^2 + \tilde{\gamma}^2),$$

where

$$\underline{v} = \min\left\{\frac{1}{\gamma_m}, \frac{4v_*}{K\gamma_m(\gamma_M - \gamma_m)^2}\right\}, \tag{3.71}$$

holds on $\mathcal{X}$. This and the triangle inequality $|\tilde{s}\tilde{\gamma}| \leq \frac{1}{2}\tilde{s}^2 + \frac{1}{2}\tilde{\gamma}^2$ give

$$|V_2(\tilde{s}, \tilde{\gamma})| \;=\; |\tilde{s}\tilde{\gamma}| \;\leq\; \frac{V_1(\tilde{s}, \tilde{\gamma})}{\underline{v}}.$$

(Our choice of V_2 was motivated by our desire to have the preceding estimate.) Hence, the second part of Assumption 3.2 is satisfied as well, so Theorem 3.1 applies with the constant function

$$p_2(s) \;\equiv\; \frac{1}{\underline{v}}.$$

We now explicitly build the strict Lyapunov-like function from Theorem 3.1.

Since $j = 2$ and $a_2 = 1$, we get

$$k_2(s) \;=\; \left(\frac{1}{\underline{v}} + 1\right) s,$$

hence

$$U_2(\tilde{s}, \tilde{\gamma}) = \Upsilon_1 V_1(\tilde{s}, \tilde{\gamma}) + V_2(\tilde{s}, \tilde{\gamma}),$$

where Υ_1 is the constant we defined in (3.11). Also, we can take Φ from (3.28) to be ϕ_2 to get

$$\Omega_1(s) = \left[\frac{\gamma_M^2}{v_\star} + \frac{K(\gamma_M - \gamma_m)^2}{2}\right] s \;\text{ and}$$

$$\Omega_2(s) \equiv s.$$

Hence the formula (3.30) for S becomes

$$\begin{aligned} S(\tilde{s}, \tilde{\gamma}) &= U_2(\tilde{s}, \tilde{\gamma}) + 2\left[\frac{\gamma_M^2}{v_\star} + \frac{K(\gamma_M - \gamma_m)^2}{2}\right] V_1(\tilde{s}, \tilde{\gamma}) \\ &= V_2(\tilde{s}, \tilde{\gamma}) + \left[\Upsilon_1 + \frac{2\gamma_M^2}{v_*} + K(\gamma_M - \gamma_m)^2\right] V_1(\tilde{s}, \tilde{\gamma}) \end{aligned} \tag{3.72}$$

which agrees with (3.12).

3.6.2 Continuous Time: Two Auxiliary Functions

We next consider a case where the function (3.23) constructed in Theorem 3.1 is radially unbounded and therefore is a strict Lyapunov function. We consider the system

$$\begin{cases} \dot{x}_1 = x_2 \\ \dot{x}_2 = -x_1 - x_2^3. \end{cases} \tag{3.73}$$

We use the functions

$$\begin{aligned} V_1(x) &= \tfrac{1}{4}(x_1^2+x_2^2)^2\,, \quad \mathcal{N}_1(x) = (x_1^2+x_2^2)x_2^4\,, \\ V_2(x) &= \tfrac{1}{2}(x_1^2+x_2^2)\,, \quad \mathcal{N}_2(x) = x_2^4\,, \\ V_3(x) &= \tfrac{1}{2}(x_1^2+x_2^2)x_1x_2\,, \quad \text{and} \quad \mathcal{N}_3(x) = \tfrac{1}{2}[x_1^2+x_2^2]x_1^2\,. \end{aligned} \tag{3.74}$$

Along the trajectories of (3.73), the functions V_i satisfy

$$\begin{aligned} \dot{V}_1(x) &= -\mathcal{N}_1(x), \\ \dot{V}_2(x) &= -\mathcal{N}_2(x), \quad \text{and} \\ \dot{V}_3(x) &= \tfrac{1}{2}[x_1^2+x_2^2][x_2^2-x_1^2-x_1x_2^3] - \mathcal{N}_2(x)x_1x_2. \end{aligned}$$

Therefore, the inequality $x_1x_2 \leq \frac{1}{2}x_1^2 + \frac{1}{2}x_2^2$ gives

$$\dot{V}_3(x) \leq -\mathcal{N}_3(x) + \phi_3(V_1(x))\sqrt{\mathcal{N}_2(x)}\sqrt{V_1(x)}, \tag{3.75}$$

where

$$\phi_3(r) = 1 + 3\sqrt{r}. \tag{3.76}$$

One can easily check that

$$\begin{aligned} \sum_{r=1}^{3} \mathcal{N}_r(x) &= (x_1^2+x_2^2)x_2^4 + x_2^4 + \tfrac{1}{2}[x_1^2+x_2^2]x_1^2 \\ &\geq \rho(V_1(x))V_1(x), \end{aligned} \tag{3.77}$$

where $\rho(r) \equiv 1$. Also, V_1 and V_2 are positive definite, and

$$|V_3(x)| \leq p_3(V_1(x))V_1(x)\,, \tag{3.78}$$

where $p_3(r) \equiv 1$. Therefore, Assumptions 3.1 and 3.2 are satisfied with $j = 3$, $\bar{\rho}(s) = \sqrt{s}$, $\phi_2 \equiv 1$, $a_2 = 1$, and $a_3 = 1/2$. Hence, Theorem 3.1 provides a strict Lyapunov-like function for (3.73), which turns out to be a strict Lyapunov function.

Let us construct the strict Lyapunov-like function from the theorem. Since $p_2(s) \equiv 0$, we can satisfy the conditions (3.26) on the k_i's by taking

$$k_1(s) = k_2(s) = s \quad \text{and} \quad k_3(s) = 2s.$$

The functions U_i from Lemma 3.1 are therefore

$$\begin{aligned} &U_1(x) = V_1(x)\,, \quad U_2(x) = V_1(x) + V_2(x)\,, \text{ and} \\ &U_3(x) = 2V_1(x) + V_3(x). \end{aligned} \tag{3.79}$$

Since $a_2 = 1$ and $a_3 = 1/2$, the function Φ from (3.28) is $\Phi(s) = 16(1+3\sqrt{s})^2$. Therefore, we can satisfy the conditions on the Ω_i's in (3.29) by taking

$$\begin{aligned}\Omega_3(s) &= s\ , \quad \Omega_2(s) = 32s + 128s^{3/2} + 144s^2\ , \\ \Omega_1(s) &= \Omega_2(s) + 32^2\left(49s + 105s^2 + 80s^3\right)\ , \text{ and} \\ S(x) &= \Omega_1(2U_1(x)) + \Omega_2(U_2(x)) + U_3(x).\end{aligned} \tag{3.80}$$

With these choices, we obtain

$$\dot{S}(x) \le -\tfrac{1}{4}V_1(x). \tag{3.81}$$

In conjunction with the properness and positive definiteness of S, this shows that S is a strict Lyapunov function for (3.73).

Remark 3.5. The parameters in the functions Ω_1 and Ω_2 in (3.80) are large. However, we can construct a global strict Lyapunov function for (3.73) with smaller parameters, by the following direct construction.

We have

$$\begin{aligned}\dot{U}_1(x) &= -\mathcal{N}_1(x)\ , \\ \dot{U}_2(x) &= -\left(\mathcal{N}_1(x) + \mathcal{N}_2(x)\right)\ , \text{ and} \\ \dot{U}_3(x) &\le -2\mathcal{N}_1(x) - \mathcal{N}_3(x) + \phi_3\left(V_1(x)\right)\sqrt{\mathcal{N}_2(x)}\sqrt{V_1(x)}\ .\end{aligned} \tag{3.82}$$

Therefore,

$$\begin{aligned}\dot{U}_3(x) + \dot{U}_2(x) &\le -\mathcal{N}_1(x) - \mathcal{N}_2(x) - \mathcal{N}_3(x) \\ &\quad + \phi_3(V_1(x))\sqrt{\mathcal{N}_2(x)}\sqrt{V_1(x)} \\ &\le -V_1(x) + \phi_3(V_1(x))\sqrt{\mathcal{N}_2(x)}\sqrt{V_1(x)} \\ &\le -\tfrac{1}{2}V_1(x) + \tfrac{1}{2}\phi_3^2(V_1(x))\mathcal{N}_2(x) \\ &\le -\tfrac{1}{2}V_1(x) + (1 + 9V_1(x))\mathcal{N}_2(x)\ ,\end{aligned} \tag{3.83}$$

where the second inequality is by (3.77). Let

$$\overline{S}(x) = 2U_2(x) + 8U_2^2(x) + U_3(x). \tag{3.84}$$

This function satisfies

$$\dot{\overline{S}}(x) \le -\frac{1}{2}V_1(x)\ . \tag{3.85}$$

Moreover, $\overline{S}$ is positive definite and radially unbounded, because the U_i's are bounded below by V_1. Therefore $\overline{S}$ is a strict Lyapunov function for (3.73).

3.6.3 Discrete Time Context

We illustrate our discrete time Lyapunov function construction from Theorem 3.2 using the system

$$\begin{cases} p_{k+1} = q_k \\ q_{k+1} = r_k \\ r_{k+1} = p_k - \frac{3}{4}\frac{p_k}{1+p_k^2}. \end{cases} \tag{3.86}$$

Let $x = (p, q, r)$. We check the assumptions of the theorem using

$$\begin{aligned} &\nu_1(x) = \tfrac{1}{2}\left[p^2 + q^2 + r^2\right], \quad \nu_2(x) = r^2, \quad \nu_3(x) = q^2, \\ &\mathcal{M}_1(x) = \tfrac{15}{32}\tfrac{p^2}{1+p^2}, \quad \mathcal{M}_2(x) = r^2, \quad \text{and} \quad \mathcal{M}_3(x) = q^2 . \end{aligned} \tag{3.87}$$

Notice that

$$\begin{aligned} \nu_1(x_{k+1}) - \nu_1(x_k) &= \tfrac{1}{2}[p_{k+1}^2 + q_{k+1}^2 + r_{k+1}^2] - \tfrac{1}{2}[p_k^2 + q_k^2 + r_k^2] \\ &= \tfrac{1}{2}\left[\left(p_k - \tfrac{3}{4}\tfrac{p_k}{1+p_k^2}\right)^2 - p_k^2\right] \\ &= \tfrac{1}{2}\left[-\tfrac{3}{2}\tfrac{p_k^2}{1+p_k^2} + \tfrac{9}{16}\tfrac{p_k^2}{(1+p_k^2)^2}\right] \\ &\le -\mathcal{M}_1(x_k). \end{aligned} \tag{3.88}$$

Also,

$$\begin{aligned} \nu_2(x_{k+1}) - \nu_2(x_k) &= r_{k+1}^2 - r_k^2 \\ &= -\mathcal{M}_2(x_k) + \left(1 - \tfrac{3}{4}\tfrac{1}{1+p_k^2}\right)^2 p_k^2 \\ &\le -\mathcal{M}_2(x_k) + \tfrac{32}{15}\left(1 - \tfrac{3}{4}\tfrac{1}{1+p_k^2}\right)^2 (1+p_k^2)\mathcal{M}_1(x_k) \end{aligned} \tag{3.89}$$

and

$$\nu_3(x_{k+1}) - \nu_3(x_k) = q_{k+1}^2 - q_k^2 \; = \; -\mathcal{M}_3(x_k) + \mathcal{M}_2(x_k). \tag{3.90}$$

In summary,

$$\begin{aligned} &\nu_1(x_{k+1}) - \nu_1(x_k) \le -\mathcal{M}_1(x_k) \\ &\nu_2(x_{k+1}) - \nu_2(x_k) \le -\mathcal{M}_2(x_k) + \phi_2(\nu_1(x_k))\mathcal{M}_1(x_k) \\ &\nu_3(x_{k+1}) - \nu_3(x_k) = -\mathcal{M}_3(x_k) + \phi_3(\nu_1(x_k))\mathcal{M}_2(x_k) , \end{aligned} \tag{3.91}$$

where

$$\phi_2(l) \; = \; \frac{32}{15}(1+2l) \text{ and } \phi_3(l) = l . \tag{3.92}$$

It follows that Assumption 3.3 is satisfied. Moreover, for all choices of x,

$$\begin{aligned}
&\sum_{l=1}^{3} \mathcal{M}_l(x) = \frac{15}{32}\frac{p^2}{1+p^2} + q^2 + r^2 \;\geq\; C_1(\nu_1(x))|x|^2 \,, \\
&C_2(\nu_1(x))|x|^2 \leq \nu_1(x) \;\leq\; C_3(\nu_1(x))|x|^2, \text{ and} \\
&|\nu_i(x)| \leq C_4(\nu_1(x))|x|^2 \text{ for } i = 2, 3
\end{aligned} \tag{3.93}$$

where

$$C_1(l) = \frac{15}{32(1+2l)}, \;\; C_2(l) = C_3(l) = \frac{1}{2}, \;\text{ and }\; C_4(l) = 1$$

for all $l \geq 0$. Therefore, Assumption 3.4 is satisfied as well, so Theorem 3.2 applies. Hence, we can construct a strict Lyapunov function for the system (3.86) by arguing as in the proof of Theorem 3.2.

Let us construct a strict Lyapunov function for (3.86) of the type guaranteed by the theorem. Since

$$\begin{aligned}
&2\nu_2(x_{k+1}) - 2\nu_2(x_k) + \nu_3(x_{k+1}) - \nu_3(x_k) \\
&= -\mathcal{M}_3(x_k) - \mathcal{M}_2(x_k) + \frac{64}{15}[1 + 2\nu_1(x_k)]\mathcal{M}_1(x_k),
\end{aligned} \tag{3.94}$$

the radially unbounded positive definite function

$$\overline{S}(x) = \frac{94}{15}\big[1 + 2\nu_1(x)\big]\nu_1(x) + 2\nu_2(x) + \nu_3(x). \tag{3.95}$$

satisfies

$$\begin{aligned}
\overline{S}(x_{k+1}) - \overline{S}(x_k) &\leq -2\big[1 + 2\nu_1(x_k)\big]\mathcal{M}_1(x_k) - \mathcal{M}_2(x_k) - \mathcal{M}_3(x_k) \\
&\leq -\nu_1(x_k) \,,
\end{aligned} \tag{3.96}$$

which is the desired decay condition.

3.7 Comments

The recent paper [111] provides an alternative and very general Matrosov approach for constructing strict Lyapunov-like functions. However the Lyapunov functions provided by [111] are not in general locally bounded from below by positive definite quadratic functions, even for globally asymptotically linear systems, which admit a quadratic strict Lyapunov function. The shape of Lyapunov functions, their local properties and their simplicity matter when they are used to investigate robustness properties and construct feedbacks and gains.

The differences between Assumptions 3.1 and 3.2 and the assumptions from [111] are as follows. First, while our Assumption 3.1 ensures that V_1 is positive definite but not necessarily proper, [111] assumes that a radially unbounded non-strict Lyapunov function is known. Second, our Assumption 3.1 is a restrictive version of Assumption 2 from [111]. More precisely, our Assumption 3.1 specifies the local properties of the functions that correspond to the χ_i's of Assumption 2 in [111]. Finally, our Assumption 3.2 imposes relations between the functions $\mathcal{N}_i$ and V_1, which are not required in [111]. An important feature is that we do not require the functions $V_2, \ldots, V_j$ to be non-negative.

Our treatment of (3.3) is based on [106]. Since the Matrosov constructions in [111] assume that the given non-strict Lyapunov function is globally proper on the whole Euclidean space, and since (3.6) does not satisfy this requirement, we cannot construct the required explicit strong Lyapunov function for (3.3) using the results of [111]. Notice that the strict Lyapunov-like function (3.72) that we constructed for the anaerobic digester is a simple linear combination of V_1 and V_2. By contrast, the strong Lyapunov functions provided by [111, Theorem 3] for the $j = 2$ time-invariant case have the form

$$S(x) \;=\; Q_1\big(V_1(x)\big)V_1(x) + Q_2\big(V_1(x)\big)V_2(x),$$

where Q_1 is non-negative, and where the positive definite function Q_2 needs to globally satisfy

$$Q_2(V_1) \le \phi^{-1}\left(\frac{\omega(x)}{2\rho(|x|)}\right),$$

where

$$\nabla V_2(x)f(x) \;\le\; -\mathcal{N}_2(x) + \phi\big(\mathcal{N}_1(x)\big)\rho(|x|)$$

for some $\phi \in \mathcal{K}_\infty$ and some everywhere positive non-decreasing function ρ and the positive definite function ω needs to satisfy $\mathcal{N}_1(x) + \mathcal{N}_2(x) \ge \omega(x)$ everywhere. In particular, we cannot take Q_2 to be constant to get a linear combination of the V_i's, so the construction of [111] is more complicated than the one we provide here. Similar remarks apply to the other constructions in [111]. See Chap. 8 for strict Lyapunov function constructions under more general Matrosov type conditions.

Chapter 4
Jurdjevic-Quinn Conditions

Abstract The Jurdjevic-Quinn Theorem provides a powerful framework for guaranteeing globally asymptotic stability, using a smooth feedback of arbitrarily small amplitude. It requires certain algebraic conditions on the Lie derivatives of a suitable non-strict Lyapunov function, in the directions of the vector fields that define the system. The non-strictness of the Lyapunov function is an obstacle to proving robustness, since robustness analysis typically requires *strict* Lyapunov functions.

In this chapter, we provide a method for overcoming this obstacle. It involves transforming the non-strict Lyapunov function into an explicit global CLF. This gives a strict Lyapunov function construction for closed-loop Jurdjevic-Quinn systems with feedbacks of arbitrarily small magnitude. This is valuable because (a) the non-strict Lyapunov function from the Jurdjevic-Quinn Theorem is often known explicitly and (b) our methods apply to Hamiltonian systems, which commonly arise in mechanical engineering. We illustrate our work using a two-link manipulator model, as well as an integral input-to-state stability result.

4.1 Motivation

Consider the two-link manipulator system from [5]. This is a fully actuated system obtained by viewing the robot arm as a segment with length L and mass M. Letting m denote the mass of the hand, r the position of the hand, and θ the angle of the arm, we get the Euler-Lagrange equations

$$\left\{ \begin{aligned} \left(mr^2 + M\frac{L^2}{3}\right)\ddot{\theta} + 2Mr\dot{r}\dot{\theta} &= \tau \\ m\ddot{r} - mr\dot{\theta}^2 &= F \ , \end{aligned} \right. \tag{4.1}$$

where τ and F are forces acting on the system. See Fig. 4.1.

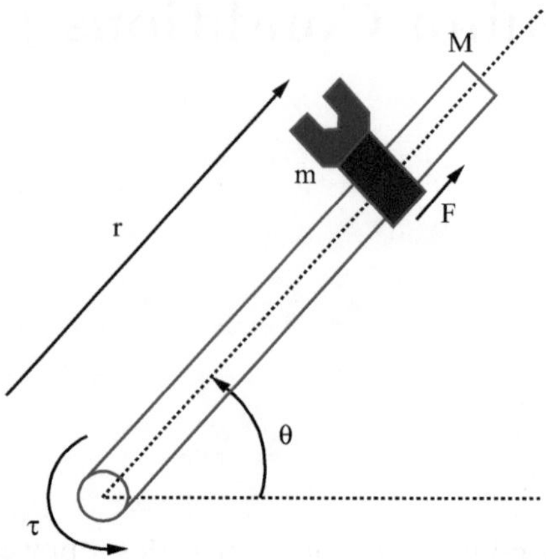

Fig. 4.1 Linear rotational actuated arm modeled by Euler-Lagrange Eq. (4.1)

It is well-known that (4.1) can be stabilized by bounded control laws. However, it is not clear how to construct a CLF for the system whose time derivative along the trajectory is made negative definite by an appropriate choice of bounded feedback. Let us show how such a CLF can be constructed.

For simplicity, we take

$$m = M = 1,\ L = \sqrt{3},\ x_1 \doteq \theta,\ x_2 \doteq \dot{\theta},\ x_3 \doteq r,\ \text{ and } x_4 \doteq \dot{r}.$$

The system (4.1) becomes

$$\begin{cases} \dot{x}_1 = x_2, & \dot{x}_2 = -\dfrac{2x_3x_2x_4}{x_3^2+1} + \dfrac{\tau}{x_3^2+1}, \\ \dot{x}_3 = x_4, & \dot{x}_4 = x_3x_2^2 + F. \end{cases} \tag{4.2}$$

We construct a globally asymptotically stabilizing feedback that is bounded by 2, and an associated CLF for (4.2). We set

$$\langle p \rangle = \frac{1}{2\sqrt{1+p^2}}$$

for all $p \in \mathbb{R}$ throughout the sequel.

Consider the positive definite and radially unbounded function

$$V(x) = \frac{1}{2}\left[(x_3^2+1)x_2^2 + x_4^2 + \sqrt{1+x_1^2} + \sqrt{1+x_3^2} - 2\right], \tag{4.3}$$

which is composed of the kinetic energy of the system with additional terms. With the change of feedback

$$\tau = -x_1\langle x_1\rangle + \tau_b \ , \ F = -x_3\langle x_3\rangle + F_b, \tag{4.4}$$

the system (4.2) takes the control affine form

$$\dot{x} = f(x) + g(x)u, \quad \text{where}$$

$$f(x) = \begin{bmatrix} x_2 \\ \frac{-2x_3x_2x_4 - x_1\langle x_1\rangle}{x_3^2+1} \\ x_4 \\ x_2^2x_3 - x_3\langle x_3\rangle \end{bmatrix}, \quad g(x) = \begin{bmatrix} 0 & 0 \\ \frac{1}{x_3^2+1} & 0 \\ 0 & 0 \\ 0 & 1 \end{bmatrix}, \quad \text{and } u = \begin{bmatrix} \tau_b \\ F_b \end{bmatrix}. \tag{4.5}$$

Next consider the vector field

$$G(x) = (0, x_1, 0, x_3)^\top$$

and the function

$$V^\sharp(x) \ = \ 40[2 + 2V(x)]^6 + L_GV(x) - 40(2^6) \ . \tag{4.6}$$

One can show that when we choose the feedbacks

$$\tau_b = -x_2\langle x_2\rangle \ \text{ and } \ F_b = -x_4\langle x_4\rangle, \tag{4.7}$$

the time derivative of $V^\sharp$ along the trajectories of the closed-loop system (4.5) satisfies

$$\dot{V}^\sharp(x) \leq -\frac{1}{2}\left[x_1^2\langle x_1\rangle + x_2^2\langle x_2\rangle + x_3^2\langle x_3\rangle + x_4^2\langle x_4\rangle\right] \ , \tag{4.8}$$

and that $V^\sharp$ is proper and positive definite; see Sect. 4.7.2 for details. The right hand side of this inequality is negative definite, and the feedback (τ, F) given by (4.4) and (4.7) is bounded in norm by 2, as desired. Also, since the feedback is 0 and continuous at the origin, the CLF (4.6) satisfies the small control property. We turn next to a general construction that leads to $V^\sharp$ as a special case.

4.2 Control Affine Case

4.2.1 Assumptions and Statement of Result

We first consider control affine systems

$$\dot{x} = f(x) + g(x)u \tag{4.9}$$

with state space $\mathcal{X} = \mathbb{R}^n$ and control set $U = \mathbb{R}^m$, where $f : \mathbb{R}^n \to \mathbb{R}^n$ and $g : \mathbb{R}^n \to \mathbb{R}^{n\times m}$ are assumed to be smooth, i.e., C^∞, and $f(0) = 0$.[1] In Sect. 4.3, we use our arguments for the control affine case to extend our results to general nonlinear systems $\dot{x} = \mathcal{F}(x, u)$. We assume the following:

Assumption 4.1 *There is a storage function $V : \mathbb{R}^n \to [0, \infty)$ such that $L_fV(x) \le 0$ everywhere. Moreover, there is a smooth scalar function ψ such that if $x \ne 0$ is such that $L_fV(x) = 0$ and $L_gV(x) = 0$ both hold, then $L_f\psi(x) < 0$.*

We refer to ψ as the *auxiliary scalar field*; see [41] or Sect. 4.4 below for general methods for constructing ψ when the Weak Jurdjevic-Quinn Conditions are satisfied. Our main result for (4.9) is:

Theorem 4.1. *Let (4.9) be such that Assumption 4.1 is satisfied. Then we can explicitly construct C^1 functions λ and Γ such that*

$$\mathcal{U}(x) = \lambda\big(V(x)\big)\psi(x) + \Gamma\big(V(x)\big) \tag{4.10}$$

is a CLF for (4.9) that satisfies the small control property. In fact, for any smooth everywhere positive function $\xi : \mathbb{R}^n \to (0, \infty)$, we can construct λ and Γ in such a way that (4.10) is a strict Lyapunov function for (4.9) in closed-loop with the feedback $u(x) = -\xi(x)L_gV(x)^\top$.

In particular, we get stabilizing feedbacks of arbitrarily small amplitude.

4.2.2 Main Lemmas

We use the following lemmas to prove Theorem 4.1. We use the notation

$$\begin{aligned} \mathcal{N}(x) &= -\min\{0, L_f\psi(x)\}\ , \\ H(x) &= -L_fV(x) + |L_gV(x)|^2\ , \text{ and} \\ S(x) &= H(x) + \mathcal{N}(x)\ . \end{aligned} \tag{4.11}$$

Lemma 4.1. *The function $S(x)$ is continuous and positive definite.*

Proof. By Assumption 4.1, both H and $\mathcal{N}$ are non-negative, so S is non-negative. On the other hand, if $S(x) = 0$, then $L_fV(x) = 0$, $L_gV(x) = 0$ and $L_f\psi(x) \ge 0$. Then Assumption 4.1 implies that $x = 0$. □

A key feature of our proof of Theorem 4.1 is that it provides explicit formulas for functions λ and Γ such that (4.10) is a CLF for (4.9). In fact, we will prove later that (4.10) is a CLF for (4.9) when

[1] The smoothness assumptions in this section can be replaced by the assumption that the relevant functions are C^k where k is large enough to make the CLF and feedback we construct C^1.

$$\begin{aligned}&\Gamma(r) = \int_0^r \gamma(s)\mathrm{d}s, \\ &\text{where } \gamma(s) = 1 + K_1'(s)s + 3\left[K_1(s) + K_1^{3/2}(s)\right]\end{aligned} \tag{4.12}$$

and K_1 and λ satisfy the requirements of the following lemma:

Lemma 4.2. *Let Assumption 4.1 hold. Then we can construct a function $\lambda \in \mathcal{K}_\infty \cap C^1$ and a C^1 increasing function $K_1 : [0,\infty) \to (0,\infty)$ such that $\lambda(s) \le K_1(s)$ everywhere,*

$$\lambda(v) \le v \;\; \forall v \ge 0\,, \quad \textit{and} \tag{4.13}$$

$$\lambda'(V(x))|\psi(x)| + \lambda(V(x)) \le K_1(V(x))\,, \tag{4.14}$$

$$|\psi(x)| \le K_1(V(x))\,, \tag{4.15}$$

$$|L_g\psi(x)|^2 \le K_1(V(x))\,, \tag{4.16}$$

and

$$\lambda(V(x))\big[1 + \max\{0, L_f\psi(x)\}\big] \le S(x)K_1(V(x)) \tag{4.17}$$

hold for all $x \in \mathbb{R}^n$.

Proof. Since S is positive definite and V is proper and positive definite, we can find a continuous positive definite function ρ_0 so that $S(x) \ge \rho_0(V(x))$ for all $x \in \mathbb{R}^n$ (by first finding a positive definite function $\tilde{\rho}$ that is increasing on $[0,1]$ and decreasing on $[1,\infty)$ such that $S(x) \ge \tilde{\rho}(|x|)$ everywhere). Hence, Lemma A.7 provides $\lambda \in \mathcal{K}_\infty \cap C^1$ such that (4.13) is satisfied and an everywhere positive increasing function $\bar{K} \in C^1$ such that

$$\lambda(V(x)) \le S(x)\bar{K}(V(x)) \;\; \forall x \in \mathbb{R}^n. \tag{4.18}$$

We can also find an increasing function $\bar{\kappa} \in C^1$ such that

$$1 + \max\{0, L_f\psi(x)\} \le \bar{\kappa}\big(V(x)\big) \;\; \forall x \in \mathbb{R}^n. \tag{4.19}$$

Combining (4.18) and (4.19) provides an increasing function $\kappa_1 \in C^1$ such that

$$\lambda(V(x))\big[1 + \max\{0, L_f\psi(x)\}\big] \le S(x)\kappa_1\big(V(x)\big) \;\; \forall x \in \mathbb{R}^n. \tag{4.20}$$

Next, one can determine everywhere positive increasing functions $\kappa_i \in C^1$ for $i = 2, 3, 4$ such that

$$\lambda'(V(x))|\psi(x)| + \lambda\big(V(x)\big) \le \kappa_2\big(V(x)\big) \tag{4.21}$$

and

$$|\psi(x)| \le \kappa_3\big(V(x)\big) \;\text{ and }\; |L_g\psi(x)|^2 \le \kappa_4\big(V(x)\big) \tag{4.22}$$

hold for all $x \in \mathbb{R}^n$. Since $\lambda' \ge 0$, the inequality $\lambda(s) \le \kappa_2(s)$ is satisfied everywhere. It follows that

$$K_1(v) = \sum_{i=1}^{4} \kappa_i(v), \tag{4.23}$$

is such that the inequalities (4.14)-(4.17) are all satisfied. □

In the sequel, all (in)equalities should be understood to hold for all $x \in \mathbb{R}^n$ unless otherwise indicated. The following lemma is a key ingredient in our proof of the Lyapunov decay condition for (4.10):

Lemma 4.3. *Let the functions λ and K_1 satisfy the requirements of Lemma 4.2. Then for all $x \in \mathbb{R}^n$, the inequality*

$$\lambda\big(V(x)\big)L_f\psi(x) \leq -\lambda\big(V(x)\big)S(x) + 2K_1\big(V(x)\big)H(x) \tag{4.24}$$

is satisfied.

Proof. According to the definition of $\mathcal{N}$, we get

$$L_f\psi(x) \;=\; -\mathcal{N}(x) + \max\{0, L_f\psi(x)\}. \tag{4.25}$$

Therefore, (4.17) from Lemma 4.2 gives

$$\begin{aligned} \lambda(V(x))L_f\psi(x) &= -\lambda(V(x))\mathcal{N}(x) + \lambda(V(x))\max\{0, L_f\psi(x)\} \\ &\leq -\lambda(V(x))\mathcal{N}(x) + S(x)K_1(V(x)). \end{aligned} \tag{4.26}$$

We consider two cases.

Case 1. $L_f\psi(x) \leq 0$. Then, (4.26) gives

$$\begin{aligned} \lambda(V(x))L_f\psi(x) &= -\lambda(V(x))\mathcal{N}(x) \\ &= -\lambda(V(x))S(x) + H(x)\lambda(V(x)). \end{aligned} \tag{4.27}$$

Case 2. $L_f\psi(x) > 0$. Then, the definition of $\mathcal{N}$ in (4.11) gives $\mathcal{N}(x) = 0$, which implies that $S(x) = H(x)$. This combined with (4.26) yields

$$\begin{aligned} \lambda(V(x))L_f\psi(x) &\leq H(x)K_1(V(x)) \\ &= -\lambda(V(x))S(x) + H(x)[K_1(V(x)) + \lambda(V(x))]. \end{aligned} \tag{4.28}$$

We deduce that in both cases,

$$\lambda(V(x))L_f\psi(x) \leq -\lambda(V(x))S(x) + H(x)\left[K_1(V(x)) + \lambda(V(x))\right]. \tag{4.29}$$

The result follows because $\lambda(s) \leq K_1(s)$ everywhere, by (4.14). □

4.2.3 Checking the CLF Properties

Returning to the proof of Theorem 4.1, let Γ be the function defined in (4.12), and let K_1 and λ be the functions provided by Lemma 4.2. We check that the resulting function $\mathcal{U}$ from (4.10) satisfies the required CLF properties. Notice that $\Gamma(v) \geq v + \int_0^v [K_1'(s)s + K_1(s)]\mathrm{d}s = v + K_1(v)v$ everywhere, so

$$\mathcal{U}(x) \geq \lambda\big(V(x)\big)\psi(x) + V(x) + K_1\big(V(x)\big)V(x) \tag{4.30}$$

holds for all $x \in \mathbb{R}^n$. From (4.13) and (4.15), we deduce that

$$\mathcal{U}(x) \geq V(x) \tag{4.31}$$

so $\mathcal{U}$ is positive definite and radially unbounded.

The time derivative of $\mathcal{U}$ along the trajectories of (4.9) is

$$\begin{aligned}\dot{\mathcal{U}}(x) &= \lambda\big(V(x)\big)\big[L_f\psi(x) + L_g\psi(x)u\big] \\ &\quad + \big[\lambda'(V(x))\psi(x) + \Gamma'(V(x))\big]\big[L_fV(x) + L_gV(x)u\big] \\ &= \lambda\big(V(x)\big)L_f\psi(x) + \big[\lambda'(V(x))\psi(x) + \Gamma'(V(x))\big]L_fV(x) \\ &\quad + \Theta(x)u\ ,\end{aligned} \tag{4.32}$$

where

$$\Theta(x) = \lambda(V(x))L_g\psi(x) + \Big\{\lambda'(V(x))\psi(x) + \gamma(V(x))\Big\}L_gV(x). \tag{4.33}$$

Using Lemma 4.3 and the definition of H in (4.11), we deduce that

$$\begin{aligned}\dot{\mathcal{U}}(x) &\leq -\lambda\big(V(x)\big)S(x) + 2K_1\big(V(x)\big[-L_fV(x) + |L_gV(x)|^2\big] \\ &\quad + \big[\lambda'\big(V(x)\big)\psi(x) + \Gamma'\big(V(x)\big)\big]L_fV(x) + \Theta(x)u \\ &= -\lambda\big(V(x)\big)S(x) + 2K_1\big(V(x)\big)|L_gV(x)|^2 + \Theta(x)u \\ &\quad + \big[\lambda'\big(V(x)\big)\psi(x) - 2K_1\big(V(x)\big) + \gamma\big(V(x)\big)\big]L_fV(x)\ .\end{aligned} \tag{4.34}$$

Recalling (4.14) and the facts that $\gamma(\ell) \geq 3K_1(\ell)$ for all $\ell \geq 0$ and $L_fV(x) \leq 0$ everywhere, we obtain

$$\dot{\mathcal{U}}(x) \leq \mho(x) + \Theta(x)u\ , \tag{4.35}$$

where

$$\mho(x) = -\lambda\big(V(x)\big)S(x) + 2K_1\big(V(x)\big)|L_gV(x)|^2\ . \tag{4.36}$$

Consider the control

$$u = -50\Theta(x)\ . \tag{4.37}$$

It suffices to show that $\mho(x) - 50\Theta^2(x)$ is negative definite because (4.35) and (4.37) combine to give $\dot{\mathcal{U}}(x) \leq \mho(x) - 50\Theta^2(x)$. To prove that $\mho(x) - 50\Theta^2(x)$ is negative definite, we proceed by contradiction. Suppose that there exists $x \neq 0$ such that $\mho(x) - 50\Theta^2(x) \geq 0$, or equivalently,

$$-\lambda(V(x))S(x) + 2K_1(V(x))|L_gV(x)|^2 - 50\Big[\lambda(V(x))L_g\psi(x) + \{\lambda'(V(x))\psi(x) + \gamma(V(x))\}L_gV(x)\Big]^2 \geq 0. \tag{4.38}$$

Then,

$$2K_1(V(x))|L_gV(x)|^2 \geq 50\Big[\lambda(V(x))L_g\psi(x) + \{\lambda'(V(x))\psi(x) + \gamma(V(x))\}L_gV(x)\Big]^2 . \tag{4.39}$$

Therefore, the general relation $|a+b+c| \geq |a| - |b| - |c|$ for any real numbers a, b, and c gives

$$\begin{aligned}\sqrt{K_1(V(x))}|L_gV(x)| &\geq 5\Big|\lambda(V(x))L_g\psi(x) \\ &\quad + \{\lambda'(V(x))\psi(x) + \gamma(V(x))\}L_gV(x)\Big| \\ &\geq -5|\lambda(V(x))L_g\psi(x)| - 5|\lambda'(V(x))\psi(x)||L_gV(x)| \\ &\quad +5\gamma(V(x))|L_gV(x)| .\end{aligned}$$

It follows that

$$5|\lambda(V(x))L_g\psi(x)| \geq \Big[-\sqrt{K_1(V(x))} - 5|\lambda'(V(x))\psi(x)| + 5\gamma(V(x))\Big]|L_gV(x)| . \tag{4.40}$$

From (4.14), we deduce that

$$\begin{aligned}5|\lambda(V(x))L_g\psi(x)| &\geq \Big[-\sqrt{K_1(V(x))} - 5K_1(V(x)) + 5\gamma(V(x))\Big]|L_gV(x)| \\ &\geq \big[-1 - 6K_1(V(x)) + 5\gamma(V(x))\big]|L_gV(x)| .\end{aligned}$$

Our choice of γ in (4.12) then gives$5|\lambda V(x))L_g\psi(x)| \geq 3\gamma(V(x))|L_gV(x)|$ and therefore

$$50K_1(V(x))\frac{|\lambda(V(x))L_g\psi(x)|^2}{9\gamma^2(V(x))} \geq 2K_1(V(x))|L_gV(x)|^2 . \tag{4.41}$$

Since (4.38) implies that $\lambda(V(x))S(x) \leq 2K_1(V(x))|L_gV(x)|^2$, we get

$$50K_1(V(x))\frac{|\lambda(V(x))L_g\psi(x)|^2}{9\gamma^2(V(x))} \geq \lambda(V(x))S(x) \tag{4.42}$$

and then

$$50K_1^2(V(x))\frac{|L_g\psi(x)|^2}{9\gamma^2(V(x))}\lambda(V(x)) \geq S(x)K_1(V(x)) . \tag{4.43}$$

From (4.16), we deduce that

$$\frac{50K_1^3(V(x))}{9\gamma^2(V(x))}\lambda(V(x)) \geq S(x)K_1(V(x)) . \tag{4.44}$$

Recalling our choice of γ in (4.12), we have

$$\frac{50K_1^3(V(x))}{9\gamma^2(V(x))} \leq \frac{50K_1^3(V(x))}{9[1+3K_1^{3/2}(V(x))]^2} \leq \frac{50}{81} . \tag{4.45}$$

It follows that $\frac{50}{81}\lambda(V(x)) \geq S(x)K_1(V(x))$. This contradicts (4.17). We conclude that $\mathcal{U}$ is a CLF for (4.9) that satisfies the small control property, when Γ is defined by (4.12).

4.2.4 Arbitrarily Small Stabilizing Feedbacks

In the previous section, we constructed a family of CLFs of the form

$$\mathcal{U}(x) = \lambda(V(x))\psi(x) + \int_0^{V(x)} \gamma(r)\mathrm{d}r \tag{4.46}$$

for the control affine system (4.9). We now show that for any smooth function $\xi : \mathbb{R}^n \to (0,\infty)$, we can choose γ in such a way that (4.46) is a strict Lyapunov function for (4.9) in closed-loop with

$$u(x) = -\xi(x)L_gV(x)^\top . \tag{4.47}$$

This will prove that for any control set $U \subseteq \mathbb{R}^m$ containing a neighborhood of the origin, (4.9) is C^1 globally asymptotically stabilizable by a feedback that takes all of its values in U.

To this end, pick any C^1 function γ such that

$$\gamma(s) \geq 1 + K_1'(s)s + 3\left[K_1(s) + K_1^{3/2}(s)\right] \quad \forall s \geq 0 \tag{4.48}$$

and

$$\gamma(V(x)) \geq \frac{4K_1(V(x))}{\xi(x)} + 2K_1(V(x)) + 2\xi(x)K_1^2(V(x)) \ \ \forall x \in \mathbb{R}^n , \tag{4.49}$$

where K_1 is the function provided by Lemma 4.2. We show that the time derivative of (4.46) along the solutions of (4.9) in closed-loop with (4.47) satisfies

$$\dot{\mathcal{U}}(x) \leq -W(x), \tag{4.50}$$

where

$$W(x) = \frac{3}{4}\lambda(V(x))S(x) + \frac{1}{2}\xi(x)\gamma(V(x))\left|L_gV(x)\right|^2. \tag{4.51}$$

Here λ is the function constructed in Lemma 4.2. The result is then immediate from the smoothness of u and the fact that $u(0) = 0$, because V, λ, and S are all positive definite.

To prove the estimate (4.50), first notice that (4.35) and (4.48) give

$$\begin{aligned}\dot{\mathcal{U}}(x) &\leq \mho(x) - \Theta(x)\xi(x)L_gV(x)^\top \\ &= \mho(x) - \xi(x)\Big[\lambda(V(x))L_g\psi(x)L_gV(x)^\top \\ &\quad +\{\lambda'(V(x))\psi(x) + \gamma(V(x))\}|L_gV(x)|^2\Big].\end{aligned} \tag{4.52}$$

Therefore, our choice of $\mho$ in (4.36) gives

$$\begin{aligned}\dot{\mathcal{U}}(x) \ \leq \ & -\lambda(V(x))S(x) + \xi(x)\lambda(V(x))|L_g\psi(x)||L_gV(x)| \\ & + \Big[2K_1(V(x)) + \xi(x)\{|\lambda'(V(x))\psi(x)| - \gamma(V(x))\}\Big]|L_gV(x)|^2.\end{aligned} \tag{4.53}$$

Recalling (4.14) and (4.16) gives

$$\begin{aligned}\dot{\mathcal{U}}(x) \leq & -\lambda(V(x))S(x) + \xi(x)\lambda(V(x))\sqrt{K_1(V(x))}|L_gV(x)| \\ & + \Big[2K_1(V(x)) + \xi(x)\{K_1(V(x)) - \gamma(V(x))\}\Big]|L_gV(x)|^2.\end{aligned} \tag{4.54}$$

The triangle inequality $c_1c_2 \leq c_1^2 + \frac{1}{4}c_2^2$ for non-negative c_1 and c_2 gives

$$\begin{aligned}&\xi(x)\lambda(V(x))\sqrt{K_1(V(x))}|L_gV(x)| \\ &\leq \ \xi^2(x)K_1^2(V(x))|L_gV(x)|^2 + \tfrac{1}{4K_1(V(x))}\lambda^2(V(x)).\end{aligned}$$

Combining with (4.54), we get

$$\begin{aligned}\dot{\mathcal{U}}(x) \leq & -\lambda(V(x))S(x) + \tfrac{1}{4K_1(V(x))}\lambda^2(V(x)) \\ & + \Big[2K_1(V(x)) + \xi(x)\{K_1(V(x)) + \xi(x)K_1^2(V(x)) \\ & -\gamma(V(x))\}\Big]|L_gV(x)|^2.\end{aligned} \tag{4.55}$$

Property (4.17) from Lemma 4.2 now gives

$$\frac{1}{4K_1(V(x))}\lambda^2(V(x)) \ \leq \ \frac{1}{4}\lambda(V(x))S(x). \tag{4.56}$$

Hence, (4.49) and (4.55) give

$$\dot{\mathcal{U}}(x) \leq -\tfrac{3}{4}\lambda(V(x))S(x) - \tfrac{1}{2}\xi(x)\gamma(V(x))|L_gV(x)|^2 \tag{4.57}$$

for all $x \in \mathbb{R}^n$, which is the desired Lyapunov decay condition. This completes the proof of Theorem 4.1. □

Remark 4.1. The simplicity of the formula for $\mathcal{U}$ depends on the choice for ξ. For example, if we pick

$$\xi(x) = \frac{1}{\sqrt{K_1(V(x))}},$$

then (4.49) becomes

$$\gamma\big(V(x)\big) \;\geq\; 2K_1\big(V(x)\big) + 6K_1^{3/2}\big(V(x)\big),$$

so we can satisfy (4.48)-(4.49) by taking

$$\gamma(s) \;=\; 1 + 3K_1(s) + K_1'(s)s + 6K_1^{3/2}(s)$$

to obtain our strict Lyapunov function for the corresponding closed-loop system.

4.3 General Case

We now use our results for the control affine system (4.9) to get analogous constructions for general nonlinear systems

$$\dot{x} = \mathcal{F}(x,u) \tag{4.58}$$

evolving on $\mathbb{R}^n$ with controls in $\mathbb{R}^m$, where $\mathcal{F}$ is assumed to be smooth. We also assume $\mathcal{F}(0,0) = 0$.

We can write

$$\begin{aligned} \mathcal{F}(x,u) &= f(x) + g(x)u + h(x,u)u, \quad \text{where} \\ f(x) = \mathcal{F}(x,0),\;\; g(x) &= \frac{\partial \mathcal{F}}{\partial u}(x,0), \quad \text{and} \\ h(x,u) &= \int_0^1 \left[\frac{\partial \mathcal{F}}{\partial u}(x,\lambda u) - \frac{\partial \mathcal{F}}{\partial u}(x,0)\right] \mathrm{d}\lambda. \end{aligned} \tag{4.59}$$

Since $\mathcal{F}$ is C^2 in u, we can find a continuous function $R : [0,\infty) \times [0,\infty) \to (0,\infty)$ that is non-decreasing in both variables such that

$$|h(x,u)u| \;\leq\; R\big(|x|,|u|\big)|u|^2$$

for all x and u. Hence, we assume in the rest of the subsection that our system has the form

$$\dot{x} = f(x) + g(x)u + r(x,u) \tag{4.60}$$

with $f(0) = 0$ and with r admitting an everywhere positive continuous function R that is non-decreasing in both variables so that

$$\big|r(x,u)\big| \leq R\big(|x|,|u|\big)|u|^2 \tag{4.61}$$

everywhere. Let Assumption 4.1 hold for the functions f and g in the system (4.60) and some functions V and ψ. Fix C^1 functions λ and K_1 satisfying the requirements of Lemma 4.2 as well as

$$\max\left\{\left|\frac{\partial V}{\partial x}(x)\right|, \left|\frac{\partial \psi}{\partial x}(x)\right|, \big|L_gV(x)\big|\right\} \leq K_1(V(x)) \tag{4.62}$$

for all $x \in \mathbb{R}^n$. We prove the following:

Theorem 4.2. *Let Assumption 4.1 hold. Let $\xi : \mathbb{R}^n \to (0,\infty)$ be any smooth function such that*

$$\xi(x) \leq \min\left\{\frac{1}{K_1(V(x))}, \frac{1}{4K_1(V(x))R(|x|,1)}\right\} \ \forall x \in \mathbb{R}^n. \tag{4.63}$$

Let γ be any continuous everywhere positive function satisfying (4.48) and (4.49) and

$$\gamma(V(x)) \geq \frac{\xi(x)R\big(|x|,1\big)K_1\big(V(x)\big)\big[V(x)+K_1(V(x))\big]}{0.5 - \xi(x)K_1(V(x))R(|x|,1)} \ \forall x \in \mathbb{R}^n. \tag{4.64}$$

Set

$$\Gamma(r) \doteq \int_0^r \gamma(s)\mathrm{d}s,$$

and let $S(x)$ be as in Lemma 4.1. Then

$$\mathcal{U}(x) = \lambda(V(x))\psi(x) + \Gamma(V(x)) \tag{4.65}$$

is a CLF for (4.60) whose time derivative along trajectories of (4.60) in closed-loop with

$$u(x) = -\xi(x)L_gV(x)^\top \tag{4.66}$$

satisfies

$$\dot{\mathcal{U}}(x) \leq -\frac{3}{4}\lambda\big(V(x)\big)S(x) \ \forall x \in \mathbb{R}^n. \tag{4.67}$$

In particular, $\mathcal{U}$ satisfies the small control property, and (4.60) can be rendered GAS to 0 with a smooth feedback $u(x)$ of arbitrary small amplitude.

Proof. Since the requirements from (4.48) and (4.49) are satisfied, we deduce from (4.57) and (4.61) that for all smooth everywhere positive functions ξ,

$$\begin{aligned}\dot{\mathcal{U}}(x) \leq& -\frac{3}{4}\lambda(V(x))S(x) - \frac{1}{2}\xi(x)\gamma(V(x))\left|L_gV(x)\right|^2 \\ &+\frac{\partial \mathcal{U}}{\partial x}(x)r\left(x, -\xi(x)L_gV(x)^\top\right) \\ \leq& -\frac{3}{4}\lambda(V(x))S(x) - \frac{1}{2}\xi(x)\gamma(V(x))\left|L_gV(x)\right|^2 \\ &+\left|\frac{\partial \mathcal{U}}{\partial x}(x)\right| R(|x|, |\xi(x)L_gV(x)|)\left|\xi(x)L_gV(x)\right|^2\end{aligned} \tag{4.68}$$

along all trajectories (4.60) when the controller u is from (4.47).

Next, observe that

$$\frac{\partial \mathcal{U}}{\partial x}(x) = \lambda'(V(x))\psi(x)\frac{\partial V}{\partial x}(x) + \lambda(V(x))\frac{\partial \psi}{\partial x}(x) + \gamma(V(x))\frac{\partial V}{\partial x}(x). \tag{4.69}$$

Recalling (4.13) and (4.14) from Lemma 4.2, as well as the bounds (4.62), we deduce that

$$\left|\frac{\partial \mathcal{U}}{\partial x}(x)\right| \leq K_1^2(V(x)) + V(x)K_1(V(x)) + \gamma(V(x))K_1(V(x)) \,. \tag{4.70}$$

Therefore, (4.68) gives

$$\begin{aligned}\dot{\mathcal{U}}(x) \leq& -\tfrac{3}{4}\lambda(V(x))S(x) - \tfrac{1}{2}\xi(x)\gamma(V(x))\left|L_gV(x)\right|^2 \\ &+\Big[K_1(V(x)) + V(x) + \gamma(V(x))\Big]R(|x|, \xi(x)K_1(V(x))) \\ &\times\xi^2(x)K_1(V(x))\left|L_gV(x)\right|^2 .\end{aligned} \tag{4.71}$$

By (4.63),

$$\xi(x) \leq \frac{1}{K_1(V(x))}. \tag{4.72}$$

Hence,

$$\begin{aligned}\dot{\mathcal{U}}(x) \leq& -\tfrac{3}{4}\lambda(V(x))S(x) - \tfrac{1}{2}\xi(x)\gamma(V(x))\left|L_gV(x)\right|^2 \\ &+R(|x|, 1)K_1(V(x))\Big[K_1(V(x)) + V(x) + \gamma(V(x))\Big] \\ &\times\xi^2(x)\left|L_gV(x)\right|^2 .\end{aligned} \tag{4.73}$$

Finally, our requirement (4.64) on γ gives

$$\dot{\mathcal{U}} \leq -\tfrac{3}{4}\lambda(V(x))S(x) \,. \tag{4.74}$$

This concludes the proof. □

4.4 Construction of the Auxiliary Scalar Field

Recall that Assumption 4.1 requires an auxiliary scalar field ψ with the following property: If $x \neq 0$ is such that $L_f V(x) = 0$ and $L_g V(x) = 0$ both hold, then $L_f \psi(x) < 0$. There are several methods for constructing ψ. In the next section, we discuss a method for Hamiltonian systems. Here we present a more general construction that applies to any control affine system

$$\dot{x} = f_0(x) + \sum_{i=1}^{m} f_i(x) u_i \tag{4.75}$$

with smooth functions $f_i : \mathbb{R}^n \to \mathbb{R}^n$ for $i = 0, 1, \ldots, m$ that satisfies the[2]

Weak Jurdjevic Quinn Conditions: There exists a smooth function $V : \mathbb{R}^n \to \mathbb{R}$ satisfying:

1. V is positive definite and radially unbounded;
2. for all $x \in \mathbb{R}^n$, $L_{f_0} V(x) \leq 0$; and
3. there exists an integer $l \geq 2$ such that the set

$$W(V) = \left\{ \begin{array}{l} x \in \mathbb{R}^n : \forall k \in \{1, \ldots, m\} \text{ and } \forall i \in \{0, \ldots, l\}, \\ L_{f_0} V(x) = L_{ad^i_{f_0}(f_k)} V(x) = 0 \end{array} \right\}$$

   equals $\{0\}$.

We construct ψ as follows, where we omit the arguments of our functions when they are clear from the context:

Proposition 4.1. *If (4.75) satisfies the Weak Jurdjevic-Quinn Conditions for some integer l and some storage function V, and if we define G by*

$$G = \sum_{i=0}^{l-1} \sum_{k=1}^{m} \lambda_{i,k} \mathrm{ad}^i_{f_0}(f_k), \tag{4.76}$$

where

$$\lambda_{i,k} = \sum_{j=i}^{l-1} (-1)^{j-i+1} L_{\mathrm{ad}^{(2j-i+1)}_{f_0}(f_k)} V \quad \forall i, k, \tag{4.77}$$

then the scalar field $\psi(x) = L_G V(x)$ satisfies the following property: If $x \in \mathbb{R}^n \setminus \{0\}$, and if $L_{f_i} V(x) = 0$ for $i = 0, 1, \ldots, m$, then $L_{f_0} \psi(x) < 0$.

---

[2] We are using slightly different notation for our control affine systems, to simplify the statement of the next proposition. Recall that for smooth vector fields $f, g : \mathbb{R}^n \to \mathbb{R}^n$, we use the notation

$$\mathrm{ad}^0_f(g) = g, \ \mathrm{ad}_f(g) = [f, g] = g_* f - f_* g, \ \text{and } \mathrm{ad}^k_f(g) = \mathrm{ad}_f(\mathrm{ad}^{k-1}_f(g)),$$

where the $*$ subscripts indicate gradients.

Proof. The proof closely follows that of [41, Theorem 4.3]. The fact that

$$[f_0, G] = \sum_{i=0}^{l-1} \sum_{k=1}^{m} \left(\mathrm{ad}^i_{f_0}(f_k) L_{f_0} \lambda_{i,k} + \lambda_{i,k} \mathrm{ad}^{i+1}_{f_0}(f_k) \right) \tag{4.78}$$

gives

$$\begin{aligned} L_{[f_0,G]} V &= \sum_{k=1}^{m} L_{f_0} \lambda_{0,k} L_{f_k} V + \sum_{k=1}^{m} \lambda_{l-1,k} L_{\mathrm{ad}^l_{f_0}(f_k)} V \\ &\quad + \sum_{i=0}^{l-2} \sum_{k=1}^{m} \left(L_{f_0} \lambda_{i+1,k} + \lambda_{i,k} \right) L_{\mathrm{ad}^{i+1}_{f_0}(f_k)} V. \end{aligned}$$

Recalling our choices (4.77) of the $\lambda_{i,k}$'s gives

$$\begin{aligned} L_{f_0} \lambda_{i+1,k} + \lambda_{i,k} &= \sum_{j=i+1}^{l-1} (-1)^{j-i} L_{f_0} L_{\mathrm{ad}^{2j-i}_{f_0}(f_k)} V \\ &\quad + \sum_{j=i}^{l-1} (-1)^{j-i+1} L_{\mathrm{ad}^{2j-i+1}_{f_0}(f_k)} V \\ &= \sum_{j=i+1}^{l-1} (-1)^{j-i} \left[L_{f_0} L_{\mathrm{ad}^{2j-i}_{f_0}(f_k)} V - L_{\mathrm{ad}^{2j-i+1}_{f_0}(f_k)} V \right] \\ &\quad - L_{\mathrm{ad}^{i+1}_{f_0}(f_k)} V, \quad i \le l-2\,. \end{aligned}$$

For any smooth vector field X and any point x where $L_{f_0} V(x) = 0$,

$$L_{[f_0,X]} V(x) \;=\; L_{f_0} L_X V(x), \tag{4.79}$$

since $\nabla L_{f_0} V(x) = 0$ at points where the non-positive function $L_{f_0} V$ is maximized. (We are using the fact that $L_{[f,g]} = L_f L_g - L_g L_f$ for smooth vector fields f and g.) Taking $X = G$ and then

$$X = \mathrm{ad}^{2j-i}_{f_0}(f_k)$$

in (4.79), we conclude that at all points x where $L_{f_0} V(x) = 0$, we have

$$\begin{aligned} L_{f_0} \psi(x) &= L_{f_0} L_G V(x) \;=\; L_{[f_0,G]} V(x) \ \text{ and} \\ L_{f_0} \lambda_{i+1,k} + \lambda_{i,k} &= -L_{\mathrm{ad}^{i+1}_{f_0}(f_k)} V \end{aligned}$$

if $i \le l-2$. By our choice of $\lambda_{l-1,k}$ from (4.77), we conclude that

$$[L_{f_i} V(x) = 0 \ \forall i = 0, 1, \ldots, m] \ \Rightarrow \ L_{f_0} \psi(x) = -\sum_{i=1}^{l} \sum_{k=1}^{m} \left[L_{\mathrm{ad}^i_{f_0}(f_k)} V(x) \right]^2 .$$

The result is now immediate from the Weak Jurdjevic-Quinn Conditions. □

4.5 Hamiltonian Systems

Theorem 4.1 covers an important class of dynamics that are governed by the *Euler-Lagrange equations*

$$\frac{d}{dt}\left(\frac{\partial L}{\partial \dot{q}}(q,\dot{q})\right) - \frac{\partial L}{\partial q}(q,\dot{q}) = \tau \tag{4.80}$$

for the motion of mechanical systems. Here $q \in \mathbb{R}^n$ represents the generalized configuration coordinates, $L = K - P$ is the difference between the kinetic energy K and the potential energy $P(q) \geq 0$, and τ is the control [183]. In many applications,

$$K(q,\dot{q}) = \frac{1}{2}\dot{q}^\top M(q)\dot{q}$$

where the inertia matrix $M(q)$ is C^1 and symmetric and positive definite for all $q \in \mathbb{R}^n$. The generalized momenta $\partial L/\partial \dot{q}$ are then given by

$$p = M(q)\dot{q}.$$

Hence, using the state $x = (q,p) \in \mathbb{R}^n \times \mathbb{R}^n$ leads to the system

$$\dot{q} = \frac{\partial H}{\partial p}(q,p)^\top = M^{-1}(q)p, \quad \dot{p} = -\frac{\partial H}{\partial q}(q,p)^\top + \tau, \tag{4.81}$$

where

$$H(q,p) = \frac{1}{2}p^\top M^{-1}(q)p + P(q) \tag{4.82}$$

is the *total energy* of the system. We refer to (4.81) as the *Hamiltonian system.* We assume that P is C^1 and positive definite.

The Hamiltonian system can be written as the control affine dynamics

$$\dot{x} = f(x) + g(x)u$$

with state space $\mathcal{X} = \mathbb{R}^{2n}$, control set $U = \mathbb{R}^n$,

$$f(x) = \begin{pmatrix} f_1(x) \\ f_2(x) \end{pmatrix}, \quad \text{where } f_1(x) = \frac{\partial H}{\partial p}(q,p)^\top \text{ and } f_2(x) = -\frac{\partial H}{\partial q}(q,p)^\top,$$

and

$$g(x) = \begin{pmatrix} O \\ I_n \end{pmatrix} \in \mathbb{R}^{2n\times n}, \tag{4.83}$$

where $O \in \mathbb{R}^{2n\times n}$ denotes the matrix whose entries are all 0. One readily checks that the time derivative of H along the trajectories of (4.81) satisfies

$$\dot{H}(q,p) = \frac{\partial H}{\partial p}(q,p)\tau \tag{4.84}$$

Therefore, if $P(q)$ is positive definite and radially unbounded, then H is a non-strict Lyapunov function. The radial unboundedness of H would follow from the continuity of the (positive) eigenvalues of the positive definite matrix $M^{-1}(q)$ as functions of q [161, Appendix A4], which implies that each compact set $\mathcal{S}$ of q values admits a constant $c_{\mathcal{S}} > 0$ such that

$$p^\top M^{-1}(q)p \geq c_{\mathcal{S}}|p|^2$$

for all $q \in \mathcal{S}$ and all $p \in \mathbb{R}^n$. However, P is not necessarily positive definite and radially unbounded. Fortunately, one can determine a real-valued function $\Lambda \in C^1$ satisfying $\Lambda(0) = 0$ such that the function

$$V(x) = H(q,p) + \Lambda(q) = \frac{1}{2}p^\top M^{-1}(q)p + P_n(q) \tag{4.85}$$

with

$$P_n(q) = P(q) + \Lambda(q) \tag{4.86}$$

is positive definite, radially unbounded and C^1. In fact, we can assume that Λ is such that

$$\left|\frac{\partial P_n}{\partial q}(q)\right| \geq |q|. \tag{4.87}$$

For simplicity, we take $\Lambda(q) = \frac{1}{2}|q|^2 - P(q)$.

Using the change of feedback

$$\tau = \tau_n - \frac{\partial \Lambda}{\partial q}(q)^\top, \tag{4.88}$$

one can then check readily that the time derivative of V along the trajectories of (4.81) satisfies

$$\begin{aligned}\dot{V}(q,p) &= \tfrac{\partial H}{\partial p}(q,p)\tau + \tfrac{\partial \Lambda}{\partial q}(q)\tfrac{\partial H}{\partial p}(q,p)^\top \\ &= p^\top M^{-1}(q)\tau + \tfrac{\partial \Lambda}{\partial q}(q)M^{-1}(q)p \\ &= p^\top M^{-1}(q)\tau_n\,.\end{aligned} \tag{4.89}$$

After the change of feedback (4.88), the system (4.81) can be rewritten as

$$\begin{cases}\dot{q} = \frac{\partial V}{\partial p}(q,p)^\top, \\ \dot{p} = -\frac{\partial V}{\partial q}(q,p)^\top + \tau_n.\end{cases} \tag{4.90}$$

Let

$$\begin{gathered} f_n(x) = \begin{pmatrix} f_{1n}(x) \\ f_{2n}(x) \end{pmatrix}, \text{ where} \\ f_{1n}(x) = \tfrac{\partial V}{\partial p}(q,p)^\top \text{ and } f_{2n}(x) = -\tfrac{\partial V}{\partial q}(q,p)^\top\,. \end{gathered} \tag{4.91}$$

We now show that Assumption 4.1 is satisfied by (4.90) with the choice

$$\psi(x) = q^\top p. \tag{4.92}$$

We have

$$L_{f_n} V(x) = 0\,, \tag{4.93}$$

$$L_g V(x) = p^\top M^{-1}(q)\,,\ \text{and} \tag{4.94}$$

$$L_{f_n}\psi(x) \;=\; q^\top f_{2n}(x) + p^\top f_{1n}(x). \tag{4.95}$$

Therefore, if $L_{f_n}V(x) = 0$ and $L_g V(x) = 0$, then $p = 0$ and therefore $L_{f_n}\psi(x) = q^\top f_{2n}(x) = -|q|^2$, so Assumption 4.1 is satisfied.

Since Assumption 4.1 is satisfied, we can construct a CLF that satisfies the small control property for the system (4.90) and therefore also for the system (4.81). In the particular case we consider, it turns out that we can determine a function $\Gamma \in C^1 \cap \mathcal{K}_\infty$ such that

$$\mathcal{U}(x) = \psi(x) + \Gamma(V(x)) \tag{4.96}$$

is a CLF for the system (4.90) that satisfies the small control property. To stipulate Γ, we first let $m_{i,j}(q)$ denote the (i,j) entry of $M^{-1}(q)$ for all $q \in \mathbb{R}^n$. The construction is as follows:

Proposition 4.2. *Fix any non-decreasing everywhere positive C^1 function Υ such that*

$$1 + ||M(q)||^4 \;\le\; \Upsilon(V(x)) \tag{4.97}$$

and

$$\frac{n^2}{2}|q| \sup\left\{\left|\frac{\partial m_{i,j}}{\partial q_k}(q)\right| : (i,j,k) \in \{1,...,n\}^3\right\} \le \sqrt{\Upsilon(V(x))} \tag{4.98}$$

hold for all $x = (q,p) \in \mathbb{R}^{2n}$. Choose a function $\alpha \in \mathcal{K}_\infty \cap C^1$ with $\alpha'(0) > 0$ such that

$$V(x) \;\ge\; \alpha(|p|^2 + |q|^2)$$

everywhere.[3] *Then with the choice*

$$\Gamma(\ell) = \frac{3}{2}\ell + 2\int_0^\ell \Upsilon(r)\mathrm{d}r + \frac{1}{2}\Upsilon(\ell)\alpha^{-1}(\ell)\,, \tag{4.99}$$

the function (4.96) is a CLF for the system (4.90) that satisfies the small control property.

---

[3] Such a function α exists because the positive definiteness of M^{-1} provides a constant $c_0 > 0$ such that $V(x) \ge c_0|x|^2$ on $\mathcal{B}_{2n}$. To construct α, first find a function $\underline{\alpha} \in \mathcal{K}_\infty \cap C^1$ such that $V(x) \ge \underline{\alpha}(|x|)$ for all $x \in \mathbb{R}^{2n}$. By reducing c_0 as needed without relabeling, we can assume that $c_0 r \le \underline{\alpha}(\sqrt{r})$ on $[0.5, 1]$. Choose a non-decreasing C^1 function $p : \mathbb{R} \to [0,1]$ such that $p(r) \equiv 0$ on $[0, 0.5]$ and $p(r) \equiv 1$ on $[1, \infty)$. We can then take $\alpha(r) = [1 - p(r)]c_0 r + p(r)\underline{\alpha}(\sqrt{r})$. In fact, $\alpha'(0) = c_0$.

Proof. Choose

$$\tau_n = -M^{-1}(q)p. \tag{4.100}$$

Then, along the trajectories of (4.90), we get

$$\begin{aligned} \dot{V}(x) &= -|p^\top M^{-1}(q)|^2 \text{ and} \\ \dot{\psi}(x) &= L_{f_n}\psi(x) - L_g\psi(x)M^{-1}(q)p. \end{aligned} \tag{4.101}$$

Therefore,

$$\begin{aligned} \dot{\psi}(x) &= q^\top f_{2n}(x) + p^\top f_{1n}(x) - L_g\psi(x)M^{-1}(q)p \\ &= -q^\top \frac{\partial V}{\partial q}(q,p)^\top + p^\top \frac{\partial V}{\partial p}(q,p)^\top - q^\top M^{-1}(q)p \\ &= -|q|^2 - \frac{1}{2}q^\top \left(p^\top \frac{\partial(M^{-1}(q)p)}{\partial q}\right)^\top + p^\top M^{-1}(q)p \\ &\quad -q^\top M^{-1}(q)p. \end{aligned} \tag{4.102}$$

On the other hand,

$$-q^\top M^{-1}(q)p \le \frac{1}{2}|q|^2 + \frac{1}{2}|M^{-1}(q)p|^2, \tag{4.103}$$

and (4.98) gives

$$\left|\frac{1}{2}q^\top \left(p^\top \frac{\partial(M^{-1}(q)p)}{\partial q}\right)^\top\right| \le \sqrt{\Upsilon(V(x))}|p|^2. \tag{4.104}$$

Therefore,

$$\begin{aligned} \dot{\psi}(x) &\le -\tfrac{1}{2}|q|^2 + \sqrt{\Upsilon(V(x))}|p|^2 \\ &\quad +p^\top M^{-1}(q)p + \tfrac{1}{2}|M^{-1}(q)p|^2 \\ &\le -\tfrac{1}{2}|q|^2 + \sqrt{\Upsilon(V(x))}|p|^2 + ||M(q)||\,|M^{-1}(q)p|^2 \\ &\quad +\tfrac{1}{2}|M^{-1}(q)p|^2 . \end{aligned} \tag{4.105}$$

Hence, (4.97) gives

$$\begin{aligned} \dot{\psi}(x) &\le -\tfrac{1}{2}|q|^2 + \sqrt{\Upsilon(V(x))}\sqrt{\Upsilon(V(x))}|M^{-1}(q)p|^2 \\ &\quad + \left[\Upsilon(V(x)) + \tfrac{1}{2}\right]|M^{-1}(q)p|^2 \\ &\le -\tfrac{1}{2}|q|^2 + \left(2\Upsilon(V(x)) + \tfrac{1}{2}\right)|M^{-1}(q)p|^2 . \end{aligned} \tag{4.106}$$

We deduce easily that the derivative of $\mathcal{U}$ defined in (4.96) along the trajectories of (4.90), in closed-loop with τ_n defined in (4.100), satisfies

$$\dot{\mathcal{U}}(x) \leq -\tfrac{1}{2}|q|^2 - |M^{-1}(q)p|^2, \tag{4.107}$$

using the fact that

$$\Gamma'(\ell) \geq \frac{3}{2} + 2\Upsilon(l). \tag{4.108}$$

Next, observe that

$$\begin{aligned} \mathcal{U}(x) &\geq -|\psi(x)| + \Gamma(V(x)) \geq -\Upsilon(V(x))|q||p| + \Gamma(V(x)) \\ &\geq -\tfrac{1}{2}\Upsilon(V(x))\alpha^{-1}(V(x)) + \Gamma(V(x)) , \end{aligned} \tag{4.109}$$

by our choice of α and the relation $|q||p| \leq \frac{1}{2}|p|^2 + \frac{1}{2}|q|^2$. Using the fact that

$$\Gamma(v) \geq v + \frac{1}{2}\Upsilon(v)\alpha^{-1}(v), \tag{4.110}$$

we get

$$\mathcal{U}(x) \geq V(x) , \tag{4.111}$$

so $\mathcal{U}$ is positive definite and radially unbounded. Moreover,

$$\dot{\mathcal{U}}(x) \leq -W(x) < 0 \;\; \forall x \neq 0 , \tag{4.112}$$

where

$$W(x) = \frac{1}{2}|q|^2 + \left|M^{-1}(q)p\right|^2 . \tag{4.113}$$

We conclude that we have determined a CLF for the system (4.81) that satisfies the small control property. Moreover, both $\mathcal{U}(x)$ and $W(x)$ are lower bounded in a neighborhood of the origin by a positive definite quadratic function. □

Remark 4.2. Systems of the form (4.90) can be globally asymptotically stabilized by using backstepping to design the controls. Therefore, backstepping provides an alternative construction of CLFs satisfying the small control property. However, this technique provides control laws that remove the term $-\frac{\partial H}{\partial q}(q,p)^\top$, which may lead to more complicated control laws with large nonlinearities when the system can be stabilized through arbitrarily small control laws.

4.6 Robustness

We saw in Theorem 4.1 how to construct a CLF $\mathcal{U}$ for

$$\dot{x} = f(x) + g(x)u \tag{4.114}$$

that has the small control property, provided Assumption 4.1 is satisfied. In fact, for each $\varepsilon > 0$, we can choose a C^1 function K_1 satisfying $|K_1(x)| \le \varepsilon$ for all $x \in \mathbb{R}^n$ such that $\mathcal{U}$ is a strict Lyapunov function for

$$\dot{x} = f(x) + g(x)K_1(x),$$

which is therefore GAS to $x = 0$.

As we saw in previous chapters, ISS is a significant generalization of the GAS [157]. Recall that for a nonlinear system $\dot{x} = F(x, d)$ with state space $\mathcal{X} = \mathbb{R}^n$ and control set $U = \mathbb{R}^m$, the ISS property says that there exist $\beta \in \mathcal{KL}$ and $\gamma \in \mathcal{K}_\infty$ such that for all measurable essentially bounded functions $\mathbf{d} : [0, \infty) \to \mathbb{R}^m$, the corresponding trajectories $x(t)$ for

$$\dot{x}(t) \;=\; F(x(t), \mathbf{d}(t)) \tag{4.115}$$

satisfy

$$|x(t)| \le \beta\big(|x(0)|, t\big) + \gamma(|\mathbf{d}|_\infty) \;\; \forall t \ge 0. \tag{ISS}$$

Here $\mathbf{d}$ represents a disturbance, and $|\cdot|_\infty$ is the essential supremum. The ISS property includes GAS to 0 for the system $\dot{x} = f(x)$, because in that case the term $\gamma(|\mathbf{d}|_\infty)$ in the ISS decay condition is not present. Therefore, given any constant $\varepsilon > 0$, it may seem reasonable to search for a feedback $K(x)$ for (4.114) (which could in principle differ from K_1) for which

$$\dot{x} = F(x, d) \doteq f(x) + g(x)[K(x) + d] \tag{4.116}$$

is ISS with respect to the disturbance d, and for which $|K(x)| \le \varepsilon$ for all $x \in \mathbb{R}^n$. Hence, we would want an arbitrarily small feedback K that renders (4.114) GAS to $x = 0$ and that has the additional property that (4.116) is ISS with respect to the disturbance d.

This objective cannot be met in general, since there is no *bounded* feedback $K(x)$ such that the one-dimensional system $\dot{x} = K(x) + d$ is ISS. Therefore, instead of using ISS to analyze Jurdjevic-Quinn systems, we use iISS [160]. Recall from Chap. 1 that for a general nonlinear system $\dot{x} = F(x, d)$ evolving on $\mathbb{R}^n \times \mathbb{R}^m$, the iISS condition says: There exist $\beta \in \mathcal{KL}$ and $\alpha, \gamma \in \mathcal{K}_\infty$ such that for all measurable essentially bounded functions $\mathbf{d} : [0, \infty) \to \mathbb{R}^m$ and corresponding trajectories $x(t)$ for (4.115), we have

$$\alpha(|x(t)|) \;\le\; \beta\big(|x(0)|, t\big) + \int_0^t \gamma(|\mathbf{d}(s)|)\mathrm{d}s \quad \forall t \ge 0. \tag{iISS}$$

See Chap. 1 or [8, 9] for the background and further motivation for iISS. To get our iISS result, we add the following assumption to our system (4.114), which we assume in addition to Assumption 4.1:

Assumption 4.2 *An everywhere positive non-decreasing smooth function* $\mathcal{D}$ *such that*

1. $\int_0^{+\infty} \frac{1}{\mathcal{D}(s)}\,\mathrm{d}s = +\infty$; *and*
2. $|L_gV(x)| \leq \mathcal{D}(V(x))$ *for all* $x \in \mathbb{R}^n$

is known.

Assumption 4.2 holds for the two-link manipulator example we introduced in Sect. 4.1, because in that case,

$$|L_gV(x)| \;\leq\; 2(V(x)+1)$$

for all $x \in \mathbb{R}^n$, so we can take

$$\mathcal{D}(s) \;=\; 2(s+1).$$

In fact, our assumptions hold for a broad class of Hamiltonian systems as well; see Remark 4.3. We claim that if Assumptions 4.1 and 4.2 both hold, then for any constant $\varepsilon > 0$ and any C^∞ function $\xi : \mathbb{R}^n \to (0,\infty)$ such that

$$|\xi(x)L_gV(x)| \;\leq\; \varepsilon \quad \forall x \in \mathbb{R}^n\,, \tag{4.117}$$

the system

$$\dot{x} = f(x) + g(x)\big[K(x) + \mathbf{d}(t)\big]$$

is iISS with the choice $K(x) = -\xi(x)L_gV(x)^\top$.

To prove this claim, we begin by applying Theorem 4.1 to $\dot{x} = f(x)+g(x)u$, with $\xi : \mathbb{R}^n \to (0,\infty)$ satisfying (4.117) for an arbitrary prescribed constant $\varepsilon > 0$. This provides a CLF $\mathcal{U}$ satisfying the small control property for (4.114), which is also a strict Lyapunov function for (4.114) in closed-loop with

$$K(x) \;=\; -\xi(x)L_gV(x)^\top.$$

Setting

$$\mathcal{F}(x) = f(x) - g(x)\xi(x)L_gV(x)^\top, \tag{4.118}$$

it follows that $W(x) \doteq -L_{\mathcal{F}}\mathcal{U}(x)$ is positive definite.

We can determine a non-decreasing everywhere positive function $A \in C^1$ such that

$$|L_g\mathcal{U}(x)| \leq A\big(V(x)\big) \;\; \forall x \in \mathbb{R}^n\,. \tag{4.119}$$

Since $\mathcal{D}$ in Assumption 4.2 is a positive non-decreasing smooth function, we can easily construct a function $\Gamma_u \in \mathcal{K}_\infty \cap C^1$ such that Γ_u' is everywhere positive and increasing and

$$A\big(V(x)\big) \leq \Gamma_u'\big(V(x)\big)\mathcal{D}\big(V(x)\big) \;\; \forall x \in \mathbb{R}^n. \tag{4.120}$$

Therefore, for all $x \in \mathbb{R}^n$,

$$|L_g\mathcal{U}(x)| \le \Gamma_u'\big(V(x)\big)\mathcal{D}\big(V(x)\big) \ . \tag{4.121}$$

Next, consider

$$\mathcal{U}_a(x) = \mathcal{U}(x) + \Gamma_u\big(V(x)\big) \ . \tag{4.122}$$

Then

$$\big|L_g\mathcal{U}_a(x)\big| \ \le \ \big|L_g\mathcal{U}(x)\big| + \Gamma_u'\big(V(x)\big)\big|L_gV(x)\big|.$$

Using Assumption 4.2 and (4.121), we obtain

$$\big|L_g\mathcal{U}_a(x)\big| \ \le \ 2\Gamma_u'\big(V(x)\big)\mathcal{D}\big(V(x)\big) \ . \tag{4.123}$$

Let

$$\mathcal{U}_*(x) = \Gamma_u^{-1}\big(\mathcal{U}_a(x)\big) \ . \tag{4.124}$$

Since $\Gamma_u^{-1} \in C^1$ and Γ_u^{-1} is increasing, we have

$$\begin{aligned} L_g\mathcal{U}_*(x) &= \{\Gamma_u^{-1}\}'\big(\mathcal{U}_a(x)\big)L_g\mathcal{U}_a(x) \\ &= \frac{1}{\Gamma_u'(\Gamma_u^{-1}(\mathcal{U}_a(x)))}L_g\mathcal{U}_a(x) \ . \end{aligned} \tag{4.125}$$

In combination with (4.123), we obtain

$$|L_g\mathcal{U}_*(x)| \ \le \ 2\frac{\Gamma_u'(V(x))\mathcal{D}(V(x))}{\Gamma_u'(\Gamma_u^{-1}(\mathcal{U}_a(x)))} \ . \tag{4.126}$$

By the definition (4.122) of $\mathcal{U}_a$, we get

$$\Gamma_u^{-1}\big(\mathcal{U}_a(x)\big) \ge V(x) \ . \tag{4.127}$$

Since Γ_u' is non-decreasing, we obtain

$$\Gamma_u'\big(\Gamma_u^{-1}(\mathcal{U}_a(x))\big) \ \ge \ \Gamma_u'(V(x)) \ , \tag{4.128}$$

so (4.126) gives

$$|L_g\mathcal{U}_*(x)| \ \le \ 2\mathcal{D}\big(V(x)\big) \ . \tag{4.129}$$

Since $\mathcal{D}$ is non-decreasing, (4.127) gives

$$\mathcal{D}\big(V(x)\big) \ \le \ \mathcal{D}\Big(\Gamma_u^{-1}\big(\mathcal{U}_a(x)\big)\Big) = \mathcal{D}\big(\mathcal{U}_*(x)\big)$$

and therefore

$$|L_g\mathcal{U}_*(x)| \ \le \ 2\mathcal{D}\big(\mathcal{U}_*(x)\big) \ . \tag{4.130}$$

Then

$$\tilde{U}(x) \ = \ \frac{1}{2}\int_0^{\mathcal{U}_*(x)} \frac{\mathrm{d}p}{\mathcal{D}(p)} \tag{4.131}$$

satisfies

$$\left|L_g\tilde{U}(x)\right| \leq 1\,. \tag{4.132}$$

The function $\tilde{U}$ is again a CLF for our dynamics (4.114) that satisfies the small control property. Moreover, (4.132) ensures that we can determine a positive definite function $\tilde{W}(x)$ such that the time derivative of $\tilde{U}$ along the trajectories of

$$\dot{x} = f(x) + g(x)[-\xi(x)L_gV(x)^\top + d] \tag{4.133}$$

satisfies

$$\dot{\tilde{U}}(x) \leq -\tilde{W}(x) + |d| \tag{4.134}$$

for all x and d. Inequality (4.134) says (see [8]) that the positive definite radially unbounded C^1 function $\tilde{U}$ is an iISS Lyapunov function for (4.133). The fact that (4.133) is iISS now follows from the standard iISS Lyapunov characterization; see Lemma 2.3 or [8, Theorem 1]. We conclude as follows:

Corollary 4.1. *Assume that the system (4.114) satisfies Assumptions 4.1 and 4.2 for some auxiliary scalar field $\psi : \mathbb{R}^n \to \mathbb{R}$ and some storage function $V : \mathbb{R}^n \to \mathbb{R}$, and let $\varepsilon > 0$ be given. Then there exists an everywhere positive function ξ such that (a) the system*

$$\dot{x} \;=\; f(x) + g(x)[K(x) + \mathbf{d}] \tag{4.135}$$

with the feedback

$$K(x) \;\doteq\; -\xi(V(x))L_gV(x)^\top \tag{4.136}$$

is iISS and (b) $|K(x)| \leq \varepsilon$ for all $x \in \mathbb{R}^n$. Moreover, if $\mathcal{U}$ is a CLF satisfying the requirements of Theorem 4.1, and if $\Gamma_u \in \mathcal{K}_\infty \cap C^1$ is such that Γ_u' is increasing and everywhere positive and satisfies $|L_g\mathcal{U}(x)| \leq \Gamma_u'(V(x))\mathcal{D}(V(x))$ everywhere, then

$$\tilde{U}(x) = \frac{1}{2}\int_0^{\Gamma_u^{-1}\left(\mathcal{U}(x)+\Gamma_u(V(x))\right)} \frac{dp}{\mathcal{D}(p)} \tag{4.137}$$

is an iISS Lyapunov function for (4.135).

Remark 4.3. Assumptions 4.1 and 4.2 are satisfied by a broad class of important systems. For example, assume that the Hamiltonian system (4.90) satisfies the conditions from Sect. 4.5 and the following additional condition:

R. There exist $\underline{\lambda}, \bar{\lambda} > 0$ such that

$$\text{spectrum}\{M^{-1}(q)\} \subseteq [\underline{\lambda}, \bar{\lambda}]$$

for all q.

Condition R. means that there are positive constants $\underline{c}$ and $\bar{c}$ such that

$$\underline{c}|p|^2 \;\leq\; p^\top M(q)p \;\leq\; \bar{c}|p|^2$$

for all q and p. This is more restrictive than merely saying that M^{-1} is everywhere positive definite, since the smallest eigenvalue $\lambda_{\min}(q)$ of $M^{-1}(q)$ could in principle be such that

$$\liminf_{|q|\to+\infty} \lambda_{\min}(q) = 0.$$

Then (4.81) satisfies our Assumptions 4.1-4.2 and so is covered by the preceding corollary. In fact, we saw in Sect. 4.5 that Assumption 4.1 holds with $x = (q, p)$ and

$$V(x) = H(q,p) + \frac{1}{2}|q|^2 - P(q),$$

and then Assumption 4.2 follows from Condition R. because

$$\begin{aligned} |L_g V(x)|^2 \;=\; \left|\tfrac{\partial H}{\partial p}(x)\right|^2 &= |p^\top M^{-1}(q)|^2 \\ &\leq \bar{\lambda}^2 |p|^2 \\ &\leq \tfrac{\bar{\lambda}^2}{\underline{\lambda}} p^\top M^{-1}(q) p \;\leq\; 2\tfrac{\bar{\lambda}^2}{\underline{\lambda}} V(x) \end{aligned}$$

for all $x = (q, p)$. Therefore, we can take

$$\mathcal{D}(s) \doteq \sqrt{2\frac{\bar{\lambda}^2}{\underline{\lambda}}(s+1)}$$

to satisfy Assumption 4.2.

4.7 Illustrations

We showed how to construct CLFs for systems

$$\dot{x} = f(x) + g(x)u \tag{4.138}$$

that have the form

$$\mathcal{U}(x) = \lambda\big(V(x)\big)\psi(x) + \Gamma\big(V(x)\big) \tag{4.139}$$

for suitable C^1 functions λ and Γ, under the Jurdjevic-Quinn Conditions. In many cases, the construction is simplified because we can take either $\Gamma(v) \equiv v$ or $\lambda \equiv 1$. For example, the Hamiltonian systems in Sect. 4.5 lead to $\lambda \equiv 1$. We now further illustrate this point in two examples. In the first example, Γ can be taken to be the identity, so we get a simple weighted sum of V and $\psi(x) = L_G V(x)$. Then we revisit the two-link manipulator, which requires a more complicated Γ but has $\lambda \equiv 1$.

4.7.1 Two-Dimensional Example

We illustrate Theorem 4.1 using the two-dimensional system

$$\begin{cases} \dot{x}_1 = x_2 \\ \dot{x}_2 = -x_1^3 + u \ . \end{cases} \tag{4.140}$$

In this case, we have

$$f(x_1, x_2) = \begin{pmatrix} x_2 \\ -x_1^3 \end{pmatrix} \text{ and } g(x_1, x_2) = \begin{pmatrix} 0 \\ 1 \end{pmatrix} . \tag{4.141}$$

Let us check that (4.140) satisfies Assumption 4.1.

1. The positive definite radially unbounded function

$$V(x_1, x_2) = \frac{1}{4}x_1^4 + \frac{1}{2}x_2^2 \tag{4.142}$$

is not a CLF for (4.140), but it satisfies $L_f V(x) = 0$ on $\mathbb{R}^2$.
2. Choosing the vector field

$$G(x_1, x_2) = \begin{pmatrix} 0 \\ x_1 \end{pmatrix} \tag{4.143}$$

gives

$$\begin{aligned} &L_g V(x_1, x_2) = x_2 \ , \ L_G V(x_1, x_2) = x_1 x_2, \text{ and} \\ &L_f L_G V(x_1, x_2) = x_2^2 - x_1^4 \ . \end{aligned} \tag{4.144}$$

If $L_g V(x_1, x_2) = 0$ and $(x_1, x_2) \neq (0, 0)$, then $x_2 = 0$ and $x_1 \neq 0$, so $L_f L_G V(x_1, 0) = -x_1^4 < 0$.

Therefore Assumption 4.1 is satisfied with (4.142) and $\psi(x) = L_G V(x)$, so Theorem 4.1 applies to the system (4.140). Let us show that with the choice

$$\delta(v) = \frac{v^2}{8(1+v)^2} , \tag{4.145}$$

the function

$$\begin{aligned} U(x) &= V(x) + \delta\big(V(x)\big) L_G V(x) \\ &= \frac{1}{4}x_1^4 + \frac{1}{2}x_2^2 + \delta\left(\frac{1}{4}x_1^4 + \frac{1}{2}x_2^2\right) x_1 x_2 \end{aligned} \tag{4.146}$$

is a CLF for the system (4.140) whose time derivative along the trajectories of (4.140) in closed-loop with

$$u = -L_gV(x)^\top = -x_2 \tag{4.147}$$

is negative definite.

To this end, we first observe that

$$\frac{1}{2}x_1^2 \le 1 + \frac{1}{4}x_1^4, \text{ so } |x_1x_2| \le 1 + V(x) \ \forall x \in \mathbb{R}^2 . \tag{4.148}$$

Therefore

$$\begin{aligned} U(x) &\ge \frac{1}{4}x_1^4 + \frac{1}{2}x_2^2 \\ &\quad - \frac{\left(\frac{1}{4}x_1^4 + \frac{1}{2}x_2^2\right)^2}{8\left(1 + \frac{1}{4}x_1^4 + \frac{1}{2}x_2^2\right)^2}\left(1 + \frac{1}{4}x_1^4 + \frac{1}{2}x_2^2\right) \\ &= \frac{1}{4}x_1^4 + \frac{1}{2}x_2^2 - \frac{\left(\frac{1}{4}x_1^4 + \frac{1}{2}x_2^2\right)^2}{8\left(1 + \frac{1}{4}x_1^4 + \frac{1}{2}x_2^2\right)} \\ &\ge \frac{1}{8}x_1^4 + \frac{1}{4}x_2^2 . \end{aligned}$$

The time derivative of $U(x)$ along the trajectories of (4.140) in closed-loop with the feedback (4.147) is

$$\begin{aligned} \dot{U} &= -x_2^2\left[1 + \frac{V(x)}{4(1+V(x))^3}x_1x_2\right] \\ &\quad + \delta(V(x))\left[-x_1^4 - x_1x_2 + x_2^2\right] \\ &\le -\frac{5}{8}x_2^2 - \delta(V(x))x_1^4 - \delta(V(x))x_1x_2 \\ &\le -\frac{3}{8}x_2^2 - \delta(V(x))x_1^4 + \delta^2(V(x))x_1^2 \\ &\le -\frac{3}{8}x_2^2 - \delta(V(x))x_1^4 + \delta(V(x))\frac{\left(\frac{1}{4}x_1^4 + \frac{1}{2}x_2^2\right)x_1^2}{8\left(1 + \frac{1}{4}x_1^4 + \frac{1}{2}x_2^2\right)} \\ &\le -\frac{1}{4}x_2^2 - \frac{1}{4}\delta(V(x))x_1^4, \end{aligned} \tag{4.149}$$

where we used (4.148) to get the first and last inequalities, and the second inequality used $(\delta(V(x))x_1 + \frac{1}{2}x_2)^2 \ge 0$. Since the right hand side of this inequality is negative definite, the result follows.

4.7.2 Two-Link Manipulator Revisited

We show how the CLF (4.6) for the two-link manipulator dynamics follows as a special case of the construction from Theorem 4.1. Recall that the dynamics is the control affine system $\dot{x} = f(x) + g(x)u$ where

$$f(x) = \begin{bmatrix} x_2 \\ \frac{-2x_3x_2x_4 - x_1\langle x_1\rangle}{x_3^2+1} \\ x_4 \\ x_2^2x_3 - x_3\langle x_3\rangle \end{bmatrix}, \quad g(x) = \begin{bmatrix} 0 & 0 \\ \frac{1}{x_3^2+1} & 0 \\ 0 & 0 \\ 0 & 1 \end{bmatrix}, \text{ and } u = \begin{bmatrix} \tau_b \\ F_b \end{bmatrix}. \quad (4.150)$$

We show that Assumption 4.1 is satisfied for the system with the choices

$$V(x) = \frac{1}{2}\left[(x_3^2+1)x_2^2 + x_4^2 + \sqrt{1+x_1^2} + \sqrt{1+x_3^2} - 2\right] \quad (4.151)$$

and $\psi = L_G V$, where

$$G(x) = (0, x_1, 0, x_3)^\top. \quad (4.152)$$

Setting

$$\langle p\rangle = \frac{1}{2\sqrt{1+p^2}}$$

for all $p \in \mathbb{R}$, simple calculations show that

$$\nabla V(x) = \left(x_1\langle x_1\rangle, x_2\left[x_3^2+1\right], x_3\langle x_3\rangle + x_2^2x_3, x_4\right).$$

Hence, along the trajectories of the system, we have

$$\dot{V}(x) = x_2\tau_b + x_4F_b,$$

and

$$L_GV(x) = \frac{\partial V}{\partial x_2}(x)x_1 + \frac{\partial V}{\partial x_4}(x)x_3 = (x_3^2+1)x_2x_1 + x_4x_3. \quad (4.153)$$

Since

$$\nabla(L_GV(x)) = \left(x_2(x_3^2+1), x_1(x_3^2+1), x_4 + 2x_1x_2x_3, x_3\right),$$

we have

$$L_fL_GV(x) = x_2^2(2x_3^2+1) + x_4^2 - x_1^2\langle x_1\rangle - x_3^2\langle x_3\rangle. \quad (4.154)$$

Notice that

$$L_fV(x) = 0 \text{ and } L_gV(x) = [x_2 \;\; x_4]$$

everywhere. Also, if $L_gV(x) = 0$, then $x_2 = x_4 = 0$, in which case we get

$$L_fL_GV(x) = -x_1^2\langle x_1\rangle - x_3^2\langle x_3\rangle.$$

It follows that if $x \neq 0$ and $L_gV(x) = 0$, then $L_fL_GV(x) < 0$. Therefore Assumption 4.1 is satisfied with

$$\psi(x) = L_GV(x), \tag{4.155}$$

so Theorem 4.1 applies.

We now derive the CLF whose existence is guaranteed by the theorem. To this end, first note that

$$a \leq 3\left\{(\sqrt{1+a}-1) + (\sqrt{1+a}-1)^2\right\} \quad \forall a \geq 0. \tag{4.156}$$

It follows from the formula for V that

$$\begin{aligned} \max\{x_1^2, x_3^2\} &\leq 3\{2V(x) + 4V^2(x)\} \text{ and} \\ \max\{x_2^2, x_4^2\} &\leq 2V(x) \quad \forall x \in \mathbb{R}^4. \end{aligned} \tag{4.157}$$

Combining the triangle inequality, (4.153), and (4.157) gives

$$\begin{aligned} |L_GV(x)| &\leq \tfrac{1}{2}x_3^4 + \tfrac{1}{4}x_1^4 + \tfrac{1}{4}x_2^4 + \tfrac{1}{2}|x|^2 \\ &\leq 288V^4(x) + 85V^2(x) + 8V(x). \end{aligned} \tag{4.158}$$

We readily conclude that the function

$$V^\sharp(x) = 40\left[2 + 2V(x)\right]^6 + L_GV(x) - 40(2^6) \tag{4.159}$$

is such that

$$V^\sharp(x) \geq 3(x_1^2 + x_2^2 + x_3^2 + x_4^2)$$

for all $x \in \mathbb{R}^4$, so $V^\sharp$ is positive definite and radially unbounded.

Moreover, its time derivative along trajectories of the system satisfies

$$\begin{aligned} \dot{V}^\sharp(x) = 480\left[2 + 2V(x)\right]^5(x_2\tau_b + x_4F_b) + x_2^2(2x_3^2 + 1) \\ +x_4^2 - x_1^2\langle x_1\rangle - x_3^2\langle x_3\rangle + x_1\tau_b + x_3F_b\ , \end{aligned} \tag{4.160}$$

since $\dot{V}(x) = x_2\tau_b + x_4F_b$. Hence, the triangle inequality gives

$$\begin{aligned} \dot{V}^\sharp(x) \leq \sqrt{1 + x_1^2\tau_b^2} + 480\left[2 + 2V(x)\right]^5 x_2\tau_b + x_2^2(2x_3^2 + 1) \\ +\sqrt{1 + x_3^2F_b^2} + 480\left[2 + 2V(x)\right]^5 x_4F_b \\ +x_4^2 - \tfrac{1}{2}x_1^2\langle x_1\rangle - \tfrac{1}{2}x_3^2\langle x_3\rangle\ . \end{aligned} \tag{4.161}$$

To show that $V^\sharp$ is a CLF for the system, we show that the right side of (4.161) is negative definite when we take

$$\tau_b = -x_2\langle x_2\rangle \text{ and } F_b = -x_4\langle x_4\rangle. \tag{4.162}$$

This will also show that $V^\sharp$ has the small control property.

To this end, we first note that with the choices (4.162), we have

$$\begin{aligned}\dot{V}^\sharp(x) \;&\leq\; T_1(x)x_2^2\langle x_2\rangle + T_2(x)x_4^2\langle x_4\rangle\\ &\quad -\tfrac{1}{2}\left[x_1^2\langle x_1\rangle + x_2^2\langle x_2\rangle + x_3^2\langle x_3\rangle + x_4^2\langle x_4\rangle\right],\end{aligned} \tag{4.163}$$

where we define the T_i's by

$$T_1(x) = \sqrt{1+x_1^2} - 480(2+2V(x))^5 + 2\sqrt{1+x_2^2}(2x_3^2+1) + \frac{1}{2}$$

and

$$T_2(x) = \sqrt{1+x_3^2} - 480(2+2V(x))^5 + 2\sqrt{1+x_4^2} + \frac{1}{2}.$$

We deduce from (4.157) that T_1 and T_2 are non-positive and therefore

$$\dot{V}^\sharp(x) \leq -\frac{1}{2}\left[x_1^2\langle x_1\rangle + x_2^2\langle x_2\rangle + x_3^2\langle x_3\rangle + x_4^2\langle x_4\rangle\right]. \tag{4.164}$$

The right hand side of this inequality is negative definite and the feedbacks resulting from (4.4) and (4.162) give the small control property.

In fact, our analysis from Sect. 4.6 shows that for any positive constant $c > 0$, the scaled feedback

$$K(x) = -c\begin{pmatrix} x_1\langle x_1\rangle + x_2\langle x_2\rangle \\ x_3\langle x_3\rangle + x_4\langle x_4\rangle \end{pmatrix} \tag{4.165}$$

renders the system iISS to actuator errors, meaning

$$\dot{x} = f(x) + g(x)[K(x) + \mathbf{d}(t)]$$

is iISS. We illustrate this point in the simulation below, where we took the feedback

$$K^\flat(x) \;=\; -0.005\begin{pmatrix} x_1\langle x_1\rangle + x_2\langle x_2\rangle \\ x_3\langle x_3\rangle + x_4\langle x_4\rangle \end{pmatrix}, \tag{4.166}$$

the disturbance

$$\mathbf{d}(t) = \begin{pmatrix} \dfrac{1}{1+0.25t^2} \\ e^{-0.25t} \end{pmatrix} \tag{4.167}$$

and the initial state $x(0) = (1,1,1,1)$. While the feedback (4.166) renders the closed-loop system GAS to 0 when the disturbance is set to 0, the state components may or may not be driven to zero when there are disturbances present. In our simulation, the angle of the link x_1 converges to zero by time $t = 1000$. However, the gripper position x_3 has an overshoot caused by the disturbance that keeps this component from converging to zero. See Figs. 4.2 and 4.3.

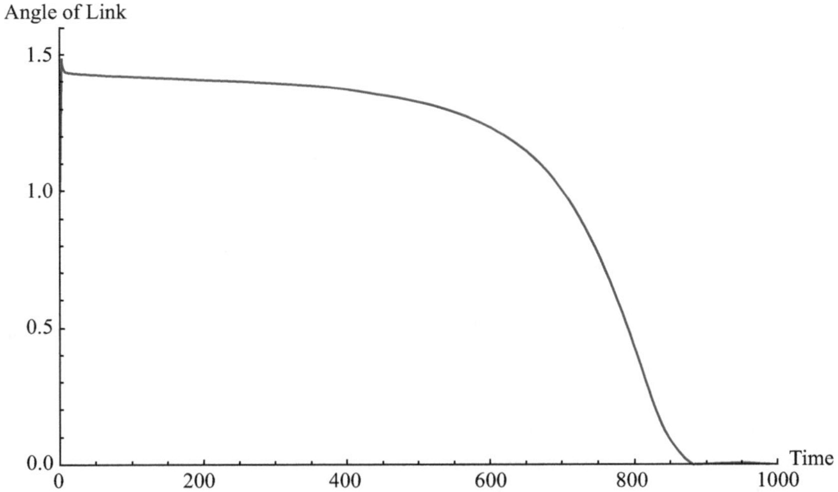

Fig. 4.2 Angle of link x_1 using feedback (4.166) and disturbance (4.167)

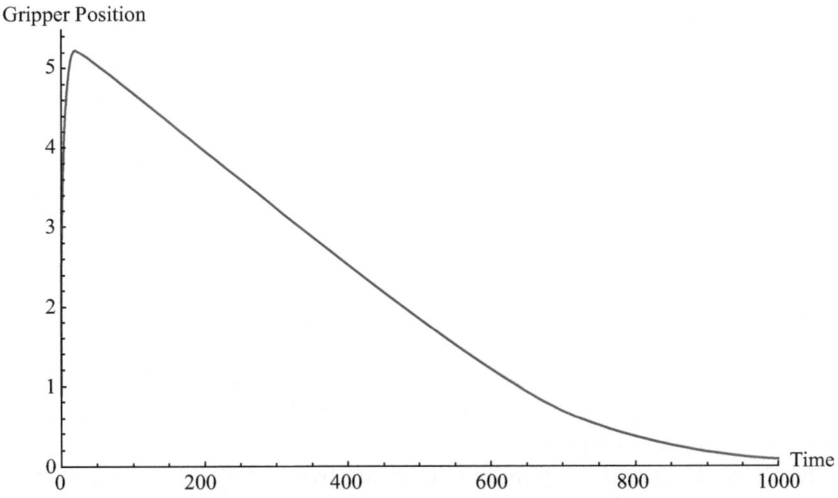

Fig. 4.3 Gripper position x_3 using feedback (4.166) and disturbance (4.167)

Remark 4.4. An important feature of the preceding analysis is that the strict Lyapunov function $V^\sharp$ has a negative definite time derivative along the closed-loop trajectories, using a bounded feedback. In fact, for each constant $\varepsilon >$ 0, our constructions from the preceding sections provide a strict Lyapunov function whose time derivative is negative definite using a feedback stabilizer $K^\flat : \mathbb{R}^n \to \varepsilon\mathcal{B}_2$ that is bounded by ε. This is done by choosing the function Γ in our strict Lyapunov function construction appropriately. Moreover, we see from (4.164) that $-\dot{V}^\sharp$ is *proper* along the closed trajectories, and the dynamics are control affine, so we can immediately use control redesign to get ISS to actuator errors, if we allow unbounded feedbacks. For example, the combined feedback

$$K^\sharp(x) = -(x_1\langle x_1\rangle + x_2\langle x_2\rangle, x_3\langle x_3\rangle + x_4\langle x_4\rangle)^\top - L_g V^\sharp(x)$$

renders the system ISS with respect to actuator errors, so we recover the ISS results for the two-link manipulator from [5].

The properness of $\dot{V}^\sharp$ is essential for the preceding control redesign argument. In general, if the time derivative of a strict Lyapunov function $\mathcal{V}$ is merely negative definite along the closed-loop trajectories of a given control affine system, then adding $-L_g\mathcal{V}$ to the feedback will not necessarily give ISS. On the other hand, we can always transform $\mathcal{V}$ into a new strict Lyapunov function $\mathcal{V}_a$ for which $-\dot{\mathcal{V}}_a$ is proper along the closed-loop trajectories (e.g., by arguing as in [157, p.440]), and then we can generate ISS with respect to actuator errors by subtracting $L_g\mathcal{V}_a$ as above.

4.8 Comments

The Jurdjevic-Quinn Method can be summarized by saying that appropriate controllability conditions and a first integral of the drift vector can be used to design smooth asymptotically stabilizing control laws. Since Jurdjevic and Quinn's original paper [68], the method has been extended in several directions [11, 41, 45, 126]. The first general result on global explicit strict Lyapunov function constructions under the Weak Jurdjevic-Quinn Conditions appears to be [40], whose results are limited to homogenous systems. Our construction of the auxiliary scalar field in Sect. 4.4 is similar to, but somewhat simpler than, the one in [41, Theorem 4.3]. This is because [41] uses a more complicated construction that guarantees that G is homogenous of degree zero, assuming the original dynamics and given non-strict Lyapunov function are both homogeneous.

The model (4.1) and accompanying figure are from [165]. There it is shown that if one takes closed-loop controllers of the form

$$\tau = -k_1\dot{\theta} - k_2(\theta - q_d) \text{ and } F = -k_3\dot{r} - k_4(r - r_d), \tag{4.168}$$

then (4.1) in closed-loop with (4.168) is not ISS with respect to (q_d, r_d). In particular, bounded signals can destabilize the system, which is called a nonlinear resonance effect. Our treatment of the two-link manipulator is based on [102], which provides an alternative CLF construction under the

Jurdjevic-Quinn conditions that differs from the one we presented in this chapter.

Chapter 5
Systems Satisfying the Conditions of LaSalle

Abstract The LaSalle Invariance Principle uses non-strict Lyapunov functions to show asymptotic stability. However, even when a system is known to be asymptotically stable, it is still desirable to be able to construct a strict Lyapunov function for the system, e.g., for robustness analysis and feedback design. In this chapter, we give two more methods for constructing strict Lyapunov functions, which apply to cases where asymptotic stability is already known from the LaSalle Invariance Principle.

The first imposes simple algebraic conditions on the higher order Lie derivatives of the non-strict Lyapunov functions, in the directions of the vector fields that define the systems. Our second method uses our continuous time Matrosov Theorem from Chap. 3. We illustrate our approach by constructing a strict Lyapunov function for an appropriate error dynamics involving the Lotka-Volterra Predator-Prey System.

5.1 Background and Motivation

As we noted in preceding chapters, Lyapunov functions are a vital tool for the analysis of, and controller design for, nonlinear systems. The two main types of Lyapunov functions are *strict* Lyapunov functions (also known as *strong* Lyapunov functions, having negative definite time derivatives along the trajectories of the system) and *non-strict* Lyapunov functions (whose time derivatives along the trajectories are negative *semi*-definite, and which are also called *weak* Lyapunov functions).

Strict Lyapunov functions are typically far more useful than non-strict ones. The key point is that in general, non-strict Lyapunov functions can only be used to prove stability, via the LaSalle Invariance Principle, while *strict* Lyapunov functions can be used to show robustness properties, such as ISS to actuator errors. Robustness is an essential feature in engineering applications, largely due to the uncertainty in dynamical models and noise entering into controllers. Many controller design methods, e.g., backstepping [75], forwarding [113, 149] and universal stabilizing controllers [158], are based on strict Lyapunov functions. In particular, if V is a global strict Lyapunov function for $\dot{x} = f(t,x)$ for which $\alpha(x) = \inf_t\{-[V_t(t,x) + V_x(t,x)f(t,x)]\}$ is radially unbounded, with f and g both locally Lipschitz, and with V, f, and g all periodic in t with the same period T, then $\dot{x} = f(t,x) + g(t,x)[K(t,x) + d]$ is ISS if we take the feedback $K(t,x) = -V_x(t,x)g(t,x)$. Consequently, when an explicit strict Lyapunov function is known, many important stabilization problems can be solved almost immediately.

In general, it is much easier to obtain non-strict Lyapunov functions than strict ones, owing to the more restrictive decay condition in the strict Lyapunov function definition. For instance, when a passive nonlinear system is stabilized by linear output feedback, the energy (i.e., storage) function can typically be used as the weak Lyapunov function. This fact is useful for electro-mechanical systems. Also, when a system is stabilized via the Jurdjevic-Quinn Theorem, non-strict Lyapunov functions are typically available, e.g., by taking the Hamiltonian for Euler-Lagrange systems; see Chap. 4 or [41, 68, 102, 127]. This has motivated a significant literature devoted to transforming non-strict Lyapunov functions into strict Lyapunov functions.

In this chapter, we present two more strict Lyapunov function constructions, both based on transforming non-strict Lyapunov functions into strict ones under suitable Lie derivative conditions. The assumptions in our first construction agree with those of [110], but they lead to simpler designs than the one in [110]. Our second result uses the Matrosov approach in Theorem 3.1. In general, Matrosov's Method can be difficult to apply because one has to find suitable auxiliary functions. Here we give simple sufficient conditions leading to a systematic design of auxiliary functions. This makes it possible to construct strict Lyapunov functions via Theorem 3.1. We illustrate our approach by constructing a strict Lyapunov function for an error dynamics involving the celebrated Lotka-Volterra System, which plays a fundamental role in bioengineering. Throughout the chapter, all (in)equalities should be understood to hold globally unless otherwise indicated, and we omit the arguments of our functions when they are clear from the context.

5.2 First Method: Iterated Lie Derivatives

Recall that if $f : \mathbb{R}^n \to \mathbb{R}^n$ is a smooth (i.e., C^∞) vector field and $V : \mathbb{R}^n \to \mathbb{R}$ is a smooth scalar function, the *Lie derivatives* of V in the direction of f are defined recursively by

$$L_f^1 V(x) \doteq L_f V(x) \doteq \tfrac{\partial V}{\partial x}(x) f(x) \;\text{ and}$$
$$L_f^k V(x) \doteq L_f\big(L_f^{k-1} V\big)(x) \;\text{ for } k \geq 2\,.$$

We refer to the functions $L_f^k V$ as *iterated Lie derivatives*. We next construct a strict Lyapunov function for the system

$$\dot{x} = f(x), \quad x \in \mathbb{R}^n \tag{5.1}$$

with f smooth and $f(0) = 0$, under appropriate Lie derivative assumptions. Specifically, assume that (5.1) admits a global non-strict Lyapunov function such that for each $p \in \mathbb{R}^n \setminus \{0\}$, there is an $i \in \mathbb{N}$ such that $L_f^i V(p) \neq 0$. If $L_f V(\phi(t, x_0)) \equiv 0$ along some trajectory $\phi(\cdot, x_0)$ of (5.1), then we can differentiate repeatedly to get

$$L_f^k V(\phi(t, x_0)) \equiv 0 \;\; \forall t \geq 0 \text{ and } k \in \mathbb{N},$$

so $x_0 = 0$. Hence, GAS follows from the LaSalle Invariance Principle. On the other hand, it is not obvious how to construct a *strict* Lyapunov function in this situation. This motivates our hypotheses in the following theorem, in which $a_i(x) \doteq (-1)^i L_f^i V(x)$ for all i:

Theorem 5.1. *Assume that there exists a smooth function $V : \mathbb{R}^n \to [0, \infty)$ such that the following conditions hold:*

1. *$V(\cdot)$ is a non-strict Lyapunov function for the system (5.1); and*
2. *there exists a positive integer $\ell \in \mathbb{N}$ such that for each $x \neq 0$, there exists an integer $i \in [1, \ell]$ (possibly depending on x) such that $L_f^i V(x) \neq 0$.*

Then we can construct explicit expressions for functions $\mathcal{F}_j$ and $\mathcal{G}$ so that

$$\begin{aligned} V^\sharp(x) &= \sum_{j=1}^{\ell-1} \mathcal{F}_j\big(V(x)\big) A_j(x) + \mathcal{G}\big(V(x)\big), \quad \textit{where} \\ A_j(x) &= \sum_{m=1}^{j} a_{m+1}(x) a_m(x) \end{aligned} \tag{5.2}$$

is a strict Lyapunov function for (5.1).

Proof. Since Condition 2. in Theorem 5.1 is satisfied for some $\ell \geq 1$, it holds for all larger integers as well, so we assume without loss of generality (to simplify the proof) that $\ell \geq 3$. Note for later use that $a_{i+1} \equiv -\dot{a}_i$ for all i, along the trajectories of (5.1).

Condition 2. from Theorem 5.1 guarantees that we can construct a positive definite continuous function ρ such that

$$a_1(x) + \sum_{m=2}^{\ell} a_m^2(x) \;\geq\; \rho(V(x)) \;\; \forall x \in \mathbb{R}^n\,, \tag{5.3}$$

e.g.,

$$\rho(r) = \min\left\{ a_1(x) + \sum_{m=2}^{\ell} a_m^2(x) : V(x) = r \right\}.$$

By minorizing ρ as necessary without relabeling and using Lemma A.7, we can assume that

$$\rho(r) = \frac{\omega(r)}{K(r)} \tag{5.4}$$

for some function $\omega \in \mathcal{K}_\infty \cap C^1$ and some increasing everywhere positive function $K \in C^1$. We can also determine an everywhere positive increasing function $\Gamma \in C^1$ such that

$$\Gamma\big(V(x)\big) \geq (\ell+2)|a_m(x)| + 1 \ \forall m \in \{1, ..., \ell+1\} \tag{5.5}$$

holds for all $x \in \mathbb{R}^n$. For example, take

$$\Gamma_0(r) = (\ell+2) \max \left\{ \sum_{m=1}^{\ell+1} |a_m(x)| + 1 : V(x) \leq r \right\},$$

and then majorize by an increasing C^1 function.

Let us introduce the following functions:

$$M_j(x) = \sum_{m=1}^{j} a_{m+1}(x) a_m(x) + \int_0^{V(x)} \Gamma(r) \mathrm{d}r, \ \ j = 1, 2, \ldots, \ell - 1 \, ; \tag{5.6}$$

$$N_0(x) = a_1(x), \ \text{ and } N_j(x) = \sum_{m=2}^{j+1} a_m^2(x) + a_1(x), \ \ j = 1, 2, \ldots \ell - 1. \tag{5.7}$$

Since $a_1(x) \geq 0$ everywhere, (5.5) gives

$$\begin{aligned} \dot{M}_1(x) &= \dot{a}_2(x) a_1(x) - a_2^2(x) - \Gamma\big(V(x)\big) a_1(x) \\ &\leq -a_2^2(x) - a_1(x) \\ &= -N_1(x) \, . \end{aligned} \tag{5.8}$$

Also, for each $j \in \{2, ..., \ell - 1\}$, we get

$$\begin{aligned} \dot{M}_j(x) &= -\sum_{m=1}^{j} a_{m+1}^2(x) + \sum_{m=1}^{j} \dot{a}_{m+1}(x) a_m(x) - \Gamma\big(V(x)\big) a_1(x) \\ &\leq -\sum_{m=1}^{j} a_{m+1}^2(x) + \sum_{m=2}^{j} |a_{m+2}(x)||a_m(x)| + |a_3(x)| a_1(x) \\ &\quad -\Gamma\big(V(x)\big) a_1(x) \\ &\leq -\sum_{m=1}^{j} a_{m+1}^2(x) + \sum_{m=2}^{j} |a_{m+2}(x)||a_m(x)| + |a_3(x)| a_1(x) \\ &\quad -\big[(\ell+2)|a_3(x)| + 1\big] a_1(x). \end{aligned} \tag{5.9}$$

From this inequality and (5.5), we deduce that for all $j \in \{2, ..., \ell - 1\}$,

$$\begin{aligned} \dot{M}_j(x) \leq & -\sum_{m=1}^{j} a_{m+1}^2(x) \\ & + \frac{\Gamma(V(x))}{\ell+2} \sum_{m=2}^{j} |a_m(x)| - \big[(\ell+1)|a_3(x)| + 1\big] a_1(x). \end{aligned} \tag{5.10}$$

It follows from the Cauchy Inequality that for all $j \in \{2, ..., \ell - 1\}$,

$$\begin{aligned} \dot{M}_j(x) & \leq -\sum_{m=1}^{j} a_{m+1}^2(x) + \Gamma(V(x))\sqrt{\sum_{m=2}^{j} a_m^2(x)} \\ & \quad -\big[(\ell+1)|a_3(x)| + 1\big] a_1(x) \\ & = -\sum_{m=2}^{j+1} a_m^2(x) - a_1(x) + \Gamma(V(x))\sqrt{\sum_{m=2}^{j} a_m^2(x)} \\ & \quad -(\ell+1)|a_3(x)| a_1(x). \end{aligned} \tag{5.11}$$

From the definitions of the functions N_j, we deduce that for all $j \in \{2, ..., \ell - 1\}$,

$$\dot{M}_j(x) \leq -N_j(x) + \Gamma(V(x))\sqrt{N_{j-1}(x)}. \tag{5.12}$$

Set

$$\Omega(v) = \frac{2\omega(v)}{\ell \Gamma^2(v) K(v)} \tag{5.13}$$

and define the positive definite functions $k_1, k_2, \ldots, k_{\ell-1} \in C^1$ by

$$k_{\ell-1}(v) = 2K(v)\omega^{2^{\ell-1}}(v) \tag{5.14}$$

and

$$k_p(v) = k_{\ell-1}(v)\Omega^{1-2^{\ell-p-1}}(v) \tag{5.15}$$

for $p = 1, 2, \ldots, \ell - 2$.

Pick a C^1 everywhere positive increasing function k_0 such that

$$k_0(V(x)) + k_0'(V(x))V(x) \geq \sum_{p=1}^{\ell-1} \left|k_p'(V(x))M_p(x)\right| + 1. \tag{5.16}$$

Let

$$S_1(x) \doteq \sum_{p=1}^{\ell-1} k_p\big(V(x)\big)M_p(x) + k_0\big(V(x)\big)V(x). \tag{5.17}$$

Then

$$\begin{aligned}\dot{S}_1(x) &= \sum_{p=1}^{\ell-1} k_p\big(V(x)\big)\dot{M}_p(x) + \left[\sum_{p=1}^{\ell-1} k_p'\big(V(x)\big)M_p(x)\right]\dot{V}(x) \\ &\quad + \big[k_0\big(V(x)\big) + k_0'\big(V(x)\big)V(x)\big]\dot{V}(x).\end{aligned} \tag{5.18}$$

It follows from (5.16) and the fact that $\dot{V}$ is non-positive everywhere that

$$\dot{S}_1(x) \;\leq\; \sum_{p=1}^{\ell-1} k_p(V(x))\dot{M}_p(x). \tag{5.19}$$

Using (5.8) and (5.12), we deduce that

$$\begin{aligned}\dot{S}_1(x) &\leq -k_1\big(V(x)\big)N_1(x) \\ &\quad + \sum_{p=2}^{\ell-1} k_p\big(V(x)\big)\left[-N_p(x) + \Gamma\big(V(x)\big)\sqrt{N_{p-1}(x)}\right] \\ &= -\sum_{p=1}^{\ell-1} k_p\big(V(x)\big)N_p(x) + \sum_{p=2}^{\ell-1} k_p\big(V(x)\big)\Gamma\big(V(x)\big)\sqrt{N_{p-1}(x)}.\end{aligned} \tag{5.20}$$

By (5.3) and (5.4), we deduce that

$$N_{\ell-1}(x) \geq \frac{\omega(V(x))}{K(V(x))}. \tag{5.21}$$

Therefore,

$$\begin{aligned}\dot{S}_1(x) &\leq -k_{\ell-1}(V(x))\frac{\omega(V(x))}{K(V(x))} - \sum_{p=1}^{\ell-2} k_p\big(V(x)\big)N_p(x) \\ &\quad + \sum_{p=1}^{\ell-2} k_{p+1}\big(V(x)\big)\Gamma\big(V(x)\big)\sqrt{N_p(x)}.\end{aligned} \tag{5.22}$$

From the triangular inequality $c_1c_2 \leq c_1^2 + \frac{1}{4}c_2^2$ for non-negative values c_1 and c_2, we deduce that

$$\begin{aligned}&\left\{\sqrt{k_p(V(x))N_p(x)}\right\}\left\{\frac{\Gamma\big(V(x)\big)k_{p+1}\big(V(x)\big)}{\sqrt{k_p(V(x))}}\right\} \\ &\leq\; k_p(V(x))N_p(x) + \frac{\Gamma^2(V(x))k_{p+1}^2(V(x))}{4k_p(V(x))}\end{aligned} \tag{5.23}$$

for $p = 1, 2, \ldots, \ell - 2$ when $x \neq 0$. Summing the inequalities in (5.23) over $p = 1, 2, \ldots, \ell - 2$ and combining with (5.22), we deduce that for $x \neq 0$,

$$\dot{S}_1(x) \leq -k_{\ell-1}\big(V(x)\big)\frac{\omega(V(x))}{K(V(x))} + \sum_{p=1}^{\ell-2} \frac{\Gamma^2(V(x))k_{p+1}^2(V(x))}{4k_p(V(x))}. \tag{5.24}$$

By our choices of the k_p's, we get

$$\begin{aligned}\dot{S}_1(x) &\leq -k_{\ell-1}\big(V(x)\big)\frac{\omega(V(x))}{K(V(x))}\\ &\quad+\sum_{p=1}^{\ell-2} \frac{\Gamma^2(V(x))k_{\ell-1}^2(V(x))\Omega^{2(1-2^{\ell-p-2})}(V(x))}{4k_{\ell-1}(V(x))\Omega^{1-2^{\ell-p-1}}(V(x))}\\ &= -k_{\ell-1}\big(V(x)\big)\frac{\omega(V(x))}{K(V(x))}\\ &\quad+(\ell-2)\frac{\Gamma^2(V(x))k_{\ell-1}(V(x))\Omega(V(x))}{4}, \quad x \neq 0\,.\end{aligned} \tag{5.25}$$

Our choice of Ω in (5.13) now gives

$$\dot{S}_1(x) \leq -k_{\ell-1}\big(V(x)\big)\frac{\omega(V(x))}{2K(V(x))} \quad \forall x \in \mathbb{R}^n\,. \tag{5.26}$$

Recalling our choice (5.14) of $k_{\ell-1}$ now gives

$$\dot{S}_1(x) \leq -\omega^{2^{\ell-1}+1}\big(V(x)\big). \tag{5.27}$$

All of the functions k_p are C^1 and the right hand side of (5.27) is negative definite. However, S_1 is not necessarily a strict Lyapunov function because S_1 is not necessarily positive definite and radially unbounded. To obtain a strict Lyapunov function, consider

$$V^\sharp(x) = V(x)S_1(x) + \kappa\big(V(x)\big)V(x)\,, \tag{5.28}$$

where $\kappa \in C^1$ is an everywhere positive function with an everywhere positive first derivative such that $\kappa(V(x)) \geq |S_1(x)| + 1$ for all $x \in \mathbb{R}^n$. Then $V^\sharp$ is positive definite and radially unbounded because $V^\sharp(x) \geq V(x)$ and

$$\begin{aligned}\dot{V}^\sharp(x) &= V(x)\dot{S}_1(x) + \dot{V}(x)S_1(x) + \big[\kappa'\big(V(x)\big)V(x) + \kappa\big(V(x)\big)\big]\dot{V}(x)\\ &\leq -\omega^{2^{\ell-1}+1}\big(V(x)\big)V(x).\end{aligned} \tag{5.29}$$

The result readily follows from the formula (5.17) for S_1, by collecting the functions involving V to form the expression for $V^\sharp$. □

5.3 Discussion and Extensions of First Method

5.3.1 Local vs. Global

While stated for systems on $\mathbb{R}^n$, we can also prove the following local version of Theorem 5.1 [110]: *Suppose that all conditions of Theorem 5.1 hold on a given neighborhood of the origin $E \subseteq \mathbb{R}^n$. Then, there exists a neighborhood of the origin E_1 with $E_1 \subseteq E$ and functions $\mathcal{F}_i$ and $\mathcal{G}$ such that (5.2) is a strict Lyapunov function for the system (3.17) on the set E_1.* The proof is similar to that of Theorem 5.1, by taking E_1 to be a suitable open sublevel set of V. Alternatively, we can prove the local version by using the construction from [110], which in general leads to a strict Lyapunov function that differs from the one we gave in Theorem 5.1.

5.3.2 Real Analytic Case

When V and f are real analytic, Theorem 5.1 remains true if its Condition 2. is replaced by the assumption that there exist positive constants $\underline{B}$ and $\bar{B}$ such that: There is an integer $\ell \in \mathbb{N}$ such that for each $x \in \{p \in \mathbb{R}^n : 0 < |p| < \underline{B}$ or $|p| > \bar{B}\}$, there is an integer $i \in [1, \ell]$ such that $L_f^i V(x) \neq 0$. This follows from the following simple observation from [110]:

Proposition 5.1. *Assume that (5.1) is GAS, f is real analytic, and Condition 1. of Theorem 5.1 holds with a real analytic function V. Then, for each compact set $E \subseteq \mathbb{R}^n$ that does not contain the origin, there exists $\ell \in \mathbb{N}$ such that each point $x \in E$ admits an index $i \in [1, \ell]$ such that $L_f^i V(x) \neq 0$.*

Hence, to apply the local version of Theorem 5.1, it suffices to check its Condition 1., and then check its Condition 2. on a set of the form $\underline{B}\mathcal{B}_n \setminus \{0\}$ for some constant $\underline{B} > 0$. Let us sketch the proof of Proposition 5.1.

Proof. We proceed in two steps.

Step 1. Fix any $x_0 \in E$. Since the system is assumed to be GAS, there must be a time $t_c > 0$ at which $L_f V(x(t_c, x_0)) \neq 0$. (This is because if no such t_c existed, then

$$V\big(x(t, x_0)\big) = V(x_0) + \int_0^t L_f V\big(x(r, x_0)\big)\mathrm{d}r \equiv V(x_0)$$

for all $t \geq 0$ would contradict the GAS property.) Since V and f are real analytic functions, so is $t \mapsto L_f V(\phi(t, x_0))$. Consider its expansion

$$L_f V\big(\phi(t, x_0)\big) = \sum_{i=0}^{\infty} L_f^{i+1} V(x_0) \frac{t^i}{i!} \tag{5.30}$$

around $t = 0$. Since $t \mapsto L_f V(\phi(t, x_0))$ is not the zero function, there must exist an integer $i = i(x_0)$ such that $L_f^i V(x_0) \neq 0$.

Step 2. Suppose that the statement of the proposition were false. Then there would exist a sequence $x_p \in E$ and a strictly increasing sequence of positive integers n_p such that

$$L_f^i V(x_p) = 0 \ \ \forall i \in [1, n_p - 1] \ , \quad \text{but} \quad L_f^{n_p} V(x_p) \neq 0 \ . \tag{5.31}$$

Since E is compact, we can assume that $x_p \to x^*$ for some non-zero $x^* \in E$. (Otherwise, we can pass to a subsequence without relabeling.) By Step 1 of the proof applied with $x_0 = x^*$, we can find an integer $J = J(x^*)$ such that $L_f^J V(x^*) \neq 0$. Since $L_f^J V$ is continuous, there exists a constant $\bar{p} \in \mathbb{N}$ such that for each $p \geq \bar{p}$, we have

$$L_f^J V(x_p) \neq 0.$$

This contradicts (5.31) once we pick p so that $n_p > J$. The result follows. □

5.3.3 Necessity vs. Sufficiency

Conditions 1. and 2. from Theorem 5.1 are not necessary for GAS of the system (5.1) [110]. To see why, consider the following example from [127]:

$$\begin{cases} \dot{x}_1 = x_2 \\ \dot{x}_2 = -x_1 - x_2 B(x_2), \end{cases} \tag{5.32}$$

where B is the smooth function

$$B(s) = \begin{cases} \exp\left(-\frac{1}{(s-1)^2}\right), & s \neq 1 \\ 0, & s = 1 \end{cases} .$$

Then Condition 1. of Theorem 5.1 is satisfied with $V(x_1, x_2) = x_1^2 + x_2^2$ since $\dot{V} = -2x_2^2 B(x_2)$, and the LaSalle Invariance Principle implies that (5.32) is GAS to zero. However, Condition 2. of Theorem 5.1 does not hold since for $x^* = (0\ 1)^\top$, we have $L_f^i V(x^*) = 0$ for all $i \in \mathbb{N}$.

5.3.4 Recovering Exponential Stability

When (5.1) is locally exponentially stable, the time derivative of (5.2) along the trajectories of (5.1) will not in general be upper bounded by a negative definite quadratic function. Moreover, it is not clear how to use (5.2) to verify local or global exponential stability. However, we can use $V^\sharp$ to get another strong Lyapunov function W that can be used to verify exponential stability. For example, if Conditions 1.-2. of Theorem 5.1 hold and (5.1) has an exponentially stable linearization, then one can construct a Lyapunov function $V^\star$ and $\alpha_1, \alpha_2, \alpha_3 \in \mathcal{K}_\infty$ such that

$$\alpha_1(|x|) \le V^\star(x) \le \alpha_2(|x|) \quad \text{and} \quad \dot{V}^\star \le -\alpha_3(|x|)$$

hold for all $x \in \mathbb{R}^n$ and, moreover, there exist positive constants δ, a, and b such that $\alpha_1(s) = as^2$ and $\alpha_3(s) = bs^2$ for all $s \in [0, \delta]$ [60, Lemma 10.1.5].

5.4 Second Method: Matrosov Conditions

We again consider a general nonlinear system

$$\dot{x} = f(x), \quad x \in \mathcal{X} \tag{5.33}$$

evolving on an open positively invariant set $\mathcal{X} \subseteq \mathbb{R}^n$ that contains the origin, where $f(0) = 0$. We use Theorem 3.1 in Chap. 3 to construct strict Lyapunov functions for (5.33). Recall that Theorem 3.1 is a continuous time Matrosov Theorem, which requires auxiliary functions, in addition to a non-strict Lyapunov function. In general, it can be difficult to find appropriate auxiliary functions to apply the Matrosov Theorem. Hence, our work sheds light on the Matrosov Theorems as well, because it gives a new mechanism for choosing auxiliary functions.

However, the most important features of our second method are that (a) the result applies to systems for which the state space is only a proper subset of $\mathbb{R}^n$ and (b) it may yield Lyapunov functions that are simpler than the ones obtained from Theorem 5.1, and that also have desirable local properties such as local boundedness from below by positive definite quadratic functions; see Sect. 5.5.

To account for the restricted state space for (5.33), we use the following definitions. A C^1 function $V : \mathcal{X} \to \mathbb{R}$ on a general open set $\mathcal{X} \subseteq \mathbb{R}^n$ containing the origin is called a *storage function* provided there exist continuous positive definite functions $\alpha_1, \alpha_2 : \mathcal{X} \to [0, \infty)$ such that the following hold:

1. for each i, $\alpha_i(x) \to +\infty$ whenever $|x| \to +\infty$ with x remaining in $\mathcal{X}$; and
2. $\alpha_1(x) \le V(x) \le \alpha_2(x)$ for all $x \in \mathcal{X}$.

Condition 1. holds vacuously when $\mathcal{X}$ is bounded. A storage function V is called a *non-strict* (resp., *strict*) *Lyapunov-like function* for (5.33) provided it is C^1 and $L_f V(x)$ is negative semi-definite (resp., negative definite). If, in addition, for each i and each $\bar{q} \in \partial\mathcal{X}$, $\alpha_i(q) \to +\infty$ when $q \to \bar{q}$ then a non-strict (resp., strict) Lyapunov-like function is called a *non-strict* (resp., *strict*) *Lyapunov function.* In the rest of this subsection, we assume:

Assumption 5.1 *There exist a smooth storage function $V_1 : \mathcal{X} \to [0, \infty)$; functions $h_1, \dots, h_m \in C^\infty(\mathbb{R}^n)$ such that $h_j(0) = 0$ for all j; everywhere positive functions $r_1, \dots, r_m \in C^\infty(\mathbb{R}^n)$ and $\rho \in C^\infty(\mathbb{R})$; and an integer $N > 0$ for which*

$$\nabla V_1(x) f(x) \le -r_1(x) h_1^2(x) - \dots - r_m(x) h_m^2(x) \tag{5.34}$$

$$\text{and} \quad \sum_{l=0}^{N-1} \sum_{j=1}^{m} \left[L_f^l h_j(x)\right]^2 \ge \rho(V_1(x)) V_1(x) \tag{5.35}$$

hold for all $x \in \mathcal{X}$. Moreover, f is defined on $\mathbb{R}^n$ and there is a function $\overline{\Gamma} \in \mathcal{K}_\infty$ such that

$$|f(x)| \le \overline{\Gamma}(|x|) \ \forall x \in \mathbb{R}^n. \tag{5.36}$$

Also, V_1 has a positive definite quadratic lower bound near the origin.

To simplify our notation, we introduce the functions

$$\begin{aligned} \mathcal{N}_1(x) &= R(x) \sum_{l=1}^{m} h_l^2(x) \\ \text{and} \ \ \mathcal{N}_i(x) &= \sum_{l=1}^{m} \left[L_f^{i-1} h_l(x)\right]^2 \end{aligned} \tag{5.37}$$

for all $i \ge 2$, where

$$R(x) = \frac{\prod_{i=1}^{m} r_i(x)}{\prod_{i=1}^{m} [r_i(x) + 1]}$$

for all $i \ge 2$. We assume that f is sufficiently smooth.

The following is shown in [105]:

Theorem 5.2. *If (5.33) satisfies Assumption 5.1, then one can determine explicit functions $k_l, \Omega_l \in \mathcal{K}_\infty \cap C^1$ and an everywhere positive continuous function ρ_0 such that*

$$S(x) = \sum_{l=1}^{N} \Omega_l\big(k_l(V_1(x)) + V_l(x)\big) \tag{5.38}$$

with the choices

$$V_i(x) = -\sum_{l=1}^{m} L_f^{i-2} h_l(x) L_f^{i-1} h_l(x)\ , \quad i = 2, \dots, N \tag{5.39}$$

satisfies $S(x) \ge V_1(x)$ and $\nabla S(x) f(x) \le -\rho_0(x) V_1(x)$ for all $x \in \mathcal{X}$. If, in addition, $\mathcal{X} = \mathbb{R}^n$, then the system (5.33) is GAS.

Proof. Sketch. Since R is everywhere positive and satisfies $R(x) \leq r_i(x)$ for all $x \in \mathbb{R}^n$ and all $i \in \{1, ..., m\}$, we get

$$\begin{aligned} &\nabla V_1(x) f(x) \leq -\mathcal{N}_1 \quad \text{by (5.34), and} \\ &\nabla V_i(x) f(x) \leq -\mathcal{N}_i + \textstyle\sum_{l=1}^m |L_f^{i-2} h_l||L_f^i h_l| \end{aligned} \tag{5.40}$$

for $i = 2, \ldots, N$ and $x \in \mathcal{X}$. In particular, we have:

$$\begin{aligned} \nabla V_2(x) f(x) &\leq -\mathcal{N}_2(x) + \sum_{l=1}^m \frac{|L_f^2 h_l(x)|}{\sqrt{R(x)}} \sqrt{\mathcal{N}_1(x)}; \\ \nabla V_i(x) f(x) &\leq -\mathcal{N}_i(x) + \left[\sum_{l=1}^m |L_f^i h_l(x)| \right] \sqrt{\mathcal{N}_{i-1}(x)} \end{aligned}$$

for $i = 3, 4, \ldots, N$. Moreover, the fact that V_1 is a storage function implies that there exists a function $\underline{\alpha} \in \mathcal{K}_\infty$ such that $V_1(x) \geq \underline{\alpha}(|x|)$ for all $x \in \mathcal{X}$.

Therefore, we can use (5.36) to determine a continuous everywhere positive function ϕ_1 such that

$$\sum_{l=1}^m \frac{|L_f^2 h_l(x)|}{\sqrt{R(x)}} \leq \phi_1(V_1(x)) \sqrt{V_1(x)} \tag{5.41}$$

and

$$\sum_{l=1}^m |L_f^i h_l(x)| \leq \phi_1(V_1(x)) \sqrt{V_1(x)} \tag{5.42}$$

for all $x \in \mathcal{X}$ and $i = 3, \ldots, N$. The construction of ϕ_1 satisfying (5.42) is as follows; the requirement (5.41) is handled in a similar way. Since $L_f^i h_l$ is sufficiently smooth for each i and l and zero at the origin, we have

$$\sum_{l=1}^m |L_f^i h_l(x)| \leq |x| \mathcal{G}_1(|x|) \leq \bar{\kappa} \sqrt{V_1(x)} \mathcal{G}_1\big(\underline{\alpha}^{-1}(V_1(x))\big)$$

for some increasing everywhere positive function $\mathcal{G}_1$ and constant $\bar{\kappa} > 0$ in some neighborhood $\mathcal{O}$ of the origin. We can also find a function $\mathcal{G}_2 \in \mathcal{K}_\infty$ such that $\sum_{l=1}^m |L_f^i h_l(x)| / (\underline{\alpha}(|x|))^{1/2} \leq \mathcal{G}_2(|x|)$ on $\mathbb{R}^n \setminus \mathcal{O}$. Hence, we can take $\phi_1(r) = 1 + \bar{\kappa} \mathcal{G}_1(\underline{\alpha}^{-1}(r)) + \mathcal{G}_2(\underline{\alpha}^{-1}(r))$.

It follows that

$$\nabla V_i(x) f(x) \leq -\mathcal{N}_i(x) + \phi_1(V_1(x)) \sqrt{\mathcal{N}_{i-1}(x)} \sqrt{V_1(x)} \tag{5.43}$$

for $i = 2, \ldots, N$. We can determine an everywhere non-negative function p_1 such that $|V_i(x)| \leq p_1(V_1(x)) V_1(x)$ for $i = 1, \ldots, N$ for all $x \in \mathcal{X}$. Hence, Theorem 3.1 constructs the necessary strict Lyapunov-like function. □

5.5 Application: Lotka-Volterra Model

5.5.1 Strict Lyapunov Function Construction

We illustrate Theorem 5.2 using the celebrated Lotka-Volterra Predator-Prey System

$$\begin{cases} \dot{\chi} = \gamma\chi\left(1 - \frac{\chi}{L}\right) - a\chi\zeta \\ \dot{\zeta} = \beta\chi\zeta - \Delta\zeta \end{cases} \tag{5.44}$$

with positive constants a, β γ, Δ, and L. System (5.44) is a simple model of one predator feeding on one prey. The population of the predator is ζ, χ is the population of the prey, and the constants are related to the birth and death rates of the predator and prey. We assume that the population levels are positive.

The time scaling, change of coordinates, and constants

$$\begin{aligned} &x(t) = \tfrac{1}{L}\chi\left(\tfrac{t}{\gamma}\right), \quad y(t) = \tfrac{a}{\beta L}\zeta\left(\tfrac{t}{\gamma}\right), \\ &\alpha = \tfrac{\beta L}{\gamma} \text{ and } d = \tfrac{\Delta}{\gamma} \end{aligned} \tag{5.45}$$

give the simpler Lotka-Volterra system

$$\begin{cases} \dot{x} = x\,(1 - x) - \alpha xy \\ \dot{y} = \alpha xy - dy. \end{cases} \tag{5.46}$$

We assume that $\alpha > d$, and we set

$$x_* = \tfrac{d}{\alpha} \text{ and } y_* = \tfrac{1}{\alpha} - \tfrac{d}{\alpha^2}. \tag{5.47}$$

Then $x_* \in (0, 1)$ and $y_* > 0$. Also, the new variables $\tilde{x} = x - x_*$ and $\tilde{y} = y - y_*$ have the dynamics

$$\begin{cases} \dot{\tilde{x}} = -[\tilde{x} + \alpha\tilde{y}](\tilde{x} + x_*) \\ \dot{\tilde{y}} = \alpha\tilde{x}(\tilde{y} + y_*)\,, \end{cases} \tag{5.48}$$

with state space $\mathcal{X} = (-x_*, \infty) \times (-y_*, \infty)$. We do our Lyapunov function construction for (5.48), so we set

$$f(\tilde{x}, \tilde{y}) = \begin{bmatrix} -[\tilde{x} + \alpha\tilde{y}](\tilde{x} + x_*) \\ \alpha\tilde{x}(\tilde{y} + y_*) \end{bmatrix}. \tag{5.49}$$

Let us check that the assumptions from Theorem 5.2 are satisfied with $m = 1$, $N = 2$, $r_1 \equiv 1$, $h_1(\tilde{x}, \tilde{y}) \doteq \tilde{x}$, and

$$V_1(\tilde{x}, \tilde{y}) = \tilde{x} - x_* \ln\left(1 + \tfrac{\tilde{x}}{x_*}\right) + \tilde{y} - y_* \ln\left(1 + \tfrac{\tilde{y}}{y_*}\right). \tag{5.50}$$

One easily checks that $V_1 : \mathcal{X} \to [0, \infty)$ is a storage function. Along the trajectories of (5.48), it has the time derivative

$$\begin{aligned}
\dot{V}_1 &= \frac{\tilde{x}}{x_* + \tilde{x}}\dot{\tilde{x}} + \frac{\tilde{y}}{y_* + \tilde{y}}\dot{\tilde{y}} \\
&= -\frac{\tilde{x}}{x_* + \tilde{x}}[\tilde{x} + \alpha\tilde{y}](\tilde{x} + x_*) + \frac{\alpha\tilde{y}}{y_* + \tilde{y}}\tilde{x}(\tilde{y} + y_*) \\
&= -\tilde{x}[\tilde{x} + \alpha\tilde{y}] + \alpha\tilde{y}\tilde{x} \;=\; -\tilde{x}^2 \,.
\end{aligned} \tag{5.51}$$

Also,

$$L_f h_1(\tilde{x}, \tilde{y}) = -[\tilde{x} + \alpha\tilde{y}](\tilde{x} + x_*).$$

Defining the $\mathcal{N}_i$'s as in (5.37), a simple argument based on the fact that V_1 becomes unbounded as $\tilde{x}$ approaches $-x_*$ or $\tilde{y}$ approaches $-y_*$ provides a constant $\underline{d} > 0$ such that

$$\sum_{i=1}^{2} \mathcal{N}_i(\tilde{x}, \tilde{y}) \geq \underline{d}\frac{V_1(\tilde{x}, \tilde{y})}{1 + V_1^2(\tilde{x}, \tilde{y})} \tag{5.52}$$

on $\mathcal{X}$; see Appendix A.3. Also, Lemma A.8 provides a positive definite quadratic lower bound for V_1 near 0. Hence, Theorem 5.2 provides the necessary strict Lyapunov function for (5.48).

We now construct the strict Lyapunov function of the type provided by the theorem. Notice that

$$\begin{aligned}
&\mathcal{N}_1(\tilde{x}, \tilde{y}) = \tfrac{1}{2}h_1^2(\tilde{x}, \tilde{y}), \;\; \mathcal{N}_2(\tilde{x}, \tilde{y}) = \big(L_f h_1(\tilde{x}, \tilde{y})\big)^2, \\
&V_2(\tilde{x}, \tilde{y}) = -h_1(\tilde{x}, \tilde{y}) L_f h_1(\tilde{x}, \tilde{y}), \;\; L_f V_1(\tilde{x}, \tilde{y}) \leq -\mathcal{N}_1(\tilde{x}, \tilde{y}),
\end{aligned}$$

and

$$\begin{aligned}
L_f V_2(\tilde{x}, \tilde{y}) &= -\big(L_f h_1(\tilde{x}, \tilde{y})\big)^2 - h_1(\tilde{x}, \tilde{y}) L_f^2 h_1(\tilde{x}, \tilde{y}) \\
&= -\mathcal{N}_2(\tilde{x}, \tilde{y}) - h_1(\tilde{x}, \tilde{y}) L_f^2 h_1(\tilde{x}, \tilde{y}).
\end{aligned} \tag{5.53}$$

Simple calculations yield

$$\begin{aligned}
L_f^2 h_1(\tilde{x}, \tilde{y}) &= -\big(\dot{\tilde{x}} + \alpha\dot{\tilde{y}}\big)\big(\tilde{x} + x_*\big) - \big[\tilde{x} + \alpha\tilde{y}\big]\dot{\tilde{x}} \\
&= -\big(x_* + 2\tilde{x} + \alpha\tilde{y}\big)\dot{\tilde{x}} - \big(x_* + \tilde{x}\big)\alpha\dot{\tilde{y}} \\
&= -\big(x_* + 2\tilde{x} + \alpha\tilde{y}\big) L_f h_1(\tilde{x}, \tilde{y}) \\
&\quad -\alpha^2\big(x_* + h_1(\tilde{x}, \tilde{y})\big) h_1(\tilde{x}, \tilde{y}) \big(\tilde{y} + y_*\big).
\end{aligned} \tag{5.54}$$

Substituting (5.54) into (5.53) gives

$$
\begin{aligned}
L_f V_2(\tilde{x}, \tilde{y}) \leq & -\mathcal{N}_2(\tilde{x}, \tilde{y}) + (x_* + 2|\tilde{x}| + \alpha|\tilde{y}|)\,|h_1(\tilde{x}, \tilde{y})|\,|L_f h_1(\tilde{x}, \tilde{y})| \\
& + \alpha^2 (x_* + |\tilde{x}|)(|\tilde{y}| + y_*) h_1^2(\tilde{x}, \tilde{y}) \\
\leq & -\mathcal{N}_2(\tilde{x}, \tilde{y}) + (x_* + 2|\tilde{x}| + \alpha|\tilde{y}|)\,|h_1(\tilde{x}, \tilde{y})|\,|L_f h_1(\tilde{x}, \tilde{y})| \\
& + \alpha^2 x_* y_* \left(1 + \tfrac{|\tilde{x}|}{x_*}\right)\left(1 + \tfrac{|\tilde{y}|}{y_*}\right) h_1^2(\tilde{x}, \tilde{y}).
\end{aligned}
$$

Next, observe that

$$
\begin{aligned}
& \left(\tfrac{1}{x_*} + \tfrac{1}{y_*}\right) V_1(\tilde{x}, \tilde{y}) \geq \\
& \tfrac{\tilde{x}}{x_*} - \ln\left(1 + \tfrac{\tilde{x}}{x_*}\right) + \tfrac{\tilde{y}}{y_*} - \ln\left(1 + \tfrac{\tilde{y}}{y_*}\right).
\end{aligned}
\tag{5.55}
$$

This, Lemma A.8, and the relation $1 + A^2 \geq \frac{1}{2}(1 + |A|)$ give

$$
\begin{aligned}
e^{\left(\frac{1}{x_*} + \frac{1}{y_*}\right) V_1(\tilde{x}, \tilde{y})} & \geq \left(\frac{e^{\frac{\tilde{x}}{x_*}}}{1 + \frac{\tilde{x}}{x_*}}\right)\left(\frac{e^{\frac{\tilde{y}}{y_*}}}{1 + \frac{\tilde{y}}{y_*}}\right) \\
& \geq \tfrac{1}{36}\left(1 + \tfrac{\tilde{x}^2}{x_*^2}\right)\left(1 + \tfrac{\tilde{y}^2}{y_*^2}\right) \\
& \geq \tfrac{1}{144}\left(1 + \tfrac{|\tilde{x}|}{x_*}\right)\left(1 + \tfrac{|\tilde{y}|}{y_*}\right).
\end{aligned}
\tag{5.56}
$$

Hence,

$$
\begin{aligned}
& |\tilde{x}| \leq 144 x_* e^{\left(\frac{1}{x_*} + \frac{1}{y_*}\right) V_1(\tilde{x}, \tilde{y})} \text{ and} \\
& |\tilde{y}| \leq 144 y_* e^{\left(\frac{1}{x_*} + \frac{1}{y_*}\right) V_1(\tilde{x}, \tilde{y})}.
\end{aligned}
$$

Setting $\mathcal{M}(r) = (289 x_* + 144 \alpha y_*)\, e^{\left(\frac{1}{x_*} + \frac{1}{y_*}\right) r}$ therefore gives

$$
\begin{aligned}
L_f V_2(\tilde{x}, \tilde{y}) \leq & -\mathcal{N}_2(\tilde{x}, \tilde{y}) \\
& + 2\mathcal{M}(V_1(\tilde{x}, \tilde{y}))\sqrt{\mathcal{N}_1(\tilde{x}, \tilde{y})}\sqrt{\mathcal{N}_2(\tilde{x}, \tilde{y})} \\
& + 288 \alpha^2 x_* y_* e^{\left(\frac{1}{x_*} + \frac{1}{y_*}\right) V_1(\tilde{x}, \tilde{y})} \mathcal{N}_1(\tilde{x}, \tilde{y}).
\end{aligned}
$$

Using the triangular inequality, we have

$$
\begin{aligned}
& \mathcal{M}(V_1)\sqrt{\mathcal{N}_1}\sqrt{\mathcal{N}_2} \\
& \leq \tfrac{1}{4}\mathcal{N}_2 + (289 x_* + 144 \alpha y_*)^2 e^{2\left(\frac{1}{x_*} + \frac{1}{y_*}\right) V_1} \mathcal{N}_1
\end{aligned}
\tag{5.57}
$$

where we omit the dependencies on $(\tilde{x}, \tilde{y})$. Therefore,

$$
L_f V_2(\tilde{x}, \tilde{y}) \leq -\frac{1}{2}\mathcal{N}_2(\tilde{x}, \tilde{y}) + \phi_1(V_1(\tilde{x}, \tilde{y}))\mathcal{N}_1(\tilde{x}, \tilde{y}), \tag{5.58}
$$

where

$$
\phi_1(r) = 2\left[(289 x_* + 144 \alpha y_*)^2 + 144 \alpha^2 x_* y_*\right] e^{2\left(\frac{1}{x_*} + \frac{1}{y_*}\right) r}.
$$

Since $V_2(\tilde{x}, \tilde{y}) = \tilde{x}[\tilde{x} + \alpha \tilde{y}](\tilde{x} + x_*)$, we easily get

$$|V_2(\tilde{x}, \tilde{y})| \leq 2(x_* + 1)(1 + \alpha)\left[\tilde{y}^4 + |\tilde{x}|^3 + \tilde{x}^2 + \tilde{y}^2\right], \tag{5.59}$$

and Lemma A.8 applied with $A = \tilde{x}/x_*$ gives

$$\left|\frac{\tilde{x}}{x_*}\right| \leq 2\left\{\left[\frac{V_1}{x_*}\right] + \left[\frac{V_1}{x_*}\right]^2\right\}^{1/2} \leq 2\left[\max\left\{\frac{1}{x_*}, \frac{1}{x_*^2}\right\}\{V_1 + V_1^2\}\right]^{1/2}$$

and similarly for y, where we omit the dependence of V_1 on $(\tilde{x}, \tilde{y})$. Combining these estimates with (5.59) and setting $\bar{d} = 1 + x_* + y_*$, simple algebra gives

$$|V_2(\tilde{x}, \tilde{y})| \leq 4(x_*+1)(1+\alpha)\textstyle\sum_{i=2}^4 \left\{2\bar{d}\sqrt{V_1 + V_1^2}\right\}^i \leq p_1(V_1(\tilde{x}, \tilde{y}))V_1(\tilde{x}, \tilde{y}),$$

where $p_1(r) = 640(x_* + 1)(\alpha + 1)\bar{d}^4(1 + r)^3$, by separately considering points where $V_1 \geq 1$ and $V_1 \leq 1$.

Then the strict Lyapunov function we get is

$$S(\tilde{x}, \tilde{y}) = V_2(\tilde{x}, \tilde{y}) + \left[p_1(V_1(\tilde{x}, \tilde{y})) + 1\right]V_1(\tilde{x}, \tilde{y}) + \textstyle\int_0^{V_1(\tilde{x}, \tilde{y})} \phi_1(r)\, dr. \tag{5.60}$$

In fact, $S(\tilde{x}, \tilde{y}) \geq V_1(\tilde{x}, \tilde{y})$ and $L_f S(\tilde{x}, \tilde{y}) \leq -\frac{1}{2}[\mathcal{N}_1(\tilde{x}, \tilde{y}) + \mathcal{N}_2(\tilde{x}, \tilde{y})]$ are satisfied everywhere.

5.5.2 Robustness to Uncertainty

We can use our strict Lyapunov function constructions to quantify the effects of uncertainty in the Lotka-Volterra dynamics. For simplicity, we only consider additive uncertainty in the death rate Δ for the predator. Using the coordinate change and constants (5.45), this means that we replace the constant d with $d + \mathbf{u}$ in the dynamics (5.46), where $\mathbf{u} : [0, \infty) \to \mathbb{R}$ is a measurable essentially bounded uncertainty, and where the constant $d > 0$ now represents the nominal value of the parameter. Later, we impose bounds on the allowable values for $|\mathbf{u}|_\infty$. We continue to use d in the formulas (5.47) for x_* and y_*; we do not introduce uncertainty in the equilibrium values.

We first define an appropriately restricted state space for the dynamics. Along the trajectories of (5.46), with d replaced by $d + \mathbf{u}$, we have $\dot{x} + \dot{y} = x(1 - x) - (d + \mathbf{u})y$. Hence, if $|\mathbf{u}|_\infty \leq d/2$, then we get $\dot{x} + \dot{y} < 0$ when $x + y > 1 + \frac{2}{d}$ (by separately considering the cases $x > 1$ and $x \leq 1$). Therefore, we restrict to disturbances satisfying $|\mathbf{u}|_\infty \leq d/2$ and the forward invariant set $\mathcal{S} = \{(x, y) \in (0, \infty)^2 : x + y \leq \mathcal{B}\}$ containing (x_*, y_*), where

$$\mathcal{B} = 1 + \frac{2}{d} + y_* \,. \tag{5.61}$$

The corresponding perturbed error dynamics is

$$\begin{cases} \dot{\tilde{x}} = -[\tilde{x} + \alpha\tilde{y}](\tilde{x} + x_*) \\ \dot{\tilde{y}} = \alpha\tilde{x}(\tilde{y} + y_*) - \mathbf{u}y \end{cases} \tag{5.62}$$

which we view as having the state space $\mathcal{X}^\flat = \{(\tilde{x}, \tilde{y}) : (x, y) \in \mathcal{S}\}$ and a control set U we will specify.

To account for the restricted state space, we use the following definitions. Given an open subset $\mathcal{D}$ of a Euclidean space that contains the origin, we say that a positive definite function $\bar{\alpha} : \mathcal{D} \to [0, \infty)$ is a *modulus with respect to* $\mathcal{D}$ provided $\bar{\alpha}(p) \to +\infty$ as $|p| \to +\infty$ or as $\mathrm{dist}(p, \partial\mathcal{D}) \to 0$ (with p remaining in $\mathcal{D}$). We say that (5.62) is ISS with respect to $\mathbf{u}$ provided there exist functions $\beta \in \mathcal{KL}$ and $\gamma \in \mathcal{K}_\infty$, and a modulus with respect to $\mathcal{X}^0 \doteq (-x_*, \infty) \times (-y_*, \infty)$, such that for each disturbance $\mathbf{u} : [0, \infty) \to U$ and each trajectory $(\tilde{x}, \tilde{y}) : [0, \infty) \to \mathcal{X}^\flat$ of (5.62) corresponding to $\mathbf{u}$, we have

$$|(\tilde{x}, \tilde{y})(t)| \leq \beta\big(\bar{\alpha}((\tilde{x}, \tilde{y})(0)), t\big) + \gamma(|\mathbf{u}|_\infty) \;\; \forall t \geq 0. \tag{5.63}$$

We define iISS for (5.62) in an analogous way; see Remark 5.1 below.

To simplify the statements of our results, we use the constants

$$K_0 = 2\left[\frac{(3+\alpha)^2}{2} + \alpha^2\right]\mathcal{B}^2, \;\; \theta = \min\left\{\frac{K_0 x_*^2}{8}, \frac{K_0 x_*^2 y_*^2 \alpha^2}{8(x_* + 2\sqrt{K_0})^2}\right\},$$

$$K = \mathcal{B}^2 \max\left\{(3+\alpha)^2 + 2\alpha^2, 2\max\{9, 3\alpha^2\}\right\},$$

$$\hat{K} = \frac{\min\left\{32x_*, x_*^2\alpha^2 y_*\right\}}{16[K + \mathcal{B}^2\max\{9, 3\alpha^2\}]}, \;\; \text{and } \bar{U} = \frac{\min\{\hat{K}, \theta\}}{4(\alpha\mathcal{B}^3 + K\mathcal{B})}.$$

We continue to use the functions V_1 and V_2 from the preceding subsection. The following is shown in [105] (but see Sect. 5.5.3 for a specific numerical example):

Theorem 5.3. *The system (5.62) is ISS with respect to disturbances* $\mathbf{u}$ *valued in the control set* $\bar{U}\mathcal{B}_1$*, and iISS with respect to disturbances* $\mathbf{u}$ *valued in* $\frac{d}{2}\mathcal{B}_1$.

The proof of Theorem 5.3 entails showing that

$$\mathcal{U}_K(\tilde{x}, \tilde{y}) = V_2(\tilde{x}, \tilde{y}) + KV_1(\tilde{x}, \tilde{y}) \tag{5.64}$$

is an iISS Lyapunov function for (5.62) when the disturbance $\mathbf{u}$ is valued in $\frac{d}{2}\mathcal{B}_1$, and that

$$\overline{\mathcal{U}_K}(\tilde{x}, \tilde{y}) = \mathcal{U}_K(\tilde{x}, \tilde{y})e^{\mathcal{U}_K(\tilde{x}, \tilde{y})} \tag{5.65}$$

is an ISS Lyapunov function for (5.62) when $\mathbf{u}$ is valued in $\bar{U}\mathcal{B}_1$, where V_1 and V_2 are as defined in Sect. 5.5. It leads to the decay estimates

$$\dot{\mathcal{U}}_K \leq -\mho\frac{\mathcal{U}_K(\tilde{x}, \tilde{y})}{1 + \mathcal{U}_K(\tilde{x}, \tilde{y})} + \overline{\mathcal{B}}|\mathbf{u}|\;, \tag{5.66}$$

where

$$\mho = \min\{\widehat{K}, \theta\}$$

(which implies that $\mathcal{U}_K$ is an iISS Lyapunov function for the Lotka-Volterra error dynamics (5.62)) when the disturbance $\mathbf{u}$ satisfies the less stringent bound $|\mathbf{u}|_\infty \leq \frac{d}{2}$ and then

$$\dot{\overline{\mathcal{U}_K}} \leq -\tfrac{\mho}{4}\overline{\mathcal{U}_K}(\tilde{x},\tilde{y}) + \overline{\mathcal{B}}|\mathbf{u}| \ . \tag{5.67}$$

along the trajectories of (5.62) when $\mathbf{u}$ is valued in $\bar{U}\mathcal{B}_1$, which gives the ISS estimate. For a summary of the robustness analysis, see Appendix A.4.

Remark 5.1. A slight variant of the iISS arguments from [8] in conjunction with (5.66) and the growth properties of $\mathcal{U}_K$ can be used to show that there exist functions $\beta \in \mathcal{KL}$ and $\gamma \in \mathcal{K}_\infty$, a constant $\bar{G} > 0$, and a modulus with respect to $\mathcal{X}^0$, such that for each disturbance $\mathbf{u} : [0,\infty) \to [-d/2, d/2]$ and each trajectory $(\tilde{x},\tilde{y}) : [0,\infty) \to \mathcal{X}^b$ of (5.62) corresponding to $\mathbf{u}$, we have

$$\gamma(|(\tilde{x},\tilde{y})(t)|) \leq \beta(\bar{\alpha}((\tilde{x},\tilde{y})(0)), t) + \bar{G}\int_0^t |\mathbf{u}(r)|\mathrm{d}r \ \ \forall t \geq 0. \tag{5.68}$$

This is less stringent than the ISS condition (5.63) because it allows the possibility that a bounded (but non-integrable) disturbance $\mathbf{u}$ could give rise to an unbounded trajectory. However, if $\mathbf{u}$ is integrable, then (5.68) guarantees boundedness of the trajectories, and it also quantifies the effects of the disturbance. We next illustrate these ideas in simulations.

5.5.3 Numerical Validation

To illustrate our findings, we simulated the dynamics (5.62) using the parameter values

$$\alpha = 2, \ \ d = 1, \ \ x_* = 0.5, \ \ \text{and} \ \ y_* = 0.25 \ , \tag{5.69}$$

corresponding to the parameter choices

$$a = \gamma = \beta = \Delta = 0.5 \ \ \text{and} \ \ L = 2 \tag{5.70}$$

in the original model. Hence, the dynamics are iISS with respect to disturbances that are bounded by 0.5. We chose the disturbance $\mathbf{u}(t) = 0.49e^{-t}$. In Figs. 5.1 and 5.2, we plot the corresponding levels of predator population ζ and the prey population χ, which are related to x and y in terms of the coordinate changes (5.45).

If

$$x(t) \to x_* = 0.5 \ \ \text{and} \ \ y(t) \to y_* = 0.25,$$

then the coordinate changes (5.45) give

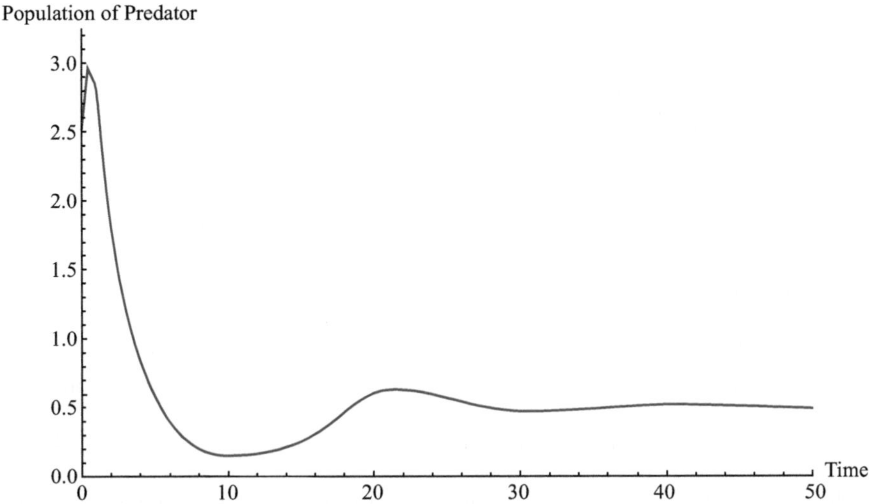

Fig. 5.1 Population of predator ζ with parameters (5.70) and $\mathbf{u}(t) = 0.49e^{-t}$

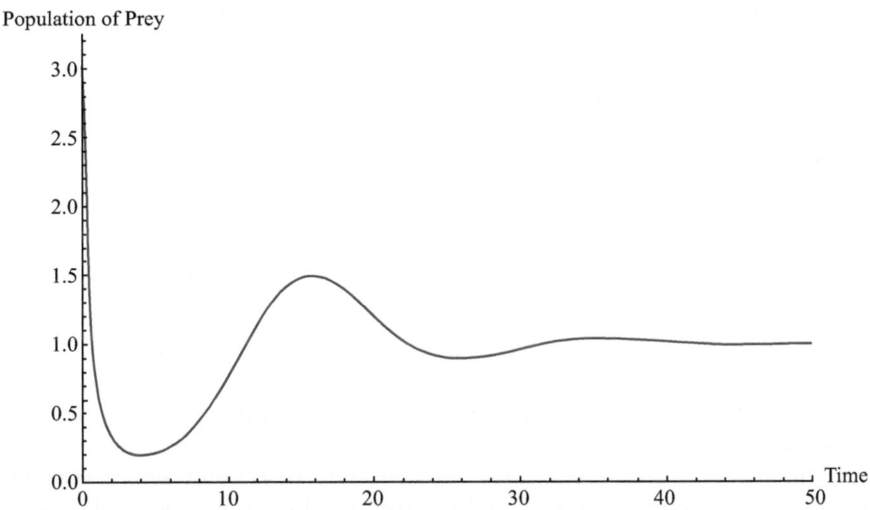

Fig. 5.2 Population of prey χ with parameters (5.70) and $\mathbf{u}(t) = 0.49e^{-t}$

$$\zeta(t) \to 0.25\frac{\beta L}{a} = 0.5 \quad \text{and} \quad \chi(t) \to 0.5L = 1, \tag{5.71}$$

which is in fact the behavior we see in the figures. This shows the robustness of the convergence in the face of the disturbance $\mathbf{u}$.

5.6 Comments

Several authors have studied ways to construct strict Lyapunov under appropriate conditions on the iterated Lie derivatives, or using non-strict Lyapunov functions. Two significant results in this direction are [5, 41]. The results of [5] deal with ISS, and [41] with controller design by using CLFs for systems that satisfy Jurdjevic-Quinn Conditions. The construction in [5] uses a weak Lyapunov function and an auxiliary Lyapunov function V_2 that satisfies certain detectability properties of the system with respect to an appropriate output $h(x)$.

More precisely, [5] assumes that there are two positive definite radially unbounded functions V_1 and V_2 and functions $\alpha_1, \alpha_2, \gamma \in \mathcal{K}_\infty$ satisfying

$$\dot{V}_1 \leq -\alpha_1(|y|) \text{ and } \dot{V}_2 \leq -\alpha_2(|x|) + \gamma(|y|) , \tag{5.72}$$

for all $x \in \mathbb{R}^n$, where $y = h(x)$. Note that V_1 in (5.72) is typically a weak Lyapunov function since $|h(x)|$ is often positive semi-definite. The function V_2 in (5.72) is an output-to-state Lyapunov function [73] that characterizes a particular form of detectability of x from the output y. The strong Lyapunov function in [5] then takes the form

$$U(x) = V_1(x) + \rho(V_2(x)),$$

where ρ is a suitable $\mathcal{K}_\infty$ function.

The main difference between our approach from Theorem 5.1 and [5] is that our conditions appear to be stronger but easier to check than those in [5]. While very general, the challenge in applying [5] stems from the need to find V_2. The auxiliary function can be found in certain useful cases, but to our knowledge there is no general procedure for finding V_2 in the context of [5]. This gives a possible advantage in checking the iterated Lie derivative condition from Theorem 5.1 and then using our construction (5.2). Another difference between [5] and our methods is that our auxiliary functions are not required to be radially unbounded or everywhere positive.

By contrast, the strict Lyapunov construction of [41] only uses the given non-strict Lyapunov function V_1 and the iterated Lie derivatives of V_1 along solutions of an auxiliary system with a scaled vector field. The results in [41] seem more direct than those of [5], but the method of [41] is in general only applicable to homogenous systems. (The translational oscillator with rotating actuator or TORA example in [41] is inhomogeneous, but [41] does not give a systematic method for inhomogenous systems.) To our knowledge, [102] provides the first general construction for CLFs for general classes of Jurdjevic Quinn systems that do not necessarily satisfy the homogeneity conditions from [41].

Conditions 1. and 2. from Theorem 5.1 agree with the assumptions from the strict Lyapunov function construction in [110, Theorem 3.1]. However, our proof of Theorem 5.1 is simpler than the arguments used in [110]. The construction in [110, Theorem 3.1] proceeds by finding a non-increasing function $\lambda : [0, \infty) \to (0, \infty)$ such that the function

$$U(x) = V(x)\left[1 + V(x) - \sum_{i=1}^{\ell-1} L^i_{f_\lambda} V(x) \cdot \left(L^{i+1}_{f_\lambda} V(x)\right)^{3^i}\right] \tag{5.73}$$

is a strict Lyapunov function for the system (5.1), where

$$f_\lambda(x) \doteq \lambda(V(x))f(x).$$

In [86, Sect. 3.3], conditions similar to Assumption 5.1 were used to conclude asymptotic stability of systems which admit a non-strict Lyapunov function, via an extension of Matrosov's Theorem. However, no strict Lyapunov functions were constructed in this earlier work.

It is possible to extend Theorem 5.1 to periodic time-varying systems, in which case we instead take

$$a_1(t, x) = -[V_t(t, x) + V_x(t, x)f(t, x)]$$

and $a_i = -\dot{a}_{i-1}$ for all $i \geq 2$ and consider the non-negative function

$$\sum_{i=2}^{\ell} a_i^2(t, x) + a_1(t, x),$$

which is allowed to be zero for some $x \neq 0$ on some intervals of positive length; see [104]. Section 5.4 is based on [104].

Our strict Lyapunov function construction for the Lotka-Volterra system is based on [104]. The Lotka-Volterra model is used extensively in mathematical biology. See [58, 79] for an extensive analysis of this model and generalizations to several predators. While there are many Lyapunov constructions for Lotka-Volterra models available (based on computing the LaSalle Invariant Set), to the best of our knowledge, the result we gave in this chapter is original and significant because we provide a *global strict Lyapunov function.*

Part III
Time-Varying Case

Chapter 6
Strictification: Basic Results

Abstract In the last three chapters, we gave general methods for constructing strict Lyapunov functions for time-invariant systems. Several of these methods have analogs for time-varying systems. In general, these involve replacing the negative semi-definite function of the state in the right side of the non-strict Lyapunov decay condition with a *product* of a negative semi-definite function of the state and a suitable time-varying parameter. We assume that the time-varying parameter satisfies a persistency of excitation (PE) condition. The challenge is then to transform non-strict Lyapunov functions satisfying this more complicated decay condition into explicit strict Lyapunov functions. In this chapter, we provide methods for solving this and related problems, including the construction of ISS Lyapunov functions for time-varying systems. We apply our work to stabilization problems for rotating rigid bodies and underactuated ships.

6.1 Background

As we noted in previous chapters, there are many situations where it is very helpful to have explicit constructions for global strict Lyapunov functions. For example, Lyapunov functions make it possible to estimate the basins of attraction for attractive equilibria, and one often needs strict Lyapunov functions for the subsystems in backstepping and forwarding. Also, a CLF satisfying the small control property can be used with Sontag's Universal Formula to obtain an explicit asymptotically stabilizing feedback that is optimal when the CLF is viewed as the value function [149, 158]. Moreover, strict Lyapunov functions are a key tool for robustness analysis.

In general, it is more difficult to construct strict Lyapunov functions for time-varying systems than it is for time-invariant systems. In fact, when the usual non-strict Lyapunov function construction methods are used for time-varying systems, the right hand sides of the Lyapunov decay condition typically end up being identically zero for some values of t, which precludes the use of the usual time-invariant strictification techniques. A simple one-dimensional example of this phenomenon is the case where $\dot{x} = -\sin^2(t)x$, where the obvious candidate Lyapunov function $V(x) = |x|^2$ gives $\dot{V} \leq -2\sin^2(t)x^2$ along trajectories. On the other hand, the right hand side is only zero for times t in the "thin" set $\{k\pi : k \in \mathbb{Z}\}$, so it is reasonable to expect the system to exhibit some stability properties. We will make this thinness notion precise later in the chapter, using the idea of PE.

This chapter provides systematic methods for transforming non-strict Lyapunov functions for time-varying systems into strict Lyapunov functions. We refer to this transformation process as *(time-varying) strictification.* The idea of using non-strict Lyapunov functions to study stability has been pursued by many authors. For example, we saw in Sect. 2.2.2 how to use time-varying versions of the LaSalle Invariance Principle in conjunction with non-strict Lyapunov functions to prove stability; see [13, 77] for more details. In [119], Narendra and Annaswamy proved that if a (possibly non-periodic) system

$$\dot{x} = f(t,x), \quad x \in \mathbb{R}^n \tag{6.1}$$

admits a uniformly proper and positive definite function V, a constant $T > 0$, and an increasing continuous function $\gamma : [0,\infty) \to [0,\infty)$ such that

1. $\frac{\partial V}{\partial t}(t,x(t)) + \frac{\partial V}{\partial x}(t,x(t))f(t,x(t)) \leq 0$; and
2. $V(t+T, x(t+T)) - V(t,x(t)) \leq -\gamma(|x(t)|)$

hold along all trajectories $x(t)$ for (6.1) for all $t \geq 0$, then (6.1) is uniformly asymptotically stable. See [1, 2] for generalizations that relax the requirement that the time derivative of V along the trajectories is negative semi-definite.

An alternative approach was pursued in [98]. The main result of [98] constructs an explicit global strict Lyapunov function for (6.1) provided one knows a storage function V, a periodic function q, and an appropriate non-negative function $\mathcal{W}(q(t),x)$ such that

$$\dot{V}(t,x) \doteq \frac{\partial V}{\partial t}(t,x) + \frac{\partial V}{\partial x}(t,x)f(t,x) \leq -\mathcal{W}(q(t),x) \tag{6.2}$$

for all $x \in \mathbb{R}^n$ and $t \geq 0$. It assumes that $\mathcal{W}(q(t),x)$ is positive definite in x for all t in suitable non-empty open intervals of time; see Sect. 6.3 for the precise hypotheses. Oftentimes, the function $\mathcal{W}(q(t),x)$ takes the form

$$\mathcal{W}(q(t),x) = p(t)W(x)\,, \tag{6.3}$$

where the non-negative bounded continuous function p is a PE parameter, meaning: *There are positive constants δ and T such that*

$$\int_t^{t+T} p(r)\mathrm{d}r \geq \delta \tag{6.4}$$

for all $t \in \mathbb{R}$. Throughout the sequel, we let $\mathcal{P}(T, \delta, \bar{p})$ denote the set of all continuous functions $p : \mathbb{R} \to [0, \infty)$ that satisfy (6.4) for some positive constants δ and T and $p(t) \leq \bar{p}$ for all $t \in \mathbb{R}$. We also set $\mathcal{P} = \cup\{\mathcal{P}(T, \delta, \bar{p}) : T, \delta, \bar{p} > 0\}$.

When a storage function satisfies an estimate of the form (6.2) with $\mathcal{W}$ having the form (6.3), standard arguments imply that the system is UGAS; see Remark 6.4. However, it is not clear how to construct global strict Lyapunov functions for (6.1) in this case. Conditions (6.2) and (6.3) are natural, because they are satisfied in many important cases. For example, consider the class of nonholonomic systems in chained form, and suppose that we wish to track a given periodic trajectory. In this case, applying the main result of [65] often gives a storage function satisfying the preceding conditions.

6.2 Motivating Example

Consider the chained form system

$$\begin{cases} \dot{x}_1 = u_1 \\ \dot{x}_2 = u_2 \\ \dot{x}_3 = x_2 u_1 \end{cases} \tag{6.5}$$

with state space $\mathcal{X} = \mathbb{R}^3$ and input space $U = \mathbb{R}^2$. We wish to find a feedback for (6.5) that stabilizes the trajectories of (6.5) to the reference trajectory $x_r(t) = (-\cos(t), 0, 0)$, in such a way that the dynamics for the error $x_e = (x_{1e}, x_{2e}, x_{3e}) = x - x_r$ is UGAS and locally exponentially stable to 0. We also want an explicit global strict Lyapunov function for the x_e dynamics.

To this end, we pick the feedback component

$$u_1 = \sin(t) - x_{1e} \tag{6.6}$$

to obtain the error dynamics

$$\begin{cases} \dot{x}_{1e} = -x_{1e} \\ \dot{x}_{2e} = u_2 \\ \dot{x}_{3e} = x_{2e}(\sin(t) - x_{1e}) . \end{cases} \tag{6.7}$$

We first consider the reduced system

$$\begin{cases} \dot{x}_{3e} = x_{2e} \sin(t) \\ \dot{x}_{2e} = u_2 \, . \end{cases} \tag{6.8}$$

We apply backstepping with the new coordinate $z_{2e} = a \sin^3(t) x_{3e} + x_{2e}$ where $a > 0$ is any constant, and then use the feedback component

$$\begin{aligned} u_2 = &-3a\sin^2(t)\cos(t)x_{3e} - a\sin^3(t)\left[-a\sin^4(t)x_{3e} + z_{2e}\sin(t)\right] \\ &-az_{2e} - \sin(t)x_{3e} \end{aligned} \tag{6.9}$$

to get the closed-loop system

$$\begin{cases} \dot{x}_{3e} = -a\sin^4(t)x_{3e} + z_{2e}\sin(t) \\ \dot{z}_{2e} = -az_{2e} - \sin(t)x_{3e} \; . \end{cases} \tag{6.10}$$

(We discuss backtepping in the next chapter.) The time derivative of

$$V(t, x_{3e}, z_{2e}) \; = \; \tfrac{1}{2}\left[x_{3e}^2 + z_{2e}^2\right] \tag{6.11}$$

along the trajectories of (6.10) satisfies

$$\dot{V}(t, x_{3e}, z_{2e}) \; = \; -a\sin^4(t)x_{3e}^2 - az_{2e}^2 \; \leq \; -2a\sin^4(t)V(t, x_{3e}, z_{2e}) \; . \tag{6.12}$$

This is the special case of the decay conditions (6.2) and (6.3) with time-varying parameter $p(t) = 2\sin^4(t)$. However, it is not clear how to transform (6.11) into a strict Lyapunov function for (6.10).

One consequence of the results of this chapter is that when $a = 1$, (6.10) admits the global strict Lyapunov function

$$U_1(t, x_{3e}, z_{2e}) \; = \; \left[4\pi(\pi+1) + P(t)\right]V(t, x_{3e}, z_{2e}) \tag{6.13}$$

where

$$P(t) \; = \; 2\int_{t-\pi}^{t}\int_{s}^{t}\sin^4(r)\mathrm{d}r\mathrm{d}s; \tag{6.14}$$

see Theorem 6.1 and Remark 6.2 with the choices $p(t) = 2\sin^4(t)$ and $T = \pi$, and see Sect. 6.4 for other strict Lyapunov function constructions for (6.10) that use different choices of a to get rate of convergence information. In fact, we can check directly that

$$\dot{U}_1(t, x_{3e}, z_{2e}) \leq -\tfrac{3\pi}{8}V(t, x_{3e}, z_{2e}) \tag{6.15}$$

along all trajectories of (6.10). Also, U_1 is a storage function.

Let us use (6.13) to get a strict Lyapunov function for the full system (6.7) in closed-loop with the feedback (6.9). We continue to take the tuning parameter $a = 1$. Changing variables as before and again using the feedback (6.9) transforms (6.7) into

$$\begin{cases} \dot{x}_{1e} = -x_{1e} \\ \dot{x}_{3e} = -\sin^4(t)x_{3e} + z_{2e}\sin(t) - x_{1e}x_{2e} \\ \dot{z}_{2e} = -z_{2e} - \sin(t)x_{3e} - x_{1e}x_{2e}\sin^3(t) \; . \end{cases} \tag{6.16}$$

Along the trajectories of (6.16), the time derivative of $U_1(t, x_{3e}, z_{2e})$ satisfies

$$\begin{aligned} \dot{U}_1(t, x_{3e}, z_{2e}) \leq &-\tfrac{3\pi}{8}V(t, x_{3e}, z_{2e}) \\ &+2\pi(3\pi+2)\left[|x_{3e}| + |z_{2e}|\right]|x_{2e}||x_{1e}| \; . \end{aligned} \tag{6.17}$$

One readily checks that

$$\begin{aligned}2\pi\left\{|x_{3e}|+|z_{2e}|\right\}\left\{|x_{2e}||x_{1e}|\right\} &\le \tfrac{\pi}{8(3\pi+2)}V(t,x_{3e},z_{2e})+32\pi(3\pi+2)x_{2e}^2x_{1e}^2\\ &\le \tfrac{\pi}{8(3\pi+2)}V(t,x_{3e},z_{2e})\\ &\quad +128\pi(3\pi+2)V(t,x_{3e},z_{2e})x_{1e}^2,\end{aligned}$$

by applying the triangular inequality $pq \le cp^2 + \frac{1}{4c}q^2$ where p and q are the terms in braces and $c > 0$ is an appropriate constant, and then using the facts that $\{|x_{3e}|+|z_{2e}|\}^2 \le 4V(t,x_{3e},z_{2e})$ and $x_{2e}^2 \le 2x_{3e}^2 + 2z_{2e}^2 = 4V(t,x_{3e},z_{2e})$ everywhere. This and (6.17) give the global inequality

$$\dot U_1(t,x_{3e},z_{2e}) \;\le\; -\frac{\pi}{4}V(t,x_{3e},z_{2e}) + 128\pi(3\pi+2)^2V(t,x_{3e},z_{2e})x_{1e}^2\,.$$

Since $V(t,x_{3e},z_{2e}) \le \frac{1}{2\pi}U_1(t,x_{3e},z_{2e})$ everywhere, it follows that the time derivative of

$$U_2(t,x_{3e},z_{2e},x_{1e}) \;=\; \ln(1+U_1(t,x_{3e},z_{2e})) + \left[64(3\pi+2)^2+0.5\right]x_{1e}^2$$

along the trajectories of (6.16) satisfies

$$\dot U_2(t,x_{3e},z_{2e},x_{1e}) \;\le\; -\frac{\pi}{4}\frac{V(t,x_{3e},z_{2e})}{1+U_1(t,x_{3e},z_{2e})} - x_{1e}^2\,. \tag{6.18}$$

Since U_2 is also a storage function, it follows from (6.18) that U_2 is a strict Lyapunov function for (6.7) in closed-loop with the feedback (6.9). Also, in a neighborhood of the origin, the right side of (6.18) is bounded from above by a negative definite quadratic function of $(x_{1e},x_{3,e},z_{2e})$. Since U_2 is bounded from above and below by positive definite quadratic functions of $(x_{1e},x_{3,e},z_{2e})$ near the origin, this shows that the feedbacks (6.6) and (6.9) globally uniformly and locally exponentially stabilize the reference trajectory $x_r(t)$. We turn next to a general result that leads to the strict Lyapunov function construction (6.13) and (6.14) as a special case. As before, all (in)equalities to follow should be understood to hold globally unless otherwise indicated.

6.3 Time-Varying Strictification Theorem

Statement of Theorem

Throughout the sequel, we understand $\gamma'(0)$ for functions γ defined on $[0,\infty)$ as a one sided derivative, and continuity of γ' at 0 to be one sided continuity. The main result from [98] is a general method for converting non-strict

Lyapunov functions, satisfying decay conditions of the form (6.2)-(6.3), into global strict Lyapunov functions for the corresponding time-varying systems (6.1). It assumes the following:

Assumption 6.1 *We are given a C^1 storage function V for (6.1), a continuous positive definite function $W(x)$, and a bounded continuous function $p : \mathbb{R} \to [0, \infty)$ such that*

$$\frac{\partial V}{\partial t}(t,x) + \frac{\partial V}{\partial x}(t,x)f(t,x) \;\leq\; -p(t)W(x) \tag{6.19}$$

hold for all $t \geq 0$ and $x \in \mathbb{R}^n$.

Assumption 6.2 *The function $p : \mathbb{R} \to [0, \infty)$ from Assumption 6.1 satisfies the PE condition (6.4) for some constants $\delta, T > 0$.*

We continue to assume that f satisfies the regularity assumptions from Chap. 1. Notice that any continuous periodic function $p : \mathbb{R} \to [0, \infty)$ that is not identically zero admits constants $T, \delta > 0$ satisfying (6.4). Assumption 6.2 also allows non-periodic p with arbitrarily large null sets, e.g., for fixed $r > 0$, set

$$p_r(t) = \max\left\{0, \frac{t}{1+|t|}\sin^3\left(\frac{t}{r}\right)\right\}.$$

Taking large r gives arbitrarily large null sets.

Theorem 6.1. *Consider the system (6.1) with state space $\mathcal{X} = \mathbb{R}^n$. Let Assumptions 6.1 and 6.2 hold. Then one can explicitly construct a function $\Gamma \in C^1 \cap \mathcal{K}_\infty$ and a positive definite C^1 function λ such that*

$$V^\sharp(t,x) \;=\; \Gamma\big(V(t,x)\big) + \lambda\big(V(t,x)\big)\int_{t-T}^{t}\int_{s}^{t} p(r)\mathrm{d}r\, \mathrm{d}s \tag{6.20}$$

is a strict Lyapunov function for the system (6.1). If V and p both have period T in t, then so does (6.20).

Proof of Theorem 6.1

We first find an everywhere positive continuous function γ and $\lambda \in \mathcal{K}_\infty \cap C^1$ such that

$$\frac{1}{4}\gamma\big(V(t,x)\big)W(x) \geq T\lambda\big(V(t,x)\big) \tag{6.21}$$

everywhere. For example, first choose $\lambda \in \mathcal{K}_\infty \cap C^1$ such that $T\lambda(\alpha_2(|x|)) \leq W(x)$ on the bounded set $\alpha_1^{-1}(1)\mathcal{B}_n$, where $\alpha_1, \alpha_2 \in \mathcal{K}_\infty$ are such that $\alpha_1(|x|) \leq V(t,x) \leq \alpha_2(|x|)$ for all $t \geq 0$ and $x \in \mathbb{R}^n$, and then choose

$$\gamma(r) = \begin{cases} \sup_{\{x:0<\alpha_1(|x|)\le r\}} \left\{ 4\max\left\{1, \frac{T\lambda(\alpha_2(|x|))}{W(x)}\right\}\right\}, & r \neq 0 \\ 4, & r = 0 . \end{cases}$$

Next note that

$$\int_{t-T}^{t}\int_{s}^{t} p(r)\mathrm{d}r\,\mathrm{d}s \;\le\; T^2\bar{p} \;=:\; P_M \quad \forall t \ge 0, \tag{6.22}$$

where $\bar{p}$ is a global bound of the PE function $p(t)$. It follows that the function $\Gamma(r) = 2P_M\lambda(r) + \int_0^r \gamma(s)\mathrm{d}s$ satisfies the following conditions for all $v \ge 0$, $t \ge 0$, and $x \in \mathbb{R}^n$:

$$\Gamma(v) \;\ge\; 2P_M\lambda(v) \;\text{ and} \tag{6.23}$$

$$\frac{1}{4}\Gamma'\big(V(t,x)\big)W(x) \;\ge\; T\lambda\big(V(t,x)\big) \; . \tag{6.24}$$

We now show that (6.20) with the preceding choices of Γ and λ satisfies the requirements of the theorem.

To check that $V^\sharp$ is a storage function, first note that our choice of P_M in (6.22) gives the inequalities

$$V^\sharp(t,x) \le \Gamma\big(V(t,x)\big) + P_M\lambda\big(V(t,x)\big). \tag{6.25}$$

Combining (6.25) with (6.23) gives

$$\Gamma\big(V(t,x)\big) \;\le\; V^\sharp(t,x) \;\le\; \frac{3}{2}\Gamma\big(V(t,x)\big) \; . \tag{6.26}$$

Hence, $V^\sharp$ is a storage function because V is a storage function and $\Gamma \in \mathcal{K}_\infty \cap C^1$.

The time derivative of $V^\sharp(t,x)$ along the trajectories of (6.1) satisfies

$$\begin{aligned} \dot{V}^\sharp(t,x) &= \Gamma'\big(V(t,x)\big)\dot{V}(t,x) + \left[-\int_{t-T}^{t} p(s)\mathrm{d}s + Tp(t)\right]\lambda\big(V(t,x)\big) \\ &\quad + P(t)\lambda'\big(V(t,x)\big)\dot{V}(t,x) \; , \end{aligned} \tag{6.27}$$

where

$$P(t) = \int_{t-T}^{t}\int_{s}^{t} p(r)\mathrm{d}r\,\mathrm{d}s. \tag{6.28}$$

Since we have $P(t) \ge 0$, $\lambda'(V(t,x)) \ge 0$, and $\dot{V}(t,x) \le 0$ everywhere, condition (6.27) gives

$$\begin{aligned} \dot{V}^\sharp(t,x) &\le -\Gamma'(V(t,x))p(t)W(x) \\ &\quad + \left[-\int_{t-T}^{t} p(s)\mathrm{d}s + Tp(t)\right]\lambda\big(V(t,x)\big) \; . \end{aligned} \tag{6.29}$$

Combined with (6.24), this gives

$$\dot{V}^\sharp(t,x) \leq -\tfrac{3}{4}\Gamma'\big(V(t,x)\big)p(t)W(x) - \left(\int_{t-T}^{t} p(s)\mathrm{d}s\right)\lambda\big(V(t,x)\big)\ . \qquad (6.30)$$

Since λ and Γ are both increasing, we conclude from Assumption 6.2 that

$$\dot{V}^\sharp(t,x) \ \leq\ -\delta\lambda\big(\alpha_1(|x|)\big)\ <\ 0\ \forall x \neq 0\ , \qquad (6.31)$$

which is the desired strict Lyapunov function decay condition. □

Remarks on Theorem 6.1

Remark 6.1. If $p(\cdot)$ is a bounded continuous everywhere non-negative function for which there exist constants $T_p > 0$ and $c_p > 0$ such that

$$p(t) + p(t - T_p)\ \geq\ c_p$$

for all $t \in \mathbb{R}$, then the formula (6.20) can be simplified by replacing the double integral

$$\int_{t-T}^{t}\int_{s}^{t} p(r)\mathrm{d}r\,\mathrm{d}s$$

in the formula (6.20) for the strict Lyapunov function with

$$\int_{t-T_p}^{t} p(r)\mathrm{d}r.$$

To see why, choose $\lambda \in C^1 \cap \mathcal{K}_\infty$ and a C^1 everywhere positive definite function γ such that (6.21) is satisfied with $T = 1$, and define Γ as before. Then the new formula for $V^\sharp$ gives

$$\begin{aligned} \dot{V}^\sharp(t,x) &\leq p(t)\big[-\Gamma'(V(t,x))W(x) + \lambda(V(t,x))\big] \\ &\quad -p(t-T_p)\lambda\big(V(t,x)\big) \\ &\leq -3p(t)\Gamma'\big(V(t,x)\big)W(x) - p(t-T_p)\lambda\big(V(t,x)\big). \end{aligned} \qquad (6.32)$$

We easily deduce the negative definiteness of the right side of (6.32) from the facts that $p(t) + p(t - T_p) \geq c_p$ for all $t \in \mathbb{R}$ and $\Gamma'(r) > 0$ for all $r > 0$.

Remark 6.2. The formula (6.20) can be simplified when $V(t,x) \equiv W(x)$. In fact, in that case, we can take

$$\lambda(r) = r,\ \ \gamma(r) \equiv 4T,\ \ \text{and}\ \ \Gamma(r) = (2P_M + 4T)r.$$

More generally, we can take $\Gamma(r) = r$ if W is proper and positive definite; see the first part of the proof of Theorem 6.2, specialized to the case where the disturbance is identically zero.

Remark 6.3. As noted in [98], we can extend Theorem 6.1 to cases where the non-strict Lyapunov function V satisfies the more general decay condition (6.2). One way to do this is to assume the following additional conditions:

1. V, q, and f are all periodic in t with the same period T;
2. $V \in C^1$ is a storage function; and
3. $\mathcal{W}$ is everywhere non-negative, and there are two constants τ_1 and τ_2 with $0 \le \tau_1 < \tau_2 \le T$ and a positive definite function W so that $\mathcal{W}(q(t), x) \ge W(x)$ for all $t \in [\tau_1, \tau_2]$ and all $x \in \mathbb{R}^n$.

Then we can again construct an explicit strict Lyapunov function for (6.1). This can be done by applying Theorem 6.1 using any continuous periodic function $p(t)$ of period T (which in general will be different from q) such that

$$\begin{array}{l} \int_0^t p(s)\mathrm{d}s > 0, \;\; p(r) \in (0,1] \;\; \forall r \in [\tau_1, \tau_2], \;\; \text{and} \\ p(r) \equiv 0 \text{ on } [0,T] \setminus [\tau_1, \tau_2]. \end{array}$$

The proof is the same as before.

Remark 6.4. Notice that Theorem 6.1 goes beyond establishing that the system (6.1) is UGAS, because it constructs an explicit strict global Lyapunov function for the system. If we merely want to establish UGAS under the assumptions of Theorem 6.1, then we can use the following argument from [139], assuming there is a $\mathcal{K}$ function α such that $W(x) \ge \alpha(|x|)$ for all $x \in \mathbb{R}^n$. The details are as follows. By arguing as in [139, Proposition 13], we can find a function $\rho \in C^1 \cap \mathcal{K}_\infty$ such that the function $\bar{V} \doteq \rho(V)$ satisfies

$$\bar{V}_t(t,x) + \bar{V}_x(t,x) f(t,x) \;\le\; -p(t)\bar{V}(t,x) \tag{6.33}$$

everywhere. We now argue as in [64]. Letting $t \ge t_0 \ge 0$ and k be the largest integer $kT \le t$, we get

$$\int_{t_0}^{t_0+t} p(s)\mathrm{d}s \;\ge\; \int_{t_0}^{t_0+kT} p(s)\mathrm{d}s \;\ge\; k\delta \;\ge\; \left(\frac{t}{T} - 1\right)\delta\,.$$

Integrating (6.33) therefore gives

$$\bar{V}\big(t+t_0, x(t+t_0, t_0, x_0)\big) \;\le\; \bar{V}(t_0, x_0)e^{-(\frac{t}{T}-1)\delta} \;\le\; \bar{V}(t_0, x_0)De^{-\delta t/T}$$

where $D = e^{\delta}$. Since $\bar{V}$ is again a storage function, this gives the UGAS condition. However, it is not clear from the preceding arguments how to construct an explicit strict Lyapunov function for (6.1), which is one of our motivations for studying strictification methods for time-varying systems.

6.4 Remarks on Rate of Convergence

In our analysis of the dynamics (6.5), we found that for any constant $a > 0$, we can find a change of coordinates so that the time derivative of

$$V(t, x_{3e}, z_{2e}) = \frac{1}{2}\left[x_{3e}^2 + z_{2e}^2\right] \tag{6.34}$$

along the trajectories of the system (6.10) in the new coordinates satisfies

$$\dot{V}(t, x_{3e}, z_{2e}) \leq -2a\sin^4(t)V(t, x_{3e}, z_{2e}) \,. \tag{6.35}$$

This can be shown to imply that (6.10) can be made exponentially stable to the origin with arbitrary fast rate of convergence. We can carry out our strictification approach in such a way that the rate of convergence information is reflected in the strict Lyapunov function decay rate. One general result in this direction is as follows.

We again consider our system (6.1) under our standing assumptions, and we assume that there exists a weak Lyapunov function such that along the trajectories of (6.1), we get

$$\dot{V}(t, x) \leq -ap(t)V(t, x), \tag{6.36}$$

with $a > 0$ a given constant and p a PE parameter that is bounded by a constant $\bar{p}$. Choose δ and T so that the PE condition (6.4) is satisfied.

Let

$$\mathcal{V}(t, x) = e^{R(t)}V(t, x), \text{ where } R(t) = \frac{a}{T}\int_{t-T}^{t}\left(\int_{l}^{t} p(m)\mathrm{d}m\right)\mathrm{d}l \,. \tag{6.37}$$

Reasoning as we did to get (6.27) gives

$$\dot{R}(t) = ap(t) - \tfrac{a}{T}\left(\int_{t-T}^{t} p(m)\mathrm{d}m\right) \tag{6.38}$$

and therefore

$$\begin{aligned}\dot{\mathcal{V}}(t, x) &= e^{R(t)}\left[\dot{V}(t, x) + \dot{R}(t)V(t, x)\right]\\ &\leq \left[-ap(t) + \dot{R}(t)\right]\mathcal{V}(t, x)\\ &\leq -\tfrac{a}{T}\left(\int_{t-T}^{t} p(m)\mathrm{d}m\right)\mathcal{V}(t, x)\\ &\leq -\tfrac{\delta}{T}a\mathcal{V}(t, x),\end{aligned} \tag{6.39}$$

where the last inequality is from (6.4). The Lyapunov function $\mathcal{V}$ decays exponentially to zero with the decay rate $\frac{\delta}{T}a$, as desired.

An immediate consequence of (6.39) and the expression (6.37) for $\mathcal{V}$ is that if there exist an integer $k \geq 1$ and positive constants c_1 and c_2 such that

$$c_1|x|^k \le V(t,x) \le c_2|x|^k \tag{6.40}$$

for all $(t,x) \in [0,\infty) \times \mathbb{R}^n$, then the solutions of (6.1) exponentially converge to zero with decay rate $\frac{\delta}{kT}a$, by the following argument. Integrating (6.39) on $[t_0, t]$ along any trajectory of (6.1) gives

$$\mathcal{V}(t, x(t)) \le e^{-\frac{\delta}{T}a(t-t_0)}\mathcal{V}(t_0, x(t_0)) \; . \tag{6.41}$$

Therefore,

$$\begin{aligned} V(t, x(t)) &\le e^{-\frac{\delta}{T}a(t-t_0)}e^{R(t_0)-R(t)}V(t_0, x(t_0)) \\ &\le e^{-\frac{\delta}{T}a(t-t_0)}e^{\frac{aT}{2}\overline{p}}V(t_0, x(t_0)) \; , \end{aligned} \tag{6.42}$$

where the last inequality used

$$0 \le R(t) \le \frac{aT}{2}\overline{p}. \tag{6.43}$$

Applying (6.40), dividing by c_1, and taking kth roots gives

$$|x(t)| \le \left[\frac{c_2 e^{\frac{aT}{2}\overline{p}}}{c_1}\right]^{\frac{1}{k}} e^{-\frac{\delta}{kT}a(t-t_0)}|x(t_0)| \, , \tag{6.44}$$

which is the desired exponential decay condition.

Remark 6.5. The preceding construction readily generalizes to cases where one has a non-strict Lyapunov V_1, a PE parameter p, and an everywhere positive C^2 function L such that $L(0) = 1$ and

$$\dot{V}_1(t,x) \le -ap(t)\frac{V_1(t,x)}{L(V_1(t,x))} \tag{6.45}$$

along all trajectories of (6.1), as follows. Without loss of generality, we may assume that L is non-decreasing. Then, for all $v \in (0,1)$,

$$\int_1^v \frac{L(m)}{m}\mathrm{d}m \;\; \le \;\; -\int_v^1 \frac{L(0)}{m}\mathrm{d}m \;\; = \;\; \ln(v). \tag{6.46}$$

We deduce that the continuous function

$$k(v) = \begin{cases} \exp\left(\int_1^v \frac{2L(m)}{m}\mathrm{d}m\right), & v > 0 \\ 0, & v = 0 \end{cases} \tag{6.47}$$

is C^1 and of class $\mathcal{K}_\infty$. Indeed, (6.46) gives $k(v) \le v^2$ for all $v \in (0,1)$, so $k'(0+) = 0$, and

$$k'(v) \;=\; 2k(v)\frac{L(v)}{v} \;\le\; 2vL(v) \;\to\; 0$$

as $v \to 0^+$.

Next, consider the function

$$V(t,x) = k(V_1(t,x)) \,. \tag{6.48}$$

It satisfies $V(t,0) = 0$ for all $t \geq 0$, and

$$\begin{aligned}
\dot{V}(t,x) &= 2\exp\left(\int_1^{V_1(t,x)} \frac{2L(m)}{m}\mathrm{d}m\right) \frac{L(V_1(t,x))}{V_1(t,x)}\dot{V}_1(t,x) \\
&\leq -2\exp\left(\int_1^{V_1(t,x)} \frac{2L(m)}{m}\mathrm{d}m\right) \frac{L(V_1(t,x))}{V_1(t,x)}ap(t)\frac{V_1(t,x)}{L(V_1(t,x))} \\
&\leq -\exp\left(\int_1^{V_1(t,x)} \frac{2L(m)}{m}\mathrm{d}m\right) ap(t) \\
&= -ap(t)V(t,x) \,, \; x \neq 0 \,,
\end{aligned} \tag{6.49}$$

which is the decay condition we had in (6.36). Hence, applying the previous result with the choice $V(t,x) = k\big(V_1(t,x)\big)$ and k defined by (6.47) gives the function

$$W(t,x) = e^{R(t)}V(t,x) \,, \tag{6.50}$$

whose time derivative along the trajectories of (6.1) satisfies

$$\dot{W}(t,x) \;\leq\; -\frac{\delta}{T}aW(t,x) \,. \tag{6.51}$$

Using the fact that $k \in \mathcal{K}_\infty$, we readily conclude that W is a strict Lyapunov (6.1). On the other hand, (6.51) does not imply that the solutions of the system converge exponentially to zero, even when V_1 satisfies inequalities of the type (6.40).

6.5 Input-to-State Stability

We next present extensions of Theorem 6.1 to systems with disturbances based on ISS, and some further extensions for systems with outputs.

Throughout this section, we will assume that our nonautonomous system

$$\dot{x} = f(t,x,u) \tag{6.52}$$

has state space $\mathcal{X} = \mathbb{R}^n$ and input set $U = \mathbb{R}^p$, and that it satisfies our usual assumptions from Chap. 1. For convenience, we also assume that f is periodic in t, which means that there exists a constant $T > 0$ such that

$f(t+T,x,u) = f(t,x,u)$ for all $t \geq 0$, $x \in \mathbb{R}^n$, and $u \in \mathbb{R}^m$. However, most of the results to follow remain true if the periodicity assumption on f is relaxed to the requirement that f is uniformly locally bounded in t, meaning, for each compact subset $K \subseteq \mathbb{R}^n \times \mathbb{R}^p$, we have

$$\sup\left\{|f(t,x,u)| : (x,u) \in K, t \geq 0\right\} \ < \ +\infty. \tag{6.53}$$

The control functions for our system (6.52) comprise the set of all measurable essentially bounded functions $\alpha : [0,\infty) \to \mathbb{R}^m$; we denote this set by $\mathcal{M}(\mathbb{R}^m)$ as before. For each $t_0 \geq 0$, $x_0 \in \mathbb{R}^n$, and $\alpha \in \mathcal{M}(\mathbb{R}^m)$, we let $I \ni t \mapsto x(t,t_0,x_0,\alpha)$ denote the unique trajectory of (6.52) for the input α satisfying $x(t_0) = x_0$ and defined on its maximal interval $I \subseteq [t_0,\infty)$. This trajectory will be denoted by $x(t)$ when this would not lead to confusion.

Recall that a C^1 function $V : [0,\infty) \times \mathbb{R}^n \to [0,\infty)$ is said to be of class UBPPD (written $V \in$ UBPPD) provided (a) it is a storage function and (b) its gradient is uniformly bounded in t, i.e., there exists a function $\alpha_3 \in \mathcal{K}_\infty$ such that

$$|\nabla V(t,x)| \leq \alpha_3(|x|) \tag{6.54}$$

for all $t \geq 0$ and $x \in \mathbb{R}^n$, where $\nabla V = (V_t, V_x)$. Given a storage function $V \in$ UBPPD and functions $\alpha_1, \alpha_2 \in \mathcal{K}_\infty$ such that

$$\alpha_1(|x|) \ \leq \ V(t,x) \ \leq \ \alpha_2(|x|) \ \ \forall t \geq 0 \text{ and } x \in \mathbb{R}^n\,, \tag{6.55}$$

we can assume that α_1 and α_2 are C^1, e.g., by taking $\alpha_2(s) = \int_0^s \alpha_3(r)\,dr$ and minorizing α_1 by a C^1 function of class $\mathcal{K}_\infty$. We continue to use the notation

$$\dot{V}(t,x,u) \doteq \frac{\partial V}{\partial t}(t,x) + \frac{\partial V}{\partial x}(t,x) f(t,x,u)$$

for functions $V \in$ UBPPD. We later use the fact that

$$s \mapsto \sup\left\{\left|\dot{V}(t,x,u)\right| : t \geq 0, |x| \leq \chi(s), |u| \leq s\right\}$$

is of class $\mathcal{K}$ for each $\chi \in \mathcal{K}_\infty$, which follows from (6.53) and (6.54).

For each element $p \in \mathcal{P}$, we can define corresponding notions of non-strict ISS and non-strict ISS Lyapunov functions, as follows:

Definition 6.1. Let $p \in \mathcal{P}$. A function $V \in$ UBPPD is called an *ISS(p) Lyapunov function* for (6.52) provided there exist $\chi \in \mathcal{K}_\infty$ and $\mu \in \mathcal{K}_\infty \cap C^1$ such that

$$|x| \geq \chi(|u|) \ \Rightarrow \ \dot{V}(t,x,u) \leq -p(t)\mu(|x|) \ \ \forall t \geq 0. \tag{6.56}$$

An ISS(p) Lyapunov function for (6.52) and $p(t) \equiv 1$ is also called a *strict ISS Lyapunov function.*

Condition (6.56) allows

$$\dot{V}(t,x,u) \ = \ 0$$

for those t where $p(t) = 0$. This corresponds to allowing V to non-strictly decrease along the solutions of (6.52).

Definition 6.2. Let $p \in \mathcal{P}$. We say that (6.52) is ISS(p), or that it is ISS with decay rate p, provided there exist $\beta \in \mathcal{KL}$ and $\gamma \in \mathcal{K}_\infty$ such that for all $t_0 \geq 0$, $x_0 \in \mathbb{R}^n$, $u_0 \in \mathcal{M}(\mathbb{R}^m)$ and $h \geq 0$,

$$|x(t_0 + h, t_0, x_0, u_0)| \leq \beta\left(|x_0|, \int_{t_0}^{t_0+h} p(s)\,\mathrm{d}s\right) + \gamma\left(|u_0|_{[t_0,t_0+h]}\right). \quad (6.57)$$

If (6.52) is ISS(p) with $p \equiv 1$, then we say that (6.52) is ISS.

Notice that ISS(p) systems are automatically forward complete. We also study dissipation-type decay conditions as follows:

Definition 6.3. Let $p \in \mathcal{P}$. A function $V \in$ UBPPD is called a *non-strict dissipative Lyapunov function* for (6.52) and p, or a *DIS(p) Lyapunov function*, provided there exist $\Omega \in \mathcal{K}_\infty$ and $\mu \in \mathcal{K}_\infty \cap C^1$ such that

$$\dot{V}(t, x, u) \leq -p(t)\mu(|x|) + \Omega(|u|) \quad (6.58)$$

for all $t \geq 0$, $x \in \mathbb{R}^n$, and $u \in \mathbb{R}^m$. A DIS(p) Lyapunov function for (6.52) and $p(t) \equiv 1$ is also called a *strict DIS Lyapunov function.*

As we saw in Sect. 6.3, the decay condition (6.58) in the special case where the disturbance u is fixed at zero implies the existence of a (possibly different) storage function that satisfies the standard decay condition

$$\dot{V}(t, x, 0) \leq -\mu(|x|),$$

which gives UGAS. However, it is not clear how to use a DIS(p) or ISS(p) Lyapunov function to construct a strict ISS Lyapunov function. In the next section, we show how such constructions can be carried out explicitly. We use the following elementary observations:

Lemma 6.1. *Let $T, \delta, \bar{p} > 0$ be constants and $p \in \mathcal{P}(T, \delta, \bar{p})$ be given. Then:*

1. $0 \leq \int_{t-T}^{t}\left(\int_s^t p(r)\,\mathrm{d}r\right)\mathrm{d}s \leq \frac{T^2\bar{p}}{2}$ *for all* $t \geq 0$*; and*
2. *the function*

$$[0, \infty) \ni h \mapsto \tilde{p}(h) = \inf_{t \geq 0} \int_t^{t+h} p(r)\,\mathrm{d}r$$

is continuous, non-decreasing, and unbounded.

The proof of Lemma 6.1 is a simple exercise.

6.6 Equivalent Characterizations of Non-strict ISS

The following theorem from [91] collects the various equivalences among ISS(p), ISS, and the corresponding non-strict and strict Lyapunov functions:

Theorem 6.2. *Let $p \in \mathcal{P}$ and f be as above. The following are equivalent:*

1. *f admits an ISS(p) Lyapunov function;*
2. *f admits a strict ISS Lyapunov function;*
3. *f admits a DIS(p) Lyapunov function;*
4. *f admits a strict DIS Lyapunov function;*
5. *f is ISS(p); and*
6. *f is ISS.*

While the main implications of Theorem 6.2 can be shown in non-explicit ways (e.g., using Lyapunov characterizations), a significant feature of our proof is the explicit construction of strict ISS Lyapunov functions

$$V^\sharp(t,x) = V(t,x) + w(V(t,x)) \int_{t-T}^{t} \left(\int_s^t p(r)\, \mathrm{d}r \right) \mathrm{d}s \tag{6.59}$$

for an appropriate function $w \in C^1 \cap \mathcal{K}_\infty$. The proof proceeds by showing the following implications: 1. $\Rightarrow$ 2. $\Rightarrow$ 4. $\Rightarrow$ 1., 3. $\Leftrightarrow$ 4., 2. $\Leftrightarrow$ 6., and 5. $\Leftrightarrow$ 6.. For completeness, we provide the parts of the proof that involve strict Lyapunov function constructions. We then discuss the other parts of the proof in remarks. Fix $T, \delta, \bar{p} > 0$ such that $p \in \mathcal{P}(T, \delta, \bar{p})$.

6.6.1 Proofs of Equivalences

Proof that 1. $\Rightarrow$ 2. Let V be an ISS(p) Lyapunov function for f. Pick functions $\alpha_1, \alpha_2 \in \mathcal{K}_\infty \cap C^1$ satisfying (6.55) and $\chi \in \mathcal{K}_\infty$ and $\mu \in \mathcal{K}_\infty \cap C^1$ satisfying (6.56). Set

$$\begin{aligned} \tilde{\alpha}_2(s) &\doteq \max\left\{\tfrac{T\bar{p}}{2}, 1\right\}(\alpha_2(s) + \mu(s) + s) \;\text{ and} \\ w(s) &\doteq \tfrac{1}{4T}\mu\big(\tilde{\alpha}_2^{-1}(s)\big). \end{aligned} \tag{6.60}$$

Then $\tilde{\alpha}_2, \tilde{\alpha}_2^{-1} \in \mathcal{K}_\infty \cap C^1$. Since $V(t,x) \le \tilde{\alpha}_2(|x|)$ for all $t \ge 0$ and $x \in \mathbb{R}^n$ and $\mu \in \mathcal{K}_\infty$, we get

$$|x| \ge \chi(|u|) \;\;\Rightarrow\;\; \dot{V}(t,x,u) \le -p(t)\mu\big(\tilde{\alpha}_2^{-1}(V(t,x))\big)\,. \tag{6.61}$$

Note too that $w \in \mathcal{K}_\infty \cap C^1$ and

$$
\begin{aligned}
0 \le w'(s) &= \frac{\mu'(\tilde{\alpha}_2^{-1}(s))}{4T\max\{\frac{T\bar{p}}{2},1\}(\alpha_2'(\tilde{\alpha}_2^{-1}(s))+\mu'(\tilde{\alpha}_2^{-1}(s))+1)} \\
&\le \frac{\mu'(\tilde{\alpha}_2^{-1}(s))}{4T\max\{\frac{T\bar{p}}{2},1\}(\mu'(\tilde{\alpha}_2^{-1}(s))+1)} \le \frac{1}{2T^2\bar{p}}
\end{aligned}
\tag{6.62}
$$

for all $s \ge 0$. Consider the UBPPD function

$$
V^\sharp(t,x) = V(t,x) + \xi(t)w\big(V(t,x)\big)\,, \tag{6.63}
$$

where

$$
\xi(t) = \int_{t-T}^{t}\left(\int_s^t p(r)\,\mathrm{d}r\right)\mathrm{d}s. \tag{6.64}
$$

Then

$$
\begin{aligned}
\dot{V}^\sharp(t,x,u) &= \big[1+\xi(t)w'\big(V(t,x)\big)\big]\dot{V}(t,x,u) \\
&\quad + \big[Tp(t) - \textstyle\int_{t-T}^t p(r)\,\mathrm{d}r\big]w\big(V(t,x)\big)\,.
\end{aligned}
\tag{6.65}
$$

When $|x| \ge \chi(|u|)$, condition (6.61) gives

$$
\begin{aligned}
\dot{V}^\sharp(t,x,u) &\le -p(t)\mu\big(\tilde{\alpha}_2^{-1}(V(t,x))\big) \\
&\quad + \left[Tp(t) - \textstyle\int_{t-T}^t p(r)\,\mathrm{d}r\right]\frac{1}{4T}\mu\big(\tilde{\alpha}_2^{-1}(V(t,x))\big) \\
&\le -\frac{3}{4}p(t)\mu\big(\tilde{\alpha}_2^{-1}(V(t,x))\big) \\
&\quad - \left(\int_{t-T}^t p(r)\,\mathrm{d}r\right)\frac{1}{4T}\mu\big(\tilde{\alpha}_2^{-1}(V(t,x))\big) \\
&\le -\frac{\delta}{4T}\mu\big(\tilde{\alpha}_2^{-1}(\alpha_1(|x|))\big) \;\; \forall t \ge 0.
\end{aligned}
$$

Since $\mu \circ \tilde{\alpha}_2^{-1} \circ \alpha_1 \in C^1 \cap \mathcal{K}_\infty$, it follows that $V^\sharp$ is a strict ISS Lyapunov function for (6.52). □

Proof that 2. ⇒ 4. Assume that f admits a strict ISS Lyapunov function V. Let $\mu, \chi \in \mathcal{K}_\infty$ satisfy condition (6.56) with $p \equiv 1$. Then the strict dissipative condition (6.58) with $p \equiv 1$ follows by choosing any $\Omega \in \mathcal{K}_\infty$ satisfying

$$
\Omega(s) \ge \max_{\{t\ge 0, |x|\le \chi(s), |u|\le s\}} \left\{\dot{V}(t,x,u) + \mu(|x|)\right\} \;\; \forall s \ge 0.
$$

Such an Ω exists by our conditions (6.53) and (6.55). Therefore, V is itself a strict DIS Lyapunov function for f. □

Proof that 4. ⇒ 1. Assume that f admits a strict DIS Lyapunov function V. Let $\mu, \Omega \in \mathcal{K}_\infty$ satisfy (6.58) with $p \equiv 1$. If $|x| \ge \chi(|u|) \doteq \mu^{-1}(2\Omega(|u|))$, then

$$\dot{V}(t,x,u) \leq -\frac{1}{2}\mu(|x|), \text{ so } \dot{V}(t,x,u) \leq -\frac{p(t)}{2\bar{p}}\mu(|x|)$$

for all $t \geq 0$. Therefore, V is also an ISS(p) Lyapunov function for f, so 1. is satisfied. □

Proof that 3. ⇔ 4. Since $p \in \mathcal{P}$ is bounded, it is immediate that 4. implies 3.. Conversely, assume $V \in$ UBPPD is a DIS(p) Lyapunov function for f and $\alpha_1, \alpha_2, \mu, \Omega \in \mathcal{K}_\infty$ satisfy (6.55) and the DIS(p) requirements. Define $\tilde{\alpha}_2, w \in \mathcal{K}_\infty \cap C^1$ and $V^\sharp$ by (6.60) and (6.63). As before, when $\tilde{\mu} = \mu \circ \tilde{\alpha}_2^{-1}$, we have

$$\dot{V}(t,x,u) \;\leq\; -p(t)\tilde{\mu}\big(V(t,x)\big) + \Omega(|u|)$$

for all $t \geq 0, x \in \mathbb{R}^n, u \in \mathbb{R}^m$. It follows from Lemma 6.1 and (6.62) that

$$1 + \xi(t)w'\big(V(t,x)\big) \in \left[1, \frac{5}{4}\right] \quad \forall t \geq 0 \text{ and } x \in \mathbb{R}^n. \tag{6.66}$$

Since $w = \frac{1}{4T}\tilde{\mu}$, we deduce from (6.65) that

$$\begin{aligned}\dot{V}^\sharp &\leq -p(t)\tilde{\mu}\big(V(t,x)\big) + \tfrac{5}{4}\Omega(|u|)\\ &\quad +Tp(t)w\big(V(t,x)\big) - \left(\textstyle\int_{t-T}^{t} p(r)\,\mathrm{d}r\right) w\big(V(t,x)\big)\\ &\leq -\delta w\big(\alpha_1(|x|)\big) + \tfrac{5}{4}\Omega(|u|).\end{aligned}$$

Since $w \circ \alpha_1 \in C^1 \cap \mathcal{K}_\infty$, it follows that $V^\sharp$ is the desired strict DIS Lyapunov function. □

6.6.2 Remarks on Proof of Equivalences

Remark 6.6. The implication 2. ⇒ 6. is standard. It follows, e.g., from [70, Theorem 4.19, p.176], generalized to allow controls in $\mathcal{M}(\mathbb{R}^m)$. The converse 6. ⇒ 2. was noted in [39, Theorem 1], and can be deduced from results in [12]. For details, see Appendix B.2.

Remark 6.7. The proof that 5. ⇔ 6. is the following straightforward consequence of Lemma 6.1. Assuming 6., there are $\beta \in \mathcal{KL}$ such that for all $t_0 \geq 0$, $x_0 \in \mathbb{R}^n$, $u_0 \in \mathcal{M}(\mathbb{R}^m)$, and $h \geq 0$,

$$\begin{aligned}|x(t_0+h, t_0, x_0, u_0)| &\leq \beta\big(|x_0|, \bar{p}h\big) + \gamma\big(|u_0|_{[t_0,t_0+h]}\big)\\ &\leq \beta\big(|x_0|, \textstyle\int_{t_0}^{t_0+h} p(s)\,\mathrm{d}s\big) + \gamma\big(|u_0|_{[t_0,t_0+h]}\big).\end{aligned}$$

Therefore, f is ISS(p) so 6. ⇒ 5.. Conversely, if f is ISS(p), then we can find $\beta \in \mathcal{KL}$ such that for all $t_0 \geq 0$, $x_0 \in \mathbb{R}^n$, $u_0 \in \mathcal{M}(\mathbb{R}^m)$, and $h \geq 0$,

$$|x(t_0+h,t_0,x_0,u_0)| \leq \beta\left(|x_0|, \int_{t_0}^{t_0+h} p(s)\,\mathrm{d}s\right) + \gamma(|u_0|_{[t_0,t_0+h]})$$
$$\leq \beta\left(|x_0|, \tilde{p}(h)\right) + \gamma(|u_0|_{[t_0,t_0+h]}),$$

where $\tilde{p}$ is the function defined in Lemma 6.1. By Lemma 6.1, $\hat{\beta}(s,t) \doteq \beta(s,\tilde{p}(t)) \in \mathcal{KL}$, so 5. $\Rightarrow$ 6., as desired.

Remark 6.8. If the functions V, α_2, μ, p are sufficiently smooth, then the particular function $\tilde{\alpha}_2$ in (6.60) we have chosen implies that the function $V^\sharp(t,x)$ is also sufficiently smooth.

Remark 6.9. Our proof of 2. $\Rightarrow$ 4. in Theorem 6.2 shows that if V is a strict ISS Lyapunov function for f, then V is also a strict DIS Lyapunov function for f. The preceding implication is no longer true if our growth requirement (6.53) on f is dropped, as illustrated by the following example from [39].

Take the one-dimensional single input system

$$\dot{x} = f(t,x,u) \doteq -x + (1+t)q(u - |x|),$$

where $q : \mathbb{R} \to \mathbb{R}$ is any C^1 function for which $q(r) \equiv 0$ for $r \leq 0$ and $q(r) > 0$ otherwise. Then $V(x) = x^2$ is a strict ISS Lyapunov function for the system since

$$|x| \geq |u| \;\Rightarrow\; \dot{V} \leq -x^2,$$

but V does not satisfy the strict DIS condition (6.58) for any choices of μ and Ω. This contrasts with the time-invariant case where strict ISS Lyapunov functions are automatically strict DIS Lyapunov functions.

6.7 Input-to-Output Stability

The ISS property estimates the decay of the state in terms of an overshoot that depends on the essential supremum of the control. However, in many applications, the current state may be difficult if not impossible to measure. Instead, only *output measurements* are available, which gives the standard model

$$\dot{x} = f(t,x,u)\,,\; y = H(x) \tag{6.67}$$

where f is as before and H is locally Lipschitz. We assume for simplicity in this section that (6.67) is forward complete and of period $T > 0$ in t.

Several generalizations of ISS for time-invariant systems with outputs are used [73, 164, 170, 171]. It is natural to generalize the ISS condition by assuming a decay of the *output* (instead of the state) with an overshoot depending as before on the sup norm of the input. This is made precise in the following definitions, which generalize the corresponding definitions for time-invariant systems from [171]. In what follows, we set

$$y(t_0+h,t_0,x_0,\mathbf{u}) = H\big(x(t_0+h,t_0,x_0,\mathbf{u})\big)$$

for all $t_0 \geq 0$, $x_0 \in \mathbb{R}^n$, $\mathbf{u} \in \mathcal{M}(\mathbb{R}^m)$, and $h \geq 0$.

Definition 6.4. We say that (6.67) is *input-to-output stable (IOS)* provided there exist $\beta \in \mathcal{KL}$ and $\gamma \in \mathcal{K}_\infty$ such that

$$\big|y(t_0+h,t_0,x_0,\mathbf{u})\big| \leq \beta\left(|x_0|,h\right) \;+\; \gamma\left(|\mathbf{u}|_{[t_0,t_0+h]}\right)$$

for all $t_0 \geq 0$, $x_0 \in \mathbb{R}^n$, $\mathbf{u} \in \mathcal{M}(\mathbb{R}^m)$ and $h \geq 0$.

The corresponding Lyapunov function notion is as follows:

Definition 6.5. A smooth $V : [0,\infty) \times \mathbb{R}^n \to [0,\infty)$ is called a *(strict) IOS Lyapunov function* for (6.67) provided there exist functions $\alpha_1, \alpha_2, \chi \in \mathcal{K}_\infty$ and $\kappa \in \mathcal{KL}$ such that

$$\alpha_1(|H(x)|) \;\leq\; V(t,x) \;\leq\; \alpha_2(|x|) \tag{6.68}$$

and

$$V(t,x) \geq \chi(|u|) \;\Rightarrow\; \dot{V}(t,x,u) \;\leq\; -\kappa\big(V(t,x),|x|\big) \tag{6.69}$$

hold for all $t \geq 0$, $x \in \mathbb{R}^n$, and $u \in \mathbb{R}^m$.

For the equivalence of the IOS property to the existence of an IOS Lyapunov function for time-invariant systems, see [171, Theorem 1.2]. Let $\mathrm{sat}\{q\}$ denote the usual projection of $q \in \mathbb{R}$ onto $[-1,+1]$, namely,

$$\mathrm{sat}(r) = \begin{cases} r, & |r| \leq 1 \\ \mathrm{sign}(r), & \text{otherwise.} \end{cases}$$

The following IOS strictification result was announced in [90]:

Theorem 6.3. *Let f and H be as above and assume $p \in \mathcal{P}(T,\delta,\bar{p})$. Let $V : [0,\infty) \times \mathbb{R}^n \to [0,\infty)$ be a C^1 function that admits $\hat{\alpha}_1, \hat{\alpha}_2, \hat{\chi} \in \mathcal{K}_\infty$ and $\hat{\kappa} \in C^1 \cap \mathcal{K}_\infty$ such that*

$$\hat{\alpha}_1\big(|H(x)|\big) \;\leq\; V(t,x) \;\leq\; \hat{\alpha}_2\big(|x|\big) \tag{6.70}$$

and

$$V(t,x) \geq \hat{\chi}(|u|) \;\Rightarrow\; \dot{V}(t,x,u) \;\leq\; -p(t)\hat{\kappa}\big(V(t,x)\big) \tag{6.71}$$

for all $x \in \mathbb{R}^n$ and $t \geq 0$. Define $w : [0,\infty) \to [0,\infty)$ by

$$w(r) = \frac{1}{T^2\bar{p} + 2T} \int_0^r \mathrm{sat}\{\hat{\kappa}'(s)\}\,\mathrm{d}s. \tag{6.72}$$

Then

$$V^\sharp(t,x) = V(t,x) + \left[\int_{t-T}^t \left(\int_s^t p(l)\mathrm{d}l\right)\mathrm{d}s\right] w\big(V(t,x)\big)$$

is a strict IOS Lyapunov function for (6.67).

Proof. Defining ξ by (6.64) as before and choosing w from (6.72) again gives (6.65). Since $Tw(r) \le \frac{1}{2}\hat{\kappa}(r)$ and $w'(r) \ge 0$ for all $r \ge 0$, it follows that if $V(t,x) \ge \hat{\chi}(|u|)$, then

$$\begin{aligned}\dot{V}^\sharp(t,x,u) &\le p(t)\left[-\hat{\kappa}\big(V(t,x)\big) + Tw\big(V(t,x)\big)\right] - w(V(t,x))\int_{t-T}^{t} p(l)\mathrm{d}l \\ &\le -\tfrac{1}{2}p(t)\hat{\kappa}\big(V(t,x)\big) - \delta w\big(V(t,x)\big) \\ &\le -\delta w\big(V(t,x)\big).\end{aligned}$$

Recalling Lemma 6.1 and the structure of $V^\sharp$, and noting that $w(r) \le \frac{r}{T^2\bar{p}}$ for all $r \ge 0$, it follows that

$$V(t,x) \;\le\; V^\sharp(t,x) \;\le\; \frac{3}{2}V(t,x) \tag{6.73}$$

for all $t \ge 0$ and $x \in \mathbb{R}^n$. Therefore, if $V^\sharp(t,x) \ge \frac{3}{2}\hat{\chi}(|u|)$, then $V(t,x) \ge \hat{\chi}(|u|)$, so (6.73) gives

$$\dot{V}^\sharp(t,x,u) \;\le\; -\delta w\big(V(t,x)\big) \;\le\; -\delta w\left(\frac{2V^\sharp(t,x)}{3}\right)$$

for all $t \ge 0$. Moreover, $\hat{\alpha}_1(|H(x)|) \le V^\sharp(t,x) \le \frac{3}{2}\hat{\alpha}_2(|x|)$ for all $t \ge 0$ and $x \in \mathbb{R}^n$, by (6.70) and (6.73). We conclude that $V^\sharp$ satisfies the strict IOS Lyapunov function requirements with

$$\alpha_1 = \hat{\alpha}_1,\quad \alpha_2 = \frac{3}{2}\hat{\alpha}_2,\quad \chi = \frac{3}{2}\hat{\chi} \text{ and } \kappa(r,s) \doteq \delta\frac{w(2r/3)}{(1+s)}$$

which proves the theorem. □

6.8 Illustrations

6.8.1 Rotating Rigid Body

We construct a strict ISS Lyapunov function for a tracking problem for a rotating rigid body. Following Lefeber [78, p.31], we only consider the dynamics of the velocities, which, after a change of feedback gives

$$\begin{cases}\dot{\omega}_1 = \delta_1 + u_1 \\ \dot{\omega}_2 = \delta_2 + u_2 \\ \dot{\omega}_3 = \omega_1\omega_2\ ,\end{cases} \tag{6.74}$$

where δ_1 and δ_2 are the inputs and u_1 and u_2 are the disturbances. We consider the reference state trajectory

$$\omega_r(t) = (\omega_{1r}, \omega_{2r}, \omega_{3r})(t) = (\sin(t), 0, 0) . \tag{6.75}$$

The substitution

$$\tilde{\omega}_i(t) = \omega_i(t) - \omega_{ir}(t)$$

transforms (6.74) into the error equations

$$\begin{cases} \dot{\tilde{\omega}}_1 = \delta_1 + u_1 - \cos(t) \\ \dot{\tilde{\omega}}_2 = \delta_2 + u_2 \\ \dot{\tilde{\omega}}_3 = [\tilde{\omega}_1 + \sin(t)]\tilde{\omega}_2 . \end{cases} \tag{6.76}$$

By applying the backstepping approach as in [65], or through direct calculations, one shows that the time derivative of the class UBPPD function

$$V(t, \tilde{\omega}) = \frac{1}{2}\left[\tilde{\omega}_1^2 + (\tilde{\omega}_2 + \sin(t)\tilde{\omega}_3)^2 + \tilde{\omega}_3^2\right] \tag{6.77}$$

with $\tilde{\omega} = (\tilde{\omega}_1, \tilde{\omega}_2, \tilde{\omega}_3)^\top$ along the trajectories of (6.76) in closed-loop with the control laws

$$\begin{aligned} \delta_1(t, \tilde{\omega}) &= -\tilde{\omega}_1 - \tilde{\omega}_2\tilde{\omega}_3 + \cos(t) \\ \delta_2(t, \tilde{\omega}) &= -\left[1 + \sin(t)\tilde{\omega}_1 + \sin^2(t)\right]\tilde{\omega}_2 \\ &\quad -(2\sin(t) + \cos(t))\tilde{\omega}_3 \end{aligned} \tag{6.78}$$

satisfies

$$\begin{aligned} \dot{V} &= -\tilde{\omega}_1^2 - [\tilde{\omega}_2 + \sin(t)\tilde{\omega}_3]^2 - \sin^2(t)\tilde{\omega}_3^2 \\ &\quad +\tilde{\omega}_1 u_1 + [\tilde{\omega}_2 + \sin(t)\tilde{\omega}_3]\, u_2 \\ &\le -\tfrac{1}{2}\tilde{\omega}_1^2 - \tfrac{1}{2}[\tilde{\omega}_2 + \sin(t)\tilde{\omega}_3]^2 - \sin^2(t)\tilde{\omega}_3^2 \\ &\quad +\tfrac{1}{2}(u_1^2 + u_2^2) \\ &\le -p(t)\tilde{\mu}(V(t, \tilde{\omega})) + \Omega(|u|) \end{aligned} \tag{6.79}$$

with $u = (u_1, u_2)^\top \in \mathbb{R}^2$,

$$p(t) = \sin^2(t), \quad \tilde{\mu}(s) = s, \quad \text{and} \quad \Omega(s) = \frac{1}{2}s^2.$$

Therefore, V is a DIS(p) Lyapunov function for (6.76) in closed-loop with the control laws (6.78). In this case, $p \in \mathcal{P}(\pi, \pi/2, 1)$. Setting

$$T = \pi \quad \text{and} \quad w(s) = \frac{1}{8T}\tilde{\mu}(s) = \frac{s}{8\pi}$$

and defining ξ by (6.64), it follows that (6.66) also holds. Therefore, our proof of Theorem 6.2 shows that

$$\begin{aligned} V^\sharp(t,\tilde{\omega}) &= V(t,\tilde{\omega}) + \left[\int_{t-T}^{t}\left(\int_s^t p(r)\,\mathrm{d}r\right)\mathrm{d}s\right] w(V(t,\tilde{\omega})) \\ &= \left[1 + \tfrac{\pi}{32} - \tfrac{1}{32}\sin(2t)\right] V(t,\tilde{\omega}) \end{aligned}$$

is a strict DIS Lyapunov function and also a strict ISS Lyapunov function for the system (6.76) in closed-loop with the control laws (6.78).

6.8.2 Stabilization of Underactuated Ships

We next consider a control problem arising from the dynamic positioning of a ship that has no side thruster, but does have two independent main thrusters. The thrusters are located at a distance from the center line in order to provide both surge force and yaw moment. We find feedback laws that stabilize both the position variables and the orientation around a periodic trajectory, using only the two available controls. We also seek a strict Lyapunov function for the corresponding closed-loop error dynamics. To keep our illustration simple, we select a very simple family of reference trajectories.

6.8.2.1 Ship Model

Following [42], the dynamic equations of the ship are

$$\begin{cases} \dot{u} = \frac{m_{22}}{m_{11}} vr - \frac{d_{11}}{m_{11}} u + \frac{1}{m_{11}} \tau_1 \\ \dot{v} = -\frac{m_{11}}{m_{22}} ur - \frac{d_{22}}{m_{22}} v \\ \dot{r} = \frac{m_{11}-m_{22}}{m_{33}} uv - \frac{d_{33}}{m_{33}} r + \frac{1}{m_{33}} \tau_3 \ . \end{cases} \tag{6.80}$$

The variables u, v and r are the velocities in surge, sway and yaw respectively. The constant parameters $m_{ii} > 0$ are given by the ship inertia and added mass effects, and the parameters $d_{ii} > 0$ are related to the hydrodynamic damping. The available controls are the surge control force τ_1 and the yaw control moment τ_3. However, we have no available control in the sway direction, and the problem of controlling the ship in three degrees-of-freedom is therefore an underactuated control problem.

When modeling the ship, the dynamics associated with the motion in heave, roll, and pitch and terms of second order at the origin in the hydrodynamic terms are assumed to be negligible. We also assume that the inertia and damping matrices are diagonal. This is true for ships having port/starboard and fore/aft symmetry. Most ships have port/starboard symmetry. Non-symmetric fore/aft of the ship implies that the off diagonal terms of the inertia matrix are non-zero, i.e., $m_{23} \neq 0$ and $m_{32} \neq 0$, as well as $d_{23} \neq 0$ and $d_{32} \neq 0$ in the damping matrix. These off-diagonal terms will be small compared with the diagonal elements m_{ii} and d_{ii} for most ships. Non-symmetric fore/aft will also give some extra cross-terms, due to Coriolis and centripetal forces. Control design in the general case where the off-diagonal terms are also taken into account is relatively simple for a fully actuated ship, while it is still a topic of future research for the underactuated ship.

The kinematics of the ship are described by

$$\begin{cases} \dot{x} = \cos(\psi)u - \sin(\psi)v \\ \dot{y} = \sin(\psi)u + \cos(\psi)v \\ \dot{\psi} = r \end{cases} \tag{6.81}$$

where (x, y) and ψ give the position and orientation of the ship in the earth fixed frame, respectively. To obtain simpler, polynomial equations, we use the global coordinate transformation from [132], namely,

$$\begin{aligned} z_1 &= \cos(\psi)x + \sin(\psi)y \\ z_2 &= -\sin(\psi)x + \cos(\psi)y \\ z_3 &= \psi \end{aligned} \tag{6.82}$$

which yields

$$\begin{cases} \dot{z}_1 = u + z_2 r \\ \dot{z}_2 = v - z_1 r \\ \dot{z}_3 = r\ . \end{cases} \tag{6.83}$$

To summarize, the overall system is

$$\begin{cases} \dot{z}_1 = u + z_2 r \\ \dot{z}_2 = v - z_1 r \\ \dot{z}_3 = r \\ \dot{u} = \frac{m_{22}}{m_{11}} vr - \frac{d_{11}}{m_{11}} u + \frac{1}{m_{11}} \tau_1 \\ \dot{v} = -\frac{m_{11}}{m_{22}} ur - \frac{d_{22}}{m_{22}} v \\ \dot{r} = \frac{m_{11} - m_{22}}{m_{33}} uv - \frac{d_{33}}{m_{33}} r + \frac{1}{m_{33}} \tau_3\ . \end{cases} \tag{6.84}$$

6.8.2.2 Trajectory Tracking Problem

We solve the problem of tracking the state reference trajectory

$$(z_{1p}, z_{2p}, z_{3p}, u_p, v_p, r_p)(t) = \big(0, 0, -\varepsilon \sin(t), 0, 0, -\varepsilon \cos(t)\big)\ , \tag{6.85}$$

where ε is an arbitrary positive real number, and we also find an explicit global strict Lyapunov function for the corresponding closed-loop system.

We find it convenient to use the change of feedback

$$\begin{aligned} \tau_1 &= m_{11}\nu_1 - m_{22}vr + d_{11}u \\ \tau_3 &= m_{33}\nu_2 - (m_{11} - m_{22})uv + d_{33}r \ , \end{aligned} \tag{6.86}$$

where ν_1 and ν_2 are the new inputs. This gives

$$\begin{cases} \dot{z}_1 = u + z_2 r \\ \dot{z}_2 = v - z_1 r \\ \dot{z}_3 = r \\ \dot{v} = -\rho_1 ur - \rho_2 v \\ \dot{u} = \nu_1 \\ \dot{r} = \nu_2 \ , \end{cases} \tag{6.87}$$

where

$$\rho_1 = \frac{m_{11}}{m_{22}} \text{ and } \rho_2 = \frac{d_{22}}{m_{22}} \ . \tag{6.88}$$

We also use the more convenient variables

$$\begin{aligned} Z_2 &= z_2 + \tfrac{1}{\rho_2}v, \\ Z_3 &= z_3 + \varepsilon\sin(t), \text{ and} \\ \zeta &= r + \varepsilon\cos(t), \end{aligned} \tag{6.89}$$

which give the time-varying system

$$\begin{cases} \dot{z}_1 = u + \left[Z_2 - \tfrac{1}{\rho_2}v\right][\zeta - \varepsilon\cos(t)] \\ \dot{Z}_2 = -z_1[\zeta - \varepsilon\cos(t)] - \tfrac{\rho_1}{\rho_2}u[\zeta - \varepsilon\cos(t)] \\ \dot{Z}_3 = \zeta \\ \dot{v} = -\rho_1 u[\zeta - \varepsilon\cos(t)] - \rho_2 v \\ \dot{u} = \nu_1 \\ \dot{\zeta} = \nu_2 - \varepsilon\sin(t) \ . \end{cases} \tag{6.90}$$

To summarize, the problem of tracking the reference trajectory (6.85) for the system (6.84) is equivalent to the problem of globally uniformly asymptotically stabilizing the origin of (6.90). We carry out the stabilization and Lyapunov function construction for (6.90) in five steps, as follows.

Step 1. A Reduced System

We first consider the problems of (a) globally uniformly asymptotically stabilizing the origin of

$$\begin{cases} \dot{z}_1 = u_f + \left[Z_2 - \frac{1}{\rho_2}v\right][\zeta_f - \varepsilon\cos(t)] \\ \dot{Z}_2 = -z_1[\zeta_f - \varepsilon\cos(t)] - \frac{\rho_1}{\rho_2}u_f[\zeta_f - \varepsilon\cos(t)] \\ \dot{Z}_3 = \zeta_f \\ \dot{v} = -\rho_1 u_f[\zeta_f - \varepsilon\cos(t)] - \rho_2 v\ , \end{cases} \tag{6.91}$$

where u_f and ζ_f are new inputs and (b) finding a strict Lyapunov function for the closed-loop system (6.91). We later use backstepping to handle the strict Lyapunov function construction for the original dynamics (6.90).

We choose the feedbacks

$$u_f(t, z_1, Z_2, Z_3) = -\frac{\rho_2}{\rho_1}z_1 - \frac{\varepsilon\rho_2}{\rho_1}Z_2[\cos(t) + Z_3] \ \text{ and } \ \zeta_f(Z_3) = -\varepsilon Z_3\ . \tag{6.92}$$

They result in

$$\begin{cases} \dot{z}_1 = -\frac{\rho_2}{\rho_1}z_1 - \frac{\varepsilon\rho_2}{\rho_1}Z_2[\cos(t) + Z_3] - \varepsilon\left(Z_2 - \frac{1}{\rho_2}v\right)[\cos(t) + Z_3] \\ \dot{Z}_2 = -\varepsilon^2 Z_2[\cos(t) + Z_3]^2 \\ \dot{Z}_3 = -\varepsilon Z_3 \\ \dot{v} = -\varepsilon\rho_2\big[z_1 + \varepsilon Z_2(\cos(t) + Z_3)\big]\big[\cos(t) + Z_3\big] - \rho_2 v\ , \end{cases} \tag{6.93}$$

or equivalently,

$$\begin{cases} \dot{z}_1 = -\frac{\rho_2}{\rho_1}z_1 + \frac{\varepsilon}{\rho_2}\cos(t)v + h_1(t, v, Z_2, Z_3) \\ \dot{v} = -\rho_2 v - \varepsilon\rho_2\cos(t)z_1 + h_2(t, z_1, Z_2, Z_3) \\ \dot{Z}_2 = -\varepsilon^2 Z_2\big[\cos(t) + Z_3\big]^2 \\ \dot{Z}_3 = -\varepsilon Z_3\ , \end{cases} \tag{6.94}$$

where

$$\begin{aligned} h_1(t, v, Z_2, Z_3) &= \frac{\varepsilon}{\rho_2}vZ_3 - \varepsilon\big(1 + \frac{\rho_2}{\rho_1}\big)Z_2\big(\cos(t) + Z_3\big) \\ h_2(t, z_1, Z_2, Z_3) &= -\varepsilon^2\rho_2\cos(t)Z_2(\cos(t) + Z_3) \\ &\quad -\varepsilon\rho_2\big[z_1 + \varepsilon Z_2(\cos(t) + Z_3)\big]Z_3\ . \end{aligned} \tag{6.95}$$

Step 2. Strict Lyapunov Function (z_1, v)-subsystem

For the (z_1, v)-subsystem of (6.94) in the absence of h_1 and h_2, one can easily check that the positive definite quadratic function

$$V_1(z_1, v) = \frac{1}{2}\left[\rho_2^2 z_1^2 + v^2\right] \tag{6.96}$$

is a strict Lyapunov function, since along the trajectories of this subsystem, we have

$$\begin{aligned}\dot{V}_1 &= \rho_2^2 z_1\left[-\frac{\rho_2}{\rho_1}z_1 + \frac{\varepsilon}{\rho_2}v\cos(t)\right] - \rho_2 v^2 - v\varepsilon\rho_2 z_1\cos(t)\\ &= -\frac{\rho_2^3}{\rho_1}z_1^2 - \rho_2 v^2\,.\end{aligned} \tag{6.97}$$

Therefore, when h_1 and h_2 are present, we have

$$\dot{V}_1 = -\frac{\rho_2^3}{\rho_1}z_1^2 - \rho_2 v^2 + \rho_2^2 z_1 h_1(t, v, Z_2, Z_3) + v h_2(t, z_1, Z_2, Z_3)\,. \tag{6.98}$$

Using the triangular inequality, we obtain

$$\dot{V}_1 \leq -\frac{\rho_2^3}{2\rho_1}z_1^2 - \frac{\rho_2}{2}v^2 + \frac{\rho_1\rho_2}{2}h_1^2(t, v, Z_2, Z_3) + \frac{h_2^2(t, z_1, Z_2, Z_3)}{2\rho_2}\,. \tag{6.99}$$

We next construct a global strict Lyapunov function for the system (6.94).

Step 3. Strict Lyapunov Function for (Z_2, Z_3)-subsystem

We first replace V_1 with the function

$$W_1(z_1, v) = \ln\left(V_1(z_1, v) + 1\right),$$

which is a positive definite and radially unbounded function. Its time derivative along the trajectories of (6.94) satisfies

$$\begin{aligned}\dot{W}_1 \leq &-\frac{\rho_2^3}{2\rho_1(V_1(z_1, v) + 1)}z_1^2 - \frac{\rho_2}{2(V_1(z_1, v) + 1)}v^2\\ &+\frac{\rho_1\rho_2}{2}\frac{h_1^2(t, v, Z_2, Z_3)}{V_1(z_1, v) + 1} + \frac{1}{2\rho_2}\frac{h_2^2(t, z_1, Z_2, Z_3)}{V_1(z_1, v) + 1}\,.\end{aligned} \tag{6.100}$$

Since

$$\begin{aligned} \left|h_1(t,v,Z_2,Z_3)\right| &\leq \tfrac{\varepsilon}{\rho_2}|vZ_3| + \varepsilon\left(1+\tfrac{\rho_2}{\rho_1}\right)|Z_2|\left(1+|Z_3|\right) \text{ and} \\ \left|h_2(t,z_1,Z_2,Z_3)\right| &\leq \varepsilon\rho_2|z_1||Z_3| + 2\varepsilon^2\rho_2|Z_2|\left(1+|Z_3|\right)^2 \end{aligned} \tag{6.101}$$

hold everywhere, the inequalities

$$\begin{aligned} \frac{h_1^2(t,v,Z_2,Z_3)}{V_1(z_1,v)+1} &\leq 4\frac{\varepsilon^2}{\rho_2^2}Z_3^2 + 2\varepsilon^2\left(1+\frac{\rho_2}{\rho_1}\right)^2 Z_2^2\left(1+|Z_3|\right)^2 \text{ and} \\ \frac{h_2^2(t,z_1,Z_2,Z_3)}{V_1(z_1,v)+1} &\leq 4\varepsilon^2 Z_3^2 + 8\varepsilon^4\rho_2^2 Z_2^2\left(1+|Z_3|\right)^4 , \end{aligned} \tag{6.102}$$

are satisfied and therefore

$$\begin{aligned} \dot{W}_1 \leq &-\frac{\rho_2^3}{2\rho_1(V_1(z_1,v)+1)}z_1^2 - \frac{\rho_2}{2(V_1(z_1,v)+1)}v^2 + K_1(\varepsilon)Z_3^2 \\ &+K_2(\varepsilon)Z_2^2\left(1+|Z_3|\right)^4 , \end{aligned} \tag{6.103}$$

where

$$K_1(\varepsilon) = 2\varepsilon^2\frac{\rho_1+1}{\rho_2} \quad \text{and} \quad K_2(\varepsilon) = \varepsilon^2\rho_1\rho_2\left(1+\tfrac{\rho_2}{\rho_1}\right)^2 + 4\varepsilon^4\rho_2 . \tag{6.104}$$

This gives

$$\begin{aligned} \dot{W}_1 \leq &-\frac{\rho_2^3}{2\rho_1(V_1(z_1,v)+1)}z_1^2 - \frac{\rho_2}{2(V_1(z_1,v)+1)}v^2 \\ &+2K_1(\varepsilon)V_2(Z_2,Z_3) \\ &+2K_2(\varepsilon)V_2(Z_2,Z_3)\left(1+\sqrt{2V_2(Z_2,Z_3)}\right)^4 , \end{aligned} \tag{6.105}$$

where

$$V_2(Z_2,Z_3) = \frac{1}{2}\left[Z_2^2 + Z_3^2\right] . \tag{6.106}$$

We deduce that

$$\begin{aligned} \dot{W}_1 \leq &-\frac{\rho_2^3}{2\rho_1(V_1(z_1,v)+1)}z_1^2 - \frac{\rho_2}{2(V_1(z_1,v)+1)}v^2 \\ &+2K_1(\varepsilon)V_2(Z_2,Z_3) \\ &+16K_2(\varepsilon)\left[1+4V_2^2(Z_2,Z_3)\right]V_2(Z_2,Z_3) . \end{aligned} \tag{6.107}$$

Next note that V_2 as defined in (6.106) is a positive definite quadratic function. It is also a weak Lyapunov function for the (Z_2, Z_3)-subsystem of the system (6.94), because its derivative along the trajectories of (6.94) satisfies

$$\dot{V}_2 = -\varepsilon^2 Z_2^2 \left[\cos(t) + Z_3\right]^2 - \varepsilon Z_3^2 \,. \tag{6.108}$$

We next construct a strict Lyapunov function for the (Z_2, Z_3)-subsystem of (6.94) by using V_2. To this end, first note that if $|Z_3| \geq \frac{1}{2}|\cos(t)|$, then (6.108) gives

$$\dot{V}_2 \leq -\frac{\varepsilon}{4}\cos^2(t) \leq -\frac{\varepsilon^2}{4(1+\varepsilon)}\cos^2(t)\frac{V_2(Z_2,Z_3)}{1+V_2(Z_2,Z_3)} \tag{6.109}$$

while if $|Z_3| \leq \frac{1}{2}|\cos(t)|$, then

$$\begin{aligned} \dot{V}_2 &\leq -\frac{\varepsilon^2}{4} Z_2^2 \cos^2(t) - \varepsilon Z_3^2 \\ &\leq -\frac{\varepsilon^2}{4(1+\varepsilon)}\cos^2(t)\left[Z_2^2 + Z_3^2\right] \\ &\leq -\frac{\varepsilon^2}{4(1+\varepsilon)}\cos^2(t)\frac{V_2(Z_2,Z_3)}{1+V_2(Z_2,Z_3)} \,. \end{aligned} \tag{6.110}$$

Therefore, in either case,

$$\dot{V}_2 \leq -\frac{\varepsilon^2}{8(1+\varepsilon)}\cos^2(t)\frac{V_2(Z_2,Z_3)}{0.5[1+V_2(Z_2,Z_3)]} \tag{6.111}$$

for all t, Z_2, and Z_3.

We can now obtain the strict Lyapunov function for the (Z_2, Z_3)-subsystem of (6.94) using the construction from Sect. 6.4. In fact, it follows from Remark 6.5 applied with

$$T = 2\pi, \;\; p(t) = \cos^2(t), \;\; L(v) = 0.5[1+v], \;\; \text{and} \;\; a = \frac{\varepsilon^2}{8(1+\varepsilon)}$$

that the time derivative of the function

$$\begin{aligned} &W_2(t, Z_2, Z_3) \\ &= \exp\left(\tfrac{\varepsilon^2(2\pi+\sin(2t))}{32(1+\varepsilon)}\right)\exp\left(V_2(Z_2,Z_3)\right)V_2(Z_2,Z_3) \end{aligned} \tag{6.112}$$

along the trajectories of (6.94) satisfies

$$\begin{aligned} \dot{W}_2 &\leq -\frac{\varepsilon^2}{16(1+\varepsilon)} W_2(t, Z_2, Z_3) \\ &\leq -\frac{\varepsilon^2}{16(1+\varepsilon)}\exp\left(\frac{\varepsilon^2(\pi-1)}{32(1+\varepsilon)}\right)\exp\left(V_2(Z_2,Z_3)\right)V_2(Z_2,Z_3) \\ &\leq -\frac{\varepsilon^2}{16(1+\varepsilon)}\exp\left(\frac{\varepsilon^2(\pi-1)}{32(1+\varepsilon)}\right)\left[1+\tfrac{1}{2}V_2^2(Z_2,Z_3)\right]V_2(Z_2,Z_3) \,. \end{aligned} \tag{6.113}$$

Hence, W_2 is a strict Lyapunov function for the (Z_2, Z_3)-subsystem of (6.94).

Step 4. Full Reduced System

We now show that

$$W_3(t, z_1, v, Z_2, Z_3) = W_1(z_1, v) + K_3(\varepsilon) W_2(t, Z_2, Z_3) \tag{6.114}$$

is a strict Lyapunov function for the system (6.94) when we choose

$$K_3(\varepsilon) = \frac{256(1+\varepsilon)[K_1(\varepsilon) + 9K_2(\varepsilon)]}{\varepsilon^2} \exp\left(-\frac{\varepsilon^2(\pi - 1)}{32(1+\varepsilon)}\right) . \tag{6.115}$$

To this end, first note that we can find functions $\gamma_s, \gamma_l \in \mathcal{K}_\infty$ such that

$$\gamma_s\big(|(z_1, v, Z_2, Z_3)|\big) \le W_3(t, z_1, v, Z_2, Z_3) \le \gamma_l\big(|(z_1, v, Z_2, Z_3)|\big) \tag{6.116}$$

for all t, z_1, v, Z_2, and Z_3. We deduce from (6.113) and (6.107) that

$$\dot{W}_3 \le -\Omega(z_1, v, Z_2, Z_3) \tag{6.117}$$

along the trajectories of (6.94) in closed-loop with (6.92), where

$$\Omega(z_1, v, Z_2, Z_3) = \frac{\rho_2}{2(V_1(z_1, v) + 1)} \left[\frac{\rho_2^2}{\rho_1} z_1^2 + v^2\right] + 8K_2(\varepsilon)\left[Z_2^2 + Z_3^2\right]$$

is a positive definite function.

Step 5. Backstepping

We now use our results for the reduced system to find stabilizing controls and associated strict Lyapunov functions for the original system (6.90). We apply the classical backstepping approach. Omitting arguments of some of the functions, we can rewrite (6.90) as

$$\left\{\begin{aligned} \dot{z}_1 &= u_f + \left[Z_2 - \tfrac{1}{\rho_2} v\right][\zeta_f - \varepsilon\cos(t)] + \mathcal{R}_1 \\ \dot{Z}_2 &= -z_1[\zeta_f - \varepsilon\cos(t)] - \tfrac{\rho_1}{\rho_2} u_f[\zeta_f - \varepsilon\cos(t)] + \mathcal{R}_2 \\ \dot{Z}_3 &= \zeta_f + \mathcal{R}_3 \\ \dot{v} &= -\rho_1 u_f[\zeta_f - \varepsilon\cos(t)] - \rho_2 v + \mathcal{R}_4 \\ \dot{\tilde{u}} &= \nu_1 + \mathcal{R}_5 \\ \dot{\tilde{\zeta}} &= \nu_2 + \mathcal{R}_6 , \end{aligned}\right. \tag{6.118}$$

where

$$\tilde{u} = u - u_f(t, z_1, Z_2, Z_3) \ , \ \tilde{\zeta} = \zeta - \zeta_f(Z_3), \tag{6.119}$$

u_f and ζ_f are defined in (6.92), and the functions $\mathcal{R}_i$ are defined as follows:

$$\begin{aligned}
\mathcal{R}_1(Z_2, v, \tilde{u}, \tilde{\zeta}) &= \tilde{u} + \left[Z_2 - \tfrac{1}{\rho_2} v\right] \tilde{\zeta} \ , \\
\mathcal{R}_2(t, z_1, Z_2, Z_3, \tilde{u}, \tilde{\zeta}) &= -z_1\tilde{\zeta} - \tfrac{\rho_1}{\rho_2}[\tilde{u}\tilde{\zeta} + u_f\tilde{\zeta} + \tilde{u}\zeta_f - \tilde{u}\varepsilon\cos(t)] \ , \\
\mathcal{R}_3(\tilde{\zeta}) &= \tilde{\zeta} \ , \\
\mathcal{R}_4(t, z_1, Z_2, Z_3, \tilde{u}, \tilde{\zeta}) &= -\rho_1[\tilde{u}\tilde{\zeta} + u_f\tilde{\zeta} + \tilde{u}\zeta_f] + \varepsilon\cos(t)\rho_1\tilde{u} \ , \\
\mathcal{R}_5(t, z_1, Z_2, Z_3, \tilde{u}, \tilde{\zeta}) &= -\dot{u}_f \ , \text{ and} \\
\mathcal{R}_6(t, Z_3) &= -\varepsilon\sin(t) - \dot{\zeta}_f \ .
\end{aligned} \tag{6.120}$$

It easily follows from (6.117) that the time derivative of

$$W_4(t, z_1, v, Z_2, Z_3, \tilde{u}, \tilde{\zeta}) = W_3(t, z_1, v, Z_2, Z_3) + \frac{1}{2}\left[\tilde{u}^2 + \tilde{\zeta}^2\right] \tag{6.121}$$

along the solutions of (6.118) satisfies

$$\begin{aligned}
\dot{W}_4 \leq &-\Omega(z_1, v, Z_2, Z_3) + \frac{\partial W_4}{\partial z_1}\mathcal{R}_1 + \frac{\partial W_4}{\partial Z_2}\mathcal{R}_2 + \frac{\partial W_4}{\partial Z_3}\mathcal{R}_3 \\
&+\frac{\partial W_4}{\partial v}\mathcal{R}_4 + \tilde{u}[\nu_1 + \mathcal{R}_5] + \tilde{\zeta}[\nu_2 + \mathcal{R}_6] \ .
\end{aligned} \tag{6.122}$$

Hence, choosing

$$\begin{aligned}
\nu_1 = &-\tilde{u} - \mathcal{R}_5 - \frac{\partial W_4}{\partial z_1} + \frac{\partial W_4}{\partial Z_2}\frac{\rho_1}{\rho_2}[\zeta_f - \varepsilon\cos(t)] \\
&+\frac{\partial W_4}{\partial v}\rho_1[\zeta_f - \varepsilon\cos(t)] \ \text{ and} \\
\nu_2 = &-\tilde{\zeta} - \mathcal{R}_6 + \frac{\partial W_4}{\partial z_1}\left[-Z_2 + \tfrac{1}{\rho_2}v\right] \\
&+\frac{\partial W_4}{\partial Z_2}\left[z_1 + \frac{\rho_1}{\rho_2}(\tilde{u} + u_f)\right] - \frac{\partial W_4}{\partial Z_3} + \rho_1\frac{\partial W_4}{\partial v}[\tilde{u} + u_f]
\end{aligned} \tag{6.123}$$

gives

$$\dot{W}_4 \leq -\overline{\Omega}(z_1, v, Z_2, Z_3, \tilde{u}, \tilde{\zeta}) \ , \tag{6.124}$$

where

$$\overline{\Omega}(z_1, v, Z_2, Z_3, \tilde{u}, \tilde{\zeta}) = \Omega(z_1, v, Z_2, Z_3) + \tilde{u}^2 + \tilde{\zeta}^2 \ . \tag{6.125}$$

The function $\overline{\Omega}$ is a positive definite function of $(z_1, v, Z_2, Z_3, \tilde{u}, \tilde{\zeta})$ and there are functions $\alpha_s, \alpha_l \in \mathcal{K}_\infty$ such that

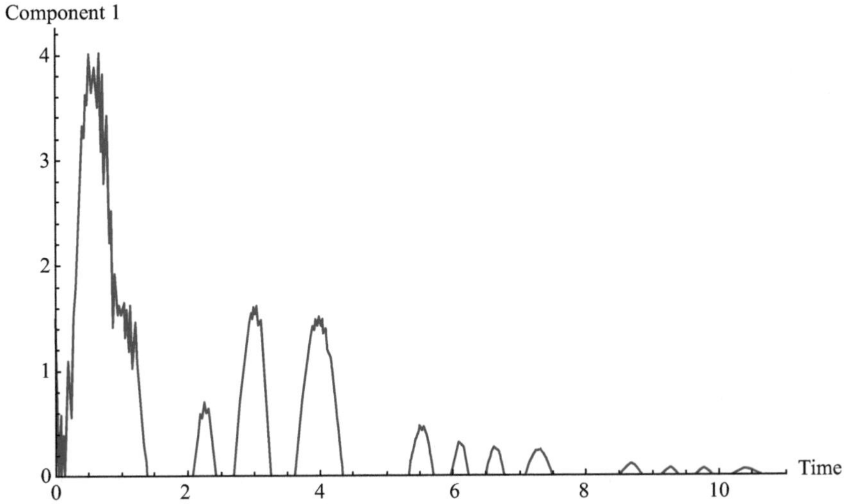

Fig. 6.1 z_1 component of (6.84)

$$\alpha_s\left(|(z_1, v, Z_2, Z_3, \tilde{u}, \tilde{\zeta})|\right) \leq W_4(t, z_1, v, Z_2, Z_3, \tilde{u}, \tilde{\zeta}) \\ \leq \alpha_l\left(|(z_1, v, Z_2, Z_3, \tilde{u}, \tilde{\zeta})|\right) . \tag{6.126}$$

Therefore W_4 is a strict Lyapunov function for the system (6.90). This completes the construction.

To validate our feedback design, we simulated (6.118) with the preceding controllers, the model parameters

$$\rho_1 = \frac{m_{11}}{m_{22}} = 0.1 \quad \text{and} \quad \rho_2 = \frac{d_{22}}{m_{22}} = 0.5, \tag{6.127}$$

the choice $\varepsilon = 0.1$, and the initial state $(1, 2, 1, 2, 1, 2)$. In Figs. 6.1-6.3, we show the corresponding trajectories for z_1, z_2, and z_3, respectively from the original transformed system (6.84). Our theory says that we should have

$$\lim_{t\to\infty} (z_1, z_2)(t) = (0, 0) \quad \text{and} \quad \lim_{t\to\infty} |z_3(t) + 0.1\sin(t)| = 0.$$

This is the type of behavior we obtained in our simulations.

6.9 Comments

The idea of transforming non-strict Lyapunov functions into strict ones has been explored by several other authors. A very different approach to strictifying was pursued by Angeli, Sontag, and Wang in [8]. There it was as-

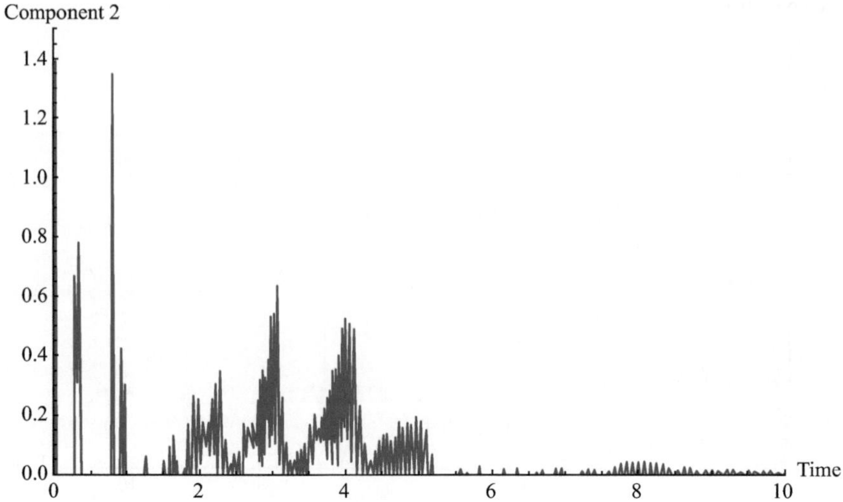

Fig. 6.2 z_2 component of (6.84)

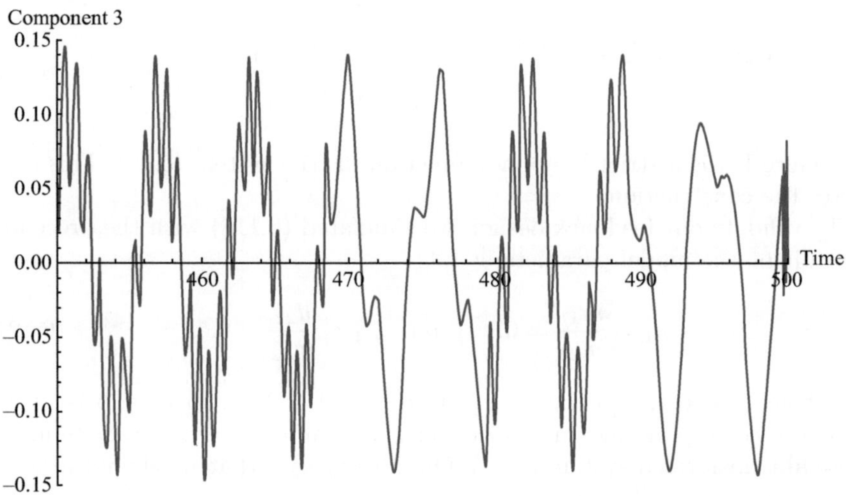

Fig. 6.3 z_3 component of (6.84)

sumed that the given classical system $\dot{x} = f(x, u)$ was *zero-output (smoothly) dissipative*, meaning there exist a smooth, proper, positive definite function $W : \mathbb{R}^n \to \mathbb{R}$ and $\sigma \in \mathcal{K}$ such that $\nabla W(x)f(x, u) \leq \sigma(|u|)$ for all $x \in \mathbb{R}^n$ and $u \in \mathbb{R}^m$. Assuming that the system is also 0-GAS, one then obtains a proper function V_0 and $k \in \mathcal{K}$ such that $V \doteq k(V_0)$ satisfies an estimate of the form

$$\nabla V(x)f(x, u) \leq -a(|x|) + b(|u|),$$

where a is only positive definite and $b \in \mathcal{K}$. The desired iISS Lyapunov function of the original system is then $V_1 \doteq W + V$. The approach in [8] is based on abstract optimal control representations of Lyapunov functions, but it does not provide ISS Lyapunov functions.

Another method for synthesizing strict Lyapunov functions involves the "change of supply rates" approach [167]. The idea is to consider supply pairs for time-invariant systems

$$\dot{x} = f(x, u). \tag{6.128}$$

A supply pair for the system (6.128) is a pair (γ, α) of $\mathcal{K}_\infty$ functions that admits a storage function V so that

$$\nabla V(x) f(x, u) \leq \gamma(|u|) - \alpha(|x|)$$

for all $x \in \mathbb{R}^n$ and $u \in \mathbb{R}^m$. The main result of [167] is the following: *If (γ, α) is a supply pair for (6.128) and $\tilde{\alpha} \in \mathcal{K}_\infty$ is such that $\tilde{\alpha}(r) = \mathrm{O}[\alpha(r)]$ as $r \to 0^+$, then there is a function $\tilde{\gamma} \in \mathcal{K}_\infty$ such that $(\tilde{\gamma}, \tilde{\alpha})$ is also a supply pair for (6.128).* As a corollary, we have the following [167]:

Corollary 6.1. *Given two ISS time-invariant systems, we can find functions $\tilde{\alpha}_1, \tilde{\alpha}_2, \tilde{\gamma} \in \mathcal{K}_\infty$ so that $(0.5\tilde{\alpha}_2, \tilde{\alpha}_1)$ is a supply pair for the first system and $(\tilde{\gamma}, \tilde{\alpha}_2)$ is a supply pair for the second system.*

Applying this corollary to ISS systems $\dot{z} = f(z, u)$ and $\dot{x} = g(x, u)$, and letting $V_1(z)$ and $V_2(x)$ denote the corresponding storage functions, it readily follows that the storage function $V(x, z) \doteq V_1(z) + V_2(x)$ satisfies

$$\dot{V}(x, z, u) \leq \tilde{\gamma}(|u|) - \frac{1}{2}\tilde{\alpha}_2(|x|) - \tilde{\alpha}_1(|z|)$$

along the trajectories of the cascade interconnection

$$\begin{cases} \dot{z} = f(z, x) \\ \dot{x} = g(x, u) \end{cases}. \tag{6.129}$$

This gives a way of building strict Lyapunov functions for certain systems of the form (6.129), assuming V_1 and V_2 are known. See [121] for analogous change of supply results for discrete time systems, and [61] for related results for time-varying interconnections satisfying appropriate small gain conditions.

Definition 6.3 is a nonlinear version of the property used in [85] to ensure UGAS of certain types of time-varying linear systems. Thus, the explicit construction of a strict DIS Lyapunov function in terms of a given DIS(p) Lyapunov function we presented extends [85], where only linear systems are studied and no strict Lyapunov function is constructed. Our proof of Theorem 6.1 closely follows [98].

There is a large literature on controlling rotating rigid bodies. See [33, 117, 118] for the background and motivation for this problem. One motivation is that a given attitude for a rigid spacecraft having two controls cannot be asymptotically stabilized by a pure state feedback, because Brockett's Necessary Condition does not hold [117]. The paper [118] gave smooth time-varying feedbacks based on center manifold theory, Lyapunov techniques, and time averaging, while [30] proved exponential stability based on a periodic switching between two control laws. Then [117] refined these results to give local exponential stability based on a cascaded high-gain control result and a single control expression. The novelty of our treatment of this model is the explicit construction of a global strict Lyapunov function for the corresponding error systems.

Our result for rotating rigid bodies is from [91]. Tracking problems for underactuated ships have been solved in several works (such as [46, 63, 134]). Also, tracking and practical stability have been proved in [133], which constructs a candidate global Lyapunov function for a suitable averaged closed-loop system. In [112], global uniform asymptotic stability of the origin of the ship model is achieved; using our strictification technique, strict Lyapunov functions can be constructed in that case as well. Our backstepping approach to the ship model is based on ideas from [31].

Chapter 7
Backstepping for Time-Varying Systems

Abstract Backstepping is one of the most popular frameworks for designing controllers for nonlinear systems. Its multiple advantages are well-known. It leads to a wide family of globally asymptotically stabilizing control laws, and it makes it possible to address robustness issues and solve adaptive control problems. This chapter begins with a review of classical backstepping for time-invariant systems. We then give several extensions that lead to time-varying strict Lyapunov functions and stabilizing feedbacks for time-varying systems. We first consider a general class of linear time-varying systems. Then we provide stronger results for linear systems in feedback form. Finally, we study nonlinear systems in feedback form and give conditions ensuring globally uniform stabilizability by bounded control laws.

7.1 Motivation: PVTOL

To motivate our results, we first consider the plane with vertical take off and landing (PVTOL) model; see, e.g., [149, Chap. 6] or Sect. 7.9 for the literature on the model. In the absence of disturbances, the equations of the PVTOL model are

$$\left\{\begin{aligned} \dot{\xi}_1 &= \xi_2 \\ \dot{\xi}_2 &= -u_1 \sin(\theta) \\ \dot{z}_1 &= z_2 \\ \dot{z}_2 &= u_1 \cos(\theta) - 1 \\ \dot{\theta} &= \omega \\ \dot{\omega} &= u_2 \,, \end{aligned}\right. \tag{7.1}$$

where ξ_1 and z_1 are the horizontal and vertical positions of the aircraft center of mass, respectively; and θ is the roll angle that the aircraft makes with the horizon. The control inputs u_1 and u_2 are the thrust (directed out from the bottom of the aircraft) and the angular acceleration (a.k.a. rolling moment), respectively. The coefficient -1 in the z_2-dynamics is the normalized gravitational acceleration.

Assume that we wish to track the following admissible trajectory for (7.1):

$$\big(\xi_{1,r}, \xi_{2,r}, z_{1,r}, z_{2,r}, \theta_r, \omega_r\big)(t) \;=\; \big(0, 0, 2\cos(3t), -6\sin(3t), 0, 0\big)\;. \tag{7.2}$$

The inputs corresponding to (7.2) are

$$u_{1,r}(t) = 1 - 18\cos(3t) \;\text{ and }\; u_{2,r}(t) = 0. \tag{7.3}$$

Using the variables $\tilde{\xi}_i = \xi_i - \xi_{i,r}(t)$ and $\tilde{z}_i = z_i - z_{i,r}(t)$ for $i = 1, 2$, $\tilde{\theta} = \theta - \theta_r(t)$, and $\tilde{\omega} = \omega - \omega_r(t)$, and the change of feedback

$$\tilde{u}_1 = u_1 - u_{1,r}(t)\;,\; \tilde{u}_2 = u_2 - u_{2,r}(t) \tag{7.4}$$

gives the error dynamics

$$\begin{cases} \dot{\tilde{\xi}}_1 = \tilde{\xi}_2 \\ \dot{\tilde{\xi}}_2 = -\big[\tilde{u}_1 + 1 - 18\cos(3t)\big]\sin(\tilde{\theta}) \\ \dot{\tilde{z}}_1 = \tilde{z}_2 \\ \dot{\tilde{z}}_2 = \big[\tilde{u}_1 + 1 - 18\cos(3t)\big]\cos(\tilde{\theta}) - 1 + 18\cos(3t) \\ \dot{\tilde{\theta}} = \tilde{\omega} \\ \dot{\tilde{\omega}} = \tilde{u}_2. \end{cases} \tag{7.5}$$

We wish to find feedback stabilizers that render (7.5) UGAS to the origin.

To this end, we first consider the auxiliary system

$$\begin{cases} \dot{\tilde{\xi}}_1 = \tilde{\xi}_2 \\ \dot{\tilde{\xi}}_2 = -\big[\tilde{u}_1 + 1 - 18\cos(3t)\big]\sin(v_2) \\ \dot{\tilde{z}}_1 = \tilde{z}_2 \\ \dot{\tilde{z}}_2 = \big[\tilde{u}_1 + 1 - 18\cos(3t)\big]\cos(v_2) - 1 + 18\cos(3t) \end{cases} \tag{7.6}$$

with $\tilde{u}_1$ and v_2 as inputs. Assume for the moment that we have constructed two control laws

$$\tilde{u}_{1s}\big(t, \tilde{\xi}_1, \tilde{\xi}_2, \tilde{z}_1, \tilde{z}_2\big) \;\text{ and }\; v_{2s}\big(t, \tilde{\xi}_1, \tilde{\xi}_2, \tilde{z}_1, \tilde{z}_2\big)$$

that have period 2π in t and that render the origin of the system (7.6) UGAS. Then a variant of classical backstepping (which we review in Sect. 7.2.3) gives a control law $\mu_s(t, \tilde{\xi}_1, \tilde{\xi}_2, \tilde{z}_1, \tilde{z}_2, \tilde{\theta})$ that also has period 2π in t such that

$$
\begin{cases}
\dot{\tilde{\xi}}_1 = \tilde{\xi}_2 \\
\dot{\tilde{\xi}}_2 = -\left[\tilde{u}_{1s}(t, \tilde{\xi}_1, \tilde{\xi}_2, \tilde{z}_1, \tilde{z}_2) + 1 - 18\cos(3t)\right] \sin(\tilde{\theta}) \\
\dot{\tilde{z}}_1 = \tilde{z}_2 \\
\dot{\tilde{z}}_2 = \left[\tilde{u}_{1s}(t, \tilde{\xi}_1, \tilde{\xi}_2, \tilde{z}_1, \tilde{z}_2) + 1 - 18\cos(3t)\right] \cos(\tilde{\theta}) - 1 + 18\cos(3t) \\
\dot{\tilde{\theta}} = \mu_s(t, \tilde{\xi}_1, \tilde{\xi}_2, \tilde{z}_1, \tilde{z}_2, \tilde{\theta})
\end{cases}
\tag{7.7}
$$

is also UGAS to the origin. Repeating this argument gives a control law

$$
\tilde{u}_{2s}(t, \tilde{\xi}_1, \tilde{\xi}_2, \tilde{z}_1, \tilde{z}_2, \tilde{\theta}, \tilde{\omega}),
$$

also having period 2π in t, such that the origin of (7.5) in closed-loop with $\tilde{u}_{1s}(t, \tilde{\xi}_1, \tilde{\xi}_2, \tilde{z}_1, \tilde{z}_2)$ and $\tilde{u}_{2s}(t, \tilde{\xi}_1, \tilde{\xi}_2, \tilde{z}_1, \tilde{z}_2, \tilde{\theta}, \tilde{\omega})$ is UGAS. However, it is by no means clear how to construct the necessary control laws $\tilde{u}_{1s}(t, \tilde{\xi}_1, \tilde{\xi}_2, \tilde{z}_1, \tilde{z}_2)$ and $v_{2s}(t, \tilde{\xi}_1, \tilde{\xi}_2, \tilde{z}_1, \tilde{z}_2)$ to stabilize (7.6). We will return to this example in Sect. 7.8, where we construct $\tilde{u}_{1s}$ and v_{2s} as a special case of a general backstepping theory for time-varying systems.

7.2 Classical Backstepping

Backstepping involves constructing stabilizing controllers for nonlinear systems having a lower triangular structure called *feedback form.* The backstepping approach is not a single technique, but rather is a collection of techniques sharing some key ideas. There is a backstepping technique based on cancelation of nonlinearities, and another involving domination of nonlinearities. We review these two methods next. Throughout the chapter, all inequalities and equalities should be understood to hold globally unless otherwise indicated, and we omit the arguments of our functions when they are clear from the context. Also, we assume that all of the functions encountered are sufficiently smooth.

7.2.1 Backstepping with Cancelation

We first recall the most important steps of backstepping by applying a basic version of backstepping with cancelation (which is also called *exact backstepping*) to the following family of time-invariant systems:

$$
\begin{cases}
\dot{x}_i = x_{i+1} + f_i(x_1, x_2, ..., x_i), \quad 1 \leq i \leq n-1 \\
\dot{x}_n = u + f_n(x_1, x_2, ..., x_n)
\end{cases}
\tag{7.8}
$$

where each $x_i \in \mathbb{R}$, $u \in \mathbb{R}$ is the input, and each function f_i is assumed to be zero at the origin and C^1. Systems of the form (7.8) are said to be in *strict feedback form*.

The key feature of (7.8) is that each $\dot{x}_i$ depends only on $x_1, x_2, ..., x_{i+1}$ and is affine in x_{i+1}. The idea behind backstepping is to consider x_2 as a "pseudo-control" (which is also frequently called a "virtual input") for the x_1-subsystem. Thus, if it were possible to simply replace x_2 with $-x_1 - f_1(x_1)$, then the x_1-subsystem would become

$$\dot{x}_1 = -x_1 \tag{7.9}$$

which has the Lyapunov function $V_1(x_1) = \frac{1}{2}x_1^2$. Since x_2 cannot be replaced with $-x_1 - f_1(x_1)$, we instead use the change of coordinates

$$\begin{aligned} z_1 &= x_1 \\ z_2 &= x_2 - \alpha_1(x_1) \end{aligned} \tag{7.10}$$

where $\alpha_1(x_1) = -x_1 - f_1(x_1)$. This change of coordinates transforms the (x_1, x_2)-subsystem of (7.8) into

$$\begin{cases} \dot{z}_1 = -z_1 + z_2 \\ \dot{z}_2 = x_3 + \overline{f}_2(z_1, z_2) \, , \end{cases} \tag{7.11}$$

where

$$\overline{f}_2(z_1, z_2) = f_2(x_1, x_2) - \alpha_1'(x_1)[x_2 + f_1(x_1)].$$

The time derivative of $V_1(z_1)$ along the trajectories of (7.11) satisfies

$$\dot{V}_1 = -z_1^2 + z_1 z_2 \, . \tag{7.12}$$

Assume $n \geq 4$. The backstepping now proceeds recursively. We view x_3 in (7.11) as a virtual input, and we use the new coordinate $z_3 = x_3 - \alpha_2(z_1, z_2)$, where $\alpha_2(z_1, z_2) = -z_1 - z_2 - \overline{f}_2(z_1, z_2)$. This gives the system

$$\begin{cases} \dot{z}_1 = -z_1 + z_2 \\ \dot{z}_2 = z_3 + \alpha_2(z_1, z_2) + \overline{f}_2(z_1, z_2) = z_3 - z_1 - z_2 \\ \dot{z}_3 = x_4 + \overline{f}_3(z_1, z_2, z_3), \end{cases} \tag{7.13}$$

where

$$\overline{f}_3(z_1, z_2, z_3) = f_3(z_1, z_2 + \alpha_1(z_1), z_3 + \alpha_2(z_1, z_2)) - \dot{\alpha}_2(z_1, z_2).$$

The time derivative of

$$V_2(z_1, z_2) = V_1(z_1) + \frac{1}{2}z_2^2 \tag{7.14}$$

along the solutions of (7.13) satisfies

$$\dot{V}_2 = -z_1^2 + z_1 z_2 + z_2(z_3 - z_1 - z_2) = -z_1^2 - z_2^2 + z_2 z_3. \tag{7.15}$$

At the i-th step, the last component of the dynamics is

$$\dot{z}_i = x_{i+1} + \overline{f}_i(z_1, ..., z_i) \tag{7.16}$$

for a suitable function $\overline{f}_i$, and we introduce the variable

$$z_{i+1} = x_{i+1} - \alpha_i(z_1, \ldots, z_i) \,, \tag{7.17}$$

where $\alpha_i(z_1, \ldots, z_i) = -z_{i-1} - z_i - \overline{f}_i(z_1, ..., z_i)$ and

$$V_i(z_1, ..., z_i) = \frac{1}{2} \sum_{r=1}^{i} z_r^2. \tag{7.18}$$

Then

$$\dot{z}_i = z_{i+1} + \alpha_i(z_1, \ldots, z_i) + \overline{f}_i(z_1, ..., z_i) = z_{i+1} - z_{i-1} - z_i \tag{7.19}$$

and the time derivative of V_i along trajectories of the $(z_1, ..., z_i)$-subsystem satisfies

$$\dot{V}_i = -\sum_{r=1}^{i} z_r^2 + z_i z_{i+1}. \tag{7.20}$$

At the last step, we have

$$\dot{z}_n = u + \overline{f}_n(z_1, ..., z_n) \,. \tag{7.21}$$

Choosing

$$u = \alpha_n(z_1, ..., z_n) \doteq -z_{n-1} - z_n - \overline{f}_n(z_1, ..., z_n) \tag{7.22}$$

and

$$V_n(z_1, ..., z_n) = \frac{1}{2} \sum_{r=1}^{n} z_r^2 \tag{7.23}$$

gives

$$\dot{z}_n = -z_{n-1} - z_n \tag{7.24}$$

and

$$\dot{V}_n = -\sum_{r=1}^{n} z_r^2. \tag{7.25}$$

Therefore, the system

$$\begin{cases} \dot{z}_1 = -z_1 + z_2 \\ \dot{z}_i = z_{i+1} - z_{i-1} - z_i, \quad i = 2, 3, \ldots, n-1 \\ \dot{z}_n = -z_{n-1} - z_n \end{cases} \tag{7.26}$$

is GAS. From the definition of the functions α_j, it follows that (7.8) in closed-loop with

$$u(x) = \alpha_n\big(\zeta_1(x), ..., \zeta_n(x)\big) \tag{7.27}$$

where $x = (x_1, .., x_n)$ and

$$\begin{aligned} \zeta_1(x) &= x_1 \\ \zeta_{i+1}(x) &= x_{i+1} - \alpha_i\big(\zeta_1(x), \ldots, \zeta_i(x)\big), \quad i = 1, 2, \ldots, n-1 \end{aligned} \tag{7.28}$$

is GAS.

7.2.2 Backstepping with Domination

The control law $u(x)$ in (7.27) depends explicitly on the nonlinear functions $f_i(x_1, x_2, ..., x_i)$ because

$$\alpha_i(z_1, \ldots, z_i) = -z_{i-1} - z_i - \overline{f}_i(z_1, ..., z_i)$$

for each i. Consequently, when the functions f_i are unknown, the technique does not apply. In [75, pp. 84-85], it is explained how backstepping can be adapted to the case where the functions f_i are replaced by

$$f_i(t, x_1, x_2, ..., x_i, u) = \varphi_i(x_1, ..., x_i)^\top \Delta(t, x, u) \ , \tag{7.29}$$

where $\varphi_i(x_1, ..., x_i)$ is a $(p \times 1)$ vector of known smooth nonlinear functions, and $\Delta(t, x, u)$ is a globally bounded $(p \times 1)$ smooth vector of uncertain nonlinearities.

We next provide a variant of [75, pp. 84-85] that constructs a state feedback to prove UGAS of the uncertain system

$$\begin{cases} \dot{x}_1 = x_2 + \varphi_1(x_1)^\top \Delta_1(t, x, u) \\ \dot{x}_i = x_{i+1} + \varphi_i(x_1, ..., x_i)^\top \Delta_i(t, x, u), \quad i = 2, 3, \ldots, n-1 \\ \dot{x}_n = u + \varphi_n(x_1, ..., x_n)^\top \Delta_n(t, x, u) \end{cases} \tag{7.30}$$

with state space $\mathbb{R}^n$ in feedback form. We do not require the functions Δ_i to be bounded. Rather, we assume that they satisfy

$$\big|\Delta_i(t, x, u)\big| \le \Delta_M \sqrt{\sum_{r=1}^{i} x_r^2} \quad \text{for} \quad i = 1, 2, \ldots, n \tag{7.31}$$

for some known positive constant Δ_M. Let $\overline{\varphi}$ be an everywhere positive, increasing function such that

$$|\varphi_i(x_1, ..., x_i)| \leq \overline{\varphi}\left(\sqrt{\sum_{r=1}^{i} x_r^2}\right) \tag{7.32}$$

for each $i \in \{1, 2, \dots, n\}$ and $x \in \mathbb{R}^n$. Our backstepping involves a change of variables, followed by the construction of an appropriate set of dominating functions.

7.2.2.1 Change of Variables

We introduce the notation $\xi_i = (x_1, ..., x_i)$ for $i = 1, 2, \dots, n$. Given arbitrary positive constants c_i and everywhere positive functions $\kappa_i \in C^n$, we use the variables

$$\begin{aligned} z_1 &= x_1 \\ z_i &= x_i - \alpha_{i-1}(\xi_{i-1}) \qquad \forall i \geq 2 \ , \end{aligned} \tag{7.33}$$

where

$$\begin{aligned} \alpha_1(\xi_1) &= -\big[c_1 + \kappa_1(\xi_1)\big] z_1 \qquad \text{and} \\ \alpha_i(\xi_i) &= -\big[c_i + \kappa_i(\xi_i)\big] z_i - z_{i-1} + \sum_{r=1}^{i-1} \frac{\partial \alpha_{i-1}}{\partial x_r}(\xi_{i-1}) x_{r+1} \end{aligned} \tag{7.34}$$

for $i = 2, \dots, n$, and we let $u = \alpha_n(\xi_n)$. We specify the functions κ_i later.

Elementary calculations yield

$$\left\{ \begin{aligned} \dot{z}_1 &= z_2 + \alpha_1(x_1) + \varphi_1(x_1)^\top \Delta_1(t, x, u) \\ \dot{z}_i &= z_{i+1} + \alpha_i(\xi_i) + \varphi_i(\xi_i)^\top \Delta_i(t, x, u) \\ &\quad - \sum_{r=1}^{i-1} \frac{\partial \alpha_{i-1}}{\partial x_r}(\xi_{i-1}) \dot{x}_r, \quad i = 2, 3, \dots, n-1 \\ \dot{z}_n &= \alpha_n(\xi_n) + \varphi_n(\xi_n)^\top \Delta_n(t, x, u) - \sum_{r=1}^{n-1} \frac{\partial \alpha_{n-1}}{\partial x_r}(\xi_{n-1}) \dot{x}_r \ , \end{aligned} \right. \tag{7.35}$$

or equivalently,

$$\left\{ \begin{aligned} \dot{z}_1 &= -\big[c_1 + \kappa_1(x_1)\big] z_1 + z_2 + \varphi_1(x_1)^\top \Delta_1(t, x, u) \\ \dot{z}_i &= -\big[c_i + \kappa_i(\xi_i)\big] z_i - z_{i-1} + z_{i+1} + \varphi_i(\xi_i)^\top \Delta_i(t, x, u) \\ &\quad - \sum_{r=1}^{i-1} \frac{\partial \alpha_{i-1}}{\partial x_r}(\xi_{i-1}) \varphi_r(\xi_r)^\top \Delta_r(t, x, u), \quad i = 2, 3, \dots, n-1 \\ \dot{z}_n &= -\big[c_n + \kappa_n(\xi_n)\big] z_n - z_{n-1} + \varphi_n(\xi_n)^\top \Delta_n(t, x, u) \\ &\quad - \sum_{r=1}^{n-1} \frac{\partial \alpha_{n-1}}{\partial x_r}(\xi_{n-1}) \varphi_r(\xi_r)^\top \Delta_r(t, x, u) \ . \end{aligned} \right. \tag{7.36}$$

The time derivative of the function

$$V_n(z_1, ..., z_n) = \frac{1}{2}\sum_{i=1}^{n} z_i^2 \tag{7.37}$$

along the trajectories of (7.36) is

$$\begin{aligned}\dot{V}_n = & -\sum_{i=1}^{n}\left[c_i + \kappa_i(\xi_i)\right]z_i^2 + z_1\varphi_1(x_1)^\top \Delta_1(t,x,u)\\ & + \sum_{i=2}^{n} z_i\left[\varphi_i(\xi_i)^\top \Delta_i(t,x,u) - \sum_{r=1}^{i-1}\frac{\partial\alpha_{i-1}}{\partial x_r}(\xi_{i-1})\varphi_r(\xi_r)^\top \Delta_r(t,x,u)\right].\end{aligned}$$

From (7.31), we deduce that

$$\begin{aligned}\dot{V}_n \le & -\sum_{i=1}^{n}\left[c_i + \kappa_i(\xi_i)\right]z_i^2 + \Delta_M|z_1||\varphi_1(x_1)||x_1|\\ & + \sum_{i=2}^{n}|z_i|\left[\Delta_M|\varphi_i(\xi_i)||\xi_i| + \Delta_M\sum_{r=1}^{i-1}\left|\frac{\partial\alpha_{i-1}}{\partial x_r}(\xi_{i-1})\right||\varphi_r(\xi_r)||\xi_r|\right].\end{aligned}$$

Using the inequality $|\xi_i| \ge |\xi_r|$ for all $r \in \{1, ..., i\}$ and (7.32) gives

$$\dot{V}_n \le -\sum_{i=1}^{n}[c_i + \kappa_i(\xi_i)]z_i^2 + \Delta_M\sum_{i=1}^{n}|z_i||\xi_i|\overline{\varphi}(|\xi_i|)\Gamma_i(\xi_i)\,, \tag{7.38}$$

where $\Gamma_1(\xi_1) = 1$ and

$$\Gamma_i(\xi_i) = 1 + \sum_{r=1}^{i-1}\left|\frac{\partial\alpha_{i-1}}{\partial x_r}(\xi_{i-1})\right| \quad \text{for } i = 2, \ldots, n\,. \tag{7.39}$$

If the everywhere positive functions κ_i are such that

$$\sum_{i=1}^{n}\kappa_i(\xi_i)z_i^2 \ge \Delta_M\sum_{i=1}^{n}|z_i||\xi_i|\overline{\varphi}(|\xi_i|)\Gamma_i(\xi_i)\,, \tag{7.40}$$

then we obtain the desirable inequality

$$\dot{V}_n \le -\sum_{i=1}^{n}c_i z_i^2 \tag{7.41}$$

which implies the GAS of the system because V_n is a positive definite quadratic function and the right side of (7.41) is negative definite. It remains to construct positive functions κ_i that satisfy (7.40), which we do next.

7.2.2.2 Construction of the Dominating Functions κ_i's

We now construct everywhere positive functions κ_i that satisfy (7.40), by induction.

Induction Assumption. For each $k \in \{1, ..., n\}$, there are k functions $\kappa_i : \mathbb{R}^i \to [1, \infty)$ of class C^n such that

$$\tfrac{k}{n} \sum_{i=1}^{k} \kappa_i(\xi_i) z_i^2 \geq \Delta_M \sum_{i=1}^{k} |z_i| |\xi_i| \overline{\varphi}(|\xi_i|) \Gamma_i(\xi_i). \tag{7.42}$$

Step 1. The result holds for $k = 1$ because we can choose an everywhere positive function $\kappa_1 \in C^n$ such that

$$\tfrac{1}{n} \kappa_1(z_1) z_1^2 \geq \Delta_M z_1^2 \overline{\varphi}(|z_1|). \tag{7.43}$$

Step $k+1$. Assume that the induction assumption is satisfied at step k. Choose an everywhere positive function $\kappa_{k+1} \in C^n$ such that

$$\tfrac{1}{4n} \kappa_{k+1}(\xi_{k+1}) \geq \tfrac{2n\Delta_M^2}{\kappa_{k+1}(\xi_{k+1})} \overline{\varphi}^2(|\xi_{k+1}|) \Gamma_{k+1}^2(\xi_{k+1}) \,. \tag{7.44}$$

The induction assumption gives

$$\begin{aligned} \frac{k+1}{n} \sum_{i=1}^{k+1} \kappa_i(\xi_i) z_i^2 &= \frac{k}{n} \sum_{i=1}^{k+1} \kappa_i(\xi_i) z_i^2 + \frac{1}{n} \sum_{i=1}^{k+1} \kappa_i(\xi_i) z_i^2 \\ &\geq \Delta_M \sum_{i=1}^{k} |z_i| |\xi_i| \overline{\varphi}(|\xi_i|) \Gamma_i(\xi_i) + \frac{1}{n} \sum_{i=1}^{k+1} \kappa_i(\xi_i) z_i^2 \\ &= \Delta_M \sum_{i=1}^{k+1} |z_i| |\xi_i| \overline{\varphi}(|\xi_i|) \Gamma_i(\xi_i) \\ &\quad + \frac{1}{n} \sum_{i=1}^{k+1} \kappa_i(\xi_i) z_i^2 \\ &\quad - \Delta_M |z_{k+1}| |\xi_{k+1}| \overline{\varphi}(|\xi_{k+1}|) \Gamma_{k+1}(\xi_{k+1}). \end{aligned} \tag{7.45}$$

Using the triangular inequality $ab \leq \frac{1}{4}a^2 + b^2$ for suitable nonnegative values a and b, we deduce that

$$\begin{aligned} &|z_{k+1}| |\xi_{k+1}| \overline{\varphi}(|\xi_{k+1}|) \Gamma_{k+1}(\xi_{k+1}) \\ &\leq \frac{\kappa_{k+1}(\xi_{k+1}) z_{k+1}^2}{4n\Delta_M} + \frac{\Delta_M n}{\kappa_{k+1}(\xi_{k+1})} |\xi_{k+1}|^2 \overline{\varphi}^2\big(|\xi_{k+1}|\big) \Gamma_{k+1}^2(\xi_{k+1}) \end{aligned}$$

and therefore

$$
\begin{aligned}
\tfrac{k+1}{n}\sum_{i=1}^{k+1}\kappa_i(\xi_i)z_i^2 \geq \Delta_M \sum_{i=1}^{k+1}|z_i||\xi_i|\overline{\varphi}\big(|\xi_i|\big)\Gamma_i(\xi_i) + \frac{1}{n}\sum_{i=1}^{k}\kappa_i(\xi_i)z_i^2 \\
+\tfrac{3}{4n}\kappa_{k+1}(\xi_{k+1})z_{k+1}^2 \\
-\tfrac{n\Delta_M^2}{\kappa_{k+1}(\xi_{k+1})}|\xi_{k+1}|^2\overline{\varphi}^2\big(|\xi_{k+1}|\big)\Gamma_{k+1}^2(\xi_{k+1})\,.
\end{aligned}
\tag{7.46}
$$

Since $x_{k+1} = z_{k+1} + \alpha_k(\xi_k)$ for all $k \geq 1$, we get

$$
|\xi_{k+1}|^2 = |\xi_k|^2 + (z_{k+1} + \alpha_k(\xi_k))^2 \;\leq\; 2z_{k+1}^2 + |\xi_k|^2 + 2\alpha_k^2(\xi_k),
$$

so our choice (7.44) of κ_{k+1} gives

$$
\begin{aligned}
\tfrac{k+1}{n}\sum_{i=1}^{k+1}\kappa_i(\xi_i)z_i^2 \geq \Delta_M \sum_{i=1}^{k+1}|z_i||\xi_i|\overline{\varphi}\big(|\xi_i|\big)\Gamma_i(\xi_i) + \frac{1}{n}\sum_{i=1}^{k}\kappa_i(\xi_i)z_i^2 \\
+\tfrac{1}{2n}\kappa_{k+1}(\xi_{k+1})z_{k+1}^2 \\
-\tfrac{2n\Delta_M^2(|\xi_k|^2+|\alpha_k(\xi_k)|^2)}{\kappa_{k+1}(\xi_{k+1})}\overline{\varphi}^2\big(|\xi_{k+1}|\big)\Gamma_{k+1}^2(\xi_{k+1})\,.
\end{aligned}
\tag{7.47}
$$

One can easily prove that there is a function $\mho$, depending on the functions $\kappa_1,\ldots,\ \kappa_k$ but not on κ_{k+1}, such that

$$
|\xi_k|^2 + 2\big|\alpha_k(\xi_k)\big|^2 \;\leq\; \mho(|\xi_k|)\sum_{i=1}^{k} z_i^2, \tag{7.48}
$$

by induction on the components of ξ_k. Therefore,

$$
\begin{aligned}
&\tfrac{n\Delta_M^2(|\xi_k|^2+2|\alpha_k(\xi_k)|^2)}{\kappa_{k+1}(\xi_{k+1})}\overline{\varphi}^2\big(|\xi_{k+1}|\big)\Gamma_{k+1}^2(\xi_{k+1}) \\
\leq\;& \tfrac{n\Delta_M^2\mho(|\xi_k|)\sum_{i=1}^{k} z_i^2}{\kappa_{k+1}(\xi_{k+1})}\overline{\varphi}^2\big(|\xi_{k+1}|\big)\Gamma_{k+1}^2(\xi_{k+1})\,.
\end{aligned}
\tag{7.49}
$$

Since $\overline{\varphi}$ and Γ_{k+1} are also independent of κ_{k+1}, we can enlarge κ_{k+1} sufficiently so that

$$
\tfrac{2n\Delta_M^2(|\xi_k|^2+|\alpha_k(\xi_k)|^2)}{\kappa_{k+1}(\xi_{k+1})}\overline{\varphi}^2\big(|\xi_{k+1}|\big)\Gamma_{k+1}^2(\xi_{k+1}) \leq \tfrac{1}{n}\sum_{i=1}^{k}\kappa_i(\xi_i)z_i^2\,. \tag{7.50}
$$

Combining this inequality with (7.47), we obtain

$$
\frac{k+1}{n}\sum_{i=1}^{k+1}\kappa_i(\xi_i)z_i^2 \geq \Delta_M \sum_{i=1}^{k+1}|z_i||\xi_i|\overline{\varphi}\big(|\xi_i|\big)\Gamma_i(\xi_i). \tag{7.51}
$$

This concludes the construction of the functions κ_i, which establishes (7.41). This proves the domination result.

7.2.3 Further Extensions

In the system (7.8), each $\dot{x}_i$ depends only on $x_1, x_2, ..., x_{i+1}$ and is affine in x_{i+1}. This assumption can be relaxed. For example, we can extend the result to systems

$$\begin{cases} \dot{x}_i = g_i(x_1, x_2, ...x_i)h_i(x_{i+1}) + f_i(x_1, x_2, ..., x_i), & 1 \le i \le n-1 \\ \dot{x}_n = g_n(x_1, x_2, ...x_n)h_n(u) + f_n(x_1, x_2, ..., x_n) , \end{cases} \tag{7.52}$$

where each $x_i \in \mathbb{R}$, $u \in \mathbb{R}$ is the input, each function f_i is assumed to be zero at the origin, each function g_i is everywhere positive or everywhere negative, and each real-valued function h_i is a diffeomorphism satisfying $h_i(0) = 0$. The extension proceeds by choosing new coordinates $z_1, z_2, \dots, z_n$ (which are different from, but analogous to, the ones we chose in Sect. 7.2.1) that give the system (7.26). In the first step, we take $z_1 = x_1$ and $z_2 = g_1(x_1)h_1(x_2) + x_1 + f_1(x_1)$ to get $\dot{z}_1 = -z_1 + z_2$ and $\dot{z}_2 = \mathcal{M}_2(z_1, z_2)h_2(x_3) + \tilde{f}_2(z_1, z_2) = z_3 - z_1 - z_2$ for appropriate functions $\mathcal{M}_2$ and $\tilde{f}_2$ with $\mathcal{M}_2$ being nowhere zero, and in general, $\dot{z}_i = \mathcal{M}_i(z_1, z_2, \dots, z_i)h_i(x_{i+1}) + \tilde{f}_i(z_1, z_2, \dots, z_i) = z_{i+1} - z_{i-1} - z_i$ for $i = 1, 2, \dots, n-1$ for suitable functions $\mathcal{M}_i$ and $\tilde{f}_i$. We next give another backstepping result, to help the reader understand later sections.

Consider a nonlinear time-varying system

$$\begin{cases} \dot{x} = f_x(t, x, z) \\ \dot{z} = g(t, x, z)h(u) + f_z(t, x, z) \end{cases} \tag{7.53}$$

that is periodic with a given period $T > 0$ in t, where $x \in \mathbb{R}^{n_x}$, $z \in \mathbb{R}$, $u \in \mathbb{R}$, h is a diffeomorphism satisfying $h(0) = 0$, and the function g is such that there exists an everywhere positive continuous function γ_p such that

$$\gamma_p(x, z) \le g(t, x, z) \tag{7.54}$$

for all t, x, and z. We assume that f_x, g, h, and f_z are C^1, and that there exists a function $z_s(t, x)$ that is periodic of period T in t such that $z_s(t, 0) \equiv 0$, and such that the system

$$\dot{x} = f_x\big(t, x, z_s(t, x)\big) \tag{7.55}$$

is UGAS to 0. Finally, we assume that a strict Lyapunov function V_1 is known for the closed-loop system (7.55), with V_1 having period T in t. This gives known functions $\alpha_1, \alpha_2 \in \mathcal{K}_\infty$ and a known positive definite function $W_1(x)$ such that

$$\alpha_1(|x|) \le V_1(t, x) \le \alpha_2(|x|) \tag{7.56}$$

and

$$\tfrac{\partial V_1}{\partial t}(t, x) + \tfrac{\partial V_1}{\partial x}(t, x) f(x, z_s(t, x)) \le -W_1(x) \tag{7.57}$$

for all $(t, x) \in \mathbb{R} \times \mathbb{R}^n$.

Then

$$V_2(t, x, z) = V_1(t, x) + \tfrac{1}{2}[z - z_s(t, x)]^2 \tag{7.58}$$

admits functions $\alpha_3, \alpha_4 \in \mathcal{K}_\infty$ such that

$$\alpha_3(|(x, z)|) \leq V_2(t, x, z) \leq \alpha_4(|(x, z)|) \tag{7.59}$$

for all $t \in \mathbb{R}$ and all $(x, z) \in \mathbb{R}^n \times \mathbb{R}$. Also, its time derivative along the trajectories of (7.53) satisfies

$$\begin{aligned}
\dot{V}_2 &= \tfrac{\partial V_1}{\partial t}(t, x) + \tfrac{\partial V_1}{\partial x}(t, x) f_x(t, x, z) - \mathcal{M}(t, x, z) \\
&\quad + [z - z_s(t, x)][g(t, x, z)h(u) + f_z(t, x, z)] \\
&= \tfrac{\partial V_1}{\partial t}(t, x) + \tfrac{\partial V_1}{\partial x}(t, x) f_x(t, x, z_s(t, x)) - \mathcal{M}(t, x, z) \\
&\quad + \tfrac{\partial V_1}{\partial x}(t, x)\big[f_x(t, x, z) - f_x(t, x, z_s(t, x))\big] \\
&\quad + [z - z_s(t, x)][g(t, x, z)h(u) + f_z(t, x, z)] \\
&= -W_1(x) - \mathcal{M}(t, x, z) \\
&\quad + [z - z_s(t, x)]\big[\tfrac{\partial V_1}{\partial x}(t, x) F(t, x, z) + g(t, x, z)h(u) + f_z(t, x, z)\big]
\end{aligned} \tag{7.60}$$

where

$$F(t, x, z) = \int_0^1 \frac{\partial f_x}{\partial z}(t, x, m(z - z_s(t, x)) + z_s(t, x))\mathrm{d}m$$

and $\mathcal{M}(t, x, z) = [z - z_s(t, x)]\dot{z}_s(t, x)$.

Since g is everywhere positive, the control law

$$\begin{aligned}
&u_s(t, x, z) \\
&= h^{-1}\left(\frac{-[z - z_s(t, x)] - \frac{\partial V_1}{\partial x}(t, x)F(t, x, z) - f_z(t, x, z) + \dot{z}(t, x)}{g(t, x, z)}\right)
\end{aligned} \tag{7.61}$$

is well defined and yields

$$\dot{V}_2 = -W_2(t, x, z)\ , \ \text{ where } W_2(t, x, z) = W_1(x) + [z - z_s(t, x)]^2. \tag{7.62}$$

Since W_2 is periodic in t, we can find a positive definite function $\underline{\alpha}$ such that $W_2(t, x, z) \geq \underline{\alpha}(|(x, z)|)$ everywhere, which gives the UGAS for (7.53).

7.3 Backstepping for Nonautonomous Systems

When adapting the backstepping approach to nonlinear time-varying systems, it is natural to consider the special case

$$\begin{cases} \dot{x} = \mathcal{F}(t, x, z) \\ \dot{z} = p(t)u + h(t, x, z) \end{cases} \tag{7.63}$$

of (7.53), where $x \in \mathbb{R}^{n_x}$, $z \in \mathbb{R}$, $u \in \mathbb{R}$ is the input, $p(t)$ is a bounded function, and $\mathcal{F}(t, x, z)$ and $h(t, x, z)$ satisfy

$$\mathcal{F}(t, 0, 0) = 0 \quad \text{and} \quad h(t, 0, 0) = 0$$

for all t. There are several cases where strict Lyapunov function methods lead to control laws that render (7.63) UGAS to the origin. We discuss these cases next.

7.3.1 Chained Form Systems

One motivation for studying (7.63) involves the system

$$\begin{cases} \dot{\xi}_4 = \xi_3 v_1 \\ \dot{\xi}_3 = \xi_2 v_1 \\ \dot{\xi}_2 = v_2 \\ \dot{\xi}_1 = v_1 \end{cases} \tag{7.64}$$

in chained form of order 4 with inputs v_1 and v_2. Assume that we want ξ_1 to asymptotically track the function $\sin(t)$ while ξ_2, ξ_3, and ξ_4 converge to zero. This is the problem of tracking the reference trajectory

$$(\xi_{1r}, \xi_{2r}, \xi_{3r}, \xi_{4r})(t) \;=\; \big(\sin(t), 0, 0, 0\big).$$

The time-varying change of variables

$$x_1 = \xi_1 - \xi_{1r}(t) \tag{7.65}$$

and the change of feedback

$$v_1 = \cos(t) + u_1 \tag{7.66}$$

result in

$$\begin{cases} \dot{\xi}_4 = \xi_3(\cos(t) + u_1) \\ \dot{\xi}_3 = \xi_2(\cos(t) + u_1) \\ \dot{\xi}_2 = v_2 \\ \dot{x}_1 = u_1. \end{cases} \tag{7.67}$$

The system (7.67) can be globally uniformly asymptotically stabilized provided one knows (a) a control $v_2(t, \xi_2, \xi_3, \xi_4)$ that is periodic of period 2π in t that renders

$$\begin{cases} \dot{\xi}_4 = \xi_3 \cos(t) \\ \dot{\xi}_3 = \xi_2 \cos(t) \\ \dot{\xi}_2 = v_2 \end{cases} \tag{7.68}$$

UGAS and (b) a strict Lyapunov function ν_1 for the corresponding closed-loop system that also has period 2π in t.

Indeed, assume that the control law and strict Lyapunov function ν_1 are known. Then, there exists a positive definite function $\mathcal{W}_1(\xi_2, \xi_3, \xi_4)$ such that the time derivative of ν_1 along the trajectories of (7.68) in closed-loop with $v_{2s}(t, \xi_2, \xi_3, \xi_4)$ satisfies

$$\dot{\nu}_1 \leq -\mathcal{W}_1(\xi_2, \xi_3, \xi_4) \; . \tag{7.69}$$

Consequently, the time derivative of

$$\nu_2(t, x_1, \xi_2, \xi_3, \xi_4) \;=\; \nu_1(t, \xi_2, \xi_3, \xi_4) + \frac{1}{2}x_1^2$$

along the trajectories of (7.67) in closed-loop with $v_{2s}(t, \xi_2, \xi_3, \xi_4)$ satisfies

$$\begin{aligned} \dot{\nu}_2 \leq & -\mathcal{W}_1(\xi_2, \xi_3, \xi_4) \\ & + \left[\tfrac{\partial \nu_1}{\partial \xi_4}(t, \xi_2, \xi_3, \xi_4)\xi_3 + \tfrac{\partial \nu_1}{\partial \xi_3}(t, \xi_2, \xi_3, \xi_4)\xi_2 + x_1\right] u_1 \; . \end{aligned} \tag{7.70}$$

The choice

$$\begin{aligned} & u_1(t, x_1, \xi_2, \xi_3, \xi_4) = \\ & -\left[\tfrac{\partial \nu_1}{\partial \xi_4}(t, \xi_2, \xi_3, \xi_4)\xi_3 + \tfrac{\partial \nu_1}{\partial \xi_3}(t, \xi_2, \xi_3, \xi_4)\xi_2 + x_1\right] \end{aligned} \tag{7.71}$$

results in

$$\dot{\nu}_2 \leq -\mathcal{W}_2(t, x_1, \xi_2, \xi_3, \xi_4) \; , \tag{7.72}$$

where

$$\begin{aligned} & \mathcal{W}_2(t, x_1, \xi_2, \xi_3, \xi_4) = \\ & \mathcal{W}_1(\xi_2, \xi_3, \xi_4) + \left[\tfrac{\partial \nu_1}{\partial \xi_4}(t, \xi_2, \xi_3, \xi_4)\xi_3 + \tfrac{\partial \nu_1}{\partial \xi_3}(t, \xi_2, \xi_3, \xi_4)\xi_2 + x_1\right]^2 \; . \end{aligned} \tag{7.73}$$

Using the periodicity of the relevant functions, we can easily prove that $\mathcal{W}_2$ is bounded from above and below by positive definite functions of x_1, ξ_2, ξ_3, and ξ_4. It follows that the origin of (7.67) in closed-loop with $v_{2s}(t, \xi_2, \xi_3, \xi_4)$ and $u_1(t, x_1, \xi_2, \xi_3, \xi_4)$ defined in (7.71) is UGAS.

Therefore, it suffices to stabilize the system (7.68) and build a corresponding strict Lyapunov function for the closed-loop system. To globally uniformly asymptotically stabilize (7.68), it suffices to do backstepping for systems of the form (7.63). Indeed, if we can construct a globally asymptotically stabilizing 2π periodic feedback for the special case

$$\begin{cases} \dot{x}_4 = x_3 \cos(t) \\ \dot{x}_3 = U \cos(t) \end{cases} \tag{7.74}$$

of (7.63) with input U, then the argument from Sect. 7.2.3 provides a control law that renders (7.68) UGAS to the origin.

7.3.2 Feedback Systems

A more general motivation for studying the systems (7.63) arises from systems in feedback form. Solving local tracking problems for feedback systems frequently involves designing exponentially stable controllers for linear systems of the form (7.63). To understand why, consider the simple family of systems

$$\begin{cases} \dot{\xi}_1 = \mathcal{H}_1(\xi_2) \\ \dot{\xi}_2 = \mathcal{H}_2(\xi_3) \\ \dot{\xi}_3 = u \ , \end{cases} \tag{7.75}$$

where the functions $\mathcal{H}_i$ are not necessarily differomorphisms. Dynamics of the form (7.75) are said to be in feedback form or feedback systems.

Assume that there exists a bounded periodic trajectory $(\xi_{1,r}, \xi_{2,r}, \xi_{3,r})$ such that

$$\begin{cases} \dot{\xi}_{1,r}(t) = \mathcal{H}_1(\xi_{2,r}(t)) \\ \dot{\xi}_{2,r}(t) = \mathcal{H}_2(\xi_{3,r}(t) \ . \end{cases} \tag{7.76}$$

Then the dynamics for the error variables

$$\tilde{\xi}_j = \xi_j - \xi_{j,r}(t) \ , \ j = 1, ..., 3 \tag{7.77}$$

has the form

$$\begin{cases} \dot{\tilde{\xi}}_1 = \mathcal{H}_1(\tilde{\xi}_2 + \xi_{2,r}(t)) - \mathcal{H}_1(\xi_{2,r}(t)) \\ \dot{\tilde{\xi}}_2 = \mathcal{H}_2(\tilde{\xi}_3 + \xi_{3,r}(t)) - \mathcal{H}_2(\xi_{3,r}(t)) \\ \dot{\tilde{\xi}}_3 = u - \dot{\xi}_{3,r}(t) \ . \end{cases} \tag{7.78}$$

The linear approximation of (7.78) at the origin is

$$\begin{cases} \dot{\tilde{\xi}}_1 = \mathcal{H}_1'(\xi_{2,r}(t))\tilde{\xi}_2 \\ \dot{\tilde{\xi}}_2 = \mathcal{H}_2'(\xi_{3,r}(t))\tilde{\xi}_3 \\ \dot{\tilde{\xi}}_3 = u. \end{cases} \tag{7.79}$$

This system can be stabilized if the time-varying chain of integrators

$$\begin{cases} \dot{x}_1 = \mathcal{H}_1'(\xi_{2,r}(t))x_2 \\ \dot{x}_2 = \mathcal{H}_2'(\xi_{3,r}(t))U \end{cases} \tag{7.80}$$

can be globally uniformly asymptotically stabilized, and (7.80) is also of the form (7.63). In fact, once we can stabilize (7.80), the argument from Sect. 7.2.3 provides a control law that renders the system (7.79) UGAS to the origin, as well as a strict Lyapunov function for the corresponding closed-loop dynamics, assuming (7.80) and its stabilizer have the same period.

7.3.3 Feedforward Systems

Another motivation for studying the systems (7.63) arises from feedforward systems. As in the case of feedback systems, solving tracking problems for feedforward systems often involves building exponentially stable controllers for linear systems of the form (7.63). To understand why, consider the Euler-Lagrange feedforward system

$$\begin{cases} \dot{\xi}_1 = \xi_2 \\ \dot{\xi}_2 = -\xi_1 + \epsilon \sin(\xi_3) \\ \dot{\xi}_3 = \xi_4 \\ \dot{\xi}_4 = v \end{cases} \tag{7.81}$$

with input v. For definiteness, we take $\epsilon = \frac{3}{4}$. This is the so-called translational oscillator with rotating actuator (TORA) system [56]. One can readily check that the trajectory

$$(\xi_{1,r}, \xi_{2,r}, \xi_{3,r}, \xi_{4,r})(t) = \left(\sin\left(\frac{t}{2}\right), \frac{1}{2}\cos\left(\frac{t}{2}\right), \frac{t}{2}, \frac{1}{2}\right) \tag{7.82}$$

satisfies

$$\begin{cases} \dot{\xi}_{1,r} = \xi_{2,r} \\ \dot{\xi}_{2,r} = -\xi_{1,r} + \frac{3}{4}\sin(\xi_{3,r}) \\ \dot{\xi}_{3,r} = \xi_{4,r} \\ \dot{\xi}_{4,r} = 0. \end{cases} \tag{7.83}$$

Therefore, (7.82) is an admissible trajectory of (7.81). The dynamics for the error variables $\tilde{\xi}_j = \xi_j - \xi_{j,r}(t)$ for $j = 1, \ldots, 4$ has the form

$$\begin{cases} \dot{\tilde{\xi}}_1 = \tilde{\xi}_2 \\ \dot{\tilde{\xi}}_2 = -\tilde{\xi}_1 + \frac{3}{4}\left[\sin(\tilde{\xi}_3 + \xi_{3,r}(t)) - \sin(\xi_{3,r}(t))\right] \\ \dot{\tilde{\xi}}_3 = \tilde{\xi}_4 \\ \dot{\tilde{\xi}}_4 = v. \end{cases} \tag{7.84}$$

To construct locally uniformly exponentially stabilizing control laws for the system (7.84), we consider its linear approximation

$$\begin{cases} \dot{\tilde{\xi}}_1 = \tilde{\xi}_2 \\ \dot{\tilde{\xi}}_2 = -\tilde{\xi}_1 + \frac{3}{4}\cos\left(\xi_{3,r}(t)\right)\tilde{\xi}_3 \\ \dot{\tilde{\xi}}_3 = \tilde{\xi}_4 \\ \dot{\tilde{\xi}}_4 = v \end{cases} \tag{7.85}$$

near the origin. Applying the backstepping approach to stabilize this system involves several steps. In the first step, we find a control law

$$\tilde{\xi}_{2s}(t, \tilde{\xi}_1)$$

such that

$$\dot{\tilde{\xi}}_1 = \tilde{\xi}_{2s}(t, \tilde{\xi}_1)$$

is UGAS. We then seek a stabilizing controller for the $(\tilde{\xi}_1, \tilde{\xi}_2)$-subsystem with $\tilde{\xi}_3$ as the fictitious input. Clearly, these two steps are equivalent to considering

$$\begin{cases} \dot{x}_1 = x_2 \\ \dot{x}_2 = -x_1 + \frac{3}{4}\cos(\xi_{3,r}(t))u = -x_1 + \frac{3}{4}\cos\left(\frac{t}{2}\right)u \,, \end{cases} \tag{7.86}$$

which again has the structure of (7.63).

7.3.4 Other Important Cases

If the continuous function $p(t)$ in (7.63) is bounded from below by a positive constant (or bounded from above by a negative constant), then state feedbacks for (7.63) can be designed by combining the Lyapunov results of [180, 181, 182]. However, if $p(t)$ is neither everywhere positive nor everywhere negative, and therefore can take the value 0 (which is the case for the systems (7.74) and (7.86)), then constructing globally uniformly asymptotically stabilizing feedbacks for systems (7.63) is much more difficult. In this situation, neither the cancelation method nor the domination method applies, because when $p(t) = 0$, the term $p(t)u = 0$ can neither cancel nor dominate a term different from 0.

We study two cases where this obstacle can be overcome. The first case involves time-varying linear systems where $p(t)$ is periodic and takes the value 0 at discrete instants. We then study *nonlinear* systems (7.63) whose x-subsystem with z regarded as a control can be stabilized by a virtual control having the form $z_s(t, x) = p^2(t)\mu_s(t, x)$, and whose term $h(t, x, z)$ is of the form $p(t)b(t, x, z)$. Here both μ_s and b are C^1. We then show how in some cases, *bounded control laws* can be constructed through a variant of the technique.

7.4 Linear Time-Varying Systems

Consider the linear time-varying system

$$\dot{X} = A(t)X + p(t)Bu + \lambda(t) , \tag{7.87}$$

where $u \in \mathbb{R}$, $X \in \mathbb{R}^n$, $A : \mathbb{R} \to \mathbb{R}^{n\times n}$ is continuous and bounded, $B \in \mathbb{R}^n$ is constant, $\lambda : \mathbb{R} \to \mathbb{R}^n$ is a continuous disturbance, and $p : \mathbb{R} \to \mathbb{R}$ is a periodic function.

Later, we consider the subfamily of (7.87) consisting of systems

$$\begin{cases} \dot{x}_1 = a_{1,1}(t)x_1 + p_1(t)x_2 + \lambda_1(t) \\ \dot{x}_2 = a_{2,1}(t)x_1 + a_{2,2}(t)x_2 + p_2(t)x_3 + \lambda_2(t) \\ \quad\vdots \\ \dot{x}_n = a_{n,1}(t)x_1 + a_{n,2}(t)x_2 + ... + a_{n,n}(t)x_n + p_n(t)u + \lambda_n(t) \end{cases} \tag{7.88}$$

in feedback form, where $x_i \in \mathbb{R}$, $u \in \mathbb{R}$ is the input, and the functions $\lambda_i : \mathbb{R} \to \mathbb{R}$ are continuous. Our conditions will ensure that we can construct linear time-varying feedbacks that render (7.88) ISS with respect to the disturbances λ_i.

Assumption and Technical Lemmas

Consider a function $p : \mathbb{R} \to \mathbb{R}$ that satisfies:

Assumption 7.1 *The function p is continuous and periodic of some period $T_p > 0$. The set $H = \{t \in [0, T_p] : p(t) = 0\}$ is finite and nonempty.*

Let the elements of H be denoted by $0 \le t_1 < ... < t_k \le T_p$. We use the positive constant

$$d_m = \frac{1}{4}\min\{t_2 - t_1, ..., t_k - t_{k-1}\} \tag{7.89}$$

and the sets

$$E_d = \cup_{j=1}^{k}[t_j - d, t_j + d] \cap [0, T_p] \quad \text{and} \quad F_d = \overline{[0, T_p] \setminus E_d} \ , \tag{7.90}$$

where $d \in (0, d_m]$ is a given constant. The next lemma follows because $p^2(t)$ is continuous and positive at each point of the compact set F_d:

Lemma 7.1. *Consider a function* $p : \mathbb{R} \to \mathbb{R}$ *that satisfies Assumption 7.1. Let* $d \in (0, d_m]$ *be any constant. Then*

$$C_d = \min_{s \in F_d} p^2(s) \tag{7.91}$$

is a positive real number.

Lemma 7.2. *We have*

$$\lim_{\delta \to 0^+} \int_0^{T_p} \frac{\delta}{p^2(a) + \delta} \mathrm{d}a = 0 \tag{7.92}$$

for any function $p : \mathbb{R} \to \mathbb{R}$ *that satisfies Assumption 7.1.*

Proof. Fix any constants $\epsilon > 0$ and

$$d \in \left(0, \min\left\{d_m, \frac{\epsilon}{4k}\right\}\right],$$

where d_m is defined in (7.89). Then

$$\begin{aligned} \int_0^{T_p} \frac{\delta}{p^2(a) + \delta} \mathrm{d}a &= \int_{E_d} \frac{\delta}{p^2(a) + \delta} \mathrm{d}a + \int_{F_d} \frac{\delta}{p^2(a) + \delta} \mathrm{d}a \\ &\leq 2kd + \int_{F_d} \frac{\delta}{p^2(a) + \delta} \mathrm{d}a \\ &\leq \tfrac{\epsilon}{2} + T_p \tfrac{\delta}{C_d} , \end{aligned} \tag{7.93}$$

where the last inequality used the facts that

$$d \in \left(0, \frac{\epsilon}{4k}\right] \quad \text{and} \quad p^2(a) \geq C_d$$

when $a \in F_d$. Therefore,

$$\int_0^{T_p} \frac{\delta}{p^2(a) + \delta} \mathrm{d}a \leq \epsilon \quad \forall \delta \in \left(0, \tfrac{\epsilon C_d}{2T_p}\right] \tag{7.94}$$

which proves the lemma. □

7.4.1 General Result for Linear Time-Varying Systems

Assumptions

Assume that the linear time-varying system (7.87) is such that Assumption 7.1 and the following are both satisfied:

Assumption 7.2 *There are known positive constants c_i and $\mathcal{L}$ and C^∞ functions $L_1 : \mathbb{R} \to \mathbb{R}^n$, $L_2 : \mathbb{R} \to \mathbb{R}^n$, and $Q : \mathbb{R} \to \mathbb{R}^{n\times n}$ such that $Q(t)$ is symmetric for all $t \in \mathbb{R}$; the function*

$$\overline{Q}(t,X) = X^\top Q(t)X \tag{7.95}$$

is such that

$$c_1|X|^2 \le \bar{Q}(t,X) \le c_2|X|^2 \ \forall X \in \mathbb{R}^n \tag{7.96}$$

and

$$|L_1(t)| \le \mathcal{L} \text{ and } |L_2(t)| \le c_4 \tag{7.97}$$

hold for all $t \in \mathbb{R}$; and the time derivative of $\overline{Q}(t,X)$ along the trajectories of

$$\dot{X} = A(t)X + Bv + \lambda(t) \tag{7.98}$$

in closed-loop with

$$v = L(t)\cdot X, \text{ where } L(t) = L_1(t) + p(t)L_2(t) \tag{7.99}$$

satisfies

$$\dot{\overline{Q}} \le -c_3\overline{Q}(t,X) + |\lambda(t)|^2 \ . \tag{7.100}$$

Remark 7.1. A simple application of the triangle inequality shows that if $\bar{Q}(t,x)$ takes the form (7.95) for some everywhere symmetric matrix $Q(t)$, and if there are positive constants c_i satisfying (7.96) for all $t \in \mathbb{R}$ and

$$\dot{\bar{Q}} \le -c_3\bar{Q}(t,X)$$

along all trajectories of $\dot{X} = A(t)X + Bv$ in closed-loop with (7.99), then the time derivative of

$$\bar{Q}_c \doteq \varepsilon\bar{Q}, \text{ where } \varepsilon = \tfrac{c_1c_3}{2c_2^2}$$

along trajectories of (7.98) in closed-loop with the controller (7.99) satisfies

$$\dot{\bar{Q}}_c \le -\frac{c_3}{2}\bar{Q}_c + |\lambda|^2$$

for all disturbances λ. To see why, first notice that condition (7.96) gives spectrum$\{Q(t)\} \subseteq [c_1, c_2]$ for all $t \in \mathbb{R}$ and therefore $X^\top Q(t)Q(t)X \le c_2^2\bar{Q}(t,X)/c_1$ everywhere. Therefore, along the closed-loop trajectories of (7.98), the triangle inequality gives

$$\begin{aligned}\dot{\bar{Q}}_c &\le \varepsilon\left[-c_3\bar{Q}(t,X) + 2X^\top Q(t)\lambda(t)\right]\\ &\le \varepsilon\left[-c_3\bar{Q}(t,X) + 2\left\{\tfrac{\varepsilon}{2}X^\top Q(t)Q(t)X + \tfrac{1}{2\varepsilon}|\lambda(t)|^2\right\}\right]\\ &\le -\tfrac{c_3}{2}\bar{Q}_c(t,X) + |\lambda(t)|^2.\end{aligned} \tag{7.101}$$

Therefore, by scaling $\bar{Q}$ and c_3, we can take $\lambda \equiv 0$ in Assumption 7.2.

We also assume the following:

Assumption 7.3 *The function $p(t)$ in the system (7.87) is C^∞ and satisfies Assumption 7.1.*

Statement of Theorem

Theorem 7.1. *Assume that the system (7.87) satisfies Assumptions 7.2 and 7.3. Then there exists a constant $\delta > 0$ such that the time derivative of*

$$\begin{aligned} &\widehat{Q}(t,X) = e^{R(t)}\overline{Q}(t,X), \quad \textit{where} \\ &R(t) = -\tfrac{1}{T_p}\int_{t-T_p}^{t}\left(\int_{\ell}^{t}\frac{2\delta^2|B|^2\mathcal{L}^2}{c_1(p^2(a)+\delta)^2}\mathrm{d}a\right)\mathrm{d}\ell \end{aligned} \tag{7.102}$$

along the trajectories of (7.87) in closed-loop with

$$u(t,X) = \frac{p(t)}{p^2(t)+\delta}L_1(t)\cdot X + L_2(t)\cdot X \tag{7.103}$$

satisfies

$$\dot{\widehat{Q}} \le -\frac{4c_3}{5}\widehat{Q}(t,X) + 2|\lambda(t)|^2. \tag{7.104}$$

Moreover,

$$c_1\exp\left(-\tfrac{2|B|^2\mathcal{L}^2}{c_1}\int_{t-T_p}^{t}\frac{\delta^2}{[p^2(a)+\delta]^2}\mathrm{d}a\right)|X|^2 \le \widehat{Q}(t,X) \le c_2|X|^2 \tag{7.105}$$

for all $t \in \mathbb{R}$ and $X \in \mathbb{R}^n$.

Discussion on Theorem 7.1

Remark 7.2. Assumption 7.2 is satisfied if the pair $(A(t), B)$ is stabilizable by a feedback $K(t)X$ that is C^∞ and uniformly bounded with respect to time, assuming A and K have the same period. Therefore, this assumption is not restrictive.

Remark 7.3. We will see in the proof of Theorem 7.1 that (7.104) is satisfied provided δ satisfies

$$\int_0^{T_p}\frac{\delta^2}{(p^2(a)+\delta)^2}\mathrm{d}a \le \frac{c_1c_3T_p}{10|B|^2\mathcal{L}^2}\,. \tag{7.106}$$

The proof of Lemma 7.2 shows that (7.106) is satisfied provided

$$0 \ < \ \delta \ \leq \ \frac{c_1 c_3 C_d}{20|B|^2\mathcal{L}^2} \, , \tag{7.107}$$

where

$$d = \min\left\{\frac{c_3 c_1 T_p}{40k|B|^2\mathcal{L}^2}, d_m\right\}, \tag{7.108}$$

d_m is defined in (7.89), and C_d is defined in (7.91). However, in general, much larger values for δ can be found, which is important from a practical point of view if very large controls cannot be used.

For instance, consider the case where $p(t) = \cos(t)$ and $T_p = 2\pi$. Then, Appendix A.5 gives

$$\begin{aligned} \int_0^{T_p} \frac{\delta^2}{(p^2(a)+\delta)^2} \mathrm{d}a &= \int_0^{2\pi} \frac{\delta^2}{(\cos^2(a)+\delta)^2} \mathrm{d}a \\ &= 4\delta^2 \int_0^{\frac{\pi}{2}} \frac{1}{(\cos^2(a)+\delta)^2} \mathrm{d}a \\ &\leq \frac{\pi\sqrt{\delta}(1+3\delta)}{(1+\delta)^{3/2}} \, . \end{aligned} \tag{7.109}$$

Hence, (7.106) is satisfied when

$$\delta \leq \delta_A = \frac{c_1^2 c_3^2}{225|B|^4\mathcal{L}^4} \, . \tag{7.110}$$

On the other hand, we can easily show that $d_m = \frac{\pi}{4}$ and $C_d = \sin^2(d)$. By reducing c_1, we can assume that

$$\frac{\pi c_3 c_1}{20k|B|^2\mathcal{L}^2} \leq \frac{\pi}{4} = d_m \, . \tag{7.111}$$

Assuming (7.111), the formula (7.108) for d gives

$$C_d \ = \ \sin^2(d) \ = \ \sin^2\left(\frac{c_3 c_1 T_p}{40k|B|^2\mathcal{L}^2}\right),$$

and therefore (7.107) gives

$$0 < \delta \leq \delta_B = \frac{c_1 c_3}{20|B|^2\mathcal{L}^2} \sin^2\left(\frac{\pi c_3 c_1}{20k|B|^2\mathcal{L}^2}\right) \, . \tag{7.112}$$

Frequently, we have

$$\frac{\sqrt{c_1 c_3}}{|B|\mathcal{L}} \leq 1,$$

in which case δ_A can be significantly larger than δ_B.

Remark 7.4. When Assumption 7.2 is satisfied, the decomposition of $L(t)$ in (7.99) as the sum of a function $L_1(t)$ and a function $p(t)L_2(t)$ is not unique. For instance, if $L(t) = L_1(t) + p(t)L_2(t)$, then we also have

$$L(t) = \tilde{L}_1(t) + p(t)\tilde{L}_2(t),$$

where $\tilde{L}_1(t) = L_1(t) + 5p(t)$ and $\tilde{L}_2(t) = L_2(t) - 5$. In particular, the trivial decomposition $L(t) = L_1(t) + p(t)L_2(t)$ with $L_2(t) = 0$ and $L_1(t) = L(t)$ is always possible. The flexibility in the choices of $L_1(t)$ and $L_2(t)$ allows different possible choices of the feedback (7.103).

Remark 7.5. If the function $p(t)$ satisfies a PE property of the type

$$\int_0^{T_p} p^2(a)\mathrm{d}a > 0 \tag{7.113}$$

but violates Assumption 7.1, then there might not exist a constant $\delta > 0$ such that (7.106) holds. Therefore, Assumption 7.3 cannot be replaced by the less restrictive assumption that $p(t)$ is a C^∞ function satisfying the PE property (7.113).

Proof of Theorem 7.1

To simplify the proof, we let $L(t) = L_1(t)$ and $L_2(t) = 0$. The case where $L_2 \neq 0$ can be easily handled by performing the preliminary change of control $u = u_1 + L_2(t)$ and replacing $A(t)$ with $A(t) + Bp(t)L_2^\top(t)$. The system (7.87) in closed-loop with (7.103) is

$$\begin{aligned}\dot{X} &= A(t)X + B\tfrac{p^2(t)}{p^2(t)+\delta}L(t)\cdot X + \lambda(t)\\ &= [A(t) + BL^\top(t)]X - B\frac{\delta}{p^2(t)+\delta}L(t)\cdot X + \lambda(t)\ .\end{aligned} \tag{7.114}$$

From (7.100) in Assumption 7.2, we immediately deduce that the time derivative of $\overline{Q}$ along the trajectories of the system (7.87) in closed-loop with (7.103) satisfies

$$\begin{aligned}\dot{\overline{Q}} &\leq -c_3\overline{Q}(t,X) + \left|-B\frac{\delta}{p^2(t)+\delta}L(t)\cdot X + \lambda(t)\right|^2\\ &\leq -c_3\overline{Q}(t,X) + \frac{2\delta^2}{(p^2(t)+\delta)^2}|B|^2\mathcal{L}^2|X|^2 + 2|\lambda(t)|^2\ ,\end{aligned} \tag{7.115}$$

where $\mathcal{L}$ is the constant from Assumption 7.2. It follows from (7.96) in Assumption 7.2 that

$$\dot{\overline{Q}} \leq -c_3\overline{Q}(t,X) + \frac{2\delta^2|B|^2\mathcal{L}^2}{(p^2(t)+\delta)^2}\frac{\overline{Q}(t,X)}{c_1} + 2|\lambda(t)|^2 \,. \tag{7.116}$$

On the other hand, the time derivative of the function $\widehat{Q}$ defined in (7.102) along the trajectories of the system (7.87), in closed-loop with (7.103), satisfies

$$\dot{\widehat{Q}}(t,X) = e^{R(t)}\left[\dot{\overline{Q}}(t,X) + \overline{Q}(t,X)\dot{R}(t)\right]. \tag{7.117}$$

Moreover,

$$\dot{R}(t) = -\frac{2\delta^2|B|^2\mathcal{L}^2}{c_1(p^2(t)+\delta)^2} + \frac{1}{T_p}\int_{t-T_p}^{t}\frac{2\delta^2|B|^2\mathcal{L}^2}{c_1(p^2(a)+\delta)^2}\mathrm{d}a. \tag{7.118}$$

Combining (7.116)-(7.118) yields

$$\begin{aligned}\dot{\widehat{Q}}(t,X) &\leq e^{R(t)}\Bigg[-c_3\overline{Q}(t,X) + 2|\lambda(t)|^2 \\ &\quad +\overline{Q}(t,X)\left(\frac{1}{T_p}\int_{t-T_p}^{t}\frac{2\delta^2|B|^2\mathcal{L}^2}{c_1(p^2(a)+\delta)^2}\mathrm{d}a\right)\Bigg] \\ &= e^{R(t)}\overline{Q}(t,X)\left[-c_3 + \frac{2|B|^2\mathcal{L}^2}{c_1T_p}\int_{t-T_p}^{t}\frac{\delta^2}{(p^2(a)+\delta)^2}\mathrm{d}a\right] \\ &\quad +2e^{R(t)}|\lambda(t)|^2 \,.\end{aligned} \tag{7.119}$$

Using the definition of $\widehat{Q}$ and the non-positivity of R, we get

$$\begin{aligned}\dot{\widehat{Q}}(t,X) &\leq \widehat{Q}(t,X)\left[-c_3 + \frac{2|B|^2\mathcal{L}^2}{c_1T_p}\int_{0}^{T_p}\frac{\delta^2}{(p^2(a)+\delta)^2}\mathrm{d}a\right] \\ &\quad +2|\lambda(t)|^2 \,.\end{aligned} \tag{7.120}$$

Using Lemma 7.2 and the inequality

$$\int_0^{T_p}\frac{\delta^2}{(p^2(a)+\delta)^2}\mathrm{d}a \;\leq\; \int_0^{T_p}\frac{\delta}{p^2(a)+\delta}\mathrm{d}a,$$

we can choose $\delta > 0$ so that

$$\frac{2|B|^2\mathcal{L}^2}{c_1T_p}\int_0^{T_p}\frac{\delta^2}{(p^2(a)+\delta)^2}\mathrm{d}a \;\leq\; \frac{1}{5}c_3 \,. \tag{7.121}$$

This choice yields

$$\dot{\widehat{Q}}(t,X) \;\leq\; -\frac{4c_3}{5}\widehat{Q}(t,X) + 2|\lambda(t)|^2 \,. \tag{7.122}$$

Finally, one can easily prove (7.105). This proves the theorem. □

Remark 7.6. A more restrictive condition on δ than the one in (7.107) guarantees that the time derivative of the storage function

$$\tilde{Q}(t,X) = \left[1 - \frac{1}{T_p}\int_{t-T_p}^{t}\int_{\ell}^{t}\frac{2\delta^2|B|^2\mathcal{L}^2}{c_1(p^2(a)+\delta)^2}\mathrm{d}a\,\mathrm{d}\ell\right]\overline{Q}(t,X) \tag{7.123}$$

along the trajectories of (7.87) in closed-loop with (7.103) satisfies

$$\dot{\tilde{Q}} \leq -\underline{c}\tilde{Q} + \bar{c}|\lambda|^2 \tag{7.124}$$

for suitable positive constants $\underline{c}$ and $\bar{c}$. The proof of (7.124) combines the arguments from (7.114)-(7.116) with the formula

$$\frac{d}{dt}\int_{t-T_p}^{t}\int_{\ell}^{t}\mathcal{M}(a)\,\mathrm{d}a\,\mathrm{d}\ell \;=\; T_p\mathcal{M}(t) - \int_{t-T_p}^{t}\mathcal{M}(\ell)\mathrm{d}\ell,$$

which is valid for any continuous scalar function $\mathcal{M}$. In some cases, it may be more convenient to use the Lyapunov function (7.123) instead of (7.102).

7.4.2 Linear Time-Varying Systems in Feedback Form

Notation and Assumptions

We consider the linear time-varying systems (7.88), with the following notation. Let $\Lambda_j = (\lambda_1,\ldots,\lambda_j)^\top \in \mathbb{R}^j$ and $\Lambda = \Lambda_n \in \mathbb{R}^n$. Let $\xi_j = (x_1,\ldots,x_j)^\top \in \mathbb{R}^j$ and $x = \xi_n = (x_1,\ldots,x_n)^\top \in \mathbb{R}^n$. Consider the systems

$$\begin{cases} \dot{x}_1 = a_{1,1}(t)x_1 + p_1(t)x_2 + \lambda_1(t) \\ \dot{x}_2 = a_{2,1}(t)x_1 + a_{2,2}(t)x_2 + p_2(t)x_3 + \lambda_2(t) \\ \quad\vdots \\ \dot{x}_j = a_{j,1}(t)x_1 + a_{j,2}(t)x_2 + \ldots + a_{j,j}(t)x_j + p_j(t)x_{j+1} + \lambda_j(t) \end{cases} \tag{7.125}$$

for $j = 1$ to $n-1$, which we denote in compact form by

$$\dot{\xi}_j = \mathcal{A}_j(t)\xi_{j+1} + \Lambda_j(t)\ . \tag{7.126}$$

We introduce two assumptions:

Assumption 7.4 *Each function $a_{i,j}(t)$ is C^∞ and periodic.*

Assumption 7.5 *Each function $p_i(t)$ is C^∞ and satisfies Assumption 7.1.*

We use $T_{p_i} > 0$ to denote the period of $p_i(t)$ for each i.

Statement of Main Result

Our main result for (7.88) is as follows:

Theorem 7.2. *Assume that (7.88) satisfies Assumptions 7.4-7.5. Then one can construct n time periodic C^∞ functions $g_i(t)$, a time periodic everywhere symmetric C^∞ matrix $H(t)$, and constants $h_i > 0$ such that*

$$h_1 I_n \le H(t) \le h_2 I_n \ \ \forall t \in \mathbb{R}\,, \tag{7.127}$$

and such that the time derivative of the function

$$\mathcal{V}(t,x) = x^\top H(t) x \tag{7.128}$$

along the trajectories of the system (7.88) in closed-loop with the feedback

$$u(t,x) \;=\; g_1(t)x_1 + ... + g_n(t)x_n \tag{7.129}$$

satisfies

$$\dot{\mathcal{V}}(t,x) \;\le\; -\mathcal{V}(t,x) + 2|\Lambda(t)|^2\,. \tag{7.130}$$

Remark 7.7. An immediate consequence of (7.130) is that the system (7.88) in closed-loop with the feedback (7.129) is globally ISS with respect to Λ. Moreover, the explicit formula for $\mathcal{V}$ yields the explicit ISS estimate

$$|x(t)| \;\le\; \sqrt{\frac{h_2}{h_1}} e^{-0.5(t+t_0)}|x(t_0)| + \frac{2|\Lambda|_\infty}{\sqrt{h_1}} \tag{7.131}$$

for all $t \ge t_0 \ge 0$ along the closed-loop trajectories.

Proof of Theorem 7.2

The proof proceeds by induction. We define the *step j subsystems* by

$$\begin{cases} \dot{\xi}_{j-1} = \mathcal{A}_{j-1}(t)\xi_j + \Lambda_{j-1}(t) \\ \dot{x}_j = \displaystyle\sum_{r=1}^{j} a_{j,r}(t)x_r + p_j(t)w_j + \lambda_j(t) \end{cases} \tag{7.132}$$

if $j > 1$ and

$$\dot{x}_1 = a_{1,1}(t)x_1 + p_1(t)w_1 + \lambda_1(t) \tag{7.133}$$

if $j = 1$.

Induction Hypothesis. There are j time periodic C^∞ functions $g_{i,j}(t)$, a time periodic everywhere symmetric C^∞ matrix $H_j(t)$, and positive real numbers $h_{1,j}$ and $h_{2,j}$ such that $h_{1,j}I_j \le H_j(t) \le h_{2,j}I_j$ for all $t \in \mathbb{R}$ for which the following holds: The time derivative of

$$\widehat{Q}_j(t,\xi_j) = \xi_j^\top H_j(t)\xi_j \tag{7.134}$$

along the trajectories of the step j subsystem in closed-loop with the feedback

$$w_j(t,\xi_j) \doteq g_{1,j}(t)x_1 + ... + g_{j,j}(t)x_j \tag{7.135}$$

satisfies

$$\dot{\widehat{Q}}_j(t,\xi_j) \le -\frac{5^{n-j}}{4^{n-j}}\widehat{Q}_j(t,\xi_j) + 2|\Lambda_j(t)|^2 \,. \tag{7.136}$$

Step 1. To show that the induction assumption is satisfied for $j = 1$, consider the one-dimensional system

$$\dot{x}_1 = a_{1,1}(t)x_1 + v + \lambda_1(t) \tag{7.137}$$

with v as the input. Let $\overline{Q}_1(t,x_1) = \frac{1}{2}x_1^2$ and

$$v(t,\xi_1) = -\left[a_{1,1}(t) + \left(\frac{5}{4}\right)^n\right]x_1 \,. \tag{7.138}$$

The system (7.137) in closed-loop with (7.138) is

$$\dot{x}_1(t) = -\left(\frac{5}{4}\right)^n x_1 + \lambda_1(t). \tag{7.139}$$

Along the trajectories of (7.139), the time derivative of $\overline{Q}_1(t,x_1)$ satisfies

$$\begin{aligned}\dot{\overline{Q}}_1(t,x_1) &= -\left(\tfrac{5}{4}\right)^n x_1^2 + \lambda_1(t)x_1\\ &= -\left(\tfrac{5}{4}\right)^n x_1^2 + \left\{\lambda_1(t)\left(\tfrac{4}{5}\right)^{n/2}\right\}\left\{\left(\tfrac{5}{4}\right)^{n/2}x_1\right\}\\ &\le -\left(\tfrac{5}{4}\right)^n \overline{Q}_1(t,x_1) + \lambda_1^2(t)\,,\end{aligned} \tag{7.140}$$

by the triangle inequality $c_1c_2 \le \frac{1}{2}c_1^2 + \frac{1}{2}c_2^2$ applied to the terms in braces.

We deduce that the system (7.133) satisfies Assumption 7.2 with $c_1 = c_2 = \frac{1}{2}$, $c_3 = (5/4)^n$, $L_1(t)x = v(t,\xi_1)$ as defined in (7.138), and $L_2 \equiv 0$. Moreover, Assumption 7.5 ensures that the function $p_1(t)$ satisfies Assumption 7.3. Hence, Theorem 7.1 provides a constant $\delta_1 > 0$ such that the time derivative of

$$\widehat{Q}_1(t,\xi_1) = e^{R_1(t)}x_1^2 \tag{7.141}$$

with

$$R_1(t) = -\frac{1}{T_{p1}}\int_{t-T_{p1}}^{t}\left(\int_\ell^t \frac{4\delta_1^2\mathcal{L}_1^2}{(p_1(a)^2+\delta_1)^2}\mathrm{d}a\right)\mathrm{d}\ell \tag{7.142}$$

and $\mathcal{L}_1 = \sup_t\{|a_{1,1}(t) + (5/4)^n|\}$ along the trajectories of

$$\dot{x}_1 = a_{1,1}(t)x_1 + p_1(t)w_1(t,\xi_1) + \lambda_1(t) \tag{7.143}$$

with

$$w_1(t,\xi_1) = g_{1,1}(t)x_1 \text{ and } g_{1,1}(t) = -p_1(t)\frac{a_{1,1}(t)+\left(\frac{5}{4}\right)^n}{p_1^2(t)+\delta_1} \tag{7.144}$$

satisfies

$$\dot{\widehat{Q}}_1(t,\xi_1) \leq -\left(\tfrac{5}{4}\right)^{n-1}\widehat{Q}_1(t,\xi_1) + 2\lambda_1^2(t)\ . \tag{7.145}$$

Therefore the induction assumption is satisfied at the first step.

Inductive Step. We assume that the induction assumption is satisfied at some step $j \in [1, n-1]$. Let us prove that it is satisfied at the step $j+1$. Consider the system

$$\begin{cases} \dot{\xi}_j = \mathcal{A}_j(t)\xi_{j+1} + \Lambda_j(t) \\ \dot{x}_{j+1} = \sum_{r=1}^{j+1} a_{j+1,r}(t)x_r + v + \lambda_{j+1}(t)\ , \end{cases} \tag{7.146}$$

where v is the input. We can determine a globally asymptotically stabilizing feedback for (7.146) using the following classical backstepping approach. Let $w_j(t,\xi_j)$ be the feedback provided by the induction assumption. The change of coordinates $\psi = x_{j+1} - w_j(t,\xi_j)$ gives

$$\begin{cases} \dot{x}_1 = a_{1,1}(t)x_1 + p_1(t)x_2 + \lambda_1(t) \\ \dot{x}_2 = a_{2,1}(t)x_1 + a_{2,2}(t)x_2 + p_2(t)x_3 + \lambda_2(t) \\ \vdots \\ \dot{x}_j = \sum_{r=1}^{j} a_{j,r}(t)x_r + p_j(t)[\psi + w_j(t,\xi_j)] + \lambda_j(t) \\ \dot{\psi} = \sum_{r=1}^{j+1} a_{j+1,r}(t)x_r + v + \lambda_{j+1}(t) - \dot{w}_j\ . \end{cases} \tag{7.147}$$

Therefore, the ψ-subsystem becomes

$$\begin{aligned} \dot{\psi} &= \sum_{r=1}^{j+1} a_{j+1,r}(t)x_r + v + \lambda_{j+1}(t) - \sum_{\ell=1}^{j} \dot{g}_{\ell,j}(t)x_\ell \\ &\quad - \sum_{\ell=1}^{j} g_{\ell,j}(t)\left(\sum_{r=1}^{\ell} a_{\ell,r}(t)x_r + p_\ell(t)x_{\ell+1} + \lambda_\ell(t)\right) \\ &= \sum_{r=1}^{j+1} b_r(t)x_r + v + \lambda_{j+1}(t) - \sum_{\ell=1}^{j} g_{\ell,j}(t)\lambda_\ell(t) \end{aligned} \tag{7.148}$$

where

$$b_r(t) = a_{j+1,r}(t) - \dot{g}_{r,j}(t) - \sum_{\ell=r}^{j} g_{\ell,j}(t) a_{\ell,r}(t) \\ -p_{r-1}(t) g_{r-1,j}(t) \tag{7.149}$$

for $r = 2, 3, \ldots, j$ and

$$b_r(t) = \begin{cases} a_{j+1,1}(t) - \dot{g}_{1,j}(t) - \sum_{\ell=1}^{j} g_{\ell,j}(t) a_{\ell,1}(t), \; r = 1 \\ a_{j+1,j+1}(t) - p_j(t) g_{j,j}(t), \qquad r = j+1 \,. \end{cases} \tag{7.150}$$

Let $\widehat{Q}_j$ be the function provided by the induction assumption. Then the time derivative of

$$W_{j+1}(t, \xi_j, \psi) \doteq \widehat{Q}_j(t, \xi_j) + \frac{1}{2}\psi^2 \tag{7.151}$$

along the trajectories of (7.147) satisfies

$$\dot{W}_{j+1} \leq -\left(\frac{5}{4}\right)^{n-j} \widehat{Q}_j(t, \xi_j) + 2|\Lambda_j(t)|^2 + \frac{\partial \widehat{Q}_j}{\partial x_j}(t, \xi_j) p_j(t) \psi \\ +\psi \left[\sum_{r=1}^{j+1} b_r(t) x_r + v + \lambda_{j+1}(t) - \sum_{\ell=1}^{j} g_{\ell,j}(t) \lambda_\ell(t) \right] \,. \tag{7.152}$$

Choosing

$$v(t, \xi_j, \psi) = -\left[2 + \left(\frac{5}{4}\right)^{n-j} \right] \psi - \frac{\partial \widehat{Q}_j}{\partial x_j}(t, \xi_j) p_j(t) - \sum_{r=1}^{j+1} b_r(t) x_r \tag{7.153}$$

we obtain

$$\dot{W}_{j+1} \leq -\left(\frac{5}{4}\right)^{n-j} \widehat{Q}_j(t, \xi_j) + 2|\Lambda_j(t)|^2 - \left[2 + \left(\frac{5}{4}\right)^{n-j} \right] \psi^2 \\ +\psi \left(\lambda_{j+1}(t) - \sum_{\ell=1}^{j} g_{\ell,j}(t) \lambda_\ell(t) \right) \,. \tag{7.154}$$

From the triangular inequality $c_1 c_2 \leq c_1^2 + \frac{1}{4} c_2^2$, we deduce that

$$\dot{W}_{j+1} \leq -\left(\tfrac{5}{4}\right)^{n-j} \widehat{Q}_j(t, \xi_j) + 2|\Lambda_j(t)|^2 - \left[1 + \left(\tfrac{5}{4}\right)^{n-j} \right] \psi^2 \\ +\tfrac{1}{4} \left(\lambda_{j+1}(t) - \sum_{m=1}^{j} g_{m,j}(t) \lambda_m(t) \right)^2 \,. \tag{7.155}$$

We easily deduce that

$$\dot{W}_{j+1} \leq -\left(\tfrac{5}{4}\right)^{n-j} W_{j+1}(t, \xi_j, \psi) + \kappa_{j+1} |\Lambda_{j+1}(t)|^2 \,, \tag{7.156}$$

where

$$\kappa_{j+1} = 2 + \sup_t \left(1 + \sum_{m=1}^{j} |g_{m,j}(t)|\right)^2 . \tag{7.157}$$

Therefore, the function

$$\overline{Q}_{j+1}(t, \xi_{j+1}) = \tfrac{W_{j+1}(t,\xi_j,\psi)}{1+\kappa_{j+1}} \tag{7.158}$$

satisfies

$$\dot{\overline{Q}}_{j+1} \leq -\left(\tfrac{5}{4}\right)^{n-j} \overline{Q}_{j+1}(t, \xi_{j+1}) + |\Lambda_{j+1}(t)|^2 \tag{7.159}$$

along the trajectories of (7.146).

Moreover, there exist positive constants γ_1 and γ_2 and a function $\Gamma : \mathbb{R} \to \mathbb{R}^{(j+1)\times(j+1)}$ such that

$$\overline{Q}_{j+1}(t, \xi_{j+1}) = \xi_{j+1}^{\top} \Gamma(t) \xi_{j+1} \ \text{ and } \ \gamma_1 |\xi_{j+1}|^2 \leq \overline{Q}_{j+1}(t, \xi_{j+1}) \leq \gamma_2 |\xi_{j+1}|^2 .$$

The existence of γ_1 follows from the periodicity of the functions $g_{i,j}(t)$. We deduce that the system

$$\begin{cases} \dot{\xi}_j = \mathcal{A}_j(t)\xi_{j+1} + \Lambda_j(t) \\ \dot{x}_{j+1} = \displaystyle\sum_{r=1}^{j+1} a_{j+1,r}(t) x_r + p_{j+1}(t) w + \lambda_{j+1}(t) \end{cases} \tag{7.160}$$

satisfies Assumption 7.2 with $L_2 = 0$, and $p_{j+1}(t)$ satisfies Assumption 7.3.

Therefore, Theorem 7.1 applies to the system (7.160). It follows that we can find a constant $\delta_{j+1} > 0$ such that if we set

$$\widehat{Q}_{j+1}(t, \xi_{j+1}) \doteq e^{R_{j+1}(t)} \overline{Q}_{j+1}(t, \xi_{j+1}), \tag{7.161}$$

where

$$R_{j+1}(t) = -\tfrac{1}{T_{p_{j+1}}} \int_{t-T_{p_{j+1}}}^{t} \left(\int_{\ell}^{t} \frac{2\mathcal{L}_{j+1}^2 \delta_{j+1}^2}{\gamma_1 (p_{j+1}(a)^2 + \delta_{j+1})^2} \mathrm{d}a \right) \mathrm{d}\ell \tag{7.162}$$

and

$$\begin{aligned} \mathcal{L}_{j+1} = 2 \max_t \Bigg\{ & \sum_{r=1}^{j+1} b_r^2(t) + \sum_{r=1}^{j} 8 p_j^2(t) (H_j)_{j,r}^2(t) \\ & 2\left[2 + \left(\frac{5}{4}\right)^{n-j}\right]^2 \sum_{r=1}^{j} (g_{r,j}^2 + 1) \Bigg\}, \end{aligned} \tag{7.163}$$

then the time derivative of $\widehat{Q}_{j+1}(t, \xi_{j+1})$ along the trajectories of the system (7.160) in closed-loop with

$$\begin{aligned} w_{j+1}(t,\xi_{j+1}) &= g_{1,j+1}(t)x_1 + \ldots + g_{j+1,j+1}(t)x_{j+1} \\ &= -\frac{p_{j+1}(t)}{p_{j+1}^2(t)+\delta_{j+1}}\left[\left(2+\left(\frac{5}{4}\right)^{n-j}\right)\psi \right. \\ &\qquad \left. + \frac{\partial \widehat{Q}_j}{\partial x_j}(t,\xi_j)p_j(t) + \sum_{r=1}^{j+1} b_r(t)x_r\right] \end{aligned} \tag{7.164}$$

satisfies

$$\dot{\widehat{Q}}_{j+1} \leq -\left(\frac{5}{4}\right)^{n-j-1} \widehat{Q}_{j+1}(t,\xi_{j+1}) + 2|\Lambda_{j+1}(t)|^2 \, . \tag{7.165}$$

One can easily prove that there exist a function $H_{j+1}(t)$ and positive constants $h_{1,j+1}$ and $h_{2,j+1}$ such that

$$\begin{aligned} \widehat{Q}_{j+1}(t,\xi_{j+1}) &= \xi_{j+1}^\top H_{j+1}(t)\xi_{j+1} \quad \text{and} \\ h_{1,j+1}I_{j+1} \leq H_{j+1}(t) &\leq h_{2,j+1}I_{j+1} \quad \forall t \in \mathbb{R} \, . \end{aligned} \tag{7.166}$$

Hence, the induction assumption is satisfied at the step $j+1$. We conclude by choosing $\mathcal{V}(t,x) = \widehat{Q}_n(t,x)$.

7.4.3 Illustration: Linear System with PE Coefficients

We use Theorem 7.2 to construct a stabilizing controller and a corresponding strict Lyapunov function for

$$\begin{cases} \dot{x}_1 = p(t)x_2 + \lambda_1(t) \\ \dot{x}_2 = p(t)u + \frac{1}{2}x_1 + \lambda_2(t) \, , \end{cases} \tag{7.167}$$

where $p(t) = 20\cos(t)$. This system is of the form (7.88) and since $p(t)$ is C^∞ and satisfies Assumptions 7.1, it follows that Assumptions 7.4-7.5 are also satisfied. Therefore, Theorem 7.2 applies to the system (7.167). Let us now construct the feedback and strict Lyapunov function guaranteed to exist by the theorem. First consider the auxiliary system

$$\begin{cases} \dot{x}_1 = p(t)x_2 + \lambda_1(t) \\ \dot{x}_2 = v + \frac{1}{2}x_1 + \lambda_2(t) \, , \end{cases} \tag{7.168}$$

where v is an input, and set $x = (x_1\ x_2)^\top$. When $\lambda_1 = \lambda_2 = 0$, one can apply the classical backstepping approach to obtain exponentially stabilizing linear control laws, as follows.

Step 1. Classical Backstepping

The time-varying change of coordinates

$$X_2 = x_2 + \cos^3(t)x_1 \tag{7.169}$$

transforms (7.168) into

$$\begin{cases} \dot{x}_1 = -20\cos^4(t)x_1 + 20\cos(t)X_2 + \lambda_1(t) \\ \dot{X}_2 = v + \frac{1}{2}x_1 - 3\cos^2(t)\sin(t)x_1 + 20\cos^4(t)\big[X_2 - \cos^3(t)x_1\big] \\ \qquad + \cos^3(t)\lambda_1(t) + \lambda_2(t)\,. \end{cases} \tag{7.170}$$

When $\lambda_1 \equiv 0$ and $\lambda_2 \equiv 0$, the time derivative of

$$\mathcal{G}(x_1, X_2) = \frac{1}{2}[x_1^2 + X_2^2] \tag{7.171}$$

along the trajectories of (7.170) satisfies

$$\begin{aligned} \dot{\mathcal{G}} = {} & -20\cos^4(t)x_1^2 \\ & + X_2\big[v + 20\cos(t)x_1 + \tfrac{1}{2}x_1 - 3\cos^2(t)\sin(t)x_1 \\ & + 20\cos^4(t)(X_2 - \cos^3(t)x_1)\big]. \end{aligned} \tag{7.172}$$

Choosing

$$\begin{aligned} v(t, x_1, X_2) = {} & -20\cos^2(t)X_2 - 20\cos(t)x_1 - \tfrac{1}{2}x_1 \\ & + 3\cos^2(t)\sin(t)x_1 - 20\cos^4(t)\big(X_2 - \cos^3(t)x_1\big) \end{aligned} \tag{7.173}$$

gives

$$\begin{aligned} \dot{\mathcal{G}} & = -20\cos^4(t)x_1^2 - 20\cos^2(t)X_2^2 \\ & \le -20\cos^4(t)\big[x_1^2 + X_2^2\big] \\ & \le -40\cos^4(t)\mathcal{G}(x_1, X_2). \end{aligned} \tag{7.174}$$

Let

$$\mathcal{H}(t, x_1, X_2) = \left(\int_{t-\frac{\pi}{2}}^{t} \cos^4(m)\mathrm{d}m\right)\mathcal{G}(x_1, X_2). \tag{7.175}$$

Then

$$\begin{aligned}\dot{\mathcal{H}} &= \left[\cos^4(t) - \cos^4(t - \tfrac{\pi}{2})\right] \mathcal{G}(x_1, X_2) \\ &\quad + \left(\int_{t-\frac{\pi}{2}}^{t} \cos^4(m)\mathrm{d}m\right) \dot{\mathcal{G}}(x_1, X_2) \\ &\leq \left[\cos^4(t) - \sin^4(t)\right]\mathcal{G}(x_1, X_2) \\ &\quad - \left(\int_{t-\frac{\pi}{2}}^{t} \cos^4(m)\mathrm{d}m\right) 40\cos^4(t)\mathcal{G}(x_1, X_2).\end{aligned} \tag{7.176}$$

Since

$$\int_{t-\frac{\pi}{2}}^{t} \cos^4(m)\mathrm{d}m = \frac{3\pi}{16} + \frac{\sin(2t)}{2} \tag{7.177}$$

and $\sin(2t) \geq -1$ everywhere, it follows that

$$\begin{aligned}\dot{\mathcal{H}} &\leq \left[\cos^4(t) - \sin^4(t) - \left\{\frac{15\pi}{2} + 20\sin(2t)\right\}\cos^4(t)\right] \mathcal{G}(x_1, X_2) \\ &= -\left[\sin^4(t) + \left\{\frac{15\pi}{2} - 1 + 20\sin(2t)\right\}\cos^4(t)\right] \mathcal{G}(x_1, X_2) \\ &\leq -\left[\sin^4(t) + \left(\frac{15\pi - 42}{2}\right)\cos^4(t)\right] \mathcal{G}(x_1, X_2).\end{aligned} \tag{7.178}$$

Step 2. Nonzero Disturbances

It follows that when λ_1 and λ_2 are present,

$$\begin{aligned}\dot{\mathcal{H}} &\leq -\left[\sin^4(t) + \left\{\frac{15\pi - 42}{2}\right\}\cos^4(t)\right] \mathcal{G}(x_1, X_2) \\ &\quad + \left(\int_{t-\frac{\pi}{2}}^{t} \cos^4(m)\mathrm{d}m\right) x_1\lambda_1(t) \\ &\quad + \left(\int_{t-\frac{\pi}{2}}^{t} \cos^4(m)\mathrm{d}m\right) X_2\left[\cos^3(t)\lambda_1(t) + \lambda_2(t)\right]\end{aligned} \tag{7.179}$$

along the trajectories of (7.170).

Using (7.177) and the global inequalities

$$\frac{15\pi - 42}{2} \geq 1 \quad \text{and} \quad \sin^4(t) + \cos^4(t) \geq \frac{1}{2},$$

we get

$$\begin{aligned}\dot{\mathcal{H}} &\leq -\frac{1}{2}\mathcal{G}(x_1, X_2) + \left(\frac{3\pi}{16} + \frac{\sin(2t)}{2}\right)|x_1\lambda_1(t)| \\ &\quad + \left(\frac{3\pi}{16} + \frac{\sin(2t)}{2}\right)\left|X_2[\cos^3(t)\lambda_1(t) + \lambda_2(t)]\right| \\ &\leq -\frac{1}{4}[x_1^2 + X_2^2] \\ &\quad + \left(\frac{3\pi}{16} + \frac{1}{2}\right)|x_1||\lambda_1(t)| + \left(\frac{3\pi}{16} + \frac{1}{2}\right)|X_2|\left(|\lambda_1(t)| + |\lambda_2(t)|\right).\end{aligned} \tag{7.180}$$

From the triangular inequality $c_1c_2 \leq 2c_1^2 + \frac{1}{8}c_2^2$ for suitable non-negative values c_1 and c_2, we get

$$\dot{\mathcal{H}} \leq -\frac{1}{8}\left[x_1^2 + X_2^2\right] + 2\left(\frac{3\pi}{16} + \frac{1}{2}\right)^2 \lambda_1^2(t) + 2\left(\frac{3\pi}{16} + \frac{1}{2}\right)^2 \left(|\lambda_1(t)| + |\lambda_2(t)|\right)^2.$$

Next, observing that

$$\begin{aligned}\frac{1}{8}\left[x_1^2 + X_2^2\right] &= \frac{1}{4}\mathcal{G}(x_1, X_2) \\ &= \frac{\mathcal{H}(t, x_1, X_2)}{\frac{3\pi}{4} + 2\sin(2t)} \geq \frac{\mathcal{H}(t, x_1, X_2)}{\frac{3\pi}{4} + 2}\end{aligned} \tag{7.181}$$

gives

$$\begin{aligned}\dot{\mathcal{H}} &\leq -\frac{\mathcal{H}(t, x_1, X_2)}{\frac{3\pi}{4} + 2} + 2\left(\frac{3\pi}{16} + \frac{1}{2}\right)^2 \lambda_1^2(t) \\ &\quad + 2\left(\tfrac{3\pi}{16} + \tfrac{1}{2}\right)^2 \left\{|\lambda_1(t)| + |\lambda_2(t)|\right\}^2 \\ &\leq -\frac{\mathcal{H}(t, x_1, X_2)}{\frac{3\pi}{4} + 2} + 6\left(\frac{3\pi}{16} + \frac{1}{2}\right)^2 \left[\lambda_1^2(t) + \lambda_2^2(t)\right].\end{aligned} \tag{7.182}$$

We now return to the original coordinates. The feedback

$$v^\sharp(t, x_1, x_2) = v\big(t, x_1, x_2 + \cos^3(t)x_1\big)$$

with v defined in (7.173) admits the decomposition

$$v^\sharp(t, x_1, x_2) = L_1(t) \cdot x + 20\cos(t)L_2(t) \cdot x\,, \tag{7.183}$$

where $L_1(t) \cdot x = -\frac{1}{2}x_1$ and

$$L_2(t) \cdot x = -\left[\cos(t) + \cos^3(t)\right]x_2 + \left[-\cos^4(t) - 1 + \frac{3}{20}\cos(t)\sin(t)\right]x_1\,.$$

Next we consider the function

$$\overline{Q}(t, x_1, x_2) = \frac{1}{12\left(\frac{3\pi}{16} + \frac{1}{2}\right)^2}\mathcal{H}\big(t, x_1, x_2 + \cos^3(t)x_1\big) . \tag{7.184}$$

By separately considering the possibilities

$$|x_1| \geq \frac{1}{\sqrt{5}}|x_2| \quad \text{and} \quad |x_1| \leq \frac{1}{\sqrt{5}}|x_2|,$$

our choice (7.169) of X_2 gives $x_1^2 + X_2^2 \geq \frac{1}{6}|x|^2$ everywhere. Also,

$$\overline{Q}(t, x) \;\geq\; \frac{1}{24\left(\frac{3\pi}{16} + \frac{1}{2}\right)^2}\left(\frac{3\pi}{16} - \frac{1}{2}\right)\left[x_1^2 + (x_2 + \cos^3(t)x_1)^2\right] \tag{7.185}$$

everywhere. One can then prove that the time derivative of $\overline{Q}(t, x)$ along the trajectories of (7.168) in closed-loop with the feedback $v^\sharp(t, x)$ satisfies

$$\dot{\overline{Q}} \;\leq\; -c_3\overline{Q}(t, x) + |\lambda(t)|^2, \quad \text{and} \quad c_1|x|^2 \;\leq\; \overline{Q}(t, x) \tag{7.186}$$

where $x = (x_1, x_2)$,

$$c_1 \;=\; \frac{\frac{3\pi}{16} - \frac{1}{2}}{144\left(\frac{3\pi}{16} + \frac{1}{2}\right)^2} \quad \text{and} \quad c_3 \;=\; \frac{1}{\frac{3\pi}{4} + 2} . \tag{7.187}$$

We deduce from Theorem 7.1 and Remark 7.3 that the feedback

$$\begin{aligned} u &= \frac{p(t)}{p^2(t) + \delta}L_1(t)x + L_2(t)x \\ &= -\frac{20\cos(t)}{400\cos^2(t) + \delta}\tfrac{1}{2}x_1 - \left[\cos(t) + \cos^3(t)\right]x_2 \\ &\quad + \left[-\cos^4(t) - 1 + \tfrac{3}{20}\cos(t)\sin(t)\right]x_1 \end{aligned} \tag{7.188}$$

with δ such that

$$\int_0^{2\pi} \frac{\delta^2}{(400\cos^2(t) + \delta)^2}\mathrm{d}t \;\leq\; \frac{2\pi}{5}\frac{\frac{3\pi}{16} - \frac{1}{2}}{144\left(\frac{3\pi}{8} + 1\right)^2}\frac{1}{\frac{3\pi}{4} + 2} \tag{7.189}$$

renders (7.167) ISS with respect to λ; see (7.106). Inequality (7.189) holds if

$$\int_0^{\frac{\pi}{2}} \frac{\left(\frac{\delta}{20}\right)^2}{(\cos^2(t) + \frac{\delta}{400})^2}\mathrm{d}t \;\leq\; \frac{\pi}{23040\left(\frac{3\pi}{8} + 1\right)^3} . \tag{7.190}$$

Therefore, we can construct an upper bound for the admissible values of $\delta > 0$ using the proof of Lemma 7.2. We leave the construction to the reader as a simple exercise.

7.5 Nonlinear Time-Varying Systems

7.5.1 Assumptions and Notation

We consider nonlinear time-varying systems of the form (7.63). Throughout the section, we assume that all of our functions are sufficiently smooth and:

Assumption 7.6 *There is a known continuous function $b(t,x,z)$ such that $h(t,x,z) = p(t)b(t,x,z)$ holds for all $(t,x,z) \in \mathbb{R} \times \mathbb{R}^n \times \mathbb{R}$.*

Therefore, the system we consider is

$$\begin{cases} \dot{x} = \mathcal{F}(t,x,z) \\ \dot{z} = p(t)[u + b(t,x,z)] \, . \end{cases} \tag{7.191}$$

Assumption 7.7 *The functions $|p(t)|$ and $|\dot{p}(t)|$ are uniformly bounded by a positive real number P and two positive numbers T and γ such that*

$$\int_t^{t+T} p^2(s)\mathrm{d}s \geq \gamma \quad \forall t \in \mathbb{R} \tag{7.192}$$

are known. Also, $p \in C^1$.

Assumption 7.8 *There are known functions V and $\alpha_i \in \mathcal{K}_\infty$, a positive definite function W, and a function $\mu_s \in C^1$ such that*

$$|\mu_s(t,x)| \leq \alpha_4(|x|) \, , \tag{7.193}$$

$$\alpha_1(|x|) \leq V(t,x) \leq \alpha_2(|x|) \, , \; \left| \frac{\partial V}{\partial x}(t,x) \right| \leq \alpha_3(|x|), \tag{7.194}$$

$$\text{and } \; \frac{\partial V}{\partial t}(t,x) + \frac{\partial V}{\partial x}(t,x)\mathcal{F}(t,x,z_s(t,x)) \leq -W(x) \tag{7.195}$$

with

$$z_s(t,x) = p^2(t)\mu_s(t,x)$$

hold for all $t \in \mathbb{R}$ and $x \in \mathbb{R}^n$. Also, z_s has period T in t.

Assumption 7.9 *There exists an everywhere positive non-decreasing function $\mathcal{C}$ such that*

$$\frac{\partial V}{\partial x}(t,x)\left[\mathcal{F}(t,x,a_1) - \mathcal{F}(t,x,a_2)\right] \leq \frac{1}{2}W(x) + \mathcal{C}\left([a_1-a_2]^2\right)(a_1-a_2)^2 \tag{7.196}$$

for all $t \in \mathbb{R}$, $x \in \mathbb{R}^n$, $a_1 \in \mathbb{R}$, and $a_2 \in \mathbb{R}$.

7.5.2 Main Result and Remarks

Our main result for this subsection is the following:

Theorem 7.3. *Assume that the system (7.191) satisfies Assumptions 7.6-7.9 for some constant T. Then for any positive constant Υ, the system is globally uniformly asymptotically stabilizable by the feedback*

$$\begin{aligned} u_s(t,x,z) = & -\Upsilon p(t)[z - z_s(t,x)] - b(t,x,z) + 2\dot{p}(t)\mu_s(t,x) \\ & +p(t)\left[\frac{\partial \mu_s}{\partial t}(t,x) + \frac{\partial \mu_s}{\partial x}(t,x)\mathcal{F}(t,x,z)\right] . \end{aligned} \tag{7.197}$$

A global strict Lyapunov function for the corresponding closed-loop system is

$$U(t,x,z) = V(t,x) + K\left(\left[\frac{T}{\Upsilon} + \int_{t-T}^{t}\int_{\ell}^{t} p^2(s)\mathrm{d}s\,\mathrm{d}\ell\right] Z^2\right) , \tag{7.198}$$

where

$$Z = z - z_s(t,x) \tag{7.199}$$

and where $K \in \mathcal{K}_\infty$ is any function such that K' is non-decreasing and

$$K'(s) \geq \frac{1}{\gamma}\mathcal{C}\left(\frac{\Upsilon}{T}s\right) + \frac{1}{2\gamma} \tag{7.200}$$

for all $s \geq 0$.

Remark 7.8. Theorem 7.3 has the following important features:

1. It does not make any linear growth assumptions on $\mathcal{F}$. The only growth restriction on $\mathcal{F}$ is Assumption 7.9.
2. The PE property in Assumption 7.7 is not very restrictive; in contrast with Assumption 7.1, the function p can be equal to zero on intervals of positive length.
3. The control law z_s, its time derivative along the trajectories, and the function h must be zero when $p(t) = 0$.

Requirement 3. has no equivalent in Theorems 7.1 and 7.2. We impose it to allow nonlinearities, and to replace Assumption 7.1 by the weaker PE property from Assumption 7.7.

Proof of Theorem 7.3

The variable defined in (7.199) gives

$$\begin{cases} \dot{x} = \mathcal{F}(t, x, Z + z_s(t, x)) \\ \dot{Z} = p(t)[u + b(t, x, z)] - \dot{z}_s . \end{cases} \tag{7.201}$$

Since

$$\begin{aligned} \dot{z}_s &= 2p(t)\dot{p}(t)\mu_s(t, x) \\ &\quad + p^2(t)\left[\frac{\partial \mu_s}{\partial t}(t, x) + \frac{\partial \mu_s}{\partial x}(t, x)\mathcal{F}(t, x, Z + z_s(t, x))\right] , \end{aligned} \tag{7.202}$$

the choice $u = u_s$ from (7.197) gives the closed-loop system

$$\begin{cases} \dot{x} = \mathcal{F}\big(t, x, z_s(t, x)\big) + \mathcal{F}\big(t, x, Z + z_s(t, x)\big) - \mathcal{F}\big(t, x, z_s(t, x)\big) \\ \dot{Z} = -\Upsilon p^2(t) Z. \end{cases} \tag{7.203}$$

Set

$$\aleph(t, Z) = \left[\frac{T}{\Upsilon} + \int_{t-T}^{t} \int_{\ell}^{t} p^2(s) \mathrm{d}s \, \mathrm{d}\ell\right] Z^2 . \tag{7.204}$$

Then the time derivative of

$$\mathcal{U}(t, x, Z) = V(t, x) + K\big(\aleph(t, Z)\big) \tag{7.205}$$

along the trajectories of (7.203) satisfies

$$\begin{aligned} \dot{\mathcal{U}} &= \dot{V} - 2K'\big(\aleph(t, Z)\big)\left[\tfrac{T}{\Upsilon} + \int_{t-T}^{t}\left(\int_{\ell}^{t} p^2(s)\mathrm{d}s\right)\mathrm{d}\ell\right]\Upsilon p^2(t) Z^2 \\ &\quad + K'\big(\aleph(t, Z)\big)\left[T p^2(t) - \int_{t-T}^{t} p^2(s)\mathrm{d}s\right] Z^2 \\ &\le -W(x) + \tfrac{\partial V}{\partial x}(t, x)[\mathcal{F}(t, x, Z + z_s(t, x)) - \mathcal{F}(t, x, z_s(t, x))] \\ &\quad - K'\big(\aleph(t, Z)\big) Z^2 \int_{t-T}^{t} p^2(s)\mathrm{d}s , \end{aligned} \tag{7.206}$$

by Assumption 7.8. Using Assumptions 7.7 and 7.9, we obtain

$$\dot{\mathcal{U}} \le -W(x) + \frac{1}{2}W(x) + \mathcal{C}(Z^2)Z^2 - K'\left(\aleph(t, Z)\right)\gamma Z^2 . \tag{7.207}$$

Since we assumed that K' is non-decreasing, we deduce that

$$\dot{\mathcal{U}} \le -\frac{1}{2}W(x) + \left[\mathcal{C}(Z^2) - K'\big(\frac{T}{\Upsilon}Z^2\big)\gamma\right] Z^2 . \tag{7.208}$$

Recalling (7.200), we obtain

$$\dot{\mathcal{U}} \leq -\frac{1}{2}W(x) - \frac{1}{2}Z^2.$$

Also, there are two functions $\alpha_5, \alpha_6 \in \mathcal{K}_\infty$ such that

$$\alpha_5(|(x,z)|) \leq U(t,x,z) = \mathcal{U}\big(t,x,z - z_s(t,x)\big) \leq \alpha_6(|(x,z)|)$$

for all $t \in \mathbb{R}$ and $(x,z) \in \mathbb{R}^n \times \mathbb{R}$, as desired. □

7.6 Bounded Backstepping

7.6.1 Assumptions and Statement of Result

We next show that when the following additional conditions are imposed, we can construct *bounded* stabilizing feedbacks for our systems (7.191):

Assumption 7.10 *There is a constant $\bar{B} > 0$ such that*

$$\begin{aligned} &|b(t,x,z)| \leq \bar{B}\,, \quad |\mu_s(t,x)| \leq \bar{B}\,, \quad \text{and} \\ &\left|\frac{\partial \mu_s}{\partial t}(t,x) + \frac{\partial \mu_s}{\partial x}(t,x)\mathcal{F}(t,x,z)\right| \leq \bar{B}(1+|z|)\,. \end{aligned} \tag{7.209}$$

hold for all $t \in \mathbb{R}$ and all $(x,z) \in \mathbb{R}^n \times \mathbb{R}$, where μ_s and b are from Assumptions 7.6 and 7.8.

Remark 7.9. If μ_s satisfies Assumption 7.10 and $p(t)$ satisfies Assumption 7.7, then the choices

$$\mathcal{M} = \max\left\{1, P^2\bar{B}\right\} \tag{7.210}$$

and $z_s = p^2(t)\mu_s$ give

$$|z_s(t,x)| \leq \mathcal{M} \tag{7.211}$$

for all $t \in \mathbb{R}$ and $x \in \mathbb{R}^n$.

We use the function

$$\Omega(s) \quad = \quad \text{sgn}(s)\int_0^{|s|}\left[1 + \max\left\{0, \frac{(a-2\mathcal{M})^3}{1+(a-2\mathcal{M})^2}\right\}\right] da, \tag{7.212}$$

where $\text{sgn}(s) = 1$ (resp., -1) if $s \geq 0$ (resp., $s < 0$). The function Ω has the following key properties:

1. Ω is of class C^2;
2. $\Omega(s) = s$ when $s \in [-2\mathcal{M}, 2\mathcal{M}]$; and
3. $\Omega'(z) \geq 1$ everywhere.

We prove:

Theorem 7.4. *Assume that the system (7.191) satisfies the Assumptions 7.6-7.10, and define $\mathcal{M}$ by (7.210). Then for any constant $\Upsilon > 0$, the system is globally uniformly asymptotically stabilizable by the feedback*

$$u_s(t,x,z) = -\Upsilon p(t)\frac{\Omega(z) - z_s(t,x)}{\Omega'(z)\sqrt{1+(\Omega(z)-z_s(t,x))^2}} - b(t,x,z) + \frac{2\dot{p}(t)\mu_s(t,x)}{\Omega'(z)} + \frac{p(t)}{\Omega'(z)}\left[\frac{\partial \mu_s}{\partial t}(t,x) + \frac{\partial \mu_s}{\partial x}(t,x)\mathcal{F}(t,x,z)\right]. \tag{7.213}$$

A global strict Lyapunov function for the corresponding closed-loop system is

$$U(t,x,z) = V(t,x) + K\big(\nu_p(t,\Omega(z) - z_s(t,x))\big) \tag{7.214}$$

where $K \in C^1$ is any $\mathcal{K}_\infty$ function with a non-decreasing first derivative such that

$$K'(s) \geq \frac{T}{2\gamma\Upsilon}\left[1 + 128\sqrt{1+\mathcal{M}^2}\mathcal{C}\left(128\sqrt{1+\mathcal{M}^2}\frac{s}{\sqrt{1+2s}}\right)\right] \tag{7.215}$$

for all $s \geq 0$,

$$\nu_p(t,Z) = \frac{1}{2}Z^2 + \frac{\Upsilon}{T}\left(\int_{t-T}^{t}\left(\int_s^t p^2(a)\mathrm{d}a\right)\mathrm{d}s\right)\frac{Z^2}{\sqrt{1+Z^2}}, \tag{7.216}$$

and

$$Z = \Omega(z) - z_s(t,x).$$

Moreover, the inequality

$$|u_s(t,x,z)| \leq \Upsilon P + \bar{B} + 2P\bar{B} + P\bar{B}(4\mathcal{M}+2) \tag{7.217}$$

holds for all $t \in \mathbb{R}$ and all $(x,z) \in \mathbb{R}^n \times \mathbb{R}$.

7.6.2 Technical Lemmas

We present two technical lemmas that form the basis for our proof of Theorem 7.4. Consider the one-dimensional system

$$\dot{\xi} = -q(t)\frac{\xi}{\sqrt{1+\xi^2}} \tag{7.218}$$

where q is any everywhere non-negative C^1 function.

Lemma 7.3. *Assume that there exist positive constants δ_1, δ_2, and T_q such that*

$$0 \leq q(t) \leq \delta_1 \quad \text{and} \quad \int_{t-T_q}^{t} q(s)\mathrm{d}s \geq \delta_2 \quad \forall t \in \mathbb{R}. \tag{7.219}$$

Then the time derivative of

$$\nu_q(t,\xi) \doteq \frac{1}{2}\xi^2 + \frac{1}{T_q}\left(\int_{t-T_q}^{t}\int_{s}^{t} q(a)\mathrm{d}a\,\mathrm{d}s\right)\frac{\xi^2}{\sqrt{1+\xi^2}} \tag{7.220}$$

along the trajectories of (7.218) satisfies

$$\dot{\nu}_q \;\leq\; -\frac{\delta_2}{T_q}\frac{\xi^2}{\sqrt{1+\xi^2}}\,. \tag{7.221}$$

Proof. The time derivative of ν_q along the trajectories of (7.218) satisfies

$$\begin{aligned}\dot{\nu}_q &\leq -q(t)\frac{\xi^2}{\sqrt{1+\xi^2}} + \left(q(t) - \frac{1}{T_q}\int_{t-T_q}^{t} q(a)\mathrm{d}a\right)\frac{\xi^2}{\sqrt{1+\xi^2}} \\ &= -\tfrac{1}{T_q}\left(\int_{t-T_q}^{t} q(a)\mathrm{d}a\right)\frac{\xi^2}{\sqrt{1+\xi^2}}\,.\end{aligned} \tag{7.222}$$

The lemma now follows from our choice of δ_2. □

Lemma 7.4. *Let $\mathcal{M}$ be defined by (7.210). Then for all $z \in \mathbb{R}$, $t \in \mathbb{R}$, and $x \in \mathbb{R}^n$, we have*

$$\left[z - z_s(t,x)\right]^2 \;\leq\; 64\sqrt{1+\mathcal{M}^2}\frac{(\Omega(z) - z_s(t,x))^2}{\sqrt{1+(\Omega(z)-z_s(t,x))^2}}. \tag{7.223}$$

Proof. We consider two cases.

Case 1. $|z| \leq 2\mathcal{M}$. Then $\Omega(z) = z$, so our bound (7.211) on z_s gives

$$(z - z_s(t,x))^2 \leq \sqrt{1+9\mathcal{M}^2}\frac{(\Omega(z) - z_s(t,x))^2}{\sqrt{1+(\Omega(z)-z_s(t,x))^2}}\,. \tag{7.224}$$

Case 2. $|z| \geq 2\mathcal{M}$. By (7.211), we get

$$\left(z - z_s(t,x)\right)^2 \;\leq\; \left(|z| + |z_s(t,x)|\right)^2 \;\leq\; \left(|z| + \mathcal{M}\right)^2 \;\leq\; \tfrac{5}{2}z^2\,. \tag{7.225}$$

If $2\mathcal{M} \leq |z| \leq 4\mathcal{M}$, then

$$\left[z - z_s(t,x)\right]^2 \;\leq\; 25\mathcal{M}^2\,. \tag{7.226}$$

On the other hand, since $|\Omega(z)| \geq |z|$ for all z, we have

$$\left(\Omega(z) - z_s(t,x)\right)^2 \geq (|\Omega(z)| - \mathcal{M})^2 \geq (|z| - \mathcal{M})^2 \geq \mathcal{M}^2 . \tag{7.227}$$

Since the function

$$\Theta(s) \doteq \frac{s}{\sqrt{1+s}} \tag{7.228}$$

is increasing on $[0,\infty)$, we get

$$\frac{(\Omega(z) - z_s(t,x))^2}{\sqrt{1+(\Omega(z) - z_s(t,x))^2}} \geq \frac{\mathcal{M}^2}{\sqrt{1+\mathcal{M}^2}} , \tag{7.229}$$

so (7.226) gives

$$\left[z - z_s(t,x)\right]^2 \leq 25\sqrt{1+\mathcal{M}^2}\frac{(\Omega(z) - z_s(t,x))^2}{\sqrt{1+(\Omega(z) - z_s(t,x))^2}} . \tag{7.230}$$

It remains to consider the case where $|z| \geq 4\mathcal{M}$; in that case,

$$\begin{aligned}|\Omega(z)| &= |z| + \int_{2\mathcal{M}}^{|z|} \frac{(m-2\mathcal{M})^3}{1+(m-2\mathcal{M})^2}\mathrm{d}m \\ &= |z| + \int_0^{|z|-2\mathcal{M}} \frac{m^3}{1+m^2}\mathrm{d}m .\end{aligned} \tag{7.231}$$

It follows that

$$\begin{aligned}\left|\Omega(z) - z_s(t,x)\right| &\geq |z| - \mathcal{M} + \int_0^{\frac{1}{2}|z|} \frac{m^3}{1+m^2}\mathrm{d}m \\ &\geq \tfrac{1}{2}|z| + \int_0^{\frac{1}{2}|z|} \frac{m^3}{1+m^2}\mathrm{d}m \\ &\geq \int_0^{\frac{1}{2}|z|} \frac{1+m^2+m^3}{1+m^2}\mathrm{d}m \geq \int_0^{\frac{1}{2}|z|} \frac{1}{2}(1+m)\mathrm{d}m\end{aligned} \tag{7.232}$$

and therefore

$$\left|\Omega(z) - z_s(t,x)\right| \geq \tfrac{1}{4}|z| + \tfrac{1}{16}z^2 . \tag{7.233}$$

Recalling that (7.228) is increasing, we deduce that

$$\begin{aligned}\frac{(\Omega(z) - z_s(t,x))^2}{\sqrt{1+(\Omega(z) - z_s(t,x))^2}} &\geq \frac{\left(\frac{1}{4}|z| + \frac{1}{16}z^2\right)^2}{\sqrt{1+\left(\frac{1}{4}|z| + \frac{1}{16}z^2\right)^2}} \\ &= \frac{\left(\frac{1}{4} + \frac{1}{16}|z|\right)^2}{\sqrt{1+\left(\frac{1}{4}|z| + \frac{1}{16}z^2\right)^2}}z^2 .\end{aligned} \tag{7.234}$$

Using the inequality

$$z^2 \geq \frac{1}{2}\big(z - z_s(t,x)\big)^2,$$

(which is valid because $|z| \geq 4\mathcal{M}$, and therefore $\frac{1}{2}z^2 \geq \frac{1}{2}\mathcal{M}^2 + \mathcal{M}|z|$), we obtain

$$\frac{(\Omega(z) - z_s(t,x))^2}{\sqrt{1 + (\Omega(z) - z_s(t,x))^2}} \geq \frac{(\frac{1}{4} + \frac{1}{16}|z|)^2}{2\sqrt{1 + (\frac{1}{4}|z| + \frac{1}{16}z^2)^2}}\big(z - z_s(t,x)\big)^2. \quad (7.235)$$

Moreover, since

$$\Theta(r^2) \geq \frac{r}{2} \text{ on } [1, \infty),$$

and since our choice (7.210) of $\mathcal{M}$ gives $\mathcal{M} \geq 1$, we get

$$\begin{aligned} \frac{\big(\frac{1}{4} + \frac{1}{16}|z|\big)^2}{\sqrt{1 + \big(\frac{1}{4}|z| + \frac{1}{16}z^2\big)^2}} &= \Theta\left(\left[\frac{1}{4}|z| + \frac{1}{16}z^2\right]^2\right)\frac{1}{z^2} \\ &\geq \frac{1}{2z^2}\left(\frac{|z|}{4} + \frac{1}{16}z^2\right) \geq \frac{1}{32} \end{aligned}$$

when $z \neq 0$. It follows that when $|z| \geq 4\mathcal{M}$, we have

$$(z - z_s(t,x))^2 \leq 64\frac{(\Omega(z) - z_s(t,x))^2}{\sqrt{1 + |\Omega(z) - z_s(t,x)|^2}}. \quad (7.236)$$

Finally, from (7.236), (7.230) and (7.224), we deduce that (7.223) is satisfied in all three cases. This completes the proof of Lemma 7.4. □

7.6.3 Proof of Bounded Backstepping Theorem

The inequality (7.193) in Assumption 7.8 implies that for any function K of class $\mathcal{K}_\infty$, there are two functions $\alpha_5, \alpha_6 \in \mathcal{K}_\infty$ such that

$$\alpha_5(|(x,z)|) \leq U(t,x,z) \leq \alpha_6(|(x,z)|) \quad (7.237)$$

for all $t \in \mathbb{R}$ and $(x,z) \in \mathbb{R}^n \times \mathbb{R}$. Also, the time-varying change of coordinates

$$Z = \Omega(z) - z_s(t,x) \quad (7.238)$$

transforms the system (7.191) into

$$\begin{cases} \dot{x} = \mathcal{F}\big(t, x, \Omega^{-1}(Z + z_s(t,x))\big) \\ \dot{Z} = \Omega'(z)p(t)[u + b(t,x,z)] - \dot{z}_s(t,x) \ . \end{cases} \tag{7.239}$$

This system in closed-loop with $u_s(t,x,z)$ defined in (7.213) yields

$$\begin{aligned} \dot{Z} &= -\Upsilon p^2(t)\frac{\Omega(z) - z_s(t,x)}{\sqrt{1 + [\Omega(z) - z_s(t,x)]^2}} + 2\dot{p}(t)p(t)\mu_s(t,x) \\ &\quad + p^2(t)\left[\frac{\partial \mu_s}{\partial t}(t,x) + \frac{\partial \mu_s}{\partial x}(t,x)\mathcal{F}(t,x,z)\right] - \dot{z}_s(t,x) \\ &= -\Upsilon p^2(t)\tfrac{Z}{\sqrt{1+Z^2}} \ . \end{aligned} \tag{7.240}$$

Therefore, we have the closed-loop system

$$\begin{cases} \dot{x} = \mathcal{F}\big(t, x, \Omega^{-1}(Z + z_s(t,x))\big) \\ \dot{Z} = -\Upsilon p^2(t)\tfrac{Z}{\sqrt{1+Z^2}} \ . \end{cases} \tag{7.241}$$

According to Assumption 7.8, the time derivative of V along the trajectories of (7.241) satisfies

$$\begin{aligned} \dot{V} &\leq -W(x) \\ &\quad + \tfrac{\partial V}{\partial x}(t,x)\left[\mathcal{F}(t,x,\Omega^{-1}(Z + z_s(t,x))) - \mathcal{F}(t,x,z_s(t,x))\right] . \end{aligned} \tag{7.242}$$

Using Assumption 7.9, we deduce that

$$\begin{aligned} \dot{V} &\leq -\tfrac{1}{2}W(x) \\ &\quad + \mathcal{C}\left([\Omega^{-1}(Z + z_s) - z_s]^2\right)[\Omega^{-1}(Z + z_s) - z_s]^2 \ , \end{aligned} \tag{7.243}$$

where we omit the dependence of z_s on (t,x).

Next notice that (7.223) gives

$$\left(\Omega^{-1}(Z + z_s(t,x)) - z_s(t,x)\right)^2 \leq 64\sqrt{1 + \mathcal{M}^2}\frac{Z^2}{\sqrt{1 + Z^2}}. \tag{7.244}$$

Combining this inequality and (7.243), we obtain

$$\begin{aligned} \dot{V} &\leq -\tfrac{1}{2}W(x) \\ &\quad + 64\sqrt{1 + \mathcal{M}^2}\,\mathcal{C}\left(64\sqrt{1 + \mathcal{M}^2}\frac{Z^2}{\sqrt{1 + Z^2}}\right)\frac{Z^2}{\sqrt{1 + Z^2}}. \end{aligned} \tag{7.245}$$

On the other hand, Lemma 7.3 with the choice $q(t) = \Upsilon p^2(t)$ implies that the time derivative of $\nu_p(t,Z)$ along the trajectories of (7.241) satisfies

$$\dot{\nu}_p(t,Z) \leq -\frac{\gamma \Upsilon}{T} \frac{Z^2}{\sqrt{1+Z^2}}\,, \tag{7.246}$$

where γ is the constant in Assumption 7.7. It follows that the time derivative of U defined in (7.214) along the trajectories of (7.241) satisfies

$$\begin{aligned}\dot{U} \leq -\tfrac{1}{2}W(x) + \Bigg[& 64\sqrt{1+\mathcal{M}^2}\,\mathcal{C}\left(64\sqrt{1+\mathcal{M}^2}\frac{Z^2}{\sqrt{1+Z^2}}\right) \\ & -K'(\nu_p(t,Z))\frac{\gamma\Upsilon}{T}\Bigg]\frac{Z^2}{\sqrt{1+Z^2}}\,.\end{aligned} \tag{7.247}$$

Since K' is non-decreasing and $\nu_p(t,Z) \geq \frac{1}{2}Z^2$, we have

$$\begin{aligned}\dot{U} \leq -\tfrac{1}{2}W(x) + \Bigg[& 64\sqrt{1+\mathcal{M}^2}\,\mathcal{C}\left(64\sqrt{1+\mathcal{M}^2}\frac{Z^2}{\sqrt{1+Z^2}}\right) \\ & -K'\left(\frac{1}{2}Z^2\right)\frac{\gamma\Upsilon}{T}\Bigg]\frac{Z^2}{\sqrt{1+Z^2}}\,.\end{aligned} \tag{7.248}$$

From (7.215), it follows immediately that

$$\dot{U} \;\leq\; -\frac{1}{2}\left[W(x) + \frac{Z^2}{\sqrt{1+Z^2}}\right]\,. \tag{7.249}$$

Therefore, the proof of Theorem 7.4 will be complete once we establish (7.217). It is easy to prove that the first three terms in the right hand side of (7.213) are bounded by ΥP, $\bar{B}$, and $2P\bar{B}$, respectively. Also, the definition of Ω gives

$$\Omega'(|z|) \;\geq\; \frac{1+|z|}{4\mathcal{M}+2} \quad \forall z \in \mathbb{R},$$

by separately considering the cases $|z| \geq 4\mathcal{M}+1$ and $|z| \leq 4\mathcal{M}+1$ (because if $|z| \geq 4\mathcal{M}+1$, then $\Omega'(|z|) \geq 1 + \frac{1}{2}(|z| - 2\mathcal{M}) \geq 1 + \frac{|z|}{4}$ and $\Omega'(|z|) \geq 1$ everywhere). This property combined with the last inequality of Assumption 7.10 bounds the last term of the right hand side of (7.213) by $P\bar{B}(4\mathcal{M}+2)$. This concludes the proof of Theorem 7.4.

7.7 Two-Dimensional Example

The two-dimensional system

$$\begin{cases} \dot{x} = \cos(t) z \\ \dot{z} = \cos(t) u \end{cases} \tag{7.250}$$

satisfies Assumptions 7.6 and 7.7 with $b \equiv 0$, $p(t) = \cos(t)$, $\gamma = \pi$, $P = 1$, and $T = 2\pi$. Choosing $\mu_s(t,x) = -\cos(t)x$, $z_s(t,x) = -\cos^3(t)x$, and

$$V(t,x) = \exp\left(\int_{t-\frac{\pi}{2}}^{t} \cos^2(s)\mathrm{d}s\right)\frac{1}{2}x^2 , \tag{7.251}$$

one can check readily that Assumption 7.8 is satisfied. In particular,

$$\begin{aligned} \frac{\partial V}{\partial t}(t,x) + \frac{\partial V}{\partial x}(t,x)\big[-\cos^4(t)x\big] &= V(t,x)\big[\cos(2t) - 2\cos^4(t)\big] \\ &= -\frac{1}{2}V(t,x)\big[1+\cos^2(2t)\big] \\ &\le -\frac{1}{2}V(t,x) \ \le \ -\frac{1}{4}x^2 \end{aligned} \tag{7.252}$$

and $\left|\frac{\partial V}{\partial x}(t,x)\cos(t)\right| \le e^{\frac{\pi}{2}}|x|$ hold for all $(t,x) \in \mathbb{R}^2$. Hence,

$$\frac{\partial V}{\partial x}(t,x)\cos(t)(a_1 - a_2) \ \le \ \frac{1}{8}|x|^2 + 2(a_1-a_2)^2 e^{\pi} , \tag{7.253}$$

by the triangle inequality. We easily deduce that Assumption 7.9 is satisfied with $W(x) = \frac{1}{4}x^2$ and $\mathcal{C}(s) = 2e^{\pi}$ for all $s \in \mathbb{R}$.

Therefore, Theorem 7.3 applies. It follows that the control law

$$\begin{aligned} u_s(t,x,z) = &-\cos(t)[z + \cos^3(t)x] \\ &+2\sin(t)\cos(t)x + \cos(t)\left[\sin(t)x - \cos^2(t)z\right] \end{aligned} \tag{7.254}$$

globally uniformly asymptotically stabilizes the system (7.250). Taking

$$K(s) \ = \ \left(\frac{2}{\pi}e^{\pi} + \frac{1}{2\pi}\right)s,$$

a global strict Lyapunov function for the system (7.250) in closed-loop with (7.254) is

$$\begin{aligned} \mathcal{U}(t,x,z) = &\exp\left(\int_{t-\frac{\pi}{2}}^{t}\cos^2(s)\mathrm{d}s\right)\frac{1}{2}x^2 \\ &+\frac{4e^{\pi}+1}{2\pi}\left[2\pi + \int_{t-2\pi}^{t}\left(\int_{\ell}^{t}\cos^2(s)\mathrm{d}s\right)d\ell\right][z+\cos^3(t)x]^2 \\ = &\exp\left(\frac{\pi}{4} + \frac{1}{2}\sin(2t)\right)\frac{1}{2}x^2 \\ &+\frac{4e^{\pi}+1}{2\pi}\left[2\pi + \frac{\pi}{2}\sin(2t) + \pi^2\right][z+\cos^3(t)x]^2 . \end{aligned}$$

7.8 PVTOL Revisited

We now use our results to construct the necessary control laws $\tilde{u}_{1s}$ and v_{2s} to stabilize (7.6). This will complete our stabilizing feedback construction for the PVTOL model from Sect. 7.1. We prove the following:

Theorem 7.5. *Choose any positive constants ε and Υ such that*

$$0 < \varepsilon \leq \frac{1}{54^2} \quad \text{and} \quad \Upsilon \leq \frac{\tan\left(\frac{3}{2}\right)}{108}. \tag{7.255}$$

Then the feedbacks

$$\tilde{u}_{1s} = \frac{[1 - 18\cos(3t)][1 - \cos(v_{2s})] - \tilde{z}_1 - \tilde{z}_2}{\cos(v_{2s})} \quad \text{and} \tag{7.256}$$

$$\begin{aligned} v_{2s}\left(t, \tilde{\xi}_1, \tilde{\xi}_2\right) = \arctan\Bigg(& -\Upsilon p(t) \frac{\Omega(\tilde{\xi}_2) + \varepsilon p^2(t) \frac{\tilde{\xi}_1}{\sqrt{1+\tilde{\xi}_1^2}}}{\Omega'(\tilde{\xi}_2)\sqrt{1 + \left(\Omega(\tilde{\xi}_2) + \varepsilon p^2(t) \frac{\tilde{\xi}_1}{\sqrt{1+\tilde{\xi}_1^2}}\right)^2}} \\ & - \varepsilon \frac{\tilde{\xi}_1}{\sqrt{1+\tilde{\xi}_1^2}} \frac{2\dot{p}(t)}{\Omega'(\tilde{\zeta}_2)} - \varepsilon \frac{p(t)}{\Omega'(\tilde{\xi}_2)(1+\tilde{\xi}_1^2)\sqrt{1+\tilde{\xi}_1^2}} \tilde{\xi}_2 \Bigg) \end{aligned} \tag{7.257}$$

with

$$\Omega(s) = \operatorname{sgn}(s) \int_0^{|s|} \left[1 + \max\left\{0, \frac{(a-2)^3}{1+(a-2)^2}\right\}\right] \mathrm{d}a \tag{7.258}$$

and $p(t) = -1 + 18\cos(3t)$ render (7.6) UGAS to the origin.

The rest of this section is devoted to the proof of Theorem 7.5. We will presently show that $|v_{2s}| \leq \frac{3}{2}$ everywhere. Assuming this to be true for the moment, we get $\cos(v_{2s}) \geq \cos(\frac{3}{2}) > 0$ and therefore we can select (7.256) in (7.6) to get

$$\begin{cases} \dot{\tilde{\xi}}_1 = \tilde{\xi}_2 \\ \dot{\tilde{\xi}}_2 = [-1 + 18\cos(3t) + \tilde{z}_1 + \tilde{z}_2]\tan(v_{2s}) \\ \dot{\tilde{z}}_1 = \tilde{z}_2 \\ \dot{\tilde{z}}_2 = -\tilde{z}_1 - \tilde{z}_2. \end{cases} \tag{7.259}$$

Since the $\tilde{z}$-subsystem of (7.259) is globally exponentially stable and since $|v_{2s}| \leq \frac{3}{2}$, this leads us to consider the problem of finding a control law u bounded by $\tan\left(\frac{3}{2}\right)$ and an iISS Lyapunov function $\bar{U}$ for the system

$$\begin{cases} \dot{\tilde{\xi}}_1 = \tilde{\xi}_2 \\ \dot{\tilde{\xi}}_2 = [-1 + 18\cos(3t) + d]u \end{cases} \tag{7.260}$$

with disturbance d. Later we use the iISS Lyapunov function to prove UGAS of the full closed-loop system (7.259).

7.8.1 Analysis of Reduced System

7.8.1.1 Zero Disturbances Case

To find the iISS Lyapunov function $\bar{U}$ for (7.260), we use the simplifying notation $x = \tilde{\xi}_1$ and $z = \tilde{\xi}_2$. Moreover, for the time being, let $d = 0$. Then we obtain the two-dimensional system

$$\left\{ \begin{array}{l} \dot{x} = z \\ \dot{z} = p(t)u . \end{array} \right. \tag{7.261}$$

This system is of the form (7.191). Therefore, to determine stabilizing bounded controls for (7.261), we use Theorem 7.4. Before applying Theorem 7.4 to (7.261), we show that this system satisfies Assumptions 7.6-7.10.

It satisfies Assumption 7.6 with $b \equiv 0$, and it satisfies Assumption 7.7 with $P = 54$, $T = 2\pi$ and $\gamma = 324\pi$. We choose

$$\mu_s(t,x) = -\varepsilon \frac{x}{\sqrt{1+x^2}} \quad \text{and} \quad z_s(t,x) = -\varepsilon p^2(t) \frac{x}{\sqrt{1+x^2}} \tag{7.262}$$

where $\varepsilon > 0$ is such that (7.255) holds. Let $V(t,x) = \sqrt{1+\nu(t,x)} - 1$, where

$$\begin{aligned} \nu(t,x) &= \frac{1}{2}x^2 + \frac{1}{2\pi}\left(\int_{t-2\pi}^{t}\left(\int_{s}^{t} \varepsilon p^2(a)\mathrm{d}a\right)\mathrm{d}s\right)\frac{x^2}{\sqrt{1+x^2}} \\ &= \frac{1}{2}x^2 + \varepsilon \mathcal{S}(t)\frac{x^2}{\sqrt{1+x^2}} \end{aligned} \tag{7.263}$$

and

$$\mathcal{S}(t) = 163\pi + 27\sin(6t) - 12\sin(3t) \,. \tag{7.264}$$

According to Lemma 7.3, we have

$$\frac{\partial \nu}{\partial t}(t,x) + \frac{\partial \nu}{\partial x}(t,x) z_s(t,x) \ \le\ -162\varepsilon \frac{x^2}{\sqrt{1+x^2}} \,. \tag{7.265}$$

It follows that

$$\begin{aligned} \frac{\partial V}{\partial t}(t,x) + \frac{\partial V}{\partial x}(t,x) z_s(t,x) &\le -81\varepsilon \frac{x^2}{\sqrt{1+x^2}\sqrt{1+\nu(t,x)}} \\ &\le -W(x) \,, \end{aligned} \tag{7.266}$$

where

$$W(x) = 81\varepsilon \frac{x^2}{1+x^2} , \tag{7.267}$$

because $|\mathcal{S}(t)| \leq 691$ for all $t \in \mathbb{R}$, and ε satisfies (7.255). This allows us to prove that Assumption 7.8 is satisfied. In addition,

$$\begin{aligned} \left|\frac{\partial V}{\partial x}(t,x)\right| &= \left|\frac{x\left[1+\varepsilon\mathcal{S}(t)\frac{2+x^2}{(1+x^2)\sqrt{1+x^2}}\right]}{2\sqrt{1+\nu(t,x)}}\right| \\ &\leq \frac{[1+691\varepsilon]}{\sqrt{1+\frac{1}{2}x^2}}|x| \;\leq\; \frac{2\sqrt{2}}{\sqrt{1+x^2}}|x| \,. \end{aligned} \tag{7.268}$$

This easily gives

$$\left(\frac{\partial V}{\partial x}(t,x)\right)^2 \leq \frac{8}{1+x^2}x^2 = \frac{8}{81\varepsilon}W(x) \,. \tag{7.269}$$

Therefore, we deduce from the triangular inequality that

$$\frac{\partial V}{\partial x}(t,x)(a_1-a_2) \leq \frac{1}{2}W(x) + \frac{4}{81\varepsilon}(a_1-a_2)^2 \tag{7.270}$$

which implies that Assumption 7.9 is satisfied with $\mathcal{C} \equiv \frac{4}{81\varepsilon}$. Finally, one can easily prove that Assumption 7.10 is satisfied with $\bar{B} = \varepsilon$ and therefore Theorem 7.4 applies to the system (7.261).

From Theorem 7.4 and the fact that $\mathcal{M} = 1$ (because (7.255) is satisfied), it follows that for any $\Upsilon > 0$ satisfying (7.255), the control law

$$\begin{aligned} u_s(t,x,z) = &-\Upsilon p(t)\frac{\Omega(z)+\varepsilon p^2(t)\frac{x}{\sqrt{1+x^2}}}{\Omega'(z)\sqrt{1+\left(\Omega(z)+\varepsilon p^2(t)\frac{x}{\sqrt{1+x^2}}\right)^2}} \\ &-\varepsilon\frac{x}{\sqrt{1+x^2}}\frac{2\dot{p}(t)}{\Omega'(z)} - \varepsilon\frac{p(t)}{\Omega'(z)(1+x^2)\sqrt{1+x^2}}z \end{aligned} \tag{7.271}$$

with Ω defined by (7.258) globally uniformly asymptotically stabilizes the origin of the system (7.261). Moreover, the proof of Theorem 7.4 implies that if we take $\nu_p(t,x)$ as defined in (7.216), namely,

$$\begin{aligned} \nu_p(t,Z) &= \frac{1}{2}Z^2 + \frac{\Upsilon}{2\pi}\left(\int_{t-2\pi}^{t}\left(\int_s^t p^2(a)\mathrm{d}a\right)\mathrm{d}s\right)\frac{Z^2}{\sqrt{1+Z^2}} \\ &= \frac{1}{2}Z^2 + \Upsilon\mathcal{S}(t)\frac{Z^2}{\sqrt{1+Z^2}}, \end{aligned} \tag{7.272}$$

$\mathcal{S}(t)$ defined in (7.264), and $K \in \mathcal{K}_\infty$ such that

$$K'(s) \geq \frac{2\pi}{2\gamma\Upsilon}\left[1+128\sqrt{1+\mathcal{M}^2}\mathcal{C}\left(128\sqrt{1+\mathcal{M}^2}\frac{s}{\sqrt{1+2s}}\right)\right] \\ = \frac{1}{324\Upsilon}\left[1+\frac{512\sqrt{2}}{81\varepsilon}\right] =: \underline{K}, \tag{7.273}$$

then the time derivative of

$$U(t,x,z) = \sqrt{1+\nu(t,x)}-1+K\left(\nu_p\left(t,\Omega(z)+\varepsilon p^2(t)\tfrac{x}{\sqrt{1+x^2}}\right)\right) \tag{7.274}$$

along the trajectories of (7.261), in closed-loop with (7.271), satisfies

$$\dot{U} \leq -\tfrac{1}{2}\left[W(x)+\tfrac{Z^2}{\sqrt{1+Z^2}}\right], \tag{7.275}$$

where $Z = \Omega(z)-z_s(t,x)$. Let us choose $K(s) = \underline{K}s$. By (7.217), the function u_s defined in (7.271) satisfies $|u_s(t,x,z)| \leq 54\Upsilon+433\varepsilon$. Noting that Υ satisfies (7.255) and observing that (7.255) implies that

$$\varepsilon \leq \frac{\tan\left(\frac{3}{2}\right)}{54^2},$$

we get

$$|u_s(t,x,z)| \leq \tan\left(\frac{3}{2}\right). \tag{7.276}$$

We therefore take $v_{2s} = \arctan(u_s)$.

7.8.1.2 Nonzero Disturbances Case

Returning to the system (7.260) when d is present, we immediately deduce from the previous analysis that the time derivative of

$$U(t,\tilde{\xi}_1,\tilde{\xi}_2) = \sqrt{1+\nu(t,\tilde{\xi}_1)}-1+\underline{K}\nu_p\left(t,\varrho(t,\tilde{\xi}_1,\tilde{\xi}_2)\right) \tag{7.277}$$

with

$$\varrho(t,\tilde{\xi}_1,\tilde{\xi}_2) = \Omega(\tilde{\xi}_2)+\varepsilon p^2(t)\frac{\tilde{\xi}_1}{\sqrt{1+\tilde{\xi}_1^2}} \tag{7.278}$$

along the solutions of (7.260) in closed-loop with $u_s(t,\tilde{\xi}_1,\tilde{\xi}_2)$ defined in (7.271) satisfies

$$\dot{U} \leq -\frac{1}{2}W(\tilde{\xi}_1)-\frac{1}{2}\hat{\varrho}(t,\tilde{\xi}_1,\tilde{\xi}_2)+\frac{\partial U}{\partial\tilde{\xi}_2}(t,\tilde{\xi}_1,\tilde{\xi}_2)du_s(t,\tilde{\xi}_1,\tilde{\xi}_2), \tag{7.279}$$

where

$$\hat{\varrho} \doteq \frac{\varrho^2}{\sqrt{1+\varrho^2}}.$$

We have

$$\begin{aligned} \frac{\partial U}{\partial \tilde{\xi}_2}(t,\tilde{\xi}_1,\tilde{\xi}_2) &= \underline{K}\frac{\partial \nu_p}{\partial Z}\left(t,\varrho\left(t,\tilde{\xi}_1,\tilde{\xi}_2\right)\right)\Omega'\left(\tilde{\xi}_2\right) \\ &= \underline{K}\left[1+\Upsilon\mathcal{S}(t)\frac{2+\varrho^2(t,\tilde{\xi}_1,\tilde{\xi}_2)}{(1+\varrho^2(t,\tilde{\xi}_1,\tilde{\xi}_2))^{\frac{3}{2}}}\right] \\ &\quad \times\varrho(t,\tilde{\xi}_1,\tilde{\xi}_2)\Omega'(\tilde{\xi}_2)\,. \end{aligned} \tag{7.280}$$

Since $\mathcal{S}(t)| \leq 691$ everywhere, we deduce that

$$\left|\frac{\partial U}{\partial \tilde{\xi}_2}(t,\tilde{\xi}_1,\tilde{\xi}_2)\right| \;\leq\; \mathcal{M}_1|\varrho(t,\tilde{\xi}_1,\tilde{\xi}_2)\Omega'(\tilde{\xi}_2)| \tag{7.281}$$

where $\mathcal{M}_1 = 2\underline{K}(1+691\Upsilon)$, and therefore

$$\left|\frac{\partial U}{\partial \tilde{\xi}_2}(t,\tilde{\xi}_1,\tilde{\xi}_2)du_s(t,\tilde{\xi}_1,\tilde{\xi}_2)\right| \;\leq\; \mathcal{M}_2|\varrho(t,\tilde{\xi}_1,\tilde{\xi}_2)\Omega'(\tilde{\xi}_2)||d| \tag{7.282}$$

where $\mathcal{M}_2 = \mathcal{M}_1\tan(3/2)$. Next, using (7.255) one can easily prove that $|\varrho(t,\tilde{\xi}_1,\tilde{\xi}_2)| \leq |\Omega(\tilde{\xi}_2)| + 19^2\varepsilon \leq |\Omega(\tilde{\xi}_2)| + 1$.

It follows that

$$\left|\frac{\partial U}{\partial \tilde{\xi}_2}(t,\tilde{\xi}_1,\tilde{\xi}_2)du_s(t,\tilde{\xi}_1,\tilde{\xi}_2)\right| \leq \mathcal{M}_2\left[|\Omega(\tilde{\xi}_2)|+1\right]\Omega'(\tilde{\xi}_2)|d|\,. \tag{7.283}$$

This inequality combined with (7.279) yields

$$\dot{U} \;\leq\; -\frac{1}{2}W(\tilde{\xi}_1) - \frac{1}{2}\hat{\varrho}(t,\tilde{\xi}_1,\tilde{\xi}_2) + \mathcal{M}_2\left[|\Omega(\tilde{\xi}_2)|+1\right]\Omega'(\tilde{\xi}_2)|d|\,. \tag{7.284}$$

Therefore, the time derivative of

$$\overline{U}(t,\tilde{\xi}_1,\tilde{\xi}_2) = \ln\left(1+U(t,\tilde{\xi}_1,\tilde{\xi}_2)\right) \tag{7.285}$$

along the closed-loop trajectories of (7.260) satisfies

$$\dot{\overline{U}} \leq \frac{-\frac{1}{2}W(\tilde{\xi}_1) - \frac{1}{2}\hat{\varrho}(t,\tilde{\xi}_1,\tilde{\xi}_2) + \mathcal{M}_2[|\Omega(\tilde{\xi}_2)|+1]\Omega'(\tilde{\xi}_2)|d|}{1+U(t,\tilde{\xi}_1,\tilde{\xi}_2)}\,. \tag{7.286}$$

Next, observe that

$$|\Omega'(\tilde{\xi}_2)| \;=\; 1+\max\left\{0,\frac{(|\tilde{\xi}_2|-2\mathcal{M})^3}{1+(|\tilde{\xi}_2|-2\mathcal{M})^2}\right\} \;\leq\; 1+|\tilde{\xi}_2| \;\leq\; 1+|\Omega(\tilde{\xi}_2)|\,.$$

It follows that

$$\dot{\overline{U}} \le \frac{-\frac{1}{2}W(\tilde{\xi}_1) - \frac{1}{2}\hat{\varrho}(t, \tilde{\xi}_1, \tilde{\xi}_2) + \mathcal{M}_2[|\Omega(\tilde{\xi}_2)| + 1]^2|d|}{1 + U(t, \tilde{\xi}_1, \tilde{\xi}_2)}, \tag{7.287}$$

and therefore

$$\begin{aligned} \dot{\overline{U}} &\le \frac{-\frac{1}{2}W(\tilde{\xi}_1) - \frac{1}{2}\hat{\varrho}(t,\tilde{\xi}_1,\tilde{\xi}_2) + \mathcal{M}_2\left[\left|\Omega(\tilde{\xi}_2) + \varepsilon p^2(t)\frac{\tilde{\xi}_1}{\sqrt{1+\tilde{\xi}_1^2}}\right| + \varepsilon p^2(t) + 1\right]^2|d|}{1+U(t,\tilde{\xi}_1,\tilde{\xi}_2)} \\ &\le \frac{-\frac{1}{2}W(\tilde{\xi}_1) - \frac{1}{2}\hat{\varrho}(t,\tilde{\xi}_1,\tilde{\xi}_2) + \mathcal{M}_2\left[\left|\varrho(t,\tilde{\xi}_1,\tilde{\xi}_2)\right| + 361\varepsilon + 1\right]^2|d|}{1+U(t,\tilde{\xi}_1,\tilde{\xi}_2)} \\ &\le \frac{-\frac{1}{2}W(\tilde{\xi}_1) - \frac{1}{2}\hat{\varrho}(t,\tilde{\xi}_1,\tilde{\xi}_2) + 2\mathcal{M}_2[\varrho^2(t,\tilde{\xi}_1,\tilde{\xi}_2) + 4]|d|}{\sqrt{1+\nu(t,\tilde{\xi}_1)} + K\left(\frac{1}{2}\varrho^2(t,\tilde{\xi}_1,\tilde{\xi}_2) + \Upsilon\mathcal{S}(t)\frac{\varrho^2(t,\tilde{\xi}_1,\tilde{\xi}_2)}{\sqrt{1+\varrho^2(t,\tilde{\xi}_1,\tilde{\xi}_2)}}\right)}. \end{aligned} \tag{7.288}$$

It follows that one can determine a constant $\mathcal{M}_3$ such that

$$\begin{aligned} \dot{\overline{U}} \le \; & \frac{-\frac{1}{2}W(\tilde{\xi}_1) - \frac{1}{2}\hat{\varrho}(t, \tilde{\xi}_1, \tilde{\xi}_2)}{\sqrt{1 + \nu(t, \tilde{\xi}_1)} + K\left(\frac{1}{2}\varrho^2(t, \tilde{\xi}_1, \tilde{\xi}_2) + \Upsilon\mathcal{S}(t)\frac{\varrho^2(t,\tilde{\xi}_1,\tilde{\xi}_2)}{\sqrt{1+\varrho^2(t,\tilde{\xi}_1,\tilde{\xi}_2)}}\right)} \\ & + \mathcal{M}_3|d|. \end{aligned} \tag{7.289}$$

This implies that $\overline{U}$ is the desired iISS Lyapunov function.

7.8.2 UGAS of Full System

Standard arguments (analogous to those in [8] but generalized to time-varying periodic systems) now provide $\underline{\alpha} \in \mathcal{K}_\infty$, $\beta \in \mathcal{KL}$, and a constant $\bar{M} > 0$ such that for each $k \in \mathbb{N} \cup \{0\}$ and $t_0 \ge 0$ and each trajectory $\tilde{\xi}(t)$ of (7.260) with initial time t_0, we have the iISS estimate

$$\underline{\alpha}\left(|\tilde{\xi}(t + 2k\pi)|\right) \le \beta(|\tilde{\xi}(t_0 + 2k\pi)|, t - t_0) + \bar{M}\int_{t_0+2k\pi}^{t+2k\pi} |d(r)|\mathrm{d}r \tag{7.290}$$

for all $t \ge t_0$ and all exponentially decaying disturbances d. Specializing to the case where $d = \tilde{z}$ converges exponentially to zero and $k = 0$, (7.290) readily gives a $\mathcal{K}_\infty$ function $\bar{\mathcal{M}}$ such that $|\tilde{\xi}(t)| \le \bar{\mathcal{M}}(|\xi(t_0)|)$ for all $t \ge t_0 \ge 0$ along the closed-loop trajectories. Also, for each pair (ε, b) of positive constants, we can find a positive integer $\tilde{K}$ such that

$$\left|\tilde{\xi}(t + 2k\pi)\right| < \varepsilon \quad \text{when} \quad \min\{t - t_0, k\} \ge \tilde{K} \;\; \text{and} \;\; |(\tilde{\xi}(t_0), \tilde{z}(t_0))| \le b.$$

Therefore, we get the uniform global attractivity condition $|\tilde{\xi}(r)| < \varepsilon$ when $r \ge T + t_0$ and $|(\tilde{\xi}(t_0), \tilde{z}(t_0))| \le b$, where $T = \tilde{K}(1 + 2\pi)$ depends only on ε

and b. We deduce that the origin of (7.259) in closed-loop with $v_{2s}(t, \tilde{\xi}_1, \tilde{\xi}_2)$ is UGAS. This proves the theorem. □

7.8.3 Numerical Example

To validate our feedback design, we simulated (7.6) in closed-loop with the feedbacks (7.256) and (7.257), using $\varepsilon = 1/54^2$ and the initial state $(\tilde{\zeta}_1, \tilde{\zeta}_2, \tilde{z}_1, \tilde{z}_2)(0) = (0.5, 0.5, 1, 1)$. We report the corresponding error trajectories for the positions and velocities in Figs. 7.1 and 7.2. Our simulation shows the rapid convergence of the tracking error to zero and therefore validates our findings.

7.9 Comments

Backstepping is a powerful method because it applies to general classes of nonlinear systems and simultaneously constructs Lyapunov functions and stabilizing feedbacks. Some pioneering works on backstepping include [19, 138, 179]; see [75] for other important references. Introductions to backstepping can be found in several articles and textbooks. In [70, 183], results similar to the one we presented in Sect. 7.2.3 are presented. In [148, Chap. 6], backstepping with cancelation is introduced. In [149, Chap. 6], strict feedback systems (which comprise a family of systems that is slightly more restrictive than the family (7.52)) are studied. Time varying versions of backstepping are given in [181]. A first result on bounded backstepping for time-invariant systems is in [44]. An extension to time-varying systems is given in [99]. This last extension borrows some key ideas of [66]. Our approach differs from this earlier work because of our global strict Lyapunov function constructions.

The literature on the PVTOL model is sizable. Some of this work uses the more general VTOL model

$$\begin{cases} \ddot{x} = -u_1 \sin(\theta) + \varepsilon u_2 \cos(\theta) \\ \ddot{y} = u_1 \cos(\theta) + \varepsilon u_2 \sin(\theta) - 1 \\ \ddot{\theta} = u_2 \ , \end{cases}$$

where the positive parameter ε represents the sloping of the wings of the aircraft. The model appears to have originated in [56], which developed an approximate input-output linearization method that led to asymptotic stability and bounded tracking. For a nonlinear small gain approach to the model, see [174]; and see [96] for an extension of [56] based on flatness. In [149, Chap. 6], the PVTOL model is stabilized by time-invariant feedback. See also [80], which uses an optimal control approach to design state feedbacks that give

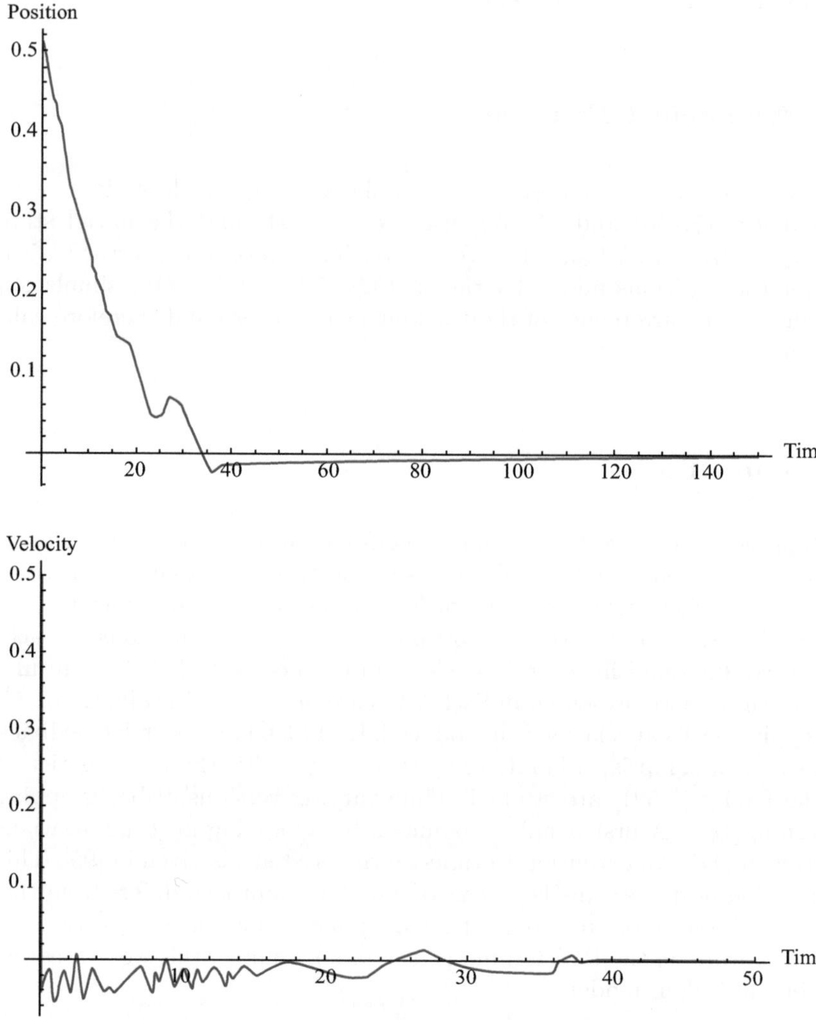

Fig. 7.1 Horizontal position and velocity components of (7.6)

robust hovering control of the PVTOL model. For internal model and output tracking approaches, see [95] and [38], respectively. Finally, see [129] for a PVTOL set up where the state is measured using a visual system that produces a delay, and [43] for state feedback designs for PVTOL models with delays in the input for cases where the velocity variables are not available for measurement. By contrast, our treatment of the PVTOL model is based on constructions of global strict Lyapunov functions.

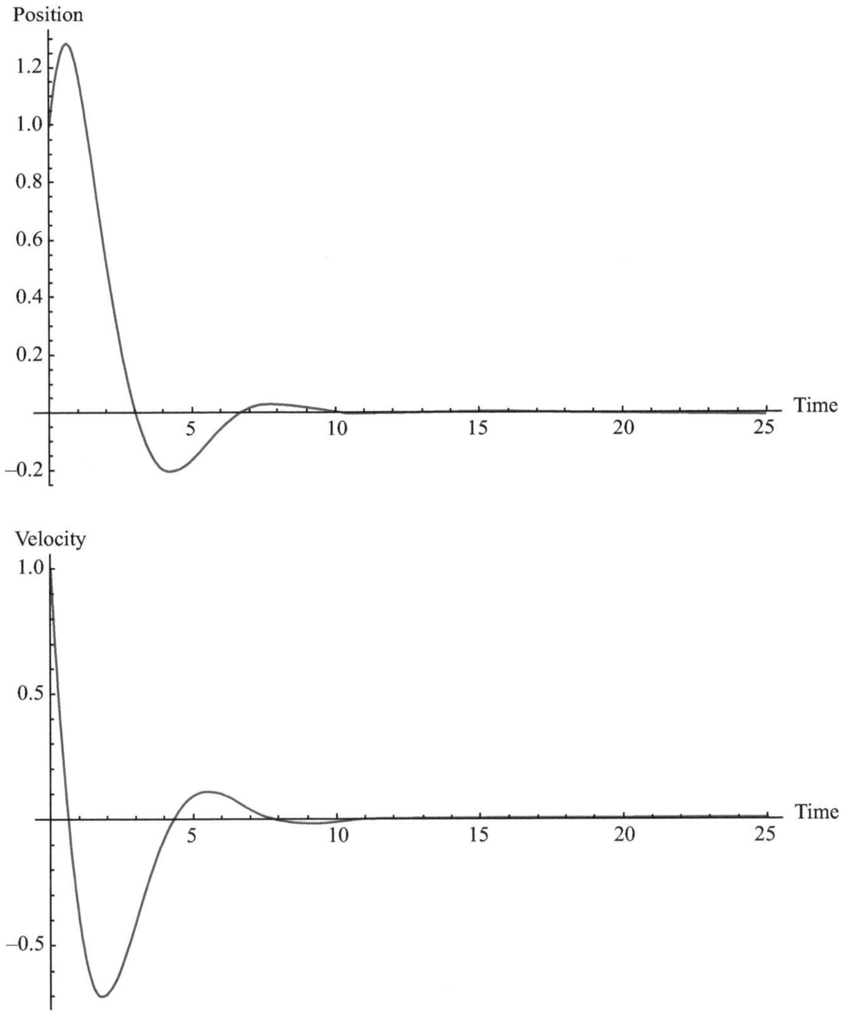

Fig. 7.2 Vertical position and velocity components of (7.6)

Chained form systems of the type (7.64) have been studied extensively. See for example [146] where they are used to control nonholonomic wheeled mobile robots and cars with multiple trailers. The TORA dynamics (7.81) has been studied by many authors; see for example [149]. The physical model consists of a platform connected to a fixed frame of reference by a spring. The platform can oscillate in the horizontal plane, and friction is assumed to be negligible. There is a rotating eccentric mass on the platform that is actuated by a DC motor. The rotating mass yields a force that can be controlled to dampen the oscillations of the platform. The control variable is the motor torque. There are several stabilizing control designs in the literature, where stability for the TORA dynamics is shown using non-strict Lyapunov functions and the LaSalle Invariant Set [149].

Chapter 8
Matrosov Conditions: General Case

Abstract In Chap. 3, we saw how to explicitly construct global, strict Lyapunov functions for time-invariant systems that satisfy Matrosov type conditions. The strict Lyapunov functions were expressed in terms of given nonstrict Lyapunov functions and the auxiliary functions from the Matrosov assumptions. The method relied on a special structure for the upper bounds on the time derivatives of the auxiliary functions.

In this chapter, we present a more general strict Lyapunov function construction for time-varying systems, under less stringent Matrosov Conditions. We apply the construction to systems that satisfy time-varying versions of the Jurdjevic-Quinn and LaSalle Conditions. We illustrate our results in a stabilization problem for a time-varying system with a sign constrained controller.

8.1 Motivation

To motivate our general Matrosov construction, we first consider a specific dynamics where we require a stabilizing controller and a corresponding global strict Lyapunov function construction, under a sign constraint on the controller (i.e., an everywhere positive or negative control). There are many results on sign restricted controllers; see, e.g., [81] for universal controllers for cases where the feedback stabilizer is constrained to be positive. The papers [34, 69, 147] provide stabilization results for families of linear systems whose inputs satisfy a sign constraint. For example, [69] proves that any stable controllable linear system $\dot{x} = Ax + Bu$ with $\det(A) \neq 0$ is globally asymptotically stabilizable by an everywhere positive (or negative) control. On the other hand, to the best of our knowledge, there are no explicit global strict Lyapunov function constructions available in the literature for the associated closed-loop systems, even for the simple system

$$\begin{cases} \dot{x}_1 = x_2 \\ \dot{x}_2 = -x_1 - |u| \; . \end{cases} \tag{8.1}$$

In this section, we construct a nowhere positive globally stabilizing controller, and a corresponding global strict Lyapunov function, for the dynamics

$$\begin{cases} \dot{x}_1 = \cos^2(t)x_2 \\ \dot{x}_2 = -\cos^2(t)x_1 + \cos^4(t)u \; . \end{cases} \tag{8.2}$$

Choosing the C^1 controller

$$u(x) = -\max\{0, x_2^3\}, \tag{8.3}$$

with $x = (x_1, x_2)$, gives the closed-loop system

$$\begin{cases} \dot{x}_1 = \cos^2(t)x_2 \\ \dot{x}_2 = -\cos^2(t)x_1 - \cos^4(t)\max\{0, x_2^3\} \; . \end{cases} \tag{8.4}$$

One can easily check that

$$V_1(x) = \frac{1}{2}|x|^2 \tag{8.5}$$

is a weak Lyapunov function for (8.4). In addition, through the LaSalle Invariance Principle and comparing the dynamics with that of a rotation, one can conclude that the origin of (8.4) is UGAS, because (8.4) is periodic [148].

However, it is by no means clear how to construct a global strict Lyapunov function for (8.4). On the other hand, it is possible to show that (8.4) admits the global strict Lyapunov function

$$\begin{aligned} V^\sharp(t,x) = {} & \left(\frac{5V_1^2(x)}{16\left[1 + V_1^2(x)\right]}\right)^3 \left\{\tilde{V}(t,x) + 5\sqrt{2}\sqrt{1 + V_1(x)}V_2(t,x)\right\} \\ & + \frac{(10\sqrt{2}+1)^4}{14}\left[(1 + V_1(x))^7 - 1\right], \end{aligned} \tag{8.6}$$

where

$$\begin{aligned} & \tilde{V}(t,x) = 6V_1(x) + V_2(t,x) + V_3(t,x) + \mathcal{C}(t,x), \\ & V_2(t,x) = x_1, \quad V_3(t,x) = \cos^4(t)x_1x_2, \quad \text{and} \\ & \mathcal{C}(t,x) = \frac{1}{2\pi}\left(\int_{t-2\pi}^{t}\int_{m}^{t}\cos^6(\ell)\mathrm{d}\ell \mathrm{d}m\right)\frac{V_1^2(x)}{1 + V_1^2(x)} \; . \end{aligned} \tag{8.7}$$

In fact, the functions V_2 and V_3 in (8.7) are the auxiliary functions in the Matrosov Conditions; see Sect. 8.7 for the derivation of (8.6). We now turn to our general procedure for constructing strict Lyapunov functions under more general Matrosov Conditions, which will include the strict Lyapunov function construction (8.6) as a special case.

8.2 Preliminaries and Matrosov Assumptions

As seen in Chap. 1, the Matrosov Theorem provides a useful framework for establishing UGAS, using a non-strict Lyapunov function and other auxiliary functions that satisfy appropriate decay conditions along the trajectories of the system. As such, Matrosov's Theorem can be viewed as a way to circumvent the need for constructing strict Lyapunov functions. However, strict Lyapunov function constructions are very useful for robustness analysis and feedback design, which motivates the search for strict Lyapunov functions under Matrosov's conditions.

In this chapter, we provide a global, explicit, strict Lyapunov function construction under Matrosov-like assumptions that are more general than those of Chap. 3. The greater generality comes from our allowing *time-varying* systems, as well as less stringent requirements on the time derivatives of the auxiliary functions. We use the following conventions and notation. Unless otherwise stated, we assume throughout the chapter that the functions encountered are sufficiently smooth. We often omit the arguments of our functions to simplify notation, and all equalities and inequalities should be understood to hold globally unless otherwise indicated. We consider the time-varying nonlinear system

$$\dot{x} = f(t, x) \tag{8.8}$$

with $t \in \mathbb{R}$ and $x \in \mathbb{R}^n$, where we assume that f is locally Lipschitz in x, uniformly in t. We always assume that (8.8) is forward complete. To simplify the statements of our results, we use the notation

$$DV \doteq \frac{\partial V}{\partial t}(t, x) + \frac{\partial V}{\partial x}(t, x) f(t, x)$$

when $V : \mathbb{R} \times \mathbb{R}^n \to \mathbb{R}$ is C^1, whenever no confusion can arise. The following definition is a slightly modified version of [148, Definition 5.14]:

Definition 8.1. A continuous function $\phi : [0, \infty) \times \mathbb{R}^n \to \mathbb{R}^p$ is *decrescent (in norm)* provided there exists a function $\Upsilon \in \mathcal{K}_\infty$ such that

$$|\phi(t, x)| \leq \Upsilon(|x|) \tag{8.9}$$

holds for all $x \in \mathbb{R}^n$ and all $t \in [0, \infty)$.

We use the following assumptions:

Assumption 8.1 *The function f in (8.8) satisfies $f(t, 0) = 0$ for all $t \in \mathbb{R}$ and is decrescent, and a non-strict Lyapunov function V_1 for (8.8) is known.*

Assumption 8.1 provides two functions $\alpha_1, \alpha_2 \in \mathcal{K}_\infty$ such that

$$\alpha_1(|x|) \le V_1(t,x) \le \alpha_2(|x|) \tag{8.10}$$

for all $x \in \mathbb{R}^n$ and all $t \in [0,\infty)$; we assume that these functions are known. We refer to the conditions of the next assumption as *Matrosov Conditions.*

Assumption 8.2 *There is an integer $j \in \mathbb{N}$, decrescent functions $V_i : [0,\infty) \times \mathbb{R}^n \to \mathbb{R}$ for $i = 2, 3, \dots, j$, positive semi-definite decrescent functions $N_i : [0,\infty) \times \mathbb{R}^n \to \mathbb{R}$ for $i = 1, \dots, j$, continuous functions $\chi_i : [0,\infty) \times \mathbb{R}^n \times \mathbb{R}^{i-1} \to \mathbb{R}$ for $i = 2, \dots, j$, and continuous positive semi-definite functions $\chi_{*i} : \mathbb{R}^n \times \mathbb{R}^{i-1} \to \mathbb{R}$ for $i = 2, \dots, j$, such that*

$$\left|\chi_i(t, x, r_1, \dots, r_{i-1})\right| \le \chi_{*i}(x, r_1, ..., r_{i-1}) \tag{8.11}$$

for all $x \in \mathbb{R}^n$, $t \in [0,\infty)$ and non-negative values $r_1, r_2, \dots, r_{i-1}$,

$$\chi_{*i}(x, 0, \dots, 0) = 0 \tag{8.12}$$

for all $x \in \mathbb{R}^n$, and

$$\begin{aligned}
DV_1 &\le -N_1(t,x) \ , \\
DV_2 &\le -N_2(t,x) + \chi_2\big(t, x, N_1(t,x)\big) \ , \\
DV_3 &\le -N_3(t,x) + \chi_3\big(t, x, N_1(t,x), N_2(t,x)\big) \ , \\
&\ \vdots \\
DV_j &\le -N_j(t,x) + \chi_j\big(t, x, N_1(t,x), \dots, N_{j-1}(t,x)\big),
\end{aligned} \tag{8.13}$$

hold for all $t \in [0,\infty)$ and all $x \in \mathbb{R}^n$, where V_1 is from Assumption 8.1.

The decrescency conditions in the preceding assumption provide a function $M_b \in \mathcal{K}_\infty$ such that

$$\sum_{i=1}^{j} N_i(t,x) + \sum_{i=1}^{j} |V_i(t,x)| \le M_b(|x|) \tag{8.14}$$

holds for all $t \in [0,\infty)$ and all $x \in \mathbb{R}^n$.

Remark 8.1. All of the results of the forthcoming sections will use Assumptions 8.1 and 8.2, as well as some other conditions. While the function V_1 from Assumption 8.1 is a non-strict Lyapunov function and therefore everywhere non-negative, the auxiliary functions $V_2, \dots, V_j$ from Assumption 8.2 do not have to be non-negative-valued. Also, the functions χ_i in the Matrosov Conditions are quite general, but they give the specific triangular form (8.13) where χ_i is independent of $N_i, \dots, N_j$ for each i.

We use the following lemmas from [111]. For their proofs, see Appendix A.1.

Lemma 8.1. *Let $n \geq 1$ and $q \geq 2$ be integers and $\chi_* : \mathbb{R}^{n+q-1} \to \mathbb{R}$ be a non-negative continuous function such that*

$$\chi_*(x, 0, ..., 0) = 0 \ \forall x \in \mathbb{R}^n. \tag{8.15}$$

Then, one can determine a continuous everywhere positive non-decreasing function ρ_* and a function $\phi_* \in \mathcal{K}_\infty$ such that

$$\chi_*(x, r_1, ..., r_{q-1}) \ \leq \ \phi_*\left(\sum_{k=1}^{q-1} r_k\right) \rho_*(|x|) \tag{8.16}$$

for all $x \in \mathbb{R}^n$ and all non-negative values $r_1, ..., r_{q-1}$.

Lemma 8.2. *Let $w_1, w_2 : \mathbb{R}^n \to \mathbb{R}$ be any continuous positive definite functions, and let $V : [0, \infty) \times \mathbb{R}^n \to \mathbb{R}$ be any storage function. Let $N \in \mathbb{N}$ be arbitrary. Then one can construct a real-valued function $L \in C^N$ such that $L(0) = 0$, $L(s) > 0$ for all $s > 0$, and*

$$L\big(V(t, x)\big) \ \leq \ w_1(x) \tag{8.17}$$

and

$$\big|L'\big(V(t, x)\big)\big| \ \leq \ w_2(x) \tag{8.18}$$

hold for all $(t, x) \in [0, \infty) \times \mathbb{R}^n$.

Lemma 8.3. *Let $\Omega : \mathbb{R}^n \to \mathbb{R}$ be a continuous function. Then, the function $\zeta : [0, \infty) \to \mathbb{R}$ defined by*

$$\zeta(r) = 1 + \int_0^{r+1} \int_0^{s_1+1} \cdots \int_0^{s_{N-1}+1} \left[\sup_{\{z \in \mathbb{R}^n : |z| \leq s_N\}} |\Omega(z)|\right] ds_N \ldots ds_1$$

is everywhere positive, of class C^N, and non-decreasing and

$$|\Omega(x)| \ \leq \ \zeta(|x|)$$

for all $x \in \mathbb{R}^n$.

8.3 One Auxiliary Function

The objective of this section is to familiarize the reader with the technique used throughout the chapter. We explicitly construct a family of strict Lyapunov functions in the simple case where (8.8) satisfies the conditions of the classical Mastrosov theorem. In this case, Assumption 8.2 is satisfied with a single auxiliary function V_2.

Theorem 8.1. *Assume that (8.8) satisfies Assumptions 8.1 and 8.2 with* $j = 2$*. Also, assume that there is a known positive definite function* ω *such that*

$$N_1(t,x) + N_2(t,x) \geq \omega(x) \tag{8.19}$$

for all $t \in [0,\infty)$ *and all* $x \in \mathbb{R}^n$*. Then, one can determine two non-negative functions* p_1 *and* p_2 *such that the function*

$$W(t,x) \doteq p_1\big(V_1(t,x)\big)V_1(t,x) + p_2\big(V_1(t,x)\big)V_2(t,x) \tag{8.20}$$

is a strict Lyapunov function for (8.8).

Proof. Let

$$S_a(t,x) = V_1(t,x) + V_2(t,x) \quad \text{and} \quad S_b(t,x) = p_2\big(V_1(t,x)\big)S_a(t,x), \tag{8.21}$$

where $p_2 \in C^1$ is a positive definite function to be specified. From Assumption 8.2,

$$DS_a = DV_1 + DV_2 \leq -N_1 - N_2 + \chi_2(t,x,N_1)\ . \tag{8.22}$$

Using (8.11) and (8.12) from Assumption 8.2 and Lemma 8.1, one can determine an explicit function $\phi \in \mathcal{K}_\infty$ and an explicit everywhere positive non-decreasing function ρ such that

$$|\chi_2(t,x,r_1)| \leq \phi(r_1)\rho(|x|) \tag{8.23}$$

for all $x \in \mathbb{R}^n$ and $r_1 \geq 0$. This inequality and (8.19) yield

$$\begin{aligned} &DS_a \leq -\omega(x) + \phi(N_1)\rho(|x|) \ \text{ and} \\ &DS_b \leq -p_2(V_1)\omega(x) + p_2(V_1)\phi(N_1)\rho(|x|) + p_2'(V_1)S_a DV_1 \end{aligned} \tag{8.24}$$

We consider the following two cases:

Case 1. $N_1 \leq p_2(V_1)$. Since ϕ is non-decreasing, it follows that the inequality $p_2(V_1)\phi(N_1)\rho(|x|) \leq p_2(V_1)\phi\big(p_2(V_1)\big)\rho(|x|)$ is satisfied.

Case 2. $N_1 \geq p_2(V_1)$. Then $p_2(V_1)\phi(N_1)\rho(|x|) \leq N_1\phi(N_1)\rho(|x|)$ is satisfied.

It follows that

$$p_2(V_1)\phi(N_1)\rho(|x|) \ \leq\ N_1\phi(N_1)\rho(|x|) + p_2(V_1)\phi\big(p_2(V_1)\big)\rho(|x|) \tag{8.25}$$

for all $x \in \mathbb{R}^n$ and $t \in [0,\infty)$. Letting α_1 be the function from (8.10) provided by Assumption 8.1 and M_b be the function satisfying (8.14), Lemma 8.2 provides a positive definite function p_2 such that

$$\begin{aligned} &p_2(V_1) \leq \inf\left\{\phi^{-1}\left(\frac{\omega(x)}{2\rho(|x|)}\right), \frac{\alpha_1(|x|)}{M_b(|x|)+1}\right\} \ \text{ and} \\ &|p_2'(V_1)| \leq \frac{1}{M_b(|x|)+1} \end{aligned} \tag{8.26}$$

for all $(t,x) \in [0,\infty) \times \mathbb{R}^n$. For such a choice, (8.25) gives

$$p_2(V_1)\phi(N_1)\rho(|x|) \;\leq\; N_1\phi(N_1)\rho(|x|) + \frac{1}{2}p_2(V_1)\omega(x) \tag{8.27}$$

for all $(t,x) \in [0,\infty) \times \mathbb{R}^n$. Combining (8.24) and (8.27), we obtain

$$DS_b \leq -\frac{1}{2}p_2(V_1)\omega(x) + N_1\phi(N_1)\rho(|x|) + p_2'(V_1)S_a DV_1 \;. \tag{8.28}$$

From (8.14) and (8.26) and the inequality $DV_1 \leq -N_1$, we obtain

$$\begin{aligned} DS_b &\leq -\tfrac{1}{2}p_2(V_1)\omega(x) + N_1\phi\big(M_b(|x|)+1\big)\rho(|x|) + |DV_1| \\ &\leq -\tfrac{1}{2}p_2(V_1)\omega(x) + \big[\phi\big(M_b(|x|)+1\big)\rho(|x|)+1\big]|DV_1| \;. \end{aligned} \tag{8.29}$$

The function $\Gamma_1(s) = \phi(M_b(s)+1)\rho(s)+1$ defined on $[0,\infty)$ is everywhere positive and non-decreasing over $[0,\infty)$. This and (8.10) imply that

$$DS_b \leq -\frac{1}{2}p_2(V_1)\omega(x) + \Gamma_1\big(\alpha_1^{-1}(V_1)\big)|DV_1| \;. \tag{8.30}$$

For an arbitrary positive integer N, Lemma 8.3 constructs a non-decreasing everywhere positive C^N function Γ_2 such that

$$\max\big\{2, \Gamma_1(\alpha_1^{-1}(r))\big\} \leq \Gamma_2(r) \;\; \forall r \geq 0 \;. \tag{8.31}$$

Hence, we obtain the inequality

$$DS_b \leq -\frac{1}{2}p_2(V_1)\omega(x) + \Gamma_2(V_1)|DV_1| \;. \tag{8.32}$$

The formula for W in (8.20) with

$$p_1(r) = \begin{cases} \frac{1}{r}\int_0^r \Gamma_2(l)\,\mathrm{d}l + p_2(r), & r \neq 0 \\ 0, & r = 0 \end{cases}$$

and p_2 satisfying (8.26) is

$$\begin{aligned} W(t,x) = &\int_0^{V_1(t,x)} \Gamma_2(l)\,\mathrm{d}l + p_2\big(V_1(t,x)\big)V_1(t,x) \\ &+ p_2\big(V_1(t,x)\big)V_2(t,x) \\ = &\int_0^{V_1(t,x)} \Gamma_2(l)\,\mathrm{d}l + S_b(t,x) \end{aligned} \tag{8.33}$$

so (8.32) implies that $DW \leq -\frac{1}{2}p_2(V_1)\omega(x)$. We deduce from (8.10) that there exists a positive definite function γ_2 such that $\frac{1}{2}p_2(V_1)\omega(x) \geq \gamma_2(x)$.

Moreover, W satisfies $W(t,x) \geq \Gamma_2(0)V_1(t,x) + p_2(V_1(t,x))V_2(t,x)$ for all $t \in [0,\infty)$ and $x \in \mathbb{R}^n$. Using (8.14) and (8.10), we obtain $W(t,x) \geq \Gamma_2(0)\alpha_1(|x|) - p_2(V_1(t,x))M_b(|x|)$. From (8.26) and (8.31),

$$\Gamma_2(0) \geq 2 \quad \text{and} \quad p_2\big(V_1(t,x)\big) \leq \frac{\alpha_1(|x|)}{M_p(|x|)+1},$$

and therefore $W(t,x) \geq \alpha_1(|x|)$. Finally, one can easily prove that W is decrescent in norm. It follows that W is a strict Lyapunov function for system (8.8). This completes the proof. □

8.4 Several Auxiliary Functions

We next extend Theorem 8.1 to the case where instead of only one auxiliary function, there are several auxiliary functions.

Theorem 8.2. *Assume that the system (8.8) satisfies Assumptions 8.1 and 8.2, and that there is a positive definite function ω such that*

$$\sum_{i=1}^{j} N_i(t,x) \geq \omega(x) \ \forall x \in \mathbb{R}^n \text{ and } t \in [0,\infty). \tag{8.34}$$

Then, one can construct non-negative functions p_i such that

$$W(t,x) = \sum_{i=2}^{j} p_i\big(V_1(t,x)\big)V_i(t,x) + p_1(V_1) \tag{8.35}$$

is a strict Lyapunov function for system (8.8).

Proof. We prove Theorem 8.2 by induction on the number of auxiliary functions in Assumption 8.2. The case of one auxiliary function follows from Theorem 8.1. Assume that the result of Theorem 8.2 holds whenever its assumptions are satisfied with $j-2$ auxiliary functions with $j \geq 3$. Let us prove that it holds when the assumptions are satisfied with $j-1$ auxiliary functions. To this end, consider a system (8.8) satisfying the assumptions of Theorem 8.2 with $j-1$ auxiliary functions $V_2, V_3, \ldots, V_j$, where $j \geq 3$. We construct a new set of $j-2$ auxiliary functions for which the assumptions of Theorem 8.2 are satisfied.

By assumption, the non-strict Lyapunov function V_1 and the $j-1$ auxiliary functions $V_2, \ldots, V_j$ are known, so we can define the function

$$S_a(t,x) \doteq \sum_{i=1}^{j} V_i(t,x) \,. \tag{8.36}$$

By Assumption 8.2 and (8.34),

$$\begin{aligned} DS_a &\leq -\sum_{i=1}^{j} N_i + \sum_{i=2}^{j} \chi_i(t, x, N_1, \dots, N_{i-1}) \\ &\leq -\omega(x) + \sum_{i=2}^{j} \chi_i(t, x, N_1, \dots, N_{i-1}) \ . \end{aligned} \tag{8.37}$$

Using the inequality (8.11) in Assumption 8.2 and Lemma 8.1, we can construct an explicit function $\phi \in \mathcal{K}_\infty$ and a non-decreasing everywhere positive function ρ such that

$$\left| \sum_{i=2}^{j} \chi_i(t, x, N_1, \dots, N_{i-1}) \right| \leq \phi\left(\sum_{i=1}^{j-1} N_i \right) \rho(|x|) \ . \tag{8.38}$$

It follows that

$$DS_a \leq -\omega(x) + \phi\left(\sum_{i=1}^{j-1} N_i \right) \rho(|x|) \ . \tag{8.39}$$

By following the proof of Theorem 8.1 verbatim from (8.24) to (8.28) except with N_1 replaced by

$$\sum_{i=1}^{j-1} N_i,$$

we can determine a positive definite function p_* and an everywhere positive increasing function $\Gamma_a \in C^1$ such that the time derivative of

$$S_b(t, x) = p_*(V_1(t, x)) S_a(t, x) \tag{8.40}$$

along the trajectories of (8.8) satisfies

$$DS_b \leq -\frac{1}{2} p_*(V_1)\omega(x) + \frac{1}{2}\left(\sum_{i=1}^{j-1} N_i \right) \Gamma_a(V_1) - \frac{1}{2}\Gamma_a(V_1) DV_1 \ . \tag{8.41}$$

Let

$$\nu_a(t, x) = S_b(t, x) + \frac{1}{2}\Gamma_a(V_1(t, x)) V_{j-1}(t, x).$$

Then,

$$\begin{aligned} D\nu_a \leq & -\tfrac{1}{2} p_*(V_1)\omega(x) + \tfrac{1}{2}\left(\sum_{i=1}^{j-1} N_i \right) \Gamma_a(V_1) - \tfrac{1}{2}\Gamma_a(V_1) DV_1 \\ & + \tfrac{1}{2}\Gamma_a'(V_1) V_{j-1} DV_1 + \tfrac{1}{2}\Gamma_a(V_1) DV_{j-1} \ . \end{aligned} \tag{8.42}$$

We can use (8.10) and (8.14) to determine an everywhere positive increasing function Γ_b such that

$$\left| -\frac{1}{2}\Gamma_a(V_1) + \frac{1}{2}\Gamma_a'(V_1)V_{j-1} \right| \leq \Gamma_b(V_1) . \tag{8.43}$$

It follows that the time derivative of

$$\nu_b(t,x) = \nu_a(t,x) + \int_0^{V_1(t,x)} \Gamma_b(l)\,\mathrm{d}l \tag{8.44}$$

along the trajectories of (8.8) satisfies

$$D\nu_b \leq -\tfrac{1}{2}p_*(V_1)\omega(x) + \tfrac{1}{2}\left(\sum_{i=1}^{j-1} N_i\right)\Gamma_a(V_1) + \tfrac{1}{2}\Gamma_a(V_1)DV_{j-1} . \tag{8.45}$$

Using Assumption 8.2, we deduce that

$$\begin{aligned} D\nu_b \leq & -\tfrac{1}{2}p_*(V_1)\omega(x) + \tfrac{1}{2}\left(\sum_{i=1}^{j-2} N_i\right)\Gamma_a(V_1) \\ & +\tfrac{1}{2}\Gamma_a(V_1)\chi_{j-1}(t,x,N_1,\ldots,N_{j-2}) . \end{aligned} \tag{8.46}$$

One can easily prove that ν_b is decrescent in norm, and use (8.14) to determine a function $M_{bn} \in \mathcal{K}_\infty$ such that

$$\sum_{i=1}^{j-1} N_i(t,x) + \frac{1}{2}p_*(V_1(t,x))\omega(x) + \sum_{i=1}^{j-2} |V_i(t,x)| + |\nu_b(t,x)| \leq M_{bn}(|x|) \tag{8.47}$$

for all $(t,x) \in [0,\infty) \times \mathbb{R}^n$. It follows that the system (8.8) satisfies the assumptions of Theorem 8.2 with the $j-2$ auxiliary functions $V_2,\ldots,V_{j-2}$ and ν_b. By our induction assumption, we can therefore explicitly construct a strict Lyapunov function for (8.8). This gives the theorem. □

8.5 Persistency of Excitation

We next weaken the requirement (8.34) by assuming that the function

$$\sum_{i=1}^{j} N_i(t,x) \tag{8.48}$$

can be equal to zero at some instants t for some choices of $x \neq 0$. Our strategy is to replace the positive definite lower bound on (8.48) with a time-varying lower bound involving PE.

Theorem 8.3. *Assume that the system (8.8) satisfies Assumptions 8.1 and 8.2 for some $j \geq 2$. Let $\bar{p} : \mathbb{R} \rightarrow \mathbb{R}$ be any continuous everywhere non-negative function for which there are positive constants τ, p_m, and p_M satisfying*

$$\int_{t-\tau}^{t} \overline{p}(l)\, \mathrm{d}l \geq p_m \quad \textit{and} \quad \overline{p}(t) \leq p_M \quad \forall t \in \mathbb{R}. \tag{8.49}$$

Assume that there is a positive definite function μ such that

$$\sum_{i=1}^{j} N_i(t,x) \;\geq\; \overline{p}(t)\mu(x) \tag{8.50}$$

for all $x \in \mathbb{R}^n$ and $t \in \mathbb{R}$. Then, one can construct non-negative functions p_i such that

$$\begin{aligned} W(t,x) = \sum_{i=2}^{j} p_i\big(V_1(t,x)\big) V_i(t,x) + p_1\big(V_1(t,x)\big) \\ + p_{j+1}\big(V_1(t,x)\big) \left(\int_{t-\tau}^{t} \int_{s}^{t} \overline{p}(l)\, \mathrm{d}l\, \mathrm{d}s \right) \end{aligned} \tag{8.51}$$

is a strict Lyapunov function for system (8.8).

Proof. We construct a function V_{j+1} such that the condition (8.34) of Theorem 8.2 is satisfied with the auxiliary functions $V_2, \ldots V_{j+1}$. Since μ is positive definite, Lemma 8.2 provides a positive definite real-valued function $\gamma \in C^1$ such that

$$\mu(x) \geq \gamma\big(V_1(t,x)\big) \quad \text{and} \quad \big|\gamma'(V_1(t,x))\big| \leq 1 \tag{8.52}$$

for all $(t,x) \in [0,\infty) \times \mathbb{R}^n$. Consider the function

$$C(t,x) = \left(\int_{t-\tau}^{t} \int_{s}^{t} \overline{p}(l)\, \mathrm{d}l\, \mathrm{d}s \right) \gamma\big(V_1(t,x)\big)\ , \tag{8.53}$$

which is decrescent in norm. The time derivative of C along the trajectories of (8.8) satisfies

$$DC = \tau\overline{p}(t)\gamma(V_1) - \left(\int_{t-\tau}^{t} \overline{p}(l)\, \mathrm{d}l \right) \gamma(V_1) + \left(\int_{t-\tau}^{t} \int_{s}^{t} \overline{p}(l)\, \mathrm{d}l\, \mathrm{d}s \right) \gamma'(V_1) DV_1.$$

Using (8.49) and (8.52), we deduce that

$$DC \;\leq\; \tau\overline{p}(t)\mu(x) - p_m\gamma(V_1(t,x)) + \tau^2 p_M |DV_1|\ . \tag{8.54}$$

Next notice that the time derivative of the function

$$V_{j+1}(t,x) \doteq C(t,x) + \tau^2 p_M V_1(t,x) + \tau \sum_{i=1}^{j} V_i(t,x) \tag{8.55}$$

along the trajectories of (8.8) satisfies

$$DV_{j+1} \le \tau\overline{p}(t)\mu(x) - p_m\gamma(V_1) + \tau^2 p_M |DV_1| + \tau^2 p_M DV_1 + \tau \sum_{i=1}^{j} DV_i \, .$$

Using the fact that DV_1 is non-positive and Assumption 8.2, we deduce from (8.50) that

$$\begin{aligned} DV_{j+1} &\le \tau\overline{p}(t)\mu(x) - p_m\gamma(V_1) - \tau \sum_{i=1}^{j} N_j \\ &\quad + \tau \sum_{i=2}^{j} \chi_i\big(t,x,N_1,\ldots,N_{i-1}\big) \\ &\le -N_{j+1} + \tau \sum_{i=2}^{j} \chi_i\big(t,x,N_1,\ldots,N_{i-1}\big) \, , \end{aligned} \tag{8.56}$$

where $N_{j+1} = p_m\gamma(V_1)$. Moreover, using (8.14) and (8.52), we can easily determine a function $M_{bn} \in \mathcal{K}_\infty$ such that

$$\sum_{i=1}^{j+1} N_i(t,x) + \sum_{i=1}^{j+1} \big|V_i(t,x)\big| \le M_{bn}(|x|) \, . \tag{8.57}$$

It readily follows that Theorem 8.2 applies, so we get a strict Lyapunov function for the system (8.8) with the features of (8.51). □

8.6 Applications

We next use our general Matrosov constructions to extend the Jurdjevic-Quinn and LaSalle results from Chapters 4 and 5 to time-varying systems. For simplicity, we assume that there is a constant $T > 0$ such that all of the functions of (t,x) in this section are periodic in t with the same period T.

8.6.1 Jurdjevic-Quinn Theorems

The Jurdjevic-Quinn approach to time-invariant systems applies to control affine dynamics $\dot{x} = f(x) + g(x)u$ that admit a storage function $V : \mathbb{R}^n \to [0,\infty)$ and a smooth scalar function ψ such that

$$\begin{aligned}&L_f V(x) \le 0 \ \ \forall x \in \mathbb{R}^n \ \text{ and} \\ &[L_f V(x) = 0 \ \text{and} \ L_g V(x) = 0 \ \text{and} \ x \neq 0] \quad \Rightarrow \quad L_f \psi(x) < 0 \ .\end{aligned} \tag{8.58}$$

To extend this work to forward complete time-varying systems

$$\dot{x} \ = \ f(t, x) + g(t, x)u \tag{8.59}$$

with $t \in \mathbb{R}$, $x \in \mathbb{R}^n$, $u \in \mathbb{R}^m$, and $f(t, 0) = 0$ for all t, we assume:

Assumption 8.3 *The functions f and g are locally Lipschitz in x, and the system $\dot{x} = f(t, x)$ satisfies Assumption 8.1 for some function $V = V_1$. Also, there exist a C^1 decrescent function $\psi : [0, \infty) \times \mathbb{R}^n \to \mathbb{R}$, a function $\phi \in \mathcal{K}_\infty$, an everywhere positive non-decreasing function $\Gamma \in C^1$, and a non-positive function $B_2(t, x)$ such that with the choice*

$$B_1(x, t) = \frac{\partial V}{\partial t}(t, x) + \frac{\partial V}{\partial x}(t, x) f(t, x), \tag{8.60}$$

we have

$$\begin{aligned}&\frac{\partial \psi}{\partial t}(t, x) + \frac{\partial \psi}{\partial x}(t, x) f(t, x) \\ &\le B_2(t, x) + \phi\left(-B_1(x, t) + \left| \frac{\partial V}{\partial x}(t, x) g(t, x) \right|^2 \right) \Gamma(|x|)\end{aligned} \tag{8.61}$$

for all $t \in [0, \infty)$ and $x \in \mathbb{R}^n$.

Remark 8.2. The inequality (8.61) is a time-varying analog of the latter condition in (8.58).

Proposition 8.1. *Assume that the system (8.59) satisfies the preceding conditions. Choose the functions*

$$V_1(t, x) = V(t, x) \ , \ V_2(t, x) = \psi(t, x) \ , \tag{8.62}$$

$$N_1(t, x) = -B_1(x, t) + \left| \frac{\partial V}{\partial x}(t, x) g(t, x) \right|^2 , \tag{8.63}$$

$$N_2(t, x) = -B_2(t, x) \ , \ \text{ and} \tag{8.64}$$

$$\chi_2(t, x, r_1) = \left| \frac{\partial \psi}{\partial x}(t, x) g(t, x) \right| \sqrt{r_1} + \phi(r_1) \Gamma(|x|) \ . \tag{8.65}$$

Then (8.59) in closed-loop with the control law

$$u = -\left[\frac{\partial V}{\partial x}(t, x) g(t, x) \right]^\top , \tag{8.66}$$

satisfies Assumption 8.2 with the choices (8.62)-(8.65).

Proof. The time derivatives of V_1 and V_2 along the trajectories of (8.59) in closed-loop with (8.66) satisfy $\dot{V}_1 = -N_1(t,x)$ and

$$\dot{V}_2 = \frac{\partial\psi}{\partial t}(t,x) + \frac{\partial\psi}{\partial x}(t,x)\left[f(t,x) - g(t,x)\left[\frac{\partial V}{\partial x}(t,x)g(t,x)\right]^\top\right] . \quad (8.67)$$

We deduce from (8.61) and (8.64) that

$$\begin{aligned}\dot{V}_2 \leq -N_2(t,x) + \phi\left(-B_1(x,t) + \left|\frac{\partial V}{\partial x}(t,x)g(t,x)\right|^2\right)\Gamma(|x|) \\ + \left|\frac{\partial\psi}{\partial x}(t,x)g(t,x)\right|\left|\frac{\partial V}{\partial x}(t,x)g(t,x)\right| .\end{aligned} \quad (8.68)$$

This allows us to conclude. □

Remark 8.3. If the functions V_1 and V_2 constructed in Proposition 8.1 also satisfy conditions of Theorem 8.3, then we can construct a strict Lyapunov function for (8.59) in closed-loop with (8.66).

8.6.2 LaSalle Type Conditions

In Chap. 5, we constructed strict Lyapunov functions for systems whose GAS can be deduced from LaSalle Invariance. We now extend the results to time-varying systems, assuming as before that there is a constant $T > 0$ such that all of the functions of (t,x) to follow are periodic in t with the same period T.

Proposition 8.2. *Assume that the system (8.8) satisfies Assumption 8.1, and set*

$$b_1(t,x) = DV_1(t,x) \quad \text{and} \quad b_{i+1}(t,x) = Db_i(t,x) \ \text{ for all } \ i \geq 1 . \quad (8.69)$$

Given any integer $j \geq 2$, define the functions

$$V_i(t,x) = -b_{i-1}(t,x)b_i(t,x) \ \text{ for all } \ i \in \{2,\ldots,j\} , \quad (8.70)$$

$$N_1(t,x) = -b_1(t,x), \ \ N_i(t,x) = b_i^2(t,x) \ \text{ for all } \ i \geq 2 , \quad (8.71)$$

and

$$\chi_i(t,x,r_1,\ldots,r_{i-1}) = \left(\sqrt{r_{i-1}} + r_1\right)|b_{i+1}(t,x)| . \quad (8.72)$$

The preceding functions satisfy the requirements of Assumption 8.2.

Proof. The definition of N_1 gives

$$DV_1(t,x) = -N_1(t,x) \leq 0 , \quad (8.73)$$

and for all $i \geq 2$,

$$
\begin{aligned}
DV_i &= -b_i^2(t,x) - b_{i-1}(t,x)b_{i+1}(t,x) \\
&\leq -N_i(t,x) + \left(\sqrt{N_{i-1}(t,x)} + N_1(t,x)\right)|b_{i+1}(t,x)| \,.
\end{aligned}
\tag{8.74}
$$

This immediately gives the requirements of Assumption 8.2. □

Remark 8.4. As in the Jurdjevic-Quinn case, if the functions V_i constructed in Proposition 8.2 also satisfy conditions of Theorem 8.3, then we can use them to construct a strict Lyapunov function.

8.7 Sign Constrained Controller

Let us show how the strict Lyapunov function construction from Sect. 8.1 follows from our general Matrosov approach. We must show that

$$
\begin{aligned}
V^\sharp(t,x) = &\left(\frac{5V_1^2(x)}{16\left[1+V_1^2(x)\right]}\right)^3 \left\{\tilde{V}(t,x) + 5\sqrt{2}\sqrt{1+V_1(x)}V_2(t,x)\right\} \\
&+\frac{(10\sqrt{2}+1)^4}{14}[(1+V_1(x))^7 - 1],
\end{aligned}
\tag{8.75}
$$

is a global strict Lyapunov function for

$$
\left\{
\begin{aligned}
\dot{x}_1 &= \cos^2(t)x_2 \\
\dot{x}_2 &= -\cos^2(t)x_1 - \cos^4(t)\max\{0, x_2^3\} \,,
\end{aligned}
\right.
\tag{8.76}
$$

where

$$
\begin{aligned}
&\tilde{V}(t,x) = 6V_1(x) + V_2(t,x) + V_3(t,x) + \mathcal{C}(t,x), \\
&V_2(t,x) = x_1, \quad V_3(t,x) = \cos^4(t)x_1x_2, \quad \text{and} \\
&\mathcal{C}(t,x) = \tfrac{1}{2\pi}\left(\int_{t-2\pi}^{t}\int_m^t \cos^6(\ell)\mathrm{d}\ell \mathrm{d}m\right)\tfrac{V_1^2(x)}{1+V_1^2(x)} \,.
\end{aligned}
\tag{8.77}
$$

It is tempting to try to build a strict Lyapunov function for (8.76) from the arguments of Sect. 8.6.2, using the non-strict Lyapunov function

$$
V_1(x) = \frac{1}{2}|x|^2. \tag{8.78}
$$

However, such an approach would not apply, because the Lie derivative of V_1 along the vector field of (8.76) is identically equal to zero on the set $\mathbb{R}\times(-\infty,0)$ and therefore all the successive derivatives of V_1 along the trajectories of (8.76) are identically equal to zero on the set $\mathbb{R}\times(-\infty,0)$. However, we can find auxiliary functions satisfying Assumptions 8.2, and then verify that the assumptions of Theorem 8.3 are satisfied, as follows.

8.7.1 Verifying the Assumptions of the Theorem

Consider the positive definite quadratic function $V_1(x)$ in (8.78). Its derivative along the trajectories of (8.76) satisfies

$$\dot{V}_1 = -N_1(t,x) , \tag{8.79}$$

where

$$N_1(t,x) = \cos^4(t) x_2^3 \max\{0, x_2\} \geq 0 . \tag{8.80}$$

We next choose the auxiliary functions

$$V_2(t,x) = x_1 \text{ and } V_3(t,x) = \cos^4(t) x_1 x_2 \tag{8.81}$$

and the non-negative functions

$$N_2(t,x) = -\cos^2(t)\min\{0,x_2\} \text{ and } N_3(t,x) = \cos^6(t) x_1^2 . \tag{8.82}$$

Along the trajectories of (8.76), we have the time derivatives

$$\begin{aligned} \dot{V}_2 &= \cos^2(t)\max\{0,x_2\} + \cos^2(t)\min\{0,x_2\} \\ &\leq -N_2(t,x) + N_1^{1/4}(t,x) , \end{aligned} \tag{8.83}$$

and

$$\begin{aligned} \dot{V}_3 &= -4\cos^3(t)\sin(t)x_1x_2 + \cos^4(t)\cos^2(t)x_2x_2 \\ &\quad + \cos^4(t)x_1(-\cos^2(t)x_1 - \cos^4(t)\max\{0,x_2^3\}) \\ &= -N_3(t,x) + [-4\cos(t)\sin(t)x_1 + \cos^4(t)x_2]\cos^2(t)x_2 \\ &\quad - \cos^8(t)x_1\max\{0,x_2^3\} . \end{aligned} \tag{8.84}$$

It follows that

$$\begin{aligned} \dot{V}_3 &\leq -N_3(t,x) + 5|x||\cos^2(t)x_2| + \cos^8(t)|x_1|\max\{0,x_2^3\} \\ &\leq -N_3(t,x) + 5\sqrt{2}\sqrt{V_1(x)}|\cos^2(t)x_2| + \sqrt{2}\sqrt{V_1(x)}N_1^{3/4}(t,x) . \end{aligned} \tag{8.85}$$

Using

$$\begin{aligned} |\cos^2(t)x_2| &= |\cos^2(t)\min\{0,x_2\} + \cos^2(t)\max\{0,x_2\}| \\ &\leq N_2(t,x) + N_1^{1/4}(t,x) \end{aligned} \tag{8.86}$$

we obtain

$$\dot{V}_3 \leq -N_3(t,x) + \chi_3\big(t,x,N_1(t,x),N_2(t,x)\big) , \tag{8.87}$$

where

$$\chi_3(t,x,r_1,r_2) = \sqrt{2}\sqrt{V_1(x)}\{5+\sqrt{r_1}\}r_1^{\frac{1}{4}} + 5\sqrt{2}\sqrt{V_1(x)}r_2 . \tag{8.88}$$

Since $\chi_3(t,x,0,0) = 0$ for all $(t,x) \in [0,\infty) \times \mathbb{R}^2$, it follows that Assumptions 8.1 and 8.2 hold.

In addition,

$$\begin{aligned}\sum_{i=1}^{3} N_i(t,x) = \cos^4(t)x_2^3 \max\{0,x_2\} \\ - \cos^2(t)\min\{0,x_2\} + \cos^6(t)x_1^2 \\ \geq \cos^6(t)\mu(x) ,\end{aligned} \tag{8.89}$$

in terms of the positive definite function

$$\begin{aligned}\mu(x) &= x_1^2 - \min\{0,x_2\} + x_2^3 \max\{0,x_2\} \\ &\geq x_1^2 + \frac{|x_2^3|}{1+|x_2^3|}\big(-\min\{0,x_2\}\big) + \frac{|x_2^3|}{1+|x_2^3|}\max\{0,x_2\} \\ &= x_1^2 + \frac{x_2^4}{1+|x_2^3|} .\end{aligned} \tag{8.90}$$

Hence, Theorem 8.3 applies and so a strict Lyapunov function can be constructed for the system (8.76).

8.7.2 Strict Lyapunov Function Construction

We now construct the strict Lyapunov function whose existence is guaranteed by the theorem. Set

$$\begin{aligned}S(t,x) &= V_1(x) + V_2(t,x) + V_3(t,x) \ \text{ and} \\ \mathcal{C}(t,x) &= \tfrac{1}{2\pi}\left(\int_{t-2\pi}^{t}\int_{m}^{t}\cos^6(\ell)\mathrm{d}\ell\mathrm{d}m\right)\tfrac{V_1^2(x)}{1+V_1^2(x)} ,\end{aligned} \tag{8.91}$$

where $V_1(x) = \frac{1}{2}|x|^2$, and V_2 and V_3 are from (8.81). The inequalities (8.83), (8.87), (8.89), and (8.90) imply that

$$
\begin{aligned}
\dot{S} \leq & -\cos^6(t)\left(x_1^2 + \frac{x_2^4}{1+|x_2^3|}\right) \\
& +\chi_3\big(t, x, N_1(t,x), N_2(t,x)\big) + N_1^{\frac{1}{4}}(t,x) \ .
\end{aligned} \tag{8.92}
$$

For all $x \in \mathbb{R}^2$, we have $2V_1^2(x) \leq x_1^4 + x_2^4$, hence also

$$
x_1^2 + \frac{x_2^4}{1+|x_2^3|} \ \geq \ \frac{x_1^4}{1+x_1^4} + \frac{x_2^4}{2+x_2^4} \ \geq \ \frac{x_1^4 + x_2^4}{2+x_1^4+x_2^4} \geq \frac{V_1^2(x)}{1+V_1^2(x)}.
$$

This gives

$$
\begin{aligned}
\dot{S} \leq & -\cos^6(t)\frac{V_1^2(x)}{1+V_1^2(x)} \\
& +\chi_3\big(t, x, N_1(t,x), N_2(t,x)\big) + N_1^{1/4}(t,x) \ .
\end{aligned} \tag{8.93}
$$

Also,

$$
\begin{aligned}
\dot{\mathcal{C}} = & \cos^6(t)\frac{V_1^2(x)}{1+V_1^2(x)} - \frac{1}{2\pi}\left(\int_{t-2\pi}^{t} \cos^6(\ell)\mathrm{d}\ell\right)\frac{V_1^2(x)}{1+V_1^2(x)} \\
& +\frac{1}{2\pi}\left(\int_{t-2\pi}^{t}\int_{m}^{t} \cos^6(\ell)\mathrm{d}\ell\mathrm{d}m\right)\frac{2V_1(x)}{(1+V_1^2(x))^2}\dot{V}_1(t,x) \ ,
\end{aligned}
$$

hence

$$
\begin{aligned}
\dot{\mathcal{C}} & \leq \cos^6(t)\frac{V_1^2(x)}{1+V_1^2(x)} - \frac{1}{2\pi}\left(\int_{t-2\pi}^{t} \cos^6(\ell)\mathrm{d}\ell\right)\frac{V_1^2(x)}{1+V_1^2(x)} \\
& \leq \cos^6(t)\frac{V_1^2(x)}{1+V_1^2(x)} - \frac{5}{16}\frac{V_1^2(x)}{1+V_1^2(x)} \ ,
\end{aligned} \tag{8.94}
$$

since $\dot{V}_1 \leq 0$ everywhere.

Let $S_1(t,x) = S(t,x) + \mathcal{C}(t,x)$ and

$$
\begin{aligned}
S_2(t,x) & = S_1(t,x) + 5\sqrt{2}\sqrt{1+V_1(x)}V_2(t,x) \\
& = V_1(x) + \mathcal{C}(t,x) + \cos^4(t)x_1x_2 \\
& \quad + \left(1 + 5\sqrt{2}\sqrt{1+V_1(x)}\right)x_1 \ .
\end{aligned} \tag{8.95}
$$

Then

$$
\dot{S}_1 \leq -\frac{5}{16}\frac{V_1^2(x)}{1+V_1^2(x)} + \chi_3\big(t, x, N_1(t,x), N_2(t,x)\big) + N_1^{1/4}(t,x) \tag{8.96}
$$

and

$$\begin{aligned}
\dot{S}_2 \leq & -\frac{5}{16}\frac{V_1^2(x)}{1+V_1^2(x)} + \chi_3\big(t,x,N_1(t,x),N_2(t,x)\big) + N_1^{1/4}(t,x) \\
& +5\sqrt{2}\frac{\dot{V}_1}{2\sqrt{1+V_1(x)}}V_2(t,x) + 5\sqrt{2}\sqrt{1+V_1(x)}\dot{V}_2 \\
\leq & -\frac{5}{16}\frac{V_1^2(x)}{1+V_1^2(x)} \\
& + \left[\sqrt{2}\sqrt{V_1(x)}\left(5+\sqrt{N_1(t,x)}\right)+1\right] N_1^{1/4}(t,x) \\
& +5\sqrt{2}\sqrt{V_1(x)}N_2(t,x) - 5\sqrt{2}\frac{N_1(t,x)}{2\sqrt{1+V_1(x)}}V_2(t,x) \\
& +5\sqrt{2}\sqrt{1+V_1(x)}\left[-N_2(t,x)+N_1^{1/4}(t,x)\right] .
\end{aligned} \tag{8.97}$$

It follows that

$$\begin{aligned}
\dot{S}_2 \leq & -\frac{5}{16}\frac{V_1^2(x)}{1+V_1^2(x)} \\
& + \left[\sqrt{2}\sqrt{V_1(x)}\left(5+\sqrt{N_1(t,x)}\right)+1\right] N_1^{1/4}(t,x) \\
& -5\sqrt{2}\frac{N_1(t,x)}{2\sqrt{1+V_1(x)}}V_2(t,x) + 5\sqrt{2}\sqrt{1+V_1(x)}N_1^{1/4}(t,x) \\
\leq & -\frac{5}{16}\frac{V_1^2(x)}{1+V_1^2(x)} + 5N_1(t,x) \\
& + \left[\sqrt{2}\left\{\sqrt{V_1(x)}\left(5+\sqrt{N_1(t,x)}\right)+5\sqrt{1+V_1(x)}\right\}+1\right] N_1^{1/4}(t,x),
\end{aligned}$$

where the last inequality followed from

$$\left|5\sqrt{2}\frac{N_1(t,x)}{2\sqrt{1+V_1(x)}}V_2(t,x)\right| \leq 5N_1(t,x)$$

and grouping terms. Using the inequality $N_1(t,x) \leq 4V_1^2(x)$, we deduce that

$$\begin{aligned}
\dot{S}_2 \leq & -\frac{5}{16}\frac{V_1^2(x)}{1+V_1^2(x)} + 5N_1(t,x) \\
& +\left[\sqrt{2}\{\sqrt{V_1(x)}(5+2V_1(x)) + 5\sqrt{1+V_1(x)}\}+1\right]N_1^{1/4}(t,x) .
\end{aligned}$$

This readily gives

$$\dot{S}_2 \leq -\frac{5}{16}\frac{V_1^2(x)}{1+V_1^2(x)} + (10\sqrt{2}+1)\{1+V_1(x)\}^{\frac{3}{2}} N_1^{1/4}(t,x) + 5N_1(t,x).$$

Let

$$S_3(t,x) = \left(\frac{5}{16}\frac{V_1^2(x)}{1+V_1^2(x)}\right)^3 \{S_2(t,x) + 5V_1(x)\}\ . \tag{8.98}$$

Then

$$\begin{aligned}\dot{S}_3 \leq & -\left(\frac{5}{16}\frac{V_1^2}{1+V_1^2}\right)^4 \\ & + \left\{\left(\frac{5}{16}\frac{V_1^2}{1+V_1^2}\right)^3\right\}\left\{(10\sqrt{2}+1)(1+V_1)^{3/2}N_1^{1/4}\right\} \\ & -6\left(\frac{5}{16}\right)^3 \frac{V_1^5}{(1+V_1^2)^4}\left(S_2+5V_1\right)N_1\ .\end{aligned} \tag{8.99}$$

By Young's Inequality

$$ab \leq \frac{3}{4}a^{\frac{4}{3}} + \frac{1}{4}b^4$$

for $a \geq 0$ and $b \geq 0$, applied to the terms in braces in (8.99), we deduce that

$$\begin{aligned}\dot{S}_3 \leq & -\frac{1}{4}\left(\frac{5}{16}\frac{V_1^2}{1+V_1^2(x)}\right)^4 + \frac{1}{4}\left((10\sqrt{2}+1)(1+V_1)^{\frac{3}{2}}N_1^{1/4}\right)^4 \\ & -6\left(\frac{5}{16}\right)^3 \frac{V_1^5}{(1+V_1^2)^4}S_2N_1\ .\end{aligned}$$

Recalling the formula (8.95) for S_2, noting that $V_1(x) + \mathcal{C}(t,x) \geq 0$, and reorganizing terms, we get

$$\begin{aligned}\dot{S}_3 \leq & -\frac{1}{4}\left(\frac{5}{16}\frac{V_1^2}{1+V_1^2}\right)^4 + \frac{1}{4}\left(10\sqrt{2}+1\right)^4\left(1+V_1\right)^6 N_1 \\ & +6\left(\frac{5}{16}\right)^3 \frac{V_1^5}{(1+V_1^2)^4}\left|\cos^4(t)x_1x_2 + \left(1+5\sqrt{2}\sqrt{1+V_1}\right)x_1\right| N_1\ .\end{aligned}$$

Using

$$\begin{aligned}& \left|\cos^4(t)x_1x_2 + (1+5\sqrt{2}\sqrt{1+V_1(x)})x_1\right| \\ & \leq V_1(x) + \left(1+5\sqrt{2}\sqrt{1+V_1(x)}\right)\sqrt{2}\sqrt{V_1(x)} \\ & \leq 13(1+V_1(x))\end{aligned} \tag{8.100}$$

we deduce that

$$
\begin{aligned}
\dot{S}_3 \le & -\frac{1}{4}\left(\frac{5}{16}\frac{V_1^2(x)}{1+V_1^2(x)}\right)^4 \\
& + \left[\frac{1}{4}\left(10\sqrt{2}+1\right)^4 (1+V_1(x))^6 + 78\left(\frac{5}{8}\right)^3 \frac{V_1^5(x)}{(1+V_1^2(x))^3}\right] N_1(t,x) . \\
\le & -\frac{1}{4}\left(\frac{5}{16}\frac{V_1^2(x)}{1+V_1^2(x)}\right)^4 + \frac{1}{2}\left(10\sqrt{2}+1\right)^4 \left(1+V_1(x)\right)^6 N_1(t,x) .
\end{aligned}
$$

Let

$$
\begin{aligned}
S_4(t,x) &= S_3(t,x) + \frac{(10\sqrt{2}+1)^4}{14}[(1+V_1(x))^7 - 1] \\
&= \left\{6V_1(x) + \mathcal{C}(t,x) + \left[1+5\sqrt{2}\left(1+\sqrt{V_1(x)}\right)\right] x_1 \right. \\
&\quad \left. + \cos^4(t)x_1x_2\right\} \left(\frac{5V_1^2(x)}{16(1+V_1^2(x))}\right)^3 \\
&\quad + \frac{(10\sqrt{2}+1)^4}{14}\left[(1+V_1(x))^7 - 1\right] .
\end{aligned}
\tag{8.101}
$$

Then

$$
\dot{S}_4 \le -\tfrac{1}{4}\left(\tfrac{5}{16}\tfrac{V_1^2(x)}{1+V_1^2(x)}\right)^4 . \tag{8.102}
$$

Also, the inequalities $(a+b)^2 \le 2a^2 + 2b^2$ and

$$
\begin{aligned}
& \left|6V_1(x) + \mathcal{C}(t,x) + \left[1+5\sqrt{2}\left(1+\sqrt{V_1(x)}\right)\right] x_1 + \cos^4(t)x_1x_2\right| \\
\le\ & 8V_1(x) + \pi + \tfrac{1}{2}\left[1+5\sqrt{2}\left(1+\sqrt{V_1(x)}\right)\right]^2 \\
\le\ & \left(1+5\sqrt{2}\right)^2 + \pi + 58V_1(x) \ \le 125\left[1+V_1(x)\right] \ \le\ 250\left[1+V_1^2(x)\right]
\end{aligned}
$$

and the formula for S_4 give

$$
\begin{aligned}
S_4(t,x) &\ge -250\left(\frac{5}{16}\right)^2 \frac{V_1^4(x)}{(1+V_1^2(x))^2}(1+V_1^2(x)) \\
&\quad + \frac{(10\sqrt{2}+1)^4}{14}\left([1+V_1(x)]^7 - 1\right) \\
&\ge -250V_1^4(x) + \frac{(10\sqrt{2}+1)^4}{14}\left([1+V_1(x)]^7 - 1\right) \ \ge\ V_1(x) .
\end{aligned}
$$

Since the right side of (8.102) is negative definite, it follows that S_4 is a strict Lyapunov function for the system (8.76). Since S_4 agrees with the function $V^\sharp$ defined in (8.75), this proves our assertions. □

8.8 Comments

The novelty of Matrosov's Theorem lies in its use of a non-strict Lyapunov function and a (not necessarily positive definite) auxiliary function to prove UGAS [97]. Different generalizations of the Matrosov Theorem involving an arbitrary number of auxiliary functions have been reported [86]. However, the proofs in [86, 97] do not construct strict Lyapunov functions. Rather, they conclude uniform asymptotic stability by directly considering the trajectories of the system. For another application of the Matrosov approach to robot manipulators, see [128].

This chapter is largely based on [111]. However, the material in Sects. 8.6 and 8.7 appears here for the first time. Sect. 8.6.2 is an extension of the main result of [110] to time-varying systems. Similar conditions were used in [86, Sect. 3.3] to conclude uniform asymptotic stability of time-varying systems.

Chapter 9
Adaptively Controlled Systems

Abstract In the preceding chapters, we saw how to transform non-strict Lyapunov functions into explicit strict Lyapunov functions for cases where the system dynamics are completely known. However, there are important cases where the system parameters are unknown, and where the objectives are to simultaneously (a) design controllers that force the trajectories to track a prescribed reference trajectory and (b) estimate the unknown parameters. In this chapter, we present a generalization of strictification that can be used to meet these two objectives. It involves constructing global strict Lyapunov functions for an augmented system that includes the tracking error and the parameter estimation error. Our strict Lyapunov function approach makes it possible to quantify the effects of other types of uncertainty in the model as well, using the input-to-state stability framework. We illustrate our results using Rössler's dynamics and Lorenz systems.

9.1 Overview of Adaptive Control

This chapter is concerned with nonlinear systems

$$\dot{x} = f(t, x, \theta, u) \tag{9.1}$$

having a vector θ of unknown constant parameters. Given a sufficiently smooth reference trajectory $x_r(t)$, the *adaptive tracking control* problem for (9.1) involves finding a dynamic feedback controller

$$u = u(t, x, \hat{\theta}), \quad \dot{\hat{\theta}} = \tau(t, x, \hat{\theta}) \tag{9.2}$$

guaranteeing that $x_r(t) - x(t) \rightarrow 0$ as $t \rightarrow +\infty$ while keeping all closed-loop signals bounded. *In general*, solving the adaptive tracking problem does not guarantee parameter identification; i.e., the parameter estimate $\hat{\theta}$ might not converge to θ, and it might not even converge to a constant vector [74]. As a result, one cannot in general prove asymptotic stability for adaptive closed-loop systems.

The PE concept that we reviewed in Sect. 6.1 has been linked to the asymptotic stability of adaptive systems [119]. Using PE, it is possible to give necessary (and sometimes sufficient) conditions for parameter identification. Specifically, the regressor matrix needs to satisfy a PE inequality along the reference trajectory; see Assumption 9.2 or [59]. In many cases, PE guarantees that tracking error convergence can only happen when the adaptation system identifies the true parameters [150]. Connections between parameter identification, PE, and uniform asymptotic stability were first shown for linear systems, but there are now also versions for nonlinear systems. On the other hand, PE is neither necessary nor sufficient for uniform asymptotic stability in general [119]. Uniformity with respect to initial times is relevant in robustness analysis, since it yields stability under persistent disturbances [54] and rate of convergence information [116].

As we noted in previous chapters, even when a controller gives UGAS, the classical Lyapunov approach does not give *explicit* strict Lyapunov functions, which are generally more useful than non-strict ones when computing stability gains or quantifying the effects of uncertainty. This motivates our search for global, explicit, strict Lyapunov functions for the error dynamics for adaptive tracking problems, under PE. One method for finding such functions would be to use the variants of Matrosov's approach from the previous chapters. While very general, the Matrosov approach requires knowledge of appropriate auxiliary functions, in order to obtain an explicit strict Lyapunov function.

In this chapter, we give explicit formulas for auxiliary functions, so the Lyapunov functions we design are completely explicit. In fact, the Lyapunov functions we obtain are far simpler than the ones that would come from applying the general Matrosov approach. In addition, our Lyapunov functions are bounded from below by positive definite quadratic functions near the origin, which is another desirable feature. We also use the notion of weighting functions, which have been used in other contexts [35, 60, 167]. The global strict Lyapunov framework can potentially generalize the UGAS proofs for adaptive systems. The results of this chapter take a first step towards this generalization, and are largely based on [100, 101].

9.2 Motivating Example

To further motivate our theory, consider the controlled Rössler dynamics

$$\begin{cases} \dot{x}_1 = \theta_1 x_1 + x_2 + w_1 \\ \dot{x}_2 = -x_1 - x_3 + w_2 \\ \dot{x}_3 = \theta_2 + x_3[x_2 - \theta_3] + w_3 \end{cases} \tag{9.3}$$

with unknown constant parameters θ_i and control vector $w = (w_1, w_2, w_3)$. For the case of no controls, this model was introduced in [144], and it has been extensively studied in the context of chaotic attractors [88]. The system (9.3) can be written in the form $\dot{x} = \omega(x)\theta + u$ by taking the change of feedback

$$u = w - \begin{bmatrix} 0 & -1 & 0 \\ 1 & 0 & 1 \\ 0 & -x_3 & 0 \end{bmatrix} x, \quad \omega(x) = \begin{bmatrix} x_1 & 0 & 0 \\ 0 & 0 & 0 \\ 0 & 1 & -x_3 \end{bmatrix}, \quad \text{and} \quad \theta = \begin{bmatrix} \theta_1 \\ \theta_2 \\ \theta_3 \end{bmatrix}. \tag{9.4}$$

We consider the task of simultaneously estimating the unknown parameter vector θ and causing the state $x(t)$ to track a prescribed reference trajectory $x_r(t)$. This leads to the augmented dynamics

$$\begin{cases} \dot{x} &= \omega(x)\theta + u_s(t, x, \hat{\theta}) \\ \dot{\hat{\theta}} &= \nu(t, x, \hat{\theta}), \end{cases} \tag{9.5}$$

where $\hat{\theta}$ is the estimate for θ. For simplicity, we take the controllers

$$\begin{aligned} u_s(t, x, \hat{\theta}) &= \dot{x}_r(t) - \omega(x)\hat{\theta} + e \quad \text{and} \\ \nu(t, x, \hat{\theta}) &= -\omega(x)^\top (x_r(t) - x). \end{aligned} \tag{9.6}$$

This leads to the closed-loop dynamics

$$\begin{cases} \dot{e} &= -\omega(x)\tilde{\theta} - e \\ \dot{\tilde{\theta}} &= \omega(x)^\top e \end{cases} \tag{9.7}$$

for the combined error

$$(e, \tilde{\theta}) = (x_r - x, \theta - \hat{\theta}),$$

since θ is constant. It is immediate that

$$V_1(e, \tilde{\theta}) = \frac{1}{2}|(e, \tilde{\theta})|^2$$

is a non-strict Lyapunov function for (9.7). However, it is by no means clear how to construct a global strict Lyapunov function for (9.7), or even whether (9.7) is UGAS for an arbitrary reference trajectory x_r.

To construct a global strict Lyapunov function for (9.7), we assume that the C^1 reference trajectory x_r has the form

$$x_r(t) = \big(x_{1r}(t), x_{2r}(t), \cos(t)\big) \tag{9.8}$$

and that it satisfies

$$\max\{|x_r|_\infty, |\dot{x}_r|_\infty\} \leq 2 \tag{9.9}$$

and

$$\int_{t-2\pi}^{t} x_{1r}^2(l)\,\mathrm{d}l \;\geq\; 0.5 \quad \forall t \in \mathbb{R}. \tag{9.10}$$

Under the preceding assumptions, we can show that (9.7) admits the global strict Lyapunov function

$$\begin{aligned} V^{\sharp}(t,e,\tilde{\theta}) = \tilde{\theta}\omega(x_r(t))^\top e + 2\int_0^{V_1(e,\tilde{\theta})} J(m)\mathrm{d}m + 2\sqrt{5}V_1(e,\tilde{\theta}) \\ +\tfrac{1}{2\pi}\tilde{\theta}\left[\int_{t-2\pi}^{t}\int_m^t \omega(x_r(\ell))^\top \omega(x_r(\ell))\mathrm{d}\ell\mathrm{d}m\right]\tilde{\theta}, \end{aligned} \tag{9.11}$$

where

$$\begin{aligned} J(m) = 8\pi\left[\sqrt{5}(2+3\sqrt{2m}) + 40\pi\sqrt{1+\left(\sqrt{2m}+2\right)^2}\right]^2 \\ +4\sqrt{5}\sqrt{1+\left(\sqrt{2m}+2\right)^2} + 0.5. \end{aligned} \tag{9.12}$$

We now turn to our general theory that includes the strict Lyapunov function construction (9.11) as a special case.

9.3 Assumptions and Main Construction

For a given vector $\theta \in \mathbb{R}^p$ of unknown constant parameters, we consider dynamical systems of the form

$$\dot{x} \;=\; \omega(x)\theta + u \quad (x, u \in \mathbb{R}^n). \tag{9.13}$$

Fix a C^1 function $x_r : \mathbb{R} \to \mathbb{R}^n$ which we call a *reference trajectory*. Let $\mathbb{R}^{n\times p}$ denote the set of all $n \times p$ real matrices. For square matrices M and N of the same size, $M \geq N$ means $M - N$ is non-negative definite. We make the following two assumptions throughout the chapter:

Assumption 9.1 *A constant* $B > 0$ *such that* $\max\{|x_r|_\infty, |\dot{x}_r|_\infty\} \leq B$ *is known.*

Assumption 9.2 *The entries* ω_{ij} *of* $\omega = [\omega_{ij}] : \mathbb{R}^n \to \mathbb{R}^{n\times p}$ *are* C^1. *Also, there are known positive constants* μ *and* T *such that*

$$\mu I_p \leq \int_{t-T}^{t} \omega(x_r(l))^\top \omega(x_r(l))\,\mathrm{d}l$$

for all $t \in \mathbb{R}$.

Assumption 9.2 is the classical *PE condition* [84]. We use the functions

$$\bar{\omega}(l) \;=\; \max\{||\omega(z)|| : |z| \leq l\}$$

and

$$\bar{\omega}_1(l) = \sup\left\{\left|\left|\frac{d}{dt}\omega(\sigma(t))\right|\right|_{\infty} : \sigma \in C^1, \max\left\{|\sigma|_\infty, |\dot{\sigma}|_\infty\right\} \leq l\right\}\right\},$$

where $||\cdot||_\infty$ is the induced matrix sup norm. Then Assumption 9.1 gives $||\frac{d}{dt}\omega(x_r(t))||_\infty \leq \bar{\omega}_1(B)$.

A function $\mathcal{F}(t,d,p)$ is *uniformly bounded* in p provided there is an everywhere positive increasing function α such that $|\mathcal{F}(t,d,p)| \leq \alpha(|p|)$. Here and in the sequel, all (in)equalities should be understood to hold globally unless otherwise indicated. We also omit the arguments of our functions when they are clear.

Fix a continuous function $K : \mathbb{R}^n \to \mathbb{R}^{n\times n}$ that has positive constants c and $\bar{K}$ such that

$$\xi^\top K(\xi)\xi \geq c|\xi|^2 \text{ and } ||K(\xi)|| \leq \bar{K} \ \ \forall \xi \in \mathbb{R}^n. \tag{9.14}$$

Let $\hat{\theta}$ denote the state of the estimator of the unknown parameter $\theta \in \mathbb{R}^p$ in (9.13), set $(e,\tilde{\theta}) = (x_r - x, \theta - \hat{\theta})$, and choose the augmented dynamics

$$\begin{cases} \dot{x} = \omega(x)\theta + u_s(t,x,\hat{\theta}) \\ \dot{\hat{\theta}} = \nu(t,x,\hat{\theta}). \end{cases} \tag{9.15}$$

For simplicity, we choose the adaptive controller

$$\begin{aligned} u_s(t,x,\hat{\theta}) &= \dot{x}_r(t) - \omega(x)\hat{\theta} + K(e)e \\ \nu(t,x,\hat{\theta}) &= -\omega(x)^\top(x_r(t) - x) \end{aligned} \tag{9.16}$$

but see Sect. 9.7 for more general K, u_s, and ν. We then have the closed-loop error dynamics

$$\begin{cases} \dot{e} = -\omega(x)\tilde{\theta} - K(e)e \\ \dot{\tilde{\theta}} = \omega(x)^\top e\,, \end{cases} \tag{9.17}$$

since θ is constant. We will take the non-strict Lyapunov function

$$V_1(e,\tilde{\theta}) \ = \ \frac{1}{2}|(e,\tilde{\theta})|^2. \tag{9.18}$$

We also set

$$\begin{aligned} &V_4 = V_2 + V_3, \text{ where } V_2(t,e,\tilde{\theta}) = \tilde{\theta}^\top\omega(x_r(t))^\top e \text{ and} \\ &V_3(t,\tilde{\theta}) = \frac{1}{T}\tilde{\theta}^\top\left[\int_{t-T}^t\int_m^t \omega(x_r(l))^\top\omega(x_r(l))\,\mathrm{d}l\,\mathrm{d}m\right]\tilde{\theta}. \end{aligned} \tag{9.19}$$

Recalling the constants from Assumptions 9.1 and 9.2, we also use the functions

$$
\begin{aligned}
P_5(l) &= \tfrac{2}{c}\textstyle\int_0^l P_4(m)\,\mathrm{d}m + \bar{\omega}(B)l, \quad \text{where} \\
P_4(l) &= \tfrac{T}{2\mu}[P_0 + P_2(l) + P_3(l)]^2 + P_1(l) + \tfrac{c}{2}, \\
P_3(l) &= T\bar{\omega}\left(\sqrt{2l} + B\right)\bar{\omega}^2(B), \\
P_2(l) &= \sqrt{2lnp}\,\sup_{i,j}\left\{\left|\tfrac{\partial\omega_{ij}(q)}{\partial q}\right| : |q| \le \sqrt{2l} + B\right\}\bar{\omega}(B), \\
P_1(l) &= \bar{\omega}(B)\bar{\omega}\big(\sqrt{2l} + B\big), \quad \text{and} \\
P_0 &= \max\{2\bar{\omega}_1(B), 2\bar{K}\bar{\omega}(B)\}.
\end{aligned}
\tag{9.20}
$$

Note that $P_5 \in C^1$ on $[0, \infty)$, and that

$$
||\omega(x)|| \le \bar{\omega}(|e| + B) \quad \text{and} \quad ||\omega(x_r(t))|| \cdot ||\omega(x)|| \le P_1(V_1(e, \tilde{\theta})) \tag{9.21}
$$

for all $t \in \mathbb{R}$, $x \in \mathbb{R}^n$ and $\tilde{\theta} \in \mathbb{R}^p$ when $e = x_r(t) - x$. Also, the constant B depends only on x_r, and $\bar{\omega}$ and $\bar{\omega}_1$ depend only on ω and x_r, so the following construction from [101] is a *global* one:

Theorem 9.1. *Under the preceding assumptions, we can transform the non-strict Lyapunov function (9.18) into the explicit, global, strict Lyapunov function*

$$
V_5(t, e, \tilde{\theta}) = V_4(t, e, \tilde{\theta}) + P_5\big(V_1(e, \tilde{\theta})\big) \tag{9.22}
$$

for (9.17) which is therefore UGAS. Also, (9.17) is LES to 0.

Proof. Sketch. Since

$$
|V_2(t, e, \tilde{\theta})| \le \bar{\omega}(B)|\tilde{\theta}||e| \le \bar{\omega}(B)V_1 \quad \text{and} \quad P_4(l) \ge \frac{c}{2} \tag{9.23}
$$

everywhere, and since P_4 is non-decreasing, our formula (9.22) readily gives

$$
\begin{aligned}
V_5 &\ge V_2 + \frac{2}{c}\int_0^{V_1} P_4(l)\,\mathrm{d}l + \bar{\omega}(B)V_1 \\
&\ge \frac{1}{2}|e|^2 + \tfrac{1}{2}|\tilde{\theta}|^2 \;=:\; \alpha_1(|(e, \tilde{\theta})|) \\
V_5 &\le \bar{\omega}(B)|\tilde{\theta}||e| + \frac{T}{2}|\tilde{\theta}|^2\bar{\omega}^2(B) + \frac{2}{c}\int_0^{V_1} P_4(m)\mathrm{d}m + \bar{\omega}(B)V_1 \\
&\le \bar{\omega}(B)|\tilde{\theta}||e| + \frac{T}{2}|\tilde{\theta}|^2\bar{\omega}^2(B) + \frac{1}{2}\left[\frac{2}{c}P_4(V_1) + \bar{\omega}(B)\right]|(e, \tilde{\theta})|^2 \\
&\le \left[\bar{\omega}(B)\{1 + \bar{\omega}(B)T\} + \frac{2}{c}P_4(|(e, \tilde{\theta})|^2)\right]|(e, \tilde{\theta})|^2 \\
&=: \alpha_2(|(e, \tilde{\theta})|)
\end{aligned}
\tag{9.24}
$$

everywhere. Hence, V_5 is uniformly proper and positive definite. Our conditions (9.14) on K and (9.17) readily give

$$\begin{aligned}\dot{V}_1 &= -e^\top K(e)e \;\le\; -c|e|^2 \quad \text{and} \\ \dot{V}_2 &= \tilde{\theta}^\top \omega(x_r(t))^\top[-\omega(x)\tilde{\theta} - K(e)e] + \tilde{\theta}^\top \tfrac{d[\omega(x_r(t))]}{dt}^\top e \\ &\quad + e^\top \omega(x)\omega(x_r(t))^\top e.\end{aligned} \tag{9.25}$$

Here and in the sequel, dots indicate time derivatives along the trajectories of (9.17). By (9.21), we have the global inequality

$$e^\top \omega(x)\omega\big(x_r(t)\big)^\top e \;\le\; P_1(V_1)|e|^2.$$

Also,

$$\max\left\{-\tilde{\theta}^\top \omega\big(x_r(t)\big)^\top K(e)e,\; \tilde{\theta}^\top \frac{d[\omega(x_r(t))]}{dt}^\top e\right\} \;\le\; \frac{1}{2}P_0|\tilde{\theta}||e|\,. \tag{9.26}$$

Moreover,

$$||\omega(x) - \omega(x_r(t))|| \;\le\; |e|\sqrt{np}\max_{i,j}\left\{\left|\frac{\partial \omega_{ij}(q)}{\partial x}\right| : |q| \le \sqrt{2V_1} + B\right\}$$

gives the estimate

$$-\tilde{\theta}^\top \omega(x_r(t))^\top \big[\omega(x) - \omega(x_r(t))\big]\tilde{\theta} \;\le\; P_2(V_1)|\tilde{\theta}||e|. \tag{9.27}$$

(We used

$$||A|| \;\le\; \sqrt{np}\max_{i,j}|a_{ij}|$$

for any $A = [a_{ij}] \in \mathbb{R}^{n\times p}$, plus the mean value theorem.) Therefore, (9.25) gives

$$\begin{aligned}\dot{V}_2 &\le -\tilde{\theta}^\top \omega(x_r(t))^\top \omega(x)\tilde{\theta} + P_0|\tilde{\theta}||e| + P_1(V_1)|e|^2 \\ &\le -\tilde{\theta}^\top \omega(x_r(t))^\top \omega(x_r(t))\tilde{\theta} + \big[P_0 + P_2(V_1)\big]|\tilde{\theta}||e| + P_1(V_1)|e|^2\,.\end{aligned} \tag{9.28}$$

By (9.17) and the PE condition in Assumption 9.2, we get

$$\begin{aligned}\dot{V}_3 &= \tfrac{2}{T}\tilde{\theta}^\top \left[\int_{t-T}^t \int_m^t \omega(x_r(l))^\top \omega(x_r(l))\,\mathrm{d}l\,\mathrm{d}m\right] \omega(x)^\top e \\ &\quad + \tilde{\theta}^\top \omega(x_r(t))^\top \omega(x_r(t))\tilde{\theta} \\ &\quad - \tfrac{1}{T}\tilde{\theta}^\top \left[\int_{t-T}^t \omega(x_r(l))^\top \omega(x_r(l))\,\mathrm{d}l\right]\tilde{\theta} \\ &\le \tfrac{2}{T}\tilde{\theta}^\top \left[\int_{t-T}^t \int_m^t \omega(x_r(l))^\top \omega(x_r(l))\,\mathrm{d}l\,\mathrm{d}m\right] \omega(x)^\top e \\ &\quad + \tilde{\theta}^\top \omega(x_r(t))^\top \omega(x_r(t))\tilde{\theta} - \tfrac{\mu}{T}|\tilde{\theta}|^2.\end{aligned} \tag{9.29}$$

By Assumption 9.1 and the relations (9.23) and (9.21),

$$\frac{2}{T}\tilde{\theta}^\top \left[\int_{t-T}^{t}\int_{m}^{t} \omega(x_r(l))^\top \omega(x_r(l))\,\mathrm{d}l\,\mathrm{d}m\right] \omega(x)^\top e \;\leq\; P_3(V_1)|\tilde{\theta}||e|. \tag{9.30}$$

Combining the preceding inequalities and canceling terms, we obtain

$$\begin{aligned}
\dot{V}_4 &\leq -\tilde{\theta}^\top \omega(x_r(t))^\top \omega(x_r(t))\tilde{\theta} \\
&\quad + \big[P_0 + P_2(V_1)\big]|\tilde{\theta}||e| + P_1(V_1)|e|^2 \\
&\quad + P_3(V_1)|\tilde{\theta}||e| + \tilde{\theta}^\top \omega(x_r(t))^\top \omega(x_r(t))\tilde{\theta} - \tfrac{\mu}{T}|\tilde{\theta}|^2 \\
&= \big\{[P_0 + P_2(V_1) + P_3(V_1)]|e|\big\}|\tilde{\theta}| \\
&\quad + P_1(V_1)|e|^2 - \tfrac{\mu}{T}|\tilde{\theta}|^2.
\end{aligned} \tag{9.31}$$

Applying the inequality

$$a|\tilde{\theta}| \;\leq\; \frac{T}{2\mu}a^2 + \frac{\mu}{2T}|\tilde{\theta}|^2,$$

where a is the term in braces in (9.31), gives

$$\begin{aligned}
\dot{V}_4 &\leq \tfrac{T}{2\mu}\big[P_0 + P_2(V_1) + P_3(V_1)\big]^2|e|^2 + \tfrac{\mu}{2T}|\tilde{\theta}|^2 + P_1(V_1)|e|^2 - \tfrac{\mu}{T}|\tilde{\theta}|^2 \\
&\leq P_4(V_1)|e|^2 - \tfrac{\mu}{2T}|\tilde{\theta}|^2.
\end{aligned} \tag{9.32}$$

Since $\dot{V}_1 \leq -c|e|^2$ everywhere, (9.32) and our choice of P_5 in (9.20) give

$$\begin{aligned}
\dot{V}_5 &= \dot{V}_4 + \big[\tfrac{2}{c}P_4(V_1) + \bar{\omega}(B)\big]\dot{V}_1 \;\leq\; \dot{V}_4 + \tfrac{2}{c}P_4(V_1)\dot{V}_1 \\
&\leq \dot{V}_4 - 2P_4(V_1)|e|^2 \;\leq\; -P_4\big(V_1(e,\tilde{\theta})\big)|e|^2 - \tfrac{\mu}{2T}|\tilde{\theta}|^2\,.
\end{aligned} \tag{9.33}$$

By (9.24), V_5 is uniformly proper and positive definite. Since $P_4(l) \geq c/2$ for all l, we conclude that V_5 is a global strict Lyapunov function for (9.17). The local exponential stability follows from (9.24). □

9.4 Robustness

9.4.1 Statement of ISS Theorem

We illustrate the usefulness of our strict Lyapunov function constructions by showing that when ω has affine growth, the *perturbed* error dynamics

$$\begin{cases} \dot{e} = -\omega(x)[\tilde{\theta} + \delta(t)] - K(e)e \\ \dot{\tilde{\theta}} = \omega(x)^\top e, \end{cases} \tag{9.34}$$

obtained by replacing θ with $\theta + \delta(t)$ in (9.13) and using (9.16), is ISS with respect to suitably bounded uncertainties $\delta(t)$. We assume that $\delta(t)$ is bounded in the essential supremum by a constant $\bar{\delta} > 0$ that we specify shortly, and that Assumptions 9.1 and 9.2 hold for some positive constants B, μ and T.

We further assume that there are constants $\omega_M \geq \max\{1, \bar{\omega}(B)\}$ and $\eta > 0$ such that the following *affine growth condition* holds:

$$||\omega(x)|| \leq \omega_M + \eta|x| \quad \text{and} \quad \left|\tfrac{\partial \omega_{ij}}{\partial x}(x)\right| \leq \omega_M \;\; \forall x, i, j. \tag{9.35}$$

Hence $\bar{\omega}_1(B) \leq \omega_M \sqrt{np} B$. We prove that (9.34) is ISS by explicitly constructing an ISS Lyapunov function, which will lead to explicit formulas for the functions $\beta \in \mathcal{KL}$ and $\gamma \in \mathcal{K}_\infty$ for the ISS estimate; see Remark 9.2. We take $K(e) \equiv cI_n$, and the constants

$$\begin{aligned} \Delta_1 &= \tfrac{T}{\mu}\omega_M^6 \left[2\max\{c, (1+B)\sqrt{np}\} + T\{1 + \eta(B+2)\}\right]^2 \\ &\quad + \eta\omega_M(B+2) + \omega_M^2 + \tfrac{c}{2}, \\ C_1 &= \tfrac{2T}{\mu}\omega_M^4 \left[\sqrt{np} + \eta T\right]^2 + 2\eta\omega_M + 0.5\sqrt{c}, \quad \text{and} \\ \Delta_2 &= \omega_M + \tfrac{2\Delta_1}{c}, \end{aligned} \tag{9.36}$$

where the constant $c \geq 1$ and the disturbance bound $\bar{\delta}$ are assumed to satisfy

$$\bar{\delta} \leq \min\left\{\tfrac{C_1}{8\omega_M \eta}, \tfrac{c}{8(\omega_M + \eta[1+B])}, \tfrac{c}{C_1(\omega_M + \eta[1+B])} \min\left\{0.9\Delta_1, \tfrac{\mu}{3T}\right\}\right\}. \tag{9.37}$$

For any constant $\bar{\delta} > 0$, we can choose the constant $c > 0$ so that (9.37) is satisfied, so any disturbance bound $\bar{\delta}$ can be accommodated. The following is shown in [101]:

Theorem 9.2. *Let the preceding assumptions hold. Then (9.34) with uncertainties $\delta : \mathbb{R} \to \bar{\delta}\mathcal{B}_p$ bounded by $\bar{\delta}$ admits the ISS Lyapunov function*

$$\begin{aligned} V_5(t, e, \tilde{\theta}) &= \tilde{\theta}^\top \omega(x_r(t))^\top e + \Delta_2 V_1(e, \tilde{\theta}) + \tfrac{C_1}{c} V_1^2(e, \tilde{\theta}) \\ &\quad + \tfrac{1}{T}\tilde{\theta}^\top \left[\int_{t-T}^t \int_m^t \omega(x_r(l))^\top \omega(x_r(l))\, \mathrm{d}l\, \mathrm{d}m\right] \tilde{\theta} \end{aligned} \tag{9.38}$$

and so is ISS with respect to the uncertainty $\delta : \mathbb{R} \to \bar{\delta}\mathcal{B}_p$.

9.4.2 Proof of ISS Theorem

We only sketch the proof; see [101] for details. We use the notation from Theorem 9.1. We may assume that $\omega_M = \bar{\omega}(B) \geq 1$, by enlarging the function

$\bar{\omega}$ without relabeling. Then the definitions (9.20) of the functions P_i give

$$\begin{aligned} &P_0(l) \le 2\omega_M \max\{c, \sqrt{npB}\}, \\ &P_1(l) \le \omega_M^2 + \eta\omega_M \left(\sqrt{2l} + B\right), \\ &P_2(l) \le \omega_M^2 \sqrt{2lnp}, \\ &P_3(l) \le T\omega_M P_1(l), \ \text{and} \\ &P_4(l) \le \Delta_1 + C_1 l, \end{aligned} \tag{9.39}$$

by (9.35). By enlarging P_4 as necessary without relabeling, we assume that $P_4(l) = \Delta_1 + C_1 l$ in the sequel. Therefore, V_5 from Theorem 9.1 takes the form (9.38). One easily checks that Theorem 9.1 remains true when P_4 is enlarged in this way, by our proof of the theorem. Using Assumption 9.1 and (9.33)-(9.35), we get

$$\begin{aligned} \dot{V}_5 &\le -\left[\Delta_1 + C_1 V_1(e, \tilde{\theta})\right] |e|^2 - \tfrac{\mu}{2T} |\tilde{\theta}|^2 \\ &\quad + (\omega_M + \eta B + \eta|e|) \left| \tfrac{\partial V_5}{\partial e}(t, e, \tilde{\theta}) \right| |\delta(t)| \ \text{and} \\ \left| \tfrac{\partial V_5}{\partial e}(t, e, \tilde{\theta}) \right| &\le |\omega(x_r(t))\tilde{\theta}| + \left[\Delta_2 + 2\tfrac{C_1}{c} V_1(e, \tilde{\theta})\right] |e| \\ &\le \omega_M |\tilde{\theta}| + \left[\Delta_2 + 2\tfrac{C_1}{c} V_1(e, \tilde{\theta})\right] |e|. \end{aligned} \tag{9.40}$$

We consider two cases.
Case 1. $|e| \ge 1$. In this case, dropping $-\Delta_1|e|^2 - \frac{\mu}{2T}|\tilde{\theta}|^2$ in (9.40) gives

$$\begin{aligned} \dot{V}_5 \le &-\frac{C_1}{4}(|e|^2 + |\tilde{\theta}|^2) - \frac{C_1}{3} V_1 |e|^2 - \frac{C_1}{12}|e|^4 \\ &+ \omega_M(\omega_M + \eta B)|\tilde{\theta}||\delta(t)| \\ &+ \omega_M \eta |e||\tilde{\theta}||\delta(t)| + \{|e|^2\}\{[\omega_M + \eta(1+B)]\Delta_2|\delta(t)|\} \\ &+ 2\frac{C_1}{c}(\omega_M + \eta[B+1]) V_1 |e|^2 |\delta(t)|. \end{aligned} \tag{9.41}$$

Applying the relations $|\tilde{\theta}||\delta(t)| \le \frac{1}{2\varepsilon_1}|\tilde{\theta}|^2 + \frac{\varepsilon_1}{2}|\delta(t)|^2$ and $|\tilde{\theta}||e| \le \frac{1}{2}|\tilde{\theta}|^2 + \frac{1}{2}|e|^2$ for a suitable positive constant ε_1 to the fourth and fifth terms on the right side of (9.41), using the relation

$$ab \le \frac{C_1 a^2}{12} + \frac{3b^2}{C_1}$$

with $a = |e|^2$ on the terms in braces in (9.41), and recalling our assumption (9.37) on $\bar{\delta}$ gives

$$\dot{V}_5 \le -\frac{C_1}{40}|(e,\tilde{\theta})|^2 + \frac{5}{2C_1}\omega_M^2(\omega_M+\eta B)^2|\delta(t)|^2 \\ +\frac{3}{C_1}\left[(\omega_M+\eta B)\Delta_2+\eta\Delta_2\right]^2|\delta(t)|^2. \tag{9.42}$$

Case 2. $|e| \le 1$. In this case, (9.40) gives

$$\begin{aligned}\dot{V}_5 \le &-\left[\Delta_1 + C_1 V_1(e,\tilde{\theta})\right]|e|^2 - \frac{\mu}{2T}|\tilde{\theta}|^2 \\ &+\omega_M(\omega_M+\eta[B+1])|\tilde{\theta}||\delta(t)| \\ &+(\omega_M+\eta[B+1])\left[\Delta_2+\frac{C_1}{c}(|e|^2+|\tilde{\theta}|^2)\right]|\delta(t)| \\ \le &-\frac{\Delta_1}{10}|e|^2 - \frac{\mu}{15T}|\tilde{\theta}|^2 + \frac{5T\left\{\omega_M(\omega_M+\eta[B+1])\right\}^2}{2\mu}|\delta(t)|^2 \\ &+\Delta_2[\omega_M+\eta(B+1)]|\delta(t)|,\end{aligned} \tag{9.43}$$

where the last inequality followed from the relation $|\tilde{\theta}||\delta(t)| \le |\tilde{\theta}|^2/(2\varepsilon_2) + \varepsilon_2|\delta(t)|^2/2$ for a suitable positive constant ε_2, (9.37), and dropping the term $-C_1V_1(e,\tilde{\theta})|e|^2$. The result readily follows from (9.42) and (9.43). □

Remark 9.1. The construction from Theorem 9.2 cannot be used to prove Theorem 9.1 by simply setting the disturbance to zero. This is because its derivation is based on (9.35) which we do not require in Theorem 9.1. In other words, Theorem 9.1 applies with more general ω's that may violate (9.35).

Remark 9.2. The explicit ISS Lyapunov function (9.38) for (9.34) leads to explicit expressions for β and γ in the ISS estimate for (9.34), as follows. Define $\alpha_1, \alpha_2, \alpha_3, \alpha_4, \alpha \in \mathcal{K}_\infty$ by (9.24),

$$\alpha_3(r) = \min\left\{\tfrac{C_1}{40}, \tfrac{\mu}{15T}\right\}r^2, \quad \alpha(r) = \min\{r, \alpha_3\circ\alpha_2^{-1}(r)\}, \quad \text{and}$$

$$\begin{aligned}\alpha_4(r) = &\left\{\tfrac{5}{2C_1}\omega_M^2(\omega_M+\eta B)^2 + \tfrac{3}{C_1}\left[(\omega_M+\eta B)\Delta_2+\eta\Delta_2\right]^2 \right.\\ &\left.+\tfrac{5T\omega_M^2(\omega_M+\eta[B+1])^2}{2\mu}\right\}r^2 \\ &+\Delta_2[\omega_M+\eta(B+1)]r.\end{aligned}$$

Then

$$\alpha_1(|(e,\tilde{\theta})|) \le V_5(t,e,\tilde{\theta}) \le \alpha_2(|(e,\tilde{\theta})|) \text{ and} \\ \dot{V}_5 \le -\alpha(V_5) + \alpha_4(|\delta|_\infty) \tag{9.44}$$

along all trajectories of (9.34) when $\bar{\delta}$ satisfies (9.37) (by (9.24), (9.42), and (9.43)), and then the explicit formulas for β and γ in the ISS estimate follow by the standard argument we gave in Sect. 2.1.3.

9.5 Rössler System Revisited

We illustrate our Lyapunov function constructions using the controlled Rössler dynamics

$$\begin{cases} \dot{x}_1 = \theta_1 x_1 + x_2 + w_1 \\ \dot{x}_2 = -x_1 - x_3 + w_2 \\ \dot{x}_3 = \theta_2 + x_3[x_2 - \theta_3] + w_3 \end{cases} \tag{9.45}$$

from Sect. 9.2, which has the form $\dot{x} = \omega(x)\theta + u$ when the regressor matrix ω and the unknown parameter vector are chosen according to (9.4).

Let us show that the PE condition from Assumption 9.2 is satisfied for the class of C^1 reference trajectories

$$x_r(t) \;=\; \big(x_{1r}(t), x_{2r}(t), \cos(t)\big)$$

that satisfies conditions (9.9) and (9.10). The conditions guarantee that

$$\int_{t-2\pi}^{t} \omega(x_r(l))^\top \omega(x_r(l))\,\mathrm{d}l \;=\; \begin{bmatrix} \int_{t-2\pi}^{t} x_{1r}^2(l)\mathrm{d}l & 0 & 0 \\ 0 & 2\pi & 0 \\ 0 & 0 & \pi \end{bmatrix}.$$

Also, we can take the upper bound $\bar{\omega}(\ell) = 2\sqrt{1+\ell^2}$. Hence, Assumption 9.2 and our growth assumption (9.35) hold with $B = 2$,

$$T = 2\pi,\;\; \mu = 0.5,\;\; \eta = 1,\;\; \text{and}\;\; \omega_M = 10.$$

Therefore, for any constant $c \geq 1$, the error dynamics for the Rössler system (9.3) with the adaptive controller (9.16) with $K(e) \equiv cI_n$ admits a global strict Lyapunov function of the form (9.38) and so is UGAS.

We now show how the resulting strict Lyapunov function with the choice $c = 1$ agrees with the formula (9.11) from Sect. 9.2. One checks that we can take $\bar{\omega}_1(2) = 2$. Hence, the functions P_i from (9.20) become

$$\begin{aligned} &P_0 = 4\sqrt{5},\;\; P_1(\ell) = 4\sqrt{5}\sqrt{1 + \left[\sqrt{2\ell}+2\right]^2}, \\ &P_2(\ell) = 6\sqrt{10\ell},\;\; P_3(\ell) = 80\pi\sqrt{1 + \left[\sqrt{2\ell}+2\right]^2}, \end{aligned} \tag{9.46}$$

and

$$P_4(\ell) = 8\pi \left[\sqrt{5}\left(2 + 3\sqrt{2\ell}\right) + 40\pi\sqrt{1 + \left(\sqrt{2\ell} + 2\right)^2} \right]^2 + 4\sqrt{5}\sqrt{1 + \left[\sqrt{2\ell} + 2\right]^2} + 0.5. \tag{9.47}$$

Since P_4 agrees with the formula for J from (9.12), we readily conclude that the strict Lyapunov function for the Rössler error dynamics takes the form (9.11), as claimed. Also, Theorem 9.2 shows that (9.38) is an ISS Lyapunov function when θ is perturbed by time-varying additive uncertainty δ, but in that case c depends on the choice of the disturbance bound $\bar{\delta}$.

To illustrate the ISS property, we simulated the Rössler error dynamics with $c = 100$, i.e., the dynamics

$$\begin{aligned} \dot{e} &= -\omega(x)[\tilde{\theta} + \mathbf{d}] - 100e \\ \dot{\tilde{\theta}} &= \omega(x)^\top e\,, \\ \text{where } \omega(x) &= \begin{bmatrix} x_1 & 0 & 0 \\ 0 & 0 & 0 \\ 0 & 1 & -x_3 \end{bmatrix} \end{aligned} \tag{9.48}$$

for the combined error $(e, \tilde{\theta}) = (x_r - x, \theta - \hat{\theta})$, using the disturbance $\mathbf{d}(t) \equiv (0.05, 0.05, 0.05)^\top$, and obtained the plots in Figs. 9.1 and 9.2 on the following pages. Our simulation illustrates the robustness of the convergence of the parameter estimation error to zero in the face of the disturbance $\mathbf{d}$, with an overshoot determined by the ISS estimate, and so validates the theory.

9.6 Lorenz System

Another interesting example covered by our theory is the fully controlled Lorenz system

$$\begin{cases} \dot{x}_1 = \theta_1(x_2 - x_1) + w_1 \\ \dot{x}_2 = \theta_2 x_1 - x_2 - x_1 x_3 + w_2 \\ \dot{x}_3 = x_1 x_2 - \theta_3 x_3 + w_3 \end{cases} \tag{9.49}$$

where θ_1, θ_2, and θ_3 are unknown parameters. The change of feedback

$$u = w - \begin{bmatrix} 0 & 0 & 0 \\ 0 & 1 & x_1 \\ -x_2 & 0 & 0 \end{bmatrix} x, \tag{9.50}$$

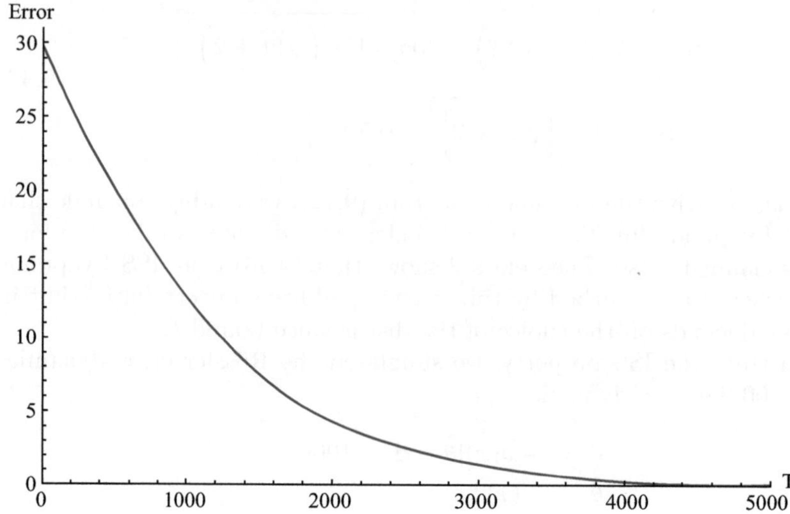

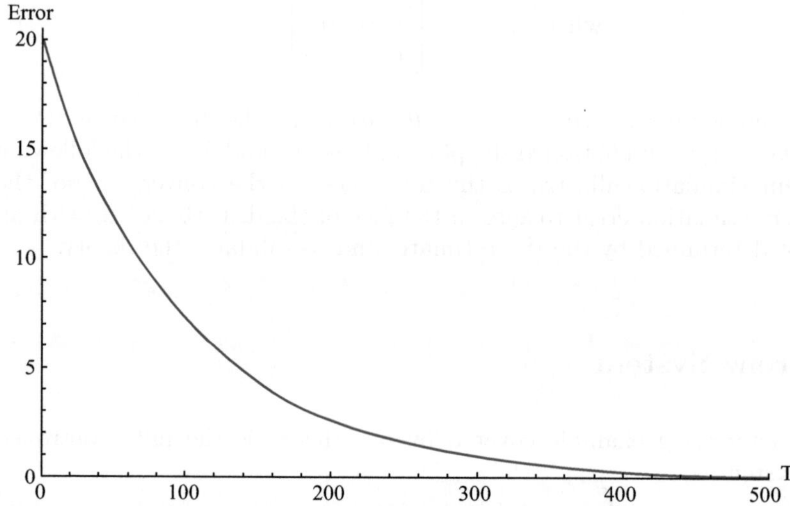

Fig. 9.1 Estimation error $\tilde{\theta}_1$ (top) and $\tilde{\theta}_2$ (bottom) for Rössler system with $K(e) \equiv 100I_3$

$$\omega(x) = \begin{bmatrix} x_2 - x_1 & 0 & 0 \\ 0 & x_1 & 0 \\ 0 & 0 & -x_3 \end{bmatrix}, \quad \text{and} \quad \theta = \begin{bmatrix} \theta_1 \\ \theta_2 \\ \theta_3 \end{bmatrix} \tag{9.51}$$

gives the system

$$\dot{x} = \omega(x)\theta + u \,. \tag{9.52}$$

Choose any periodic reference trajectory $x_r(t) = (x_{1r}(t), x_{2r}(t), x_{3r}(t))$ of period 1 such that

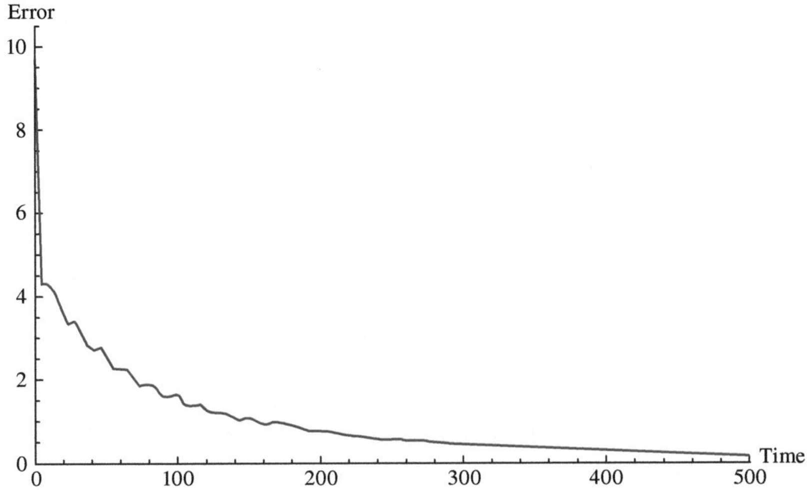

Fig. 9.2 Estimation error $\tilde{\theta}_3$ for Rössler system with $K(e) \equiv 100I_3$

$$\begin{aligned} &\textstyle\int_{t-1}^{t} x_{1r}^2(l)\,\mathrm{d}l \geq 1 \ , \ \int_{t-1}^{t} x_{3r}^2(l)\,\mathrm{d}l \geq 1 \ , \\ &\text{and } \textstyle\int_{t-1}^{t} [x_{1r}(l) - x_{2r}(l)]^2\,\mathrm{d}l \geq 1 \end{aligned} \tag{9.53}$$

hold for all $t \in \mathbb{R}$ and $\max\{|x_r|_\infty, |\dot{x}_r|_\infty\} \leq 4$.

Then

$$\textstyle\int_{t-1}^{t} \omega(x_r(l))^\top \omega(x_r(l))\,\mathrm{d}l \ \geq \ I_3 \ ,$$

so Assumption 9.2 and our growth assumption (9.35) hold. Again using the dynamic extension

$$\begin{cases} \dot{x} = \omega(x)\theta + u_s(t, x, \hat{\theta}) \\ \dot{\hat{\theta}} = \nu(t, x, \hat{\theta}) \end{cases} \tag{9.54}$$

with

$$\begin{aligned} u_s(t, x, \hat{\theta}) &= \dot{x}_r(t) - \omega(x)\hat{\theta} + e, \\ \nu(t, x, \hat{\theta}) &= -\omega(x)^\top (x_r(t) - x) \end{aligned} \tag{9.55}$$

we obtain the dynamics

$$\begin{cases} \dot{e} = -\omega(x)\tilde{\theta} - e \\ \dot{\tilde{\theta}} = \omega(x)^\top e \end{cases} \tag{9.56}$$

for the combined error $(e, \tilde{\theta}) = (x_r - x, \theta - \hat{\theta})$, since θ is constant. It is immediate that $V_1(e, \tilde{\theta}) = \frac{1}{2}|(e, \tilde{\theta})|^2$ is a non-strict Lyapunov function for (9.56), and we can construct a strict Lyapunov function for the system (9.56).

However, we will not perform this construction because it is very similar to the construction we gave for the Rössler dynamics.

9.7 Extension: More General Feedbacks

Theorem 9.1 assumes that the systems are adaptively controlled by (9.16), and that the known non-strict Lyapunov function V_1 is $\frac{1}{2}|(e, \tilde{\theta})|^2$. Let us show how these assumptions can be relaxed.

We first assume that Assumptions 9.1 and 9.2 hold, and that there exist a C^1 function $\nu(t, e, \hat{\theta})$, a bounded C^1 function $K(e)$, a uniformly proper and positive definite C^1 function V_1, a positive definite function W_1, a continuous everywhere positive increasing function P_ν, and a constant $c > 0$ such that:

1. $|\nu(t, e, \hat{\theta})| \leq P_\nu(V_1)|e|$ everywhere;
2. $\dot{V}_1(t, e, \tilde{\theta}) \leq -W_1(e)$ along all trajectories of

$$\begin{cases} \dot{e} = -\omega(x)\tilde{\theta} - K(e)e \\ \dot{\tilde{\theta}} = -\nu(t, e, \hat{\theta}) \end{cases} \tag{9.57}$$

and

3. $W_1(e) \geq c|e|^2$ everywhere,

where e and $\tilde{\theta}$ are as before. In other words, we replace $\nu = -\omega(x)^\top e$ from Theorem 9.1 with a general adaptation law that could include, for example, projection operators, least-squares estimators, and prediction-error-based estimators [59, 152]. A slight variant of the proof of Theorem 9.1 constructs a function P_5 so that (9.22) is a global strict Lyapunov function for (9.57) when 1.-3. are satisfied.

A different generalization is as follows. Let e and $\tilde{\theta}$ be as before, and let ω and x_r satisfy Assumptions 9.1 and 9.2. We now also assume:

Assumption 9.3 *There exist a (possibly unbounded) matrix function K with C^1 entries, a C^1 uniformly proper and positive definite function $V_a(t, z)$, a positive definite function $W_a(z)$, and a continuous everywhere positive function Δ so that:*

1. *$\dot{V}_a(t, z) \leq -W_a(z)$ along all trajectories of $\dot{z} = -K(z + x_r(t))z$;*
2. *all the second partial derivatives $\partial^2 V_a / \partial z_i \partial z_j$ are uniformly bounded in z; and*
3. *$W_a(z) \geq \Delta(|z|)|z|^2$ everywhere.*

Taking the augmented dynamics (9.15) as before, and choosing

$$u_s = \dot{x}_r(t) - \omega(x)\hat{\theta} + K\big(e + x_r(t)\big)e, \quad \nu = -\left[\tfrac{\partial V_a}{\partial e}(t,e)\omega(x)\right]^\top, \text{ and}$$
$$V_1(t,e,\tilde{\theta}) = V_a(t,e) + \tfrac{1}{2}|\tilde{\theta}|^2$$

guarantees that the time derivative of V_1 along the trajectories of the closed-loop error dynamics

$$\begin{cases} \dot{e} = -\omega(x)\tilde{\theta} - K\big(e + x_r(t)\big)e \\ \dot{\tilde{\theta}} = \left[\tfrac{\partial V_a}{\partial e}(t,e)\omega(x)\right]^\top \end{cases} \tag{9.58}$$

satisfies $\dot{V}_1(t,e,\tilde{\theta}) \le -W_a(e)$. Setting $V_4 = V_2 + \frac{1}{2}V_3$ with V_2 and V_3 as defined before according to (9.19), we have the following result from [101]:

Theorem 9.3. *Let Assumptions 9.1,9.2, and 9.3 hold. Then we can explicitly construct a function* $\kappa_4 \in \mathcal{K}_\infty \cap C^1$ *so that*

$$V_5(t,e,\tilde{\theta}) \;=\; \kappa_4\big(V_1(t,e,\tilde{\theta})\big) + V_4(t,e,\tilde{\theta}) \tag{9.59}$$

is a global strict Lyapunov function for the error dynamics (9.58), which is therefore UGAS.

Proof. Sketch. We only sketch the proof; see [101] for the detailed argument. First note that

$$\begin{aligned} \dot{V}_2 = &-\tilde{\theta}^\top\omega(x_r(t))^\top\omega(x_r(t))\tilde{\theta} + \left[\tfrac{\partial V_a}{\partial e}(t,e)\omega(x)\right]\omega(x_r(t))^\top e \\ &+\tilde{\theta}^\top\tfrac{d\omega(x_r(t))^\top}{dt}e \\ &-\tilde{\theta}^\top\omega(x_r(t))^\top K(e+x_r)e - \tilde{\theta}^\top\omega(x_r(t))^\top[\omega(x) - \omega(x_r(t))]\tilde{\theta}. \end{aligned} \tag{9.60}$$

Applying the mean value theorem provides an everywhere positive, increasing function κ_1 such that

$$\begin{aligned} \left[\tfrac{\partial V_a}{\partial e}(t,e)\omega(x_r(t)-e)\right]\omega(x_r(t))^\top e \;&\le\; \kappa_1(|e|)|e|^2, \\ -\tilde{\theta}^\top\omega(x_r(t))^\top K(e+x_r)e \;&\le\; \kappa_1(|e|)|\tilde{\theta}||e|, \end{aligned} \tag{9.61}$$

and $-\tilde{\theta}^\top\omega(x_r(t))^\top[\omega(x) - \omega(x_r(t))]\tilde{\theta} \le |\omega(x_r(t))\tilde{\theta}|\kappa_1(|e|)|\tilde{\theta}||e|$. We deduce from (9.60) that

$$\begin{aligned} \dot{V}_2 \le &-\tilde{\theta}^\top\omega(x_r(t))^\top\omega(x_r(t))\tilde{\theta} + \kappa_1(|e|)|e|^2 + [\bar{\omega}_1 + \kappa_1(|e|)]|\tilde{\theta}||e| \\ &+\big\{|\omega(x_r(t))\tilde{\theta}|\big\}\big\{\kappa_1(|e|)|\tilde{\theta}||e|\big\} \\ \le &-\tfrac{1}{2}\tilde{\theta}^\top\omega(x_r(t))^\top\omega(x_r(t))\tilde{\theta} + \kappa_1(|e|)|e|^2 + \left[\bar{\omega}_1 + \kappa_1(|e|)\right]|\tilde{\theta}||e| \\ &+\tfrac{1}{2}\kappa_1^2(|e|)|\tilde{\theta}|^2|e|^2, \end{aligned} \tag{9.62}$$

by applying the relation $ab \le \frac{1}{2}a^2 + \frac{1}{2}b^2$ to the terms in braces.

We can readily construct an increasing everywhere positive function κ_2 such that

$$\begin{aligned}\dot{V}_3 &\leq \tilde{\theta}^\top \omega(x_r(t))^\top \omega(x_r(t))\tilde{\theta} - \tfrac{1}{T}\tilde{\theta}^\top \left[\int_{t-T}^t \omega(x_r(s))^\top \omega(x_r(s))\mathrm{d}s\right]\tilde{\theta} \\ &\quad +\kappa_2(|e|)|\tilde{\theta}||e| \\ &\leq \tilde{\theta}^\top \omega(x_r(t))^\top \omega(x_r(t))\tilde{\theta} - \tfrac{\mu}{T}|\tilde{\theta}|^2 + 2\kappa_2(|e|)|\tilde{\theta}||e|,\end{aligned} \tag{9.63}$$

where the second inequality is by Assumption 9.2. Hence, by applying

$$ab \leq \tfrac{\mu}{4T}a^2 + \tfrac{T}{\mu}b^2$$

with $a = |\tilde{\theta}|$, we deduce from (9.62) and (9.63) that $V_4 = V_2 + \frac{1}{2}V_3$ satisfies

$$\begin{aligned}\dot{V}_4 &\leq -\tfrac{\mu}{2T}|\tilde{\theta}|^2 + \kappa_1(|e|)|e|^2 \\ &\quad +\{|\tilde{\theta}|\}\{[\bar{\omega}_1 + \kappa_1(|e|) + \kappa_2(|e|) + \tfrac{1}{2}\kappa_1^2(|e|)|e||\tilde{\theta}|]|e|\} \\ &\leq -\tfrac{\mu}{4T}|\tilde{\theta}|^2 + \kappa_1(|e|)|e|^2 \\ &\quad +\tfrac{T}{\mu}\left[\bar{\omega}_1 + \kappa_1(|e|) + \kappa_2(|e|) + \tfrac{1}{2}\kappa_1^2(|e|)|\tilde{\theta}||e|\right]^2|e|^2.\end{aligned} \tag{9.64}$$

By assuming without loss of generality that Δ is decreasing, we can construct an increasing everywhere positive continuous function κ_3 such that

$$\begin{aligned}&\kappa_3(V_1(t,e,\tilde{\theta}))W_a(e) \\ &\geq\ \kappa_1(|e|)|e|^2 + \tfrac{T}{\mu}\left[\bar{\omega}_1 + \kappa_1(|e|) + \kappa_2(|e|) + \tfrac{1}{2}\kappa_1^2(|e|)|\tilde{\theta}||e|\right]^2|e|^2.\end{aligned} \tag{9.65}$$

Consequently,

$$\dot{V}_4\ \leq\ -\frac{\mu}{4T}|\tilde{\theta}|^2 + \kappa_3(V_1(t,e,\tilde{\theta}))W_a(e).$$

One checks that $z \mapsto \inf_{t\geq 0} V_a(t,z)$ is bounded from below by a positive definite quadratic function near 0. To obtain this lower bound, let $\underline{\Delta} > 0$ be a constant lower bound for Δ on $\mathcal{B}_n$. Let $\bar{K} > 0$ be a bound for K on $(1+B)\mathcal{B}_n$. Reducing $\underline{\Delta}$, we can assume that all trajectories of $\dot{z} = -K(z + x_r(t))z$ with initial conditions $z(t_0) = z_0 \in \underline{\Delta}\mathcal{B}_n$ stay in $\mathcal{B}_n$, by Assumption 9.3. Along any such trajectory, it can be shown that

$$\frac{d}{dt}\left[V_a(t,z) - \frac{\underline{\Delta}|z|^2}{4\bar{K}}\right]\ \leq\ 0, \tag{9.66}$$

and then the lower bound on $\underline{\Delta}\mathcal{B}_n$ follows because the term in brackets in (9.66) converges down to zero as $t \to +\infty$.

Hence, we can choose $\kappa_4 \in \mathcal{K}_\infty \cap C^1$ so that $\kappa_4' \geq \kappa_3 + 1$, and so that V_5 as defined in (9.59) is uniformly proper and positive definite and satisfies

$$\dot{V}_5 \le -\frac{\mu}{4T}|\tilde{\theta}|^2 - W_a(e)$$

along all trajectories of (9.58), which is the desired decay condition. □

9.8 Comments

The two components of the adaptive tracking problem are (a) making the trajectories track a desired reference trajectory and (b) identifying the unknown model parameters. Part (a) has been well studied in the robotics literature, including cases where there are unknown parameters. One important mechanism for solving (a) is the Li-Slotine controller, which was introduced in [151] in the context of the manipulator dynamics

$$H(q)\ddot{q} + C(q, \dot{q}) + g(q) \ = \ \tau$$

for the joint displacements q. Assuming that the dynamics are linear in the unknown parameter vector a, the Li-Slotine controller is

$$\begin{aligned} \tau &= \hat{H}\ddot{q}_r + \hat{C}(q, \dot{q})\dot{q}_r + \hat{g}(q) - K_D s \\ \dot{\hat{a}} &= -\Gamma Y^\top s, \end{aligned} \tag{9.67}$$

where $\hat{H}$, $\hat{C}$, and $\hat{g}$ are obtained by replacing the unknown parameters in H, C, and g with the parameter vector estimate $\hat{a}$; Γ and K_D are appropriate positive definite matrices; the matrix Y is chosen to satisfy

$$\tilde{H}\ddot{q}_r + \tilde{C}(q, \dot{q})\dot{q}_r + \tilde{g}(q) = Y(q, \dot{q}, \dot{q}_r, \ddot{q}_r)\tilde{a};$$

$\dot{q}_r = \dot{q}_d - \Lambda\tilde{q}$ in terms of the reference trajectory q_d and a suitable positive definite matrix Λ; $s = \dot{q} - \dot{q}_r$; and $\tilde{a} = a - \hat{a}$ [151]. The Li-Slotine controller and its variants have been used extensively in robotics, the control of ships, and other applications. The nonlinear dynamics of robot manipulators provide an important example where PE guarantees asymptotic parameter error convergence when using the Li-Slotine adaptive controller [150].

Recently, PE was also shown to be both necessary and sufficient for UGAS of a class of nonlinear systems that encompasses the manipulator dynamics [84, 87]. This work generalized [150] in that it is not limited to the Slotine-Li adaptive controller when applied to the manipulator dynamics. Also, [32] proved global exponential stability for a mechanical system by constructing the regressor in the adaptive control to satisfy the PE condition. The adaptive control literature for nonlinear systems largely proceeds by analyzing the behavior of trajectories, rather than constructing explicit strict Lyapunov functions. Our global strict Lyapunov function treatment of adaptive control in this chapter closely follows [100, 101], where complete proofs of all of the theorems in this chapter can be found.

The Lorenz dynamics was introduced in [83] to model Rayleigh-Benard convection. See, e.g., [25, Chap. 13] which studies its omega limit sets.

Part IV
Systems with Multiple Time Scales

Chapter 10
Rapidly Time-Varying Systems

Abstract In the first nine chapters, we considered continuous time and discrete time systems with a single time scale. We turn next to continuous time systems with two continuous time scales, one faster than the other. Systems of this kind are called either rapidly time-varying systems or slowly time-varying systems. The presence of multiple time scales significantly complicates the problem of constructing global strict Lyapunov functions. In this chapter, we provide a systematic method for rapidly time-varying systems. Our main method involves transforming Lyapunov functions for the corresponding limiting dynamics into the desired strict Lyapunov functions for the original rapidly time-varying dynamics. We illustrate our findings using a one degree-of-freedom mass-spring system.

10.1 Motivation

Consider the following one degree-of-freedom mass-spring system from [36] with constant parameter $\alpha > 0$, which arises in the control of mechanical systems with friction:

$$\begin{cases} \dot{x}_1 = x_2 \\ \dot{x}_2 = -\sigma_1(\alpha t)x_2 - k(t)x_1 + u \\ \qquad\quad - \left\{\sigma_2(\alpha t) + \sigma_3(\alpha t)e^{-\beta_1\mu(x_2)}\right\} \operatorname{sat}(x_2) \end{cases} \tag{10.1}$$

where x_1 and x_2 are the mass position and velocity, respectively; u is a disturbance; σ_1, σ_2, and σ_3 denote everywhere positive time-varying viscous, Coulomb, and static friction related coefficients, respectively; β_1 is a positive constant corresponding to the Stribeck effect; $\mu(\cdot)$ is a positive definite function also related to the Stribeck effect; k denotes a positive time-varying spring stiffness related coefficient; and sat: $\mathbb{R} \to \mathbb{R}$ denotes any continuous function having these properties:

$$\begin{aligned}&1.\ \operatorname{sat}(0)=0;\\&2.\ \xi\operatorname{sat}(\xi)\geq 0\ \ \forall\xi\in\mathbb{R};\\&3.\ \lim_{\xi\to+\infty}\operatorname{sat}(\xi)=+1;\ \text{ and}\\&4.\ \lim_{\xi\to-\infty}\operatorname{sat}(\xi)=-1\ .\end{aligned}\tag{10.2}$$

We model the saturation as the differentiable function

$$\operatorname{sat}(x_2)=\tanh(\beta_2 x_2),\tag{10.3}$$

where β_2 is a large positive constant. Note for later use that $|\operatorname{sat}(x_2)|\leq\beta_2|x_2|$ for all $x_2\in\mathbb{R}$.

We assume that the coefficients σ_i vary in time faster than the spring stiffness coefficient so we restrict to cases where $\alpha>1$. The constant α produces a more general class of time-varying systems, in which there are two continuous time scales, one faster than the other. This captures the realistic scenario where the wear and tear due to friction is faster than the degradation of the spring stiffness. Our precise mathematical assumptions for (10.1) are: k and the σ_i's are (globally) bounded C^1 functions; μ has a bounded first derivative; and there exist constants $\tilde{\sigma}_i$ with $\tilde{\sigma}_1>0$ and $\tilde{\sigma}_i\geq 0$ for $i=2,3$, and a $o(s)$ function $s\mapsto M(s)$ such that

$$\left|\int_{t_1}^{t_2}(\sigma_i(t)-\tilde{\sigma}_i)\,\mathrm{d}t\right|\ \leq\ M(t_2-t_1),\quad i=1,2,3\tag{10.4}$$

for all $t_1,t_2\in\mathbb{R}$ satisfying $t_2>t_1$. Although the σ_i's are everywhere positive for physical reasons, we will not require their positivity in the sequel.

It is natural to ask: *Can we find a constant $\underline{\alpha}>0$ and design a class of functions $V^{[\alpha]}$ such that for each constant $\alpha\geq\underline{\alpha}$, $V^{[\alpha]}$ is an ISS Lyapunov function for (10.1)?* We will answer this question in the affirmative in Sect. 10.6.3. In fact, our construction of the $V^{[\alpha]}$'s will follow from a general constructive approach for rapidly time-varying systems from [108].

10.2 Overview of Methods

A standard method for guaranteeing stability of nonautonomous systems is the so-called *averaging method.* In averaging, the exponential stability of an appropriate *autonomous* system implies exponential stability of the original dynamics, when the time variation is sufficiently fast [70]. Such results were extended to more general rapidly time-varying systems

$$\dot{x}\ =\ f(t,\alpha t,x),\ \ t\in\mathbb{R},\ x\in\mathbb{R}^n,\ \alpha>0\tag{10.5}$$

in [136], where uniform (local) exponential stability of (10.5) was proven for large values of the constant $\alpha > 0$, assuming a suitable limiting dynamics

$$\dot{x} = \bar{f}(t, x) \tag{10.6}$$

for (10.5) is uniformly exponentially stable. We specify the choice of $\bar{f}$ below. Notice that (10.1) is a special case of (10.5) when the disturbance u is set to 0. The main arguments of [136] use partial averaging but do not lead to explicit strict Lyapunov functions for (10.5); see Sect. 10.8 for a more detailed discussion of the literature.

In this chapter, we discuss a different approach, based on [108]. Instead of averaging, we explicitly construct a family of global strict Lyapunov functions for (10.5) in terms of more readily available Lyapunov functions for (10.6), which we again assume is asymptotically stable. We consider the case where (10.6) is UGAS, in which case our conclusion is that (10.5) is UGAS (but not necessarily exponentially stable) when the constant $\alpha > 0$ is sufficiently large. The significance of the problem lies in the ubiquity of rapidly time-varying systems in a host of engineering applications (involving, e.g., suspended pendulums, Raleigh's Equations, and Duffing's Equations from [70, Chapter 10], and systems arising in identification in [136]) and the value of explicit strict Lyapunov functions in robustness analysis and controller design. The Lyapunov functions we construct are also ISS or iISS Lyapunov functions for

$$\dot{x} = f(t, \alpha t, x) + g(t, \alpha t, x)u \tag{10.7}$$

under appropriate conditions on f and g; see Remark 10.2.

10.3 Assumptions and Lemmas

Consider the systems (10.5) and (10.6) where f and $\bar{f}$ are continuous in time $t \in \mathbb{R}$, C^1 in $x \in \mathbb{R}^n$, satisfy

$$f(t, \alpha t, 0) = \bar{f}(t, 0) = 0 \quad \forall t \in \mathbb{R},\ \alpha > 0\,, \tag{10.8}$$

and are forward complete for each constant $\alpha > 0$. Throughout this chapter, we assume the following uniform growth condition on f: There exists $\rho \in \mathcal{K}_\infty$ such that $|f(t, \alpha t, x)| \leq \rho(|x|)$ everywhere, and likewise for $\bar{f}$.

Recall the UGAS property, the Converse Lyapunov Function Theorem and the classes of functions $\mathcal{KL}$, $\mathcal{K}$, $\mathcal{K}_\infty$, which we reviewed in Chapters 1 and 2. In particular, we call (10.6) *UGAS* provided there is a $\beta \in \mathcal{KL}$ such that

$$\big|\phi(t; t_0, x_0)\big| \leq \beta\big(|x_0|, t - t_0\big) \quad \forall t \geq t_0 \geq 0 \text{ and } x_0 \in \mathbb{R}^n\,, \tag{10.9}$$

where ϕ is the flow map for (10.6). We call (10.6) *uniformly globally exponentially stable (UGES)* provided there are constants $D > 1$ and $\lambda > 0$ such that (10.9) is satisfied along all of its trajectories with the choice

$$\beta(s,t) = Dse^{-\lambda t}. \tag{10.10}$$

Recall Lemma 2.2, which constructed strict Lyapunov functions for exponentially stable time-varying systems $\dot{x} = f(t,x)$ using an integral of the flow map. The following compatibility condition is also useful:

Definition 10.1. Given $\delta \in \mathcal{K}$, the dynamics (10.6) is said to be *δ-compatible* provided it admits a function $V \in C^1$, functions $\delta_1, \delta_2 \in \mathcal{K}_\infty$, and constants $\bar{c} \in (0,1)$ and $\bar{\bar{c}} > 0$ such that:

1. $V_t(t,\xi) + V_\xi(t,\xi)\,\bar{f}(t,\xi) \le -\bar{c}\,\delta^2(|\xi|)$ for all ξ and t;
2. $|V_\xi(t,\xi)| \le \delta(|\xi|)$ and $|\bar{f}(t,\xi)| \le \delta(|\xi|/2)$ for all ξ and t; and
3. $\delta(s) \le \bar{\bar{c}}s$ for all $s \ge 0$;

and $\delta_1(|\xi|) \le V(t,\xi) \le \delta_2(|\xi|)$ for all ξ and t.

Remark 10.1. The bounds on $|V_\xi|$ and $|\bar{f}|$ in Condition 2 of Definition 10.1 are asymmetric. If (10.6) satisfies the UGES requirements of Lemma 2.2 for some positive constants K, c_1, c_2, and c_3, then one easily checks that it is δ-compatible with $\delta(s) = (c_3 + 2K)s$ and $\bar{c} = (c_3 + 2K)^{-2}$, by taking the V constructed in the lemma. On the other hand, by taking different choices of δ (including cases where δ is bounded), we can also find non-UGES δ-compatible dynamics; see Sect. 10.6.1.

We also treat the nonautonomous *control system*

$$\dot{x} = F(t,x,u) \tag{10.11}$$

which we always assume to be continuous in all variables, C^1 in x with $F(t,0,0) \equiv 0$, and forward complete. Its control set is $U = \mathbb{R}^m$, and its solution for a given control $\mathbf{u} \in \mathcal{M}(\mathbb{R}^m)$ and given initial condition $x(t_0) = x_0$ is denoted by $t \mapsto \phi(t; t_0, x_0, \mathbf{u})$. Recall the definitions of ISS and iISS from Chap. 1. If (10.11) is ISS, and if the ISS estimate for (10.11) can be satisfied with a function $\beta \in \mathcal{KL}$ having the form (10.10), then we say that (10.11) is *input-to-state exponentially stable (ISES)*. We say that a function $V : [0,\infty) \times \mathbb{R}^n \to [0,\infty)$ is *uniformly positive definite* provided there is a positive definite function α_0 such that $V(t,x) \ge \alpha_0(|x|)$ everywhere. We let UPD (resp., UBPPD) indicate the class of all uniformly positive definite (resp., uniformly bounded proper and positive definite) functions $V : [0,\infty) \times \mathbb{R}^n \to [0,\infty)$.

Recall the ISS and iISS Lyapunov function definitions from Chap. 2. Since (10.11) has an ISS Lyapunov function when it is ISS (by the arguments of [169]), the proof of [8, Theorem 1] shows that if (10.11) is ISS and autonomous, then it is also iISS, but not conversely, since e.g.,

$\dot{x} = -\arctan(x) + u$ is iISS but not ISS. The next lemma follows from arguments used in [8, 39, 157]:

Lemma 10.1. *If (10.11) admits an ISS (resp., iISS) Lyapunov function, then it is ISS (resp., iISS).*

A simple application of Fubini's Theorem gives

$$\int_{t-\tau}^{t} \int_{s}^{t} p(t,l)\,\mathrm{d}l\,\mathrm{d}s = \int_{t-\tau}^{t} (r - t + \tau)p(t,r)dr \tag{10.12}$$

for each continuous function p and each constant $\tau > 0$, which implies:

Lemma 10.2. *Let $p : \mathbb{R} \times \mathbb{R} \to \mathbb{R} : (t,l) \mapsto p(t,l)$ be C^1 in t and continuous. Then*

$$\begin{aligned} &\frac{d}{dt} \int_{t-\tau}^{t} \int_{s}^{t} p(t,l)\,\mathrm{d}l\,\mathrm{d}s \\ &= \tau p(t,t) - \textstyle\int_{t-\tau}^{t} p(t,l)\,\mathrm{d}l + \displaystyle\int_{t-\tau}^{t} \int_{s}^{t} \frac{\partial p}{\partial t}(t,l)\,\mathrm{d}l\,\mathrm{d}s \end{aligned} \tag{10.13}$$

and

$$\left| \int_{t-\tau}^{t} \int_{s}^{t} p(t,l)\,\mathrm{d}l\,\mathrm{d}s \right| \le \frac{\tau^2}{2} \max_{t-\tau \le l \le t} |p(t,l)| \tag{10.14}$$

hold for all $t \in \mathbb{R}$ and all constants $\tau > 0$.

For what follows, a function $N : [0,\infty) \to [0,\infty)$ is said to be *of class $\mathcal{M}$* (written $N \in \mathcal{M}$) provided

$$\lim_{\eta \to +\infty} \eta N(\eta) = 0. \tag{10.15}$$

10.4 Main Lyapunov Function Construction

We consider the system

$$\dot{x} \;=\; f(t, \alpha t, x) + u, \quad x \in \mathbb{R}^n,\; u \in \mathbb{R}^n \tag{10.16}$$

with constant parameter $\alpha > 0$, but see Remark 10.2 for results on the more general control systems (10.7). Our main assumption will be: There exist $\delta \in \mathcal{K}$, a UGAS δ-compatible dynamics (10.6), $N \in \mathcal{M}$, and a constant $\eta_0 > 0$ such that for all $x \in \mathbb{R}^n$, all $r \in \mathbb{R}$ and all constants $\eta > \eta_0$,

$$\left| \int_{r-\frac{1}{\eta}}^{r+\frac{1}{\eta}} \{ f(l, \eta^2 l, x) - \bar{f}(l,x) \}\,\mathrm{d}l \right| \le \delta\left(\frac{|x|}{2}\right) N(\eta)\,. \tag{10.17}$$

The main result in [108] says:

Theorem 10.11. *Given a system (10.5), assume that there exist $\delta \in \mathcal{K}$, a δ-compatible UGAS system (10.6), two constants $\eta_0 > 0$ and $K > 1$, and a function $N \in \mathcal{M}$ such that (10.17) holds whenever $\eta \geq \eta_0$, $x \in \mathbb{R}^n$ and $r \in \mathbb{R}$ and such that:*

$$\begin{aligned} &\left|\tfrac{\partial \bar{f}}{\partial x}(t,x)\right| \leq K \ , \ \left|\tfrac{\partial f}{\partial x}(t,\alpha t,x)\right| \leq K \ , \ \textit{and} \\ &|f(t,\alpha t,x)| \leq \delta\left(\tfrac{|x|}{2}\right) \quad \forall t \in \mathbb{R},\ x \in \mathbb{R}^n,\ \alpha > 0. \end{aligned} \tag{10.18}$$

Then we can construct a constant $\underline{\alpha} > 0$ such that the following are true: (a) For all constants $\alpha \geq \underline{\alpha}$, (10.5) is UGAS and (10.16) is iISS. (b) If $\delta \in \mathcal{K}_\infty$, then (10.16) is ISS for all constants $\alpha \geq \underline{\alpha}$. (c) If (10.6) is UGES, then (10.5) is UGES for all constants $\alpha \geq \underline{\alpha}$ and (10.16) is ISES for all constants $\alpha \geq \underline{\alpha}$.

By (10.8), the condition

$$|f(t,\alpha t,x)| \leq \delta\left(\frac{|x|}{2}\right)$$

in (10.18) is redundant when δ is of the form $\delta(s) = \bar{r}s$ for a constant $\bar{r} > 0$, since $\bar{r}$ can always be enlarged. Two important features of Theorem 10.11 are that it applies to cases where (10.6) is UGAS but not necessarily UGES (cf. Sect. 10.6.1), and that its proof leads to explicit global strict Lyapunov functions for (10.16) in Theorem 10.12. Later we give a variant for cases where $\partial f/\partial x$ is not necessarily globally bounded. The strict Lyapunov function provided by the proof of Theorem 10.11 is as follows:

Theorem 10.12. *Let the assumptions of Theorem 10.11 be satisfied for some $\delta \in \mathcal{K}$, and $V \in C^1$ be a Lyapunov function for (10.6) satisfying the requirements of Definition 10.1. Then we can construct a constant $\underline{\alpha} > 0$ such that for all constant values $\alpha > \underline{\alpha}$,*

$$V^{[\alpha]}(t,\xi) \doteq V\left(t, \xi - \tfrac{\sqrt{\alpha}}{2}\int_{t-\frac{2}{\sqrt{\alpha}}}^{t}\int_s^t \{f(\xi,l,\alpha l) - \bar{f}(\xi,l)\}\,\mathrm{d}l\,\mathrm{d}s\right)$$

is a global strict Lyapunov function for (10.5) and an iISS Lyapunov function for (10.16). If we also have $\delta \in \mathcal{K}_\infty$, then $V^{[\alpha]}$ is also an ISS Lyapunov function for (10.16) for all constant values $\alpha > \underline{\alpha}$.

Proofs of Theorems 10.11 and 10.12

We only provide a sketch; see [108] for the complete arguments. Throughout the sequel, all inequalities and equalities should be understood to hold globally, unless we indicate otherwise. Let us first outline our method for proving the theorems. First, we give the proof of Theorem 10.11 for the special case where (10.6) is UGAS and $\delta \in \mathcal{K}_\infty$ which includes the proof of Theorem 10.12 when $\delta \in \mathcal{K}_\infty$. We then indicate the changes needed if $\delta \in \mathcal{K}$ is bounded. Finally, we specialize to the situation where (10.6) is UGES, in which case we can take $\delta(s) = \bar{r}s$ for some constant $\bar{r} > 0$, by Lemma 2.2. This will show the ISES assertion of Theorem 10.11.

Assume first that (10.6) is UGAS and $\delta \in \mathcal{K}_\infty$. We prove the ISS property for (10.16) for large constants $\alpha > 0$. Let $\eta_0 > 0$ be as in the statement of the theorem, and fix

$$\alpha = \eta^2$$

with $\eta \geq \eta_0$, $\mathbf{u} \in \mathcal{M}(\mathbb{R}^m)$, and a trajectory $x(t)$ for (10.16) and $\mathbf{u}$, for an arbitrary initial condition. Set

$$z(t) = x(t) + R_\alpha(t, x(t)), \tag{10.19}$$

where

$$R_\alpha(t,x) = -\frac{\eta}{2}\int_{t-2/\eta}^{t}\int_{s}^{t}\{f(l,\eta^2 l,x) - \bar{f}(l,x)\}\,\mathrm{d}l\,\mathrm{d}s.$$

Since we are assuming our dynamics to be forward complete, this is well defined. Taking

$$p(t,l) = f(l,\eta^2 l, x(t)) - \bar{f}(l,x(t)),$$

(10.13) multiplied through by $-\eta/2$ gives

$$\begin{aligned}\dot{z}(t) &= \bar{f}(t,z(t)) + [\bar{f}(t,x(t)) - \bar{f}(t,z(t))] + \tfrac{\eta}{2}\int_{t-\frac{2}{\eta}}^{t} p(t,l)\,\mathrm{d}l \\ &\quad -\tfrac{\eta}{2}\left\{\int_{t-\frac{2}{\eta}}^{t}\int_{s}^{t}\left(\frac{\partial f}{\partial x}(l,\eta^2 l,x(t)) - \frac{\partial \bar{f}}{\partial x}(l,x(t))\right)\mathrm{d}l\,\mathrm{d}s\right\} \\ &\quad \times\left[f(t,\eta^2 t,x(t)) + \mathbf{u}(t)\right] + \mathbf{u}(t).\end{aligned} \tag{10.20}$$

Let V, δ_1, and δ_2 satisfy the requirements of Definition 10.1. Let $\dot{V}$ denote the time derivative of $V(t,z)$ along the time-varying map $z(t)$ in (10.19). Using Condition 1. from Definition 10.1 with $\xi = z(t)$ and (10.20),

$$\begin{aligned}\dot{V} &\leq -\bar{c}\,\delta^2(|z(t)|) + V_\xi(t,z(t))\left(\bar{f}(t,x(t)) - \bar{f}(t,z(t))\right) \\ &\quad +\tfrac{\eta}{2}V_\xi(t,z(t))\int_{t-2/\eta}^{t} p(t,l)\,\mathrm{d}l \\ &\quad -\tfrac{\eta}{2}V_\xi(t,z(t))\left[\int_{t-2/\eta}^{t}\int_{s}^{t}\left(\frac{\partial f}{\partial x}(l,\eta^2 l,x(t)) - \frac{\partial \bar{f}}{\partial x}(l,x(t))\right)\mathrm{d}l\,\mathrm{d}s\right] \\ &\quad \times\left(f(t,\eta^2 t,x(t)) + \mathbf{u}(t)\right) + V_\xi(t,z(t))\,\mathbf{u}(t).\end{aligned}$$

Using (10.17)-(10.19) and Condition 2. from Definition 10.1, we get

$$\begin{aligned}\dot{V} &\leq -\bar{c}\,\delta^2(|z(t)|) + K\delta(|z(t)|)|R_\alpha(t,x(t))| + \tfrac{\eta}{2}\delta(|z(t)|)N(\eta)\,\delta(|x(t)|/2)\\ &\quad + \delta(|z(t)|)\,|\mathbf{u}(t)|\\ &\quad + \tfrac{2}{\eta}K\,\delta(|z(t)|)\{\delta(|x(t)|/2) + |\mathbf{u}(t)|\}\\ &\leq -\bar{c}\,\delta^2(|z(t)|) + \delta(|z(t)|)\,|\mathbf{u}(t)|\\ &\quad + \delta(|z(t)|)\left(\tfrac{4}{\eta}K + \tfrac{\eta}{2}N(\eta)\right)\left\{\delta\left(\tfrac{|x(t)|}{2}\right) + |\mathbf{u}(t)|\right\},\end{aligned}$$

where the last inequality used

$$|R_\alpha(t,x(t))| \;\leq\; \tfrac{\eta}{2}\int_{t-2/\eta}^{t}\int_s^t |p(t,l)|\,\mathrm{d}l\,\mathrm{d}s \;\leq\; \tfrac{2}{\eta}\delta\left(\tfrac{|x(t)|}{2}\right). \tag{10.21}$$

Also, (10.19), (10.21), and Condition 3. from Definition 10.1 give

$$|z(t)| \;\geq\; |x(t)| - \tfrac{\bar{\bar{c}}}{\eta}|x(t)| \;\geq\; \tfrac{1}{2}|x(t)| \tag{10.22}$$

when $\eta \geq \max\{2\bar{\bar{c}}, \eta_0\}$. The fact that $\delta \in \mathcal{K}$ now gives

$$\begin{aligned}\dot{V} &\leq \left(-\bar{c} + \tfrac{4}{\eta}K + \tfrac{\eta}{2}N(\eta)\right)\delta^2(|z(t)|)\\ &\quad + \left(\tfrac{4}{\eta}K + \tfrac{\eta}{2}N(\eta) + 1\right)\delta(|z(t)|)|\mathbf{u}(t)|.\end{aligned} \tag{10.23}$$

Choosing $\chi(s) = \frac{\bar{c}}{4}\delta(s/2)$ and recalling that $\bar{c} \in (0,1)$, we get

$$\begin{aligned}&|\mathbf{u}|_\infty \;\leq\; \chi(|x(t)|) \;\;\Rightarrow\;\; |\mathbf{u}|_\infty \;\leq\; \chi(2|z(t)|)\\ &\Rightarrow\;\; \dot{V} \leq \left(-\tfrac{3\bar{c}}{4} + \tfrac{8}{\eta}K + \eta N(\eta)\right)\delta^2(|z(t)|).\end{aligned} \tag{10.24}$$

Setting

$$V^{[\alpha]}(t,x) \;\doteq\; V\left(t, x + R_\alpha(t,x)\right),$$

we see that the time derivative $\dot{V} = V_t(t,z) + V_z(t,z)\dot{z}$ of $V(t,z)$ along (10.16) satisfies

$$\dot{V} = V_t^{[\alpha]}(t,x) + V_x^{[\alpha]}(t,x)\,\{f(t,\alpha t,x) + \mathbf{u}(t)\}. \tag{10.25}$$

It follows from (10.15) and (10.24) that when the constant α (and so also η) is sufficiently large, we get

$$|u| \leq \chi(|x|) \;\;\Rightarrow\;\; V_t^{[\alpha]}(t,x) + V_x^{[\alpha]}(t,x)\,[f(t,\alpha t,x) + u] \leq -\tfrac{\bar{c}}{2}\delta^2\left(\tfrac{|x|}{2}\right)$$

and the uniform proper and positive definiteness of $V^{[\alpha]}$ follows from (10.21) and (10.22). Hence, $V^{[\alpha]}$ is an ISS Lyapunov function for (10.16), so (10.16) is ISS for large α, by Lemma 10.1, as claimed. The UGAS conclusion is the special case of the preceding argument where $\mathbf{u} \equiv 0$. When δ is bounded, the iISS assertion follows from the part of the preceding argument up through (10.23), once we bound the coefficient of $|\mathbf{u}(t)|$. Then $V^{[\alpha]}$ is an iISS Lyapunov function for (10.16) for sufficiently large α, so (10.16) is iISS for large α, by Lemma 10.1.

It remains to consider the special case where (10.6) is UGES. Choose any function V that satisfies the requirements of Lemma 2.2 for (10.6) for suitable constants $c_i > 0$. Choose any trajectory $x(t)$ for (10.16) for any choice of the control $\mathbf{u} \in \mathcal{M}(\mathbb{R}^m)$, starting at an arbitrary initial state $x(t_0) = x_0$. Then

$$|R_\alpha(t, x(t))| \leq \frac{2K|x(t)|}{\eta} \quad \forall t \geq t_0 \,. \tag{10.26}$$

Defining $z(t)$ by (10.19) and arguing as before except with the preceding choice of V gives $\delta(s) = (c_3 + 2K)s$, $\bar{c} = (c_3 + 2K)^{-2}$, and therefore

$$\begin{aligned} \dot{V} \leq & \left(-\tfrac{1}{\bar{D}^2} + \tfrac{4}{\eta}K + \tfrac{\eta}{2}N(\eta)\right) \bar{D}^2|z(t)|^2 \\ & + \left[\tfrac{4}{\eta}K + \tfrac{\eta}{2}N(\eta) + 1\right] \bar{D}|z(t)||\mathbf{u}(t)|, \end{aligned}$$

where $\bar{D} = c_3 + 2K$.

Taking $\tilde{\chi}(s) = \frac{s}{6\bar{D}}$, it follows as in the UGAS case that if $|\mathbf{u}|_\infty \leq \tilde{\chi}(|x(t)|)$ for all t and η is large enough, then (10.22) gives

$$\dot{V} \leq -\frac{|z(t)|^2}{4} \leq -\frac{V(t, z(t))}{4c_2}.$$

Integrating over t and recalling the properties of c_1 and c_2 and (10.26) now gives

$$|x(t)| \leq \sqrt{\frac{4c_2}{c_1}} \left(1 + \frac{2K}{\eta}\right) |x(t_0)| \exp\left(-\frac{t - t_0}{8c_2}\right), \tag{10.27}$$

so if (10.6) is UGES, then (10.5) is also UGES when the constant $\alpha > 0$ is large enough. Also, standard arguments (e.g., from the proof of [169, Lemma 2.14]) imply that (10.16) is ISES. This completes the proof. □

Remark 10.2. A slight variant of the proof of Theorem 10.11 shows the ISS property for (10.7) under suitable growth assumptions on the matrix-valued function $g : \mathbb{R} \times \mathbb{R} \times \mathbb{R}^n \to \mathbb{R}^{n \times m}$. Some growth condition on g is needed and linear growth of g is insufficient, because $\dot{x} = -x + xu$ is not ISS. One method for extending the theorem to (10.7) with $g \in C^1$ is to add the assumption that there is a constant $c_0 > 1$ such that

$$||g(t, \alpha t, x)|| \leq c_0 + \sqrt{\delta\left(\frac{|x|}{2}\right)}$$

for all $t \in \mathbb{R}$, $x \in \mathbb{R}^n$, and $\alpha > 0$, where $\delta \in \mathcal{K}_\infty$ satisfies the requirements of Definition 10.1 for some Lyapunov function V for (10.6). Using the new $\mathcal{K}_\infty$ function

$$\chi(s) = \frac{\bar{c}\,\delta(s/2)}{4\left\{c_0 + \sqrt{\delta(s/2)}\right\}} \tag{10.28}$$

in the first part of the proof of Theorem 10.11, we conclude as in the earlier proof that (10.7) is ISS for large enough $\alpha > 0$. If $\delta \in \mathcal{K}$ is bounded, then we instead conclude that (10.7) is iISS when the rapidness parameter α is sufficiently large, using a slight variant of our earlier argument.

10.5 Alternative Strictification Result

The strict Lyapunov function construction from Theorem 10.12 is based on transforming a strict Lyapunov function for the limiting dynamics (10.6). It is natural to inquire whether we can instead use a suitable generalization of the strictification approach from Chap. 6 to transform a non-strict Lyapunov function for (10.5) into a strict Lyapunov function for (10.5). In this section, we show how such a strictification can indeed be carried out. The possible advantages of this alternative result are that:

1. it does not require (10.5) to be globally Lipschitz in the state;
2. it allows the time derivative of the non-strict Lyapunov function to be zero or even positive at some points; and
3. it does not require any knowledge of limiting dynamics.

See Sect. 10.7.2 below for details. For simplicity, we focus on systems with no controls. Our key assumption is:

Assumption 10.1 *There exist functions $V \in C^1 \cap \mathrm{UBPPD}$ and $W \in \mathrm{UPD}$, a C^1 function*

$$\Theta : [0, \infty) \times \mathbb{R}^n \to \mathbb{R},$$

a bounded continuous function $p : \mathbb{R} \to \mathbb{R}$, and positive constants c and T such that:

1. $V_t(t,x) + V_x(t,x) f(t, \alpha t, x) \leq -W(t,x) + p(\alpha t)\Theta(t,x)$;
2. $\int_{kT}^{(k+1)T} p(r)\mathrm{d}r = 0$;
3. $W(t,x) \geq c \max\{|\Theta(t,x)|, |\Theta_t(t,x) + \Theta_x(t,x) f(t, \alpha t, x)|\}$; *and*
4. $V(t,x) \geq c|\Theta(t,x)|$

for all $t \geq 0$, $x \in \mathbb{R}^n$, $\alpha > 0$, and $k \in \mathbb{Z}$.

For any choices of $s \in [\alpha t - \alpha T, \alpha t]$ and $t \geq 0$, Condition 2. of Assumption 10.1 gives

$$\int_s^{\alpha t} p(l)\mathrm{d}l \;=\; \int_s^{\bar{\tau}(s)} p(l)\mathrm{d}l + \int_{\underline{\tau}(\alpha t)}^{\alpha t} p(l)\mathrm{d}l, \tag{10.29}$$

where

$$\bar{\tau}(u) \doteq \min\{kT : k \in \mathbb{Z}, kT \geq u\} \text{ and}$$
$$\underline{\tau}(u) \doteq \max\{kT : k \in \mathbb{Z}, kT \leq u\}.$$

The proof of (10.29) is based on the facts that both endpoints of the interval $[\min\{\bar{\tau}(s), \underline{\tau}(\alpha t)\}, \max\{\bar{\tau}(s), \underline{\tau}(\alpha t)\}]$ are integer multiples of T, and that the integral of p over any interval whose endpoints are integer multiples of T is zero.

Our strictification result is as follows:

Theorem 10.13. *If Assumption 10.1 holds, then there exists a constant $\underline{\alpha} > 0$ such that for each constant $\alpha \geq \underline{\alpha}$,*

$$U^{[\alpha]}(t,x) \;\doteq\; V(t,x) - \tfrac{1}{T}\left(\int_{t-T}^{t}\int_s^t p(\alpha l)\,\mathrm{d}l\,\mathrm{d}s\right)\Theta(t,x) \tag{10.30}$$

is a global strict Lyapunov function for (10.5). In particular, (10.5) is UGAS for all constants $\alpha \geq \underline{\alpha}$.

Proof. Set

$$\dot{U}^{[\alpha]}(t,x) = U_t^{[\alpha]}(t,x) + U_x^{[\alpha]}(t,x)f(t,\alpha t,x)$$

for all $t \geq 0$, $x \in \mathbb{R}^n$, and $\alpha > 0$. Recalling Conditions 1. and 3. in Assumption 10.1 and using (10.13) with $p(t,l)$ independent of t gives

$$\begin{aligned}
\dot{U}^{[\alpha]}(t,x) &\leq -W(t,x) + p(\alpha t)\Theta(t,x) - p(\alpha t)\Theta(t,x)\\
&\quad +\tfrac{1}{T}\left(\int_{t-T}^{t} p(\alpha l)\mathrm{d}l\right)\Theta(t,x)\\
&\quad -\tfrac{1}{T}\left(\int_{t-T}^{t}\int_s^t p(\alpha l)\mathrm{d}l\,\mathrm{d}s\right)\\
&\quad \times\left(\tfrac{\partial\Theta}{\partial x}(t,x)f(t,\alpha t,x) + \tfrac{\partial\Theta}{\partial t}(t,x)\right)\\
&\leq -W(t,x) + \tfrac{1}{T}\left|\int_{t-T}^{t} p(\alpha l)\mathrm{d}l\right|\tfrac{1}{c}W(t,x)\\
&\quad +\tfrac{1}{T}\left|\int_{t-T}^{t}\int_s^t p(\alpha l)\mathrm{d}l\,\mathrm{d}s\right|\tfrac{1}{c}W(t,x)
\end{aligned} \tag{10.31}$$

along trajectories of (10.5). Taking $p_{\max}$ to be any uniform bound for $|p(l)|$ over $\mathbb{R}$, we conclude from (10.29) that

$$\left|\int_s^{\alpha t} p(l)\mathrm{d}l\right| \;\leq\; 2Tp_{\max} \quad \forall s \in [\alpha t - \alpha T, \alpha t] \tag{10.32}$$

for all $t \geq 0$ and $\alpha > 0$.

Hence,

$$\left|\int_{t-T}^{t} p(\alpha l)\mathrm{d}l\right| = \frac{1}{\alpha}\left|\int_{\alpha t-\alpha T}^{\alpha t} p(l)\mathrm{d}l\right| \leq \frac{2Tp_{\max}}{\alpha}$$

and

$$\begin{aligned}\left|\int_{t-T}^{t}\int_{s}^{t} p(\alpha l)\,\mathrm{d}l\,\mathrm{d}s\right| &= \tfrac{1}{\alpha^2}\left|\int_{\alpha t-\alpha T}^{\alpha t}\int_{s}^{\alpha t} p(l)\,\mathrm{d}l\,\mathrm{d}s\right| \\ &\leq \tfrac{T}{\alpha}\sup_{s\in[\alpha t-\alpha T,\alpha t]}\left|\int_{s}^{\alpha t} p(l)\mathrm{d}l\right| \\ &\leq \tfrac{2T^2 p_{\max}}{\alpha}\end{aligned} \tag{10.33}$$

for all $t \geq 0$. These estimates combined with (10.31) give

$$\dot{U}^{[\alpha]} \leq -\frac{W(t,x)}{2}$$

for all $t \geq 0$ and $x \in \mathbb{R}^n$, provided

$$\alpha > \frac{4(T+1)p_{\max}}{c}\,. \tag{10.34}$$

Since $W \in$ UPD, this gives the desired decay estimate. Also, (10.33) and Condition 4. in Assumption 10.1 give the uniform proper and positive definiteness of $U^{[\alpha]}$ for large enough constants $\alpha > 0$, hence the conclusion of the theorem. □

10.6 Illustrations

We illustrate Theorems 10.11 and 10.12 through several examples, starting with a case where (10.6) is UGAS but not necessarily UGES. We next consider a class of systems (10.5) from identification where the limiting dynamics (10.6) is linear and exponentially stable. For these systems, we provide formulas for strict Lyapunov functions for (10.5) that have the additional desirable property that they are also ISS Lyapunov functions for (10.7) for suitable functions g. In this situation, the strict Lyapunov functions are expressed in terms of quadratic Lyapunov functions for the limiting dynamics. Finally, we apply our results to a friction model for a mass-spring dynamics we discussed in Sect. 10.1. In all three examples, the limiting dynamics has a simple Lyapunov function, so our results give explicit Lyapunov functions for the original rapidly time-varying dynamics. The novelty of our treatment of these examples lies in our global strict Lyapunov function constructions; see Sect. 10.8 for a detailed comparison with the literature.

10.6.1 A UGAS Dynamics that Is Not UGES

Consider the following scalar example from [108]:

$$\dot{x} = f(t, \alpha t, x) = -\sigma_1(x)\big[2 + \sin\big(t + \cos(\sigma_2(x))\big)\big]\{1 + 10\sin(\alpha t)\} \tag{10.35}$$

where $\sigma_1, \sigma_2 : \mathbb{R} \to \mathbb{R}$ are C^1 functions such that σ_1 is odd,

$$\sup\big\{|\sigma_1'(x)| + |\sigma_1(x)\sigma_2'(x)| : x \in \mathbb{R}\big\} < \infty,$$

$\sigma_1 \in \mathcal{K}$ on $[0, \infty)$, and $\sigma_1''(s) \le 0$ for all $s > 0$. One easily checks that the hypotheses of Part (a) of Theorem 10.11 are satisfied, using

$$\begin{aligned}
&\bar{f}(t, x) \doteq -\sigma_1(x)\big[2 + \sin\big(t + \cos(\sigma_2(x))\big)\big],\\
&V(t, x) \equiv \bar{V}(x) \doteq \int_0^x \sigma_1(s)\mathrm{d}s,\\
&\delta(s) \doteq 33\sigma_1(2s), \quad \text{and} \quad N(\eta) \doteq 60/\eta^2 \quad \text{for large } \eta.
\end{aligned}$$

This allows cases such as $\sigma_1(s) = \sigma_2(s) = \arctan(s)$; in that case, the limiting dynamics (10.6) is UGAS but not UGES because $|\dot{x}(t)| \le 2\pi$ along all of its trajectories $x(t)$. Condition 1. from Definition 10.1 is satisfied because $\sigma_1(2s) \le 2\sigma_1(s)$ for all $s \ge 0$, which holds because $\sigma_1''(s) \le 0$ for all $s \ge 0$. Theorem 10.12 then gives the following iISS Lyapunov function for (10.16) for large $\alpha > 0$:

$$\bar{V}\left(\xi + 5\sqrt{\alpha}\,\sigma_1(\xi)\int_{t-\frac{2}{\sqrt{\alpha}}}^{t}\int_s^t \mu(\xi, l)\sin(\alpha l)\,\mathrm{d}l\,\mathrm{d}s\right), \tag{10.36}$$

where

$$\mu(\xi, l) \doteq 2 + \sin(l + \cos(\sigma_2(\xi))).$$

It is a global strict Lyapunov function for $\dot{x} = f(t, \alpha t, x)$, and it is also an ISS Lyapunov function for (10.16) in the special case where $\delta \in \mathcal{K}_\infty$, e.g., if $\sigma_1(s) = \mathrm{sgn}(s)\ln(1 + |s|)$ for $|s| \ge 1$ and $\sigma_2(s) = \arctan(s)$.

10.6.2 System Arising in Identification

We next consider the system

$$\dot{x} = f(\alpha t)\,\bar{m}(t)\,\bar{m}^\top(t)\,x + g(t, \alpha t, x)\,u, \tag{10.37}$$

with state $x \in \mathbb{R}^n$ and inputs $u \in \mathbb{R}^m$. We assume:

Assumption 10.2 *The following conditions are satisfied:*

1. $f : \mathbb{R} \to \mathbb{R}$ *is bounded and continuous and admits a* $o(s)$ *function* M *and a constant* $f^\star < 0$ *for which:*

   i. $f^\star = \lim_{T\to+\infty} \frac{1}{2T}\int_{-T}^{T} f(s)\mathrm{d}s$*; and*
   ii. $|\int_{t_1}^{t_2}[f(s) - f^\star]\mathrm{d}s| \le M(t_2 - t_1)$ *if* $t_2 \ge t_1$.

2. $\bar{m} : \mathbb{R} \to \mathbb{R}^n$ *is continuous and* $|\bar{m}(t)| = 1$ *for all* $t \in \mathbb{R}$*, and there exist constants* $\alpha', \beta', \tilde{c} > 0$ *such that for all* $t \in \mathbb{R}$*,* $\alpha > 0$*, and* $x \in \mathbb{R}^n$*, we have:*

   i. $\alpha' I_n \le \int_t^{t+\tilde{c}} \bar{m}(\tau)\bar{m}^\top(\tau)\mathrm{d}\tau \le \beta' I_n$*; and*
   ii. $||g(t, \alpha t, x)|| \le \beta'\{1 + \sqrt{|x|}\}$.

3. $g : \mathbb{R} \times \mathbb{R} \times \mathbb{R}^n \to \mathbb{R}^{n\times m} : (t, l, x) \mapsto g(t, l, x)$ *is continuous and is* C^1 *in* x*, and there exists a constant* $K > 1$ *such that*

$$\left|\frac{\partial g_{ij}}{\partial x}(t, \alpha t, x)\right| \le K$$

   for all $x \in \mathbb{R}^n$*,* $t \ge 0$*, and* $\alpha > 0$*, and each component* g_{ij} *of* g.

The particular case of (10.37) in which

$$\dot{x} = -\bar{m}(t)\,\bar{m}^\top(t)\,x$$

has been extensively studied in systems identification; see Sect. 10.8 for the historical background. However, the well-known results do not provide explicit ISS Lyapunov functions for (10.37). To provide ISS Lyapunov functions of this kind, we use the limiting dynamics

$$\dot{x} = \bar{f}(t, x) \doteq f^\star \bar{m}(t)\,\bar{m}^\top(t)\,x \tag{10.38}$$

and the following key lemma from [108]:

Lemma 10.3. *Let Assumption 10.2 hold and set*

$$P(t) = \left\{\frac{\tilde{c}}{2|f^\star|} + \frac{1}{4\alpha'}\tilde{c}^4|f^\star|\right\} I_n + \int_{t-\tilde{c}}^{t}\int_s^t \bar{m}(l)\bar{m}^\top(l)\,\mathrm{d}l\,\mathrm{d}s. \tag{10.39}$$

Then $V(t, x) = x^\top P(t)x$ *is a strict Lyapunov function for (10.38) for which* $2V/\alpha'$ *satisfies the requirements of Lemma 2.2.*

For the proof of this lemma, see Appendix A.1. We readily conclude that the corresponding rapidly varying dynamics

$$\dot{x} = f(\alpha t)\bar{m}(t)\bar{m}^\top(t)x$$

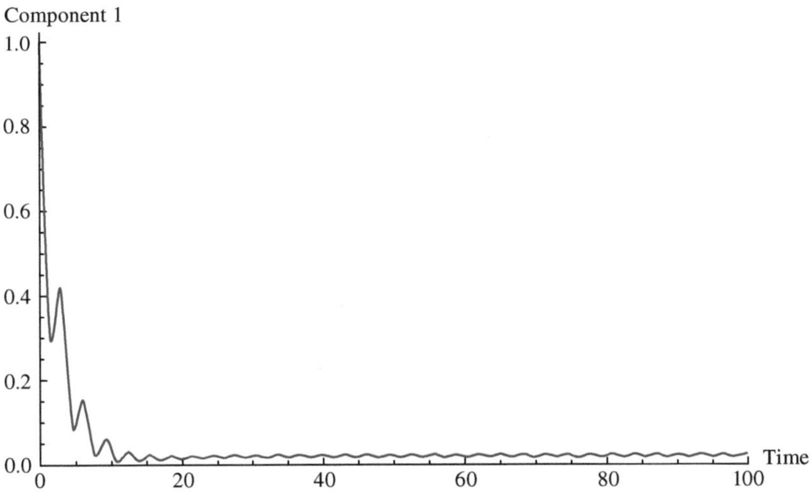

Fig. 10.1 First component of state of (10.37) with choices (10.40)

satisfies the hypotheses of Theorem 10.11 with δ of the form $\delta(s) = \bar{r}s$ for a suitable constant $\bar{r} > 0$. In fact, we can take the function V we constructed in the preceding lemma. Remark 10.2 now gives:

Corollary 10.1. *Let (10.37) satisfy Assumption 10.2 and let V be as in Lemma 10.3. Then there exists a constant $\alpha_0 > 0$ such that for each constant $\alpha > \alpha_0$,*

$$V^{[\alpha]}(t,x) \doteq V\left(t, \left[I - \frac{\sqrt{\alpha}}{2}\int_{t-\frac{2}{\sqrt{\alpha}}}^{t}\int_{s}^{t}(f(\alpha l) - f^{\star})m(l)m^{\top}(l)\,\mathrm{d}l\,\mathrm{d}s\right]x\right)$$

is an ISS Lyapunov function for (10.37).

To illustrate the ISS property, we simulated (10.37) with the choices

$$\begin{aligned} &n = m = 2,\;\; \bar{m}(t) = (\cos(t), \sin(t))^{\top},\;\; f(s) = \cos(s) - \tfrac{1}{2}, \\ &g(t, \alpha t, x) \equiv I_2,\;\; \alpha = 100,\;\; \text{and}\;\; u(t) \equiv (.005, 0)^{\top} \end{aligned} \tag{10.40}$$

and obtained the trajectories in Figs. 10.1 and 10.2. The simulations illustrate the convergence of the state components to zero, with an overshoot determined by the ISS overshoot term. Using our explicit ISS Lyapunov function, we can explicitly compute the functions $\beta \in \mathcal{KL}$ and $\gamma \in \mathcal{K}_\infty$ in the standard ISS estimate. This allows us to quantify the effects of the uncertainty u.

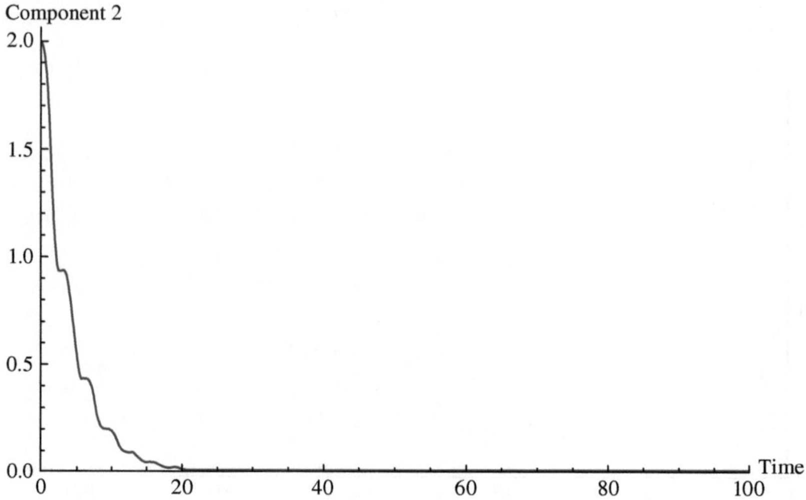

Fig. 10.2 Second component of state of (10.37) with choices (10.40)

10.6.3 Friction Example Revisited

We next show how the friction dynamics (10.1) satisfies the assumptions from Remark 10.2 with $\delta(s) = \bar{r}s$ for some constant $\bar{r}$, when the limiting dynamics (10.6) is

$$\begin{cases} \dot{x}_1 = x_2 \\ \dot{x}_2 = -\tilde{\sigma}_1 x_2 - \left\{\tilde{\sigma}_2 + \tilde{\sigma}_3 e^{-\beta_1 \mu(x_2)}\right\} \operatorname{sat}(x_2) - k(t)x_1 \, , \end{cases} \tag{10.41}$$

and the constants $\tilde{\sigma}_i$ satisfy the requirements from (10.4). Following [108], we add the assumption

$$\exists \text{ constants } k_0, \bar{k} > 0 \text{ s.t. } k_0 \le k(t) \le \bar{k} \text{ and } k'(t) \le 0 \;\; \forall t \ge 0. \tag{10.42}$$

Conditions (10.42) imply that the spring stiffness is non-increasing. To check the assumptions from Remark 10.2, set

$$\begin{aligned} S &\doteq \tilde{\sigma}_1 + (\tilde{\sigma}_2 + \tilde{\sigma}_3)\beta_2 \;\text{ and} \\ V(t,x) &\doteq A\left[k(t)x_1^2 + x_2^2\right] + x_1 x_2, \;\text{ where} \\ A &\doteq 1 + \tfrac{1}{k_0} + \tfrac{1}{\tilde{\sigma}_1}\left[1 + \tfrac{S^2}{k_0}\right]. \end{aligned} \tag{10.43}$$

Noting that $A\bar{k} \ge 1$ and $A \ge 1$, we get

$$\frac{1}{2}|x|^2 \;\le\; V(t,x) \;\le\; \frac{3}{2}A^2\bar{k}|x|^2$$

for all $x \in \mathbb{R}^2$ and $t \geq 0$. Also, the time derivative

$$\dot{V} = V_t(t,x) + V_x(t,x)\bar{f}(t,x)$$

along trajectories of (10.41) gives

$$\begin{aligned}\dot{V} \leq V_x(t,x)\bar{f}(t,x) &= [2Ak(t)x_1 + x_2]x_2 \\ &-[2Ax_2 + x_1]\{\tilde{\sigma}_1 x_2 + \left[\tilde{\sigma}_2 + \tilde{\sigma}_3 e^{-\beta_1 \mu(x_2)}\right] \mathrm{sat}(x_2) + k(t)x_1\},\end{aligned}$$

since $k' \leq 0$ everywhere. Hence, we can group terms to get

$$\begin{aligned}\dot{V} &\leq -k_0 x_1^2 - (2A\tilde{\sigma}_1 - 1)x_2^2 + S|x_1 x_2| \quad \text{(by (10.2))} \\ &\leq -b|x|^2 - \left[\tfrac{k_0}{2}x_1^2 + (A\tilde{\sigma}_1 - 1/2)x_2^2 - S|x_1 x_2|\right] \\ &= -b|x|^2 - \tfrac{k_0}{2}\left(|x_1| - \tfrac{S}{k_0}|x_2|\right)^2 + \left(\tfrac{S^2}{2k_0} + \tfrac{1}{2} - A\tilde{\sigma}_1\right)x_2^2 \\ &\leq -b|x|^2, \text{ where } b \doteq \min\{k_0/2, A\tilde{\sigma}_1 - 1/2\}.\end{aligned}$$

The preceding analysis says that V/b is a Lyapunov function for (10.41) that satisfies the conclusions of Lemma 2.2 for the limiting dynamics. The integral bound (10.17) from Theorem 10.11 follows from (10.4), using the sublinear growth of tanh, by verifying the integral bound term by term, and the other bounds from (10.18) are satisfied because sat and μ have uniformly bounded first derivatives. Hence, the proof of Theorem 10.12 implies that for large enough constants $\alpha > 0$, the dynamics (10.1) has the ISS Lyapunov function

$$V^{[\alpha]}(t,\xi) = V\left(t, \xi_1, \xi_2 + \frac{\sqrt{\alpha}}{2}\int_{t-\frac{2}{\sqrt{\alpha}}}^{t}\int_{s}^{t} \Gamma_\alpha(l,\xi)\, \mathrm{d}l\, \mathrm{d}s\right)$$

where V is the Lyapunov function (10.43) for the dynamics (10.41),

$$\Gamma_\alpha(l,\xi) \doteq \{\sigma_1(\alpha l) - \tilde{\sigma}_1\}\xi_2 + \mu_\alpha(l,\xi)\tanh(\beta_2\xi_2), \text{ and} \tag{10.44}$$

$$\mu_\alpha(l,\xi) \doteq \sigma_2(\alpha l) - \tilde{\sigma}_2 + (\sigma_3(\alpha l) - \tilde{\sigma}_3)e^{-\beta_1\mu(\xi_2)}. \tag{10.45}$$

In particular, (10.1) is ISS for large enough constant rapidness parameters $\alpha > 0$, by Remark 10.2.

Remark 10.3. The preceding analysis simplifies if σ_2 and σ_3 in (10.1) are both positive constants, since in that case, we can take the limiting dynamics

$$\begin{cases} \dot{x}_1 = x_2 \\ \dot{x}_2 = -\tilde{\sigma}_1 x_2 - \left\{\sigma_2 + \sigma_3 e^{-\beta_1\mu(x_2)}\right\}\mathrm{sat}(x_2) - k(t)x_1\,. \end{cases}$$

The ISS Lyapunov function for (10.1) can now be taken to be

$$V\left(t,\xi_1,\xi_2\left(1+\frac{\sqrt{\alpha}}{2}\int_{t-\frac{2}{\sqrt{\alpha}}}^{t}\int_{s}^{t}\{\sigma_1(\alpha l)-\tilde{\sigma}_1\}\,\mathrm{d}l\,\mathrm{d}s\right)\right)$$

with V defined by (10.43).

10.7 Further Illustrations: Strictification Approach

One easily checks that Theorem 10.13 applies to (10.35) and (10.37) without controls (using the V's from our earlier discussions of those examples), assuming (10.37) satisfies Assumption 10.2 with $\bar{m}$ having a bounded derivative and f a suitable periodic function. We next review two examples from [108] that apply Theorem 10.13 to cases that are not already covered by Theorem 10.11.

10.7.1 Systems with Unknown Functional Parameters

Consider the nonautonomous scalar dynamics

$$\dot{x}=p(\alpha t)\frac{x^2}{1+x^2}+u, \tag{10.46}$$

where the rapidly varying parameter p is unknown and u is a controller to be specified. Our only assumptions on p are that (a) there is a known constant $a_m>0$ so that $|p(l)|\leq a_m$ for all l and (b) there is a known constant $T>0$ such that Condition 2. from Assumption 10.1 holds for all $k\in\mathbb{Z}$. Let us show that the saturated state controller

$$u=-2\arctan(x) \tag{10.47}$$

makes (10.46) UGAS to the origin. For concreteness, we take $a_m=10$. The time derivative of $V(x)=\frac{1}{2}x^2$ along the trajectories of (10.46), in closed-loop with (10.47), is then

$$\dot{V}=-2x\arctan(x)+p(\alpha t)\frac{x^3}{1+x^2}. \tag{10.48}$$

Using simple calculations, we can check the assumptions of Theorem 10.13 for this closed-loop system, by taking

$$W(t,x)\equiv 2x\arctan(x),\quad \Theta(t,x)\equiv\frac{x^3}{1+x^2},$$

and a small enough constant $c > 0$. Hence, (10.47) renders (10.46) UGAS to zero, and the closed-loop system has the strict Lyapunov function

$$U^{[\alpha]}(t,x) = \frac{1}{2}x^2 - \frac{1}{T}\left(\int_{t-T}^{t}\int_{s}^{t} p(\alpha l)\,\mathrm{d}l\,\mathrm{d}s\right)\frac{x^3}{1+x^2} \tag{10.49}$$

when the constant $\alpha > 0$ is sufficiently large. A similar argument can be used to show that $u = -2\arctan(Rx)$ stabilizes (10.46) for any choice of the constant $R > 1$.

10.7.2 Dynamics that Are Not Globally Lipschitz

One easily shows that the one-dimensional dynamics

$$\dot{x} \;=\; -x^3 + 10\cos(\alpha t)\frac{x^3}{1+x^2} \tag{10.50}$$

satisfies Assumption 10.1 with

$$\begin{aligned} &V(t,x) \equiv \tfrac{x^4}{4}, \quad W(t,x) \equiv x^6, \\ &\Theta(t,x) \equiv \tfrac{x^6}{1+x^2}, \quad p(t) \doteq 10\cos(t), \quad T \doteq 2\pi, \end{aligned}$$

and a small enough constant $c > 0$. Therefore, (10.50) has the global strict Lyapunov function

$$U^{[\alpha]}(t,x) = \frac{x^4}{4} - \frac{1}{2\pi}\left(\int_{t-2\pi}^{t}\int_{s}^{t} 10\cos(\alpha l)\,\mathrm{d}l\,\mathrm{d}s\right)\frac{x^6}{1+x^2}$$

when the constant $\alpha > 0$ is large enough. On the other hand, (10.50) is not covered by Theorem 10.11, because it is not globally Lipschitz in the state.

10.8 Comments

There is a significant literature on averaging-based approaches in nonlinear control. Standard results on averaging approximate solutions of systems of the form

$$\dot{x} = \varepsilon f(t,x,\varepsilon) \tag{10.51}$$

for small positive constants ε by solving a second system that is obtained by averaging $f(t,x,\varepsilon)$ around $\varepsilon = 0$ [70]. More precisely, (10.51) is assumed to be periodic for some period T in t, and then the averaged system is $\dot{x} = \varepsilon f_{\mathrm{av}}(x)$ where

$$f_{\rm av}(x) = \frac{1}{T}\int_0^{\top} f(\tau, x, 0)\mathrm{d}\tau.$$

The averaging results in [2, 3] say that exponential stability of an appropriate *autonomous* dynamic $\dot{x}(t) = \bar{f}(x(t))$ guarantees that a given time-varying system $\dot{x}(t) = f(t, x(t))$ is exponentially stable as well, provided that the time variation of the latter is sufficiently fast. Partial averaging is one generalization of averaging. A different generalization in [135] shows that the rapid time variation assumption in averaging can be replaced by a homogeneity condition on the vector fields.

Yet another approach to averaging was pursued in [122], which proves uniform semi-global practical ISS of a general class of time-varying systems, under the assumption that the strong average of the system is ISS. A strong average of a system $\dot{x} = f(t, x, u)$ is defined to be a locally Lipschitz function $f_{\rm sa} : \mathbb{R}^n \times \mathbb{R}^m \to \mathbb{R}^n$ for which there exist a function $\beta \in \mathcal{KL}$ and a constant $T^* > 0$ such that

$$\left|\frac{1}{T}\int_t^{t+T} \left[f_{\rm sa}\big(x, w(s)\big) - f\big(s, x, w(s)\big)\right] \mathrm{d}s\right| \le \beta\big(\max\{|x|, |w|_\infty, 1\}, T\big)$$

for all $t \ge 0$, $w \in \mathcal{M}(\mathbb{R}^m)$, and $T \ge T^*$. However, this earlier work does not provide explicit strict Lyapunov functions.

The system

$$\dot{x} = -\bar{m}(t)\bar{m}^\top(t)x \tag{10.52}$$

naturally arises in adaptive identification. In that context, one is given a stable plant transfer function

$$\Pi(s) = \sum_{i=1}^{n} b_i s^{i-1}\left(s^n + \sum_{i=1}^{n} a_i s^{i-1}\right)^{-1}$$

with unknown a_i's and b_i's but known n, and no zero-pole cancelations. The problem is to find the a_i's and b_i's, based on input and output measurements for the plant. This problem is solvable when (10.52) is exponentially stable for a suitable vector $\bar{m}$; see e.g., Appendix A-I in [4]. Sufficient conditions for exponential stability of (10.52) are well-known. For example, [4, Theorem 1] establishes that the system is exponentially stable if and only if the PE in our Condition 2.i. from Assumption 10.2 holds for some positive constants α', β', and $\tilde{c}$, under the assumption that m is regulated (meaning its finite one-sided limits exist everywhere).

The paper [136] extends [2, 3] to general classes of rapidly time-varying systems, by proving a general exponential stability theorem involving the mixed partial derivatives of a Lyapunov function for the limiting dynamics. Our system (10.35) in Sect. 10.6.1 is a variant of the scalar example on [136, p.53]. The conditions on the σ_i's in (10.35) cannot be omitted even if the limiting dynamics (10.6) is UGES [136, Sect. 8.2]. For example, if we take $\sigma_1(x) = x$ and $\sigma_2(x) = x^2$, then (10.6) is UGES, but (10.35) can only shown to be *locally* exponentially stable for large $\alpha > 0$ [136]. This does not contradict our theorem because in that situation, (10.18) would be violated.

The results of [136] assume that (10.6) is uniformly locally exponentially stable, and establish exponential stability of (10.37) under our Assumption 10.2 (but with $g \equiv 0$). This generalized a result from [55, pp. 190-5] on a class of systems (10.5) satisfying certain periodicity or almost periodicity conditions. By contrast, we assumed in this chapter that (10.6) was UGAS. Although global exponential and global asymptotic stabilities are equivalent for *autonomous* systems under a coordinate change in certain dimensions, the coordinate changes are not explicit and so do not lend themselves to explicit Lyapunov function constructions [50]. The novelty of our treatment is the simple direct construction of a strict global Lyapunov function under assumptions similar to those of [136, Theorem 3].

The requirement (10.15) on $N \in \mathcal{M}$ from Theorem 10.11 can be relaxed the requirement that $\sup_\eta \eta N(\eta) < \bar{c}/4$, where $\bar{c}$ is from Condition 1. from Definition 10.1. This follows from (10.24).

Chapter 11
Slowly Time-Varying Systems

Abstract In Chap. 10, we discussed methods for building explicit global strict Lyapunov functions for rapidly time-varying systems. We turn next to the complementary problem of explicitly constructing strict Lyapunov functions for *slowly* time-varying continuous time systems. As in the case of rapidly time-varying systems, slowly time-varying systems involve two continuous time scales, one faster than the other. However, the methods for constructing strict Lyapunov functions for rapidly time-varying systems do not lend themselves to slowly time-varying systems, so our techniques in this chapter are completely different from the ones in Chap. 10. Instead of using limiting dynamics or averaging, we use a "frozen dynamics" approach, whereby Lyapunov functions for the corresponding frozen dynamics are used to build strict Lyapunov functions for the original slowly time-varying dynamics. We illustrate our results using friction and pendulum models.

11.1 Motivation

To motivate our results for slowly time-varying systems, consider the following generalized pendulum dynamics:

$$\left\{\begin{array}{l}\dot{x}_1 = x_2\\ \dot{x}_2 = -x_1 - [1 + b_2(t/\alpha)\tilde{m}(t,x)]x_2\ ,\end{array}\right. \tag{11.1}$$

where:

Assumption 11.1 *The following are satisfied:*

1. $\tilde{m} : \mathbb{R} \times \mathbb{R}^2 \to [0,1]$ *is Lipschitz in* x *and continuous; and*
2. *there are constants* $T > 0$ *and* $\underline{c} > -T$ *such that* $5\int_{t-T}^{t} b_2(l)\,\mathrm{d}l \geq \underline{c}$ *for all* $t \in \mathbb{R}$*, and* $b_2 : \mathbb{R} \to (-\infty, 0]$ *is globally bounded.*

The constant parameter $\alpha > 1$ is used to produce the slower time scale t/α.

Motivated by our work in Chap. 10, it is natural to ask the following: *Can we find a constant* $\underline{\alpha} > 0$ *and design a class of functions* $V_{\alpha}^{\sharp}$ *such that for each constant* $\alpha \geq \underline{\alpha}$, $V_{\alpha}^{\sharp}$ *is a global strict Lyapunov function for (11.1)?* We will answer this question in the affirmative in Sect. 11.6.2. While the model (11.1) has been studied by several authors, e.g., in [137], the earlier methods do not address this Lyapunov function construction problem. Moreover, the rapidly time-varying system that would arise from transforming the system does not produce a dynamics of the type covered by Chap. 10; see Remark 11.1. Rather, our construction of the $V_{\alpha}^{\sharp}$'s will follow from a general, constructive Lyapunov function theory from [103] that is specifically designed for slowly time-varying systems. We turn to this general theory next.

11.2 Overview of Methods

This chapter is devoted to the study of nonlinear slowly time-varying systems of the form

$$\dot{x} = f(t, t/\alpha, x) \tag{11.2}$$

for large values of the constant $\alpha > 1$ (but see Sect. 11.7 for the extension to systems with controls). See Sect. 11.3 for our standing assumptions on (11.2). Such systems arise in a large variety of important engineering applications [70, 137, 154], so it is important to develop methods for determining whether slowly time-varying systems are UGAS with respect to their equilibria. Even if (11.2) is known to be UGAS, it is still important to have general methods for constructing explicit closed form Lyapunov functions for (11.2), e.g., to quantify the effects of uncertainty in the model; see [5, 7, 8, 98, 102, 108] for discussions on the essentialness of explicit global strict Lyapunov functions for robustness analysis.

One well-known approach is to first show exponential stability of the corresponding "frozen dynamics"

$$\dot{x} = f(t, \tau, x) \tag{11.3}$$

for suitable values of the parameter τ, including cases where the exponent in the exponential decay estimate can take both positive and negative values for different τ values [70, 137, 154]. The stability of the frozen dynamics then leads to a proof of stability of (11.2). While useful in some applications, this standard approach is of limited value for robustness analysis, because it does not lead to explicit strict Lyapunov functions for (11.2).

The main goals of this chapter are (a) to show how to relax the exponential-like stability assumptions on (11.3) and allow cases where τ is a vector, thereby enlarging the class of dynamics to which the frozen dynamics method can be applied and (b) to show how to use a suitable class of oftentimes readily available Lyapunov functions for (11.3) to build explicit global strict

Lyapunov functions for (11.2). In general, the strict Lyapunov functions we construct for (11.2) comprise a *class* of functions, parameterized by the constant $\alpha > 0$. In some cases, we can construct strict Lyapunov functions for all values of $\alpha > 0$, while in other cases the construction is only valid when α is sufficiently large; we illustrate both of these possibilities in Sect. 11.6. We also show how to extend the results to systems with disturbances, in which case the strict Lyapunov functions we construct are ISS Lyapunov functions for the corresponding perturbed slowly time-varying dynamics.

11.3 Assumptions and Lemmas

Recall the comparison function classes $\mathcal{K}_\infty$ and $\mathcal{KL}$ we introduced in Chap. 1. For functions $r \mapsto p(r) \in \mathbb{R}^d$ with differentiable components, we let $p'(r)$ denote the vector $(p_1'(r), ..., p_d'(r))$. When ρ is defined on $[0, \infty)$, we interpret $\rho'(0)$ as a one sided derivative, and continuity at 0 as one sided continuity.

We assume that all of our uncontrolled dynamics

$$\dot{x} = h(t, x) \tag{11.4}$$

(11.4) are sufficiently smooth, forward complete and *decrescent* (a.k.a. *uniformly state bounded*), meaning there exists $\alpha_h \in \mathcal{K}_\infty$ such that $|h(t,x)| \leq \alpha_h(|x|)$ everywhere. In what follows, we often omit the arguments in our functions when they are clear from the context. Also, all (in)equalities should be interpreted to hold globally unless otherwise indicated.

11.4 Main Lyapunov Construction

For simplicity, we assume that our system (11.2) has the form

$$\dot{x} = f\big(t, p(t/\alpha), x\big) \ , \tag{11.5}$$

where $p : \mathbb{R} \to \mathbb{R}^d$ is bounded, d is any positive integer, $\alpha > 1$ is a constant called the *slowness parameter*, and the components $p_1, \ldots, p_d$ of p have bounded first derivatives. We set

$$\bar{p} \doteq \sup\{|p'(r)| : r \in \mathbb{R}\} \tag{11.6}$$

and

$$\mathcal{R}(p) \doteq \{p(t) : t \in \mathbb{R}\}. \tag{11.7}$$

Our main assumption is as follows:

Assumption 11.2 *There are known functions $\alpha_1, \alpha_2 \in \mathcal{K}_\infty$; known positive constants c_a, c_b, and T; a continuous function $q : \mathbb{R}^d \to \mathbb{R}$; and a C^1 function $V : [0,\infty) \times \mathbb{R}^d \times \mathbb{R}^n \to [0,\infty)$ satisfying the following conditions:*

1. $\alpha_1(|x|) \le V(t,\tau,x) \le \alpha_2(|x|)$;
2. $V_t(t,\tau,x) + V_x(t,\tau,x)f(t,\tau,x) \le -q(\tau)V(t,\tau,x)$;
3. $|V_\tau(t,\tau,x)| \le c_a V(t,\tau,x)$; *and*
4. $\int_{t-T}^{t} q(p(s))\mathrm{d}s \ge c_b$

hold for all $t \ge 0$, $x \in \mathbb{R}^n$, and $\tau \in \mathcal{R}(p)$.

Condition 2. is less restrictive than the standard exponential stability property of the frozen dynamics, because (a) we do not require α_1 or α_2 to be quadratic functions and (b) $q(\tau)$ can take non-positive values for some choices of τ. On the other hand, Condition 4. roughly says that q is positive on average, along the vector function $p(s)$. The following was shown in [103]:

Theorem 11.1. *If (11.5) satisfies Assumption 11.2, then for each positive constant*

$$\alpha > \frac{2Tc_a\bar{p}}{c_b}, \tag{11.8}$$

the dynamics (11.5) are UGAS and

$$V^\sharp_\alpha(t,x) \doteq \exp\left(\frac{\alpha}{T}\int_{\frac{t}{\alpha}-T}^{\frac{t}{\alpha}}\int_s^{\frac{t}{\alpha}} q(p(l))\mathrm{d}l\,\mathrm{d}s\right) V\big(t,p(t/\alpha),x\big) \tag{11.9}$$

is a global strict Lyapunov function for (11.5).

Proof. Sketch. Consider the function

$$\hat{V}(t,x) \doteq V\big(t,p(t/\alpha),x\big). \tag{11.10}$$

By Conditions 2. and 3. in Assumption 11.2 and our choice of $\bar{p}$, the time derivative of $\hat{V}$ along the trajectories of (11.5) is

$$\begin{aligned}\dot{\hat{V}} &= V_t(t,p(t/\alpha),x) + V_x(t,p(t/\alpha),x)f(t,p(t/\alpha),x)\\ &\quad + V_\tau(t,p(t/\alpha),x)\frac{p'(t/\alpha)}{\alpha}\\ &\le \left[-q\big(p(t/\alpha)\big) + \frac{c_a\bar{p}}{\alpha}\right]\hat{V}(t,x).\end{aligned} \tag{11.11}$$

Set

$$E(t,\alpha) \doteq \exp\left(\frac{\alpha}{T}\int_{\frac{t}{\alpha}-T}^{\frac{t}{\alpha}}\int_s^{\frac{t}{\alpha}} q(p(l))\,\mathrm{d}l\,\mathrm{d}s\right). \tag{11.12}$$

Then

$$V_\alpha^\sharp(t,x) \;=\; E(t,\alpha)\hat{V}(t,x)$$

everywhere, and Lemma 10.2 applied with the integrand $q(p(t))$ gives

$$\begin{aligned}&\dot{V}_\alpha^\sharp = E(t,\alpha)\left[\dot{\hat{V}} + \left\{q(p(t/\alpha)) - \tfrac{1}{T}\int_{\frac{t}{\alpha}-T}^{\frac{t}{\alpha}} q(p(l))\mathrm{d}l\right\}\hat{V}\right] \text{ and}\\ &e^{\alpha T\bar{M}/2} \;\geq\; E(t,\alpha) \;\geq\; e^{-\alpha T\bar{M}/2}\end{aligned} \tag{11.13}$$

everywhere, where $\bar{M}$ is any global bound on $q(p(t))$. Substituting (11.11) and recalling Condition 4. from Assumption 11.2 now gives

$$\begin{aligned}\dot{V}_\alpha^\sharp &\leq E(t,\alpha)\left[\frac{c_a\bar{p}}{\alpha} - \frac{1}{T}\int_{\frac{t}{\alpha}-T}^{\frac{t}{\alpha}} q(p(l))\mathrm{d}l\right]\hat{V}\\ &\leq E(t,\alpha)\left[\frac{c_a\bar{p}}{\alpha} - \frac{c_b}{T}\right]\hat{V}(t,x).\end{aligned} \tag{11.14}$$

It follows that if the constant α satisfies (11.8), then (11.14) implies

$$\dot{V}_\alpha^\sharp(t,x) \;\leq\; -\frac{c_b}{2T}e^{-\alpha T\bar{M}/2}\hat{V}(t,x) \leq -\alpha_3(|x|), \tag{11.15}$$

where the function

$$\alpha_3(s) = \frac{c_b}{2T}e^{-\alpha T\bar{M}/2}\alpha_1(s)$$

is positive definite. Since $V_\alpha^\sharp$ is also uniformly proper and positive definite, the result readily follows. □

Remark 11.1. A different approach to slowly time-varying systems $\dot{x} = f(t,t/\alpha,x)$ would be to transform them into rapidly time-varying systems and to then try to construct a Lyapunov function for the resulting rapidly time-varying systems using the methods we gave in Chap. 10. The transformation is done by taking $y(s) \doteq x(\alpha s)$, which produces the new rapidly time-varying system

$$y'(s) = g(s,\alpha s,y(s)) \doteq \alpha f(s\alpha,s,y(s)) \tag{11.16}$$

with the new rescaled time variable s. However, the system (11.16) is not of the form we saw in Chap. 10, so the earlier strict Lyapunov functions do not apply. This is one motivation for our direct construction of strict Lyapunov functions for slowly time-varying systems.

11.5 More General Decay Rates

We can relax requirements of Theorem 11.1 by assuming the following (where we continue the notation from Sect. 11.4):

Assumption 11.3 *There are known functions* $\tilde{\alpha}_1, \tilde{\alpha}_2 \in \mathcal{K}_\infty$ *a positive definite* C^1 *function* μ; *positive constants* T, $\tilde{c}_a$, *and* $\tilde{c}_b$; *a continuous function* $\tilde{q} : \mathbb{R}^d \to \mathbb{R}$; *and a* C^1 *function* $\tilde{V} : [0, \infty) \times \mathbb{R}^d \times \mathbb{R}^n \to [0, \infty)$ *satisfying*

$$\lim_{r \to +\infty} \int_1^r \frac{1}{\mu(l)} \mathrm{d}l = +\infty \tag{11.17}$$

and

1. $\tilde{\alpha}_1(|x|) \le \tilde{V}(t, \tau, x) \le \tilde{\alpha}_2(|x|)$;
2. $\tilde{V}_t(t, \tau, x) + \tilde{V}_x(t, \tau, x) f(x, t, \tau) \le -\tilde{q}(\tau)\mu(\tilde{V}(t, \tau, x))$;
3. $|\tilde{V}_\tau(t, \tau, x)| \le \tilde{c}_a \mu(\tilde{V}(t, \tau, x))$; *and*
4. $\int_{t-T}^t \tilde{q}(p(s)) \mathrm{d}s \ge \tilde{c}_b$

hold for all $x \in \mathbb{R}^n$, $t \ge 0$, *and* $\tau \in \mathcal{R}(p)$.

Assumption 11.2 is the special case of Assumption 11.3 where $\mu(l) \equiv l$. Using Theorem 11.1, one can prove the following:

Theorem 11.2. *Let (11.5) satisfy Assumption 11.3. Then we can construct a function* $k \in \mathcal{K}_\infty$ *such that the requirements of Assumption 11.2 are satisfied with* $V \doteq k(\tilde{V})$. *Hence, for each sufficiently large choice of the constant* $\alpha > 1$, *the system (11.5) is UGAS and has a global strict Lyapunov function of the form (11.9).*

This follows by taking

$$k(r) = \begin{cases} \exp\left(2B \int_1^r \frac{1}{\mu(l)} \mathrm{d}l\right), & r > 0 \\ 0, & r = 0 \end{cases} \tag{11.18}$$

with the choice $B \doteq \sup\{\mu'(s) : 0 \le s \le 1\}$. The fact that (11.18) satisfies the requirements is a consequence of

$$\lim_{r \to 0^+} \int_1^r \frac{1}{\mu(l)} \mathrm{d}l = -\infty. \tag{11.19}$$

See [103] for the full proof.

11.6 Illustrations

In general, our strict Lyapunov functions are only when the constant parameter $\alpha > 0$ in (11.5) is sufficiently large. However, in certain cases, the functions V from Assumption 11.2 are independent of the frozen parameter τ, so our proof of Theorem 11.1 shows that the strict Lyapunov function is valid for all constants $\alpha > 1$. We illustrate this phenomenon in two examples. We then provide slowly time-varying dynamics from identification theory and friction analysis where V depends on τ, and where we can therefore only guarantee UGAS for large values of α. Throughout this section, we set

$$\dot{V}(t,\tau,x) \doteq V_t(t,\tau,x) + V_x(t,\tau,x)f(t,\tau,x)$$

everywhere.

11.6.1 A Scalar Example

Consider the one-dimensional dynamics

$$\dot{x} = \frac{x}{\sqrt{1+x^2}}\left[1 - 90\cos^2\left(\frac{t}{\alpha}\right)\right] \tag{11.20}$$

and the uniformly proper and positive definite function

$$V(t,\tau,x) \equiv \bar{V}(x) \doteq e^{\sqrt{1+x^2}} - e. \tag{11.21}$$

As noted in [103], Assumption 11.2 is satisfied for this V using the frozen dynamics

$$\dot{x} = f(t,\tau,x) \doteq \frac{x}{\sqrt{1+x^2}}\left[1 - 90\tau\right], \quad 0 \le \tau \le 1. \tag{11.22}$$

To see why, notice that

$$\dot{V}(t,\tau,x) = e^{\sqrt{1+x^2}}\frac{x^2}{1+x^2} - 90\tau e^{\sqrt{1+x^2}}\frac{x^2}{1+x^2}. \tag{11.23}$$

Simple calculations give

$$\frac{2e^{\sqrt{2}}}{e-1}\bar{V}(x) \ \ge\ \frac{x^2}{1+x^2}e^{\sqrt{1+x^2}} \ \ge\ \frac{1}{2}\bar{V}(x)\,, \tag{11.24}$$

so (11.23) gives

$$\dot{V}(t,\tau,x) \le \left[\frac{2e^{\sqrt{2}}}{e-1} - 45\tau\right]\bar{V}(x). \tag{11.25}$$

Also,

$$\int_{t-\pi}^{t}\left[45\cos^2(s) - \frac{2e^{\sqrt{2}}}{e-1}\right]\mathrm{d}s \ =\ \pi\left(\frac{45}{2} - \frac{2e^{\sqrt{2}}}{e-1}\right) > 0$$

for each $t \ge 0$.

Hence, Theorem 11.1 implies that for large enough constants $\alpha > 1$, the dynamics (11.20) is UGAS, with the global strict Lyapunov function

$$\begin{aligned}&\exp\left(\frac{\alpha}{\pi}\int_{\frac{t}{\alpha}-\pi}^{\frac{t}{\alpha}}\int_{s}^{\frac{t}{\alpha}}\left[45\cos^2(l)-\frac{2e^{\sqrt{2}}}{e-1}\right]\mathrm{d}l\,\mathrm{d}s\right)\bar{V}(x)\\ &=\exp\left(45\frac{\alpha}{4}\left[\sin\left(\frac{2t}{\alpha}\right)+\pi-\frac{4\pi e^{\sqrt{2}}}{45(e-1)}\right]\right)[e^{\sqrt{1+x^2}}-e]\end{aligned} \tag{11.26}$$

where $\bar{V}$ is in (11.21). In this case V does not depend on τ, so the proof of Theorem 11.1 says that for any constant $\alpha > 0$, the system (11.20) is UGAS and has the strict Lyapunov function (11.26). Notice that (11.20) is not globally exponentially stable, because its vector field is bounded in norm by 91.

11.6.2 Pendulum Example Revisited

We now apply our construction to the pendulum example (11.1), under Assumption 11.1. For convenience, we express the inequality in the second part of Assumption 11.1 in the following equivalent way: There are constants $T, c_b > 0$ such that

$$T+5\int_{t-T}^{t} b_2(l)\,\mathrm{d}l \geq c_b \quad \forall t \in \mathbb{R}.$$

We denote the corresponding frozen dynamics by

$$f(t,\tau,x) \doteq \begin{pmatrix} x_2 \\ -x_1-[1+\tau\tilde{m}(x,t)]x_2 \end{pmatrix}.$$

To build global strict Lyapunov functions for (11.1) for large values of the constant $\alpha > 0$, we use the following simple observation from [103]:

Lemma 11.1. *The function*

$$V(x) \doteq x_1^2 + x_2^2 + x_1x_2 \tag{11.27}$$

satisfies $\nabla V(x)f(t,\tau,x) \leq -[1+5\tau]V(x)$ *for all* $x \in \mathbb{R}^n$, $t \in \mathbb{R}$, *and* $\tau \leq 0$.

Proof. Sketch. It is easily checked that

$$\nabla V(x)f(t,\tau,x) = -V(x) - 2\tau\tilde{m}(x,t)x_2^2 - \tau\tilde{m}(x,t)x_1x_2 \tag{11.28}$$

everywhere, by grouping terms. The desired estimate now readily follows because

$$V(x) \geq x_1^2 + x_2^2 - |x_1 x_2| \geq \frac{1}{2}x_1^2 + \frac{1}{2}x_2^2 \geq |x_1 x_2|$$

everywhere, using the fact that $0 \leq \tilde{m}(x,t) \leq 1$ everywhere. □

The following is an immediate consequence of Lemma 11.1, the proof of Theorem 11.1, and the fact that (11.27) only depends on x:

Theorem 11.3. *Let (11.1) satisfy Assumption 11.1. Then (11.1) has the global strict Lyapunov function*

$$V_\alpha^\sharp(t,x) \doteq \exp\left(\frac{5\alpha}{T}\int_{\frac{t}{\alpha}-T}^{\frac{t}{\alpha}}\int_s^{\frac{t}{\alpha}} b_2(l)\mathrm{d}l\,\mathrm{d}s\right)[x_1^2 + x_2^2 + x_1 x_2] \qquad (11.29)$$

for each constant $\alpha > 0$. Therefore, (11.1) is UGAS for all constants $\alpha > 0$.

11.6.3 Friction Example Revisited

We now illustrate Theorem 11.1 using the one degree-of-freedom mass-spring system from Chap. 10, which arises in the control of mechanical systems in the presence of friction. However, in contrast with Chap. 10, where the mass-spring dynamics are rapidly time-varying, here we consider the case where the dynamics are *slowly* time-varying. While slowly time-varying dynamics can be transformed into rapidly time-varying dynamics by rescaling time, doing so for the slowly time-varying mass-spring system produces a new dynamic that does not lend itself to the known methods; see Remark 11.1.

The dynamics are

$$\begin{cases} \dot{x}_1 = x_2 \\ \dot{x}_2 = -\sigma_1(t/\alpha)x_2 - k(t)x_1 \\ \qquad - \left\{\sigma_2(t/\alpha) + \sigma_3(t/\alpha)e^{-\beta_1\mu(x_2)}\right\}\mathrm{sat}(x_2)\,, \end{cases} \qquad (11.30)$$

where as in Chap. 10, x_1 and x_2 are the mass position and velocity, respectively; $\sigma_i : [0,\infty) \to (0,1]$, $i = 1,2,3$ denote positive time-varying viscous, Coulomb, and static friction related coefficients, respectively; β_1 is a positive constant corresponding to the Stribeck effect; $\mu(\cdot)$ is a positive definite function also related to the Stribeck effect; k denotes a positive time-varying spring stiffness related coefficient; and sat is again defined by (10.3), so

$$|\mathrm{sat}(x_2)| \leq \beta_2|x_2| \ \ \forall x_2 \in \mathbb{R} \qquad (11.31)$$

for a suitable constant $\beta_2 > 0$. We now assume that the friction coefficients vary in time more *slowly* than the spring stiffness coefficient, so we restrict to cases where $\alpha > 1$. We prove stability of (11.30) and construct corresponding global strict Lyapunov functions V_α when the constant $\alpha > 0$ is sufficiently large. Later in this chapter, we use our Lyapunov approach to quantify the effects of uncertainty in the model using ISS.

Our precise mathematical assumptions on (11.30) are: k and the σ_i's are C^1 with bounded derivatives; μ has a bounded first derivative; and there are positive constants c_b and T satisfying

$$\int_{t-T}^{t} \sigma_1(r)\mathrm{d}r \geq c_b \quad \forall t \geq 0. \tag{11.32}$$

We also assume this additional condition whose physical implication is that the spring stiffness is non-increasing in time: $\exists k_0, \bar{k} > 0$ such that $k_0 \leq k(t) \leq \bar{k}$ and $k'(t) \leq 0$ for all $t \geq 0$.

The frozen dynamics $\dot{x} = f(x, t, \tau)$ for (11.30) are

$$\begin{aligned} \dot{x}_1 &= x_2 \\ \dot{x}_2 &= -\tau_1 x_2 - k(t)x_1 - \left\{\tau_2 + \tau_3 e^{-\beta_1 \mu(x_2)}\right\} \mathrm{sat}(x_2)\,, \end{aligned} \tag{11.33}$$

where $\tau = (\tau_1, \tau_2, \tau_3) \in [0,1]^3$ is now a *vector* of parameters. As noted in [103], Theorem 11.1 applies with

$$p(t) = \big(\sigma_1(t), \sigma_2(t), \sigma_3(t)\big)$$

and the function

$$V(t, \tau, x) = A\big(k(t)x_1^2 + x_2^2\big) + \tau_1 x_1 x_2, \tag{11.34}$$

where

$$A = 1 + \frac{k_0}{2} + \frac{(1 + 2\beta_2)^2}{k_0}.$$

To see why, let us first check the conditions of Assumption 11.2. Since

$$A \geq \max\left\{1, \frac{1}{k_0}\right\}$$

and $\tau_1 \leq 1$, we have

$$\frac{1}{2}(x_1^2 + x_2^2) \leq V(t, \tau, x) \leq 2A^2\bar{k}|x|^2 \tag{11.35}$$

everywhere. We now compute $\dot{V}(x, t, \tau)$ for all values $\tau \in [0,1]^3$, along the trajectories of the frozen dynamics (11.33). Since $k'(t) \leq 0$ everywhere,

$$\begin{aligned} \dot{V}(t, \tau, x) &\leq V_x(t, \tau, x) f(t, \tau, x) \\ &= [2Ak(t)x_1 + \tau_1 x_2]x_2 \\ &\quad -[2Ax_2 + \tau_1 x_1]\big\{\tau_1 x_2 + k(t)x_1 + \big[\tau_2 + \tau_3 e^{-\beta_1 \mu(x_2)}\big]\,\mathrm{sat}(x_2)\big\}. \end{aligned}$$

It readily follows that

$$
\begin{aligned}
\dot{V}(t,\tau,x) &\leq -\tau_1 k_0 x_1^2 - (2A\tau_1 - \tau_1)x_2^2 + \tau_1(1+2\beta_2)|x_1 x_2| \\
&\leq -\tau_1 \frac{k_0}{2}|x|^2 \\
&\quad - \left[\tau_1 \frac{k_0}{2} x_1^2 + (A-1/2)\tau_1 x_2^2 - \tau_1(1+2\beta_2)|x_1 x_2|\right] \\
&= -\tau_1 \frac{k_0}{2}|x|^2 - \tau_1 \frac{k_0}{2}\left(|x_1| - \frac{1+2\beta_2}{k_0}|x_2|\right)^2 \\
&\quad + \left(\frac{\tau_1(1+2\beta_2)^2}{2k_0} + \frac{\tau_1}{2} - A\tau_1\right) x_2^2 \\
&\leq -\frac{\tau_1 k_0}{4A^2 \bar{k}} V(t,\tau,x)\,.
\end{aligned}
\tag{11.36}
$$

Hence, Assumption 11.2 of Theorem 11.1 is immediate from (11.32) by choosing

$$
q(\tau) = \frac{\tau_1 k_0}{4A^2 \bar{k}} \quad \text{and} \quad c_a = 1.
$$

This gives:

Corollary 11.1. *Let the preceding assumptions hold, define V as in (11.34), and set*

$$
\bar{b} = \frac{k_0}{4A^2 \bar{k}} \quad \textit{and} \quad p(t) = \big(\sigma_1(t), \sigma_2(t), \sigma_3(t)\big).
$$

Then we can construct a constant $\alpha_0 > 0$ such that for all $\alpha > \alpha_0$,

$$
V_\alpha(t,x) \doteq \exp\left(\frac{\alpha \bar{b}}{T} \int_{\frac{t}{\alpha}-T}^{\frac{t}{\alpha}} \int_s^{\frac{t}{\alpha}} \sigma_1(l)\mathrm{d}l\,\mathrm{d}s\right) V\big(t, p(t/\alpha), x\big) \tag{11.37}
$$

is a global strict Lyapunov function for the system (11.30), which is therefore UGAS to the origin.

11.6.4 Identification Dynamics Revisited

Our strict Lyapunov function design also applies to slowly time-varying dynamics of the form

$$
\dot{x} = h(t/\alpha)\bar{m}(t)\bar{m}^\top(t)x, \quad x \in \mathbb{R}^n\,. \tag{11.38}
$$

As we noted before, the particular case

$$
\dot{x} = -\bar{m}(t)\bar{m}^\top(t)x
$$

of (11.38) has been studied by several authors in the context of identification theory [108, 136]. In the more general case (11.38), we assume:

Assumption 11.4 *There are positive constants $T, \tilde{c}, \underline{\alpha}$, and $\bar{\alpha}$, with $\bar{\alpha} \geq 1$, such that:*

1. *$h : \mathbb{R} \to [-\bar{\alpha}, 0]$ is continuous with a bounded first derivative and*

$$\int_{t-T}^{t} h(r)\mathrm{d}r \leq -\underline{\alpha}$$

for all $t \in \mathbb{R}$; and

2. *$\bar{m} : \mathbb{R} \to \mathbb{R}^n$ is continuous and satisfies $|\bar{m}(t)| = 1$ and*

$$\underline{\alpha} I_n \leq \int_{t}^{t+\tilde{c}} \bar{m}(r)\bar{m}^\top(r)\mathrm{d}r \leq \bar{\alpha} I_n$$

for all $t \in \mathbb{R}$.

In this setting, for matrices $A, B \in \mathbb{R}^{n\times n}$, $B - A \geq 0$ means $B - A$ is positive semi-definite; later we allow the more general perturbed dynamics

$$\dot{x} \;=\; h(t/\alpha)\bar{m}(t)\bar{m}^\top(t)x \;+\; g(t, t/\alpha, x)u$$

for suitable matrix-valued functions g.

In Chap. 10, we saw how to build global strict explicit Lyapunov functions for the analogous rapidly time-varying system $\dot{x} = f(\alpha t)\bar{m}(t)\bar{m}^\top(t)x$ for appropriate non-positive functions f and large positive constants α. This was done using an appropriate limiting dynamics and a variant of partial averaging. However, the rapidly time-varying Lyapunov constructions do not give explicit Lyapunov functions for the slowly time-varying dynamics (11.38); see Remark 11.1. Instead, we build strict Lyapunov functions for (11.38) using the following variant of Lemma 10.3 from Chap. 10:

Lemma 11.2. *Assume there are positive constants T, $\tilde{c}$, $\underline{\alpha}$, and $\bar{\alpha}$ such that Assumption 11.4 is satisfied. Then for each constant $\tau \in [-\bar{\alpha}, 0]$, the function*

$$V(t, \tau, x) \;=\; x^\top P(t, \tau)x \tag{11.39}$$

with the choices

$$P(t, \tau) = \kappa I - \tau \int_{t-\tilde{c}}^{t} \int_{s}^{t} \bar{m}(l)\bar{m}^\top(l)\,\mathrm{d}l\,\mathrm{d}s, \tag{11.40}$$

and

$$\kappa = \frac{\tilde{c}}{2} + \frac{\bar{\alpha}^2\tilde{c}^4}{2\underline{\alpha}} + \tilde{c}^2 \tag{11.41}$$

satisfies Assumption 11.2 for the frozen dynamics

$$\dot{x} = f(t, \tau, x) = \tau\bar{m}(t)\bar{m}^\top(t)x \tag{11.42}$$

and $p(s) = h(s)$.

Proof. Sketch. We only sketch the proof; see [103] for more details. By Lemma 10.2 and (11.39), we readily have

$$\frac{\partial V}{\partial t}(t,\tau,x) = -\tau\tilde{c}x^\top \bar{m}(t)\bar{m}^\top(t)x + \tau x^\top \left[\int_{t-\tilde{c}}^{t} \bar{m}(l)\bar{m}^\top(l)\mathrm{d}l\right] x$$

and

$$\frac{\partial V}{\partial x}(t,\tau,x)f(t,\tau,x) = \\ 2\tau x^\top \left[\kappa I - \tau \textstyle\int_{t-\tilde{c}}^{t}\int_{s}^{t} \bar{m}(l)\bar{m}^\top(l)\,\mathrm{d}l\,\mathrm{d}s\right] \bar{m}(t)\bar{m}^\top(t)x\ .$$

Hence, the time derivative of (11.39) along the trajectories of (11.42) satisfies

$$\begin{aligned}\dot{V} &\le \tau\{(2\kappa-\tilde{c})(\bar{m}^\top(t)x)^2 + \underline{\alpha}|x|^2\} + \tau^2|x||\bar{m}^\top(t)x|\tilde{c}^2 \\ &= \tau\{(2\kappa-\tilde{c})(\bar{m}^\top(t)x)^2 + \underline{\alpha}|x|^2 + \tau|x||\bar{m}^\top(t)x|\tilde{c}^2\}.\end{aligned} \tag{11.43}$$

Choosing

$$\omega = \frac{\underline{\alpha}}{2\tilde{c}^2\bar{\alpha}},$$

we can use the triangle inequality to get

$$|\bar{m}^\top(t)x||x| \ \le\ \omega|x|^2 + \frac{1}{4\omega}|\bar{m}^\top(t)x|^2\ .$$

Therefore, the fact that $\tau \le 0$ gives

$$\begin{aligned}&(2\kappa-\tilde{c})|\bar{m}^\top(t)x|^2 + \underline{\alpha}|x|^2 + \tau|x||\bar{m}^\top(t)x|\tilde{c}^2 \\ &\ge \left[2\kappa - \tilde{c} + \frac{\tau\tilde{c}^2}{4\omega}\right]|\bar{m}^\top(t)x|^2 + (\underline{\alpha} + \omega\tau\tilde{c}^2)|x|^2 \\ &\ge \frac{\underline{\alpha}|x|^2}{2}\end{aligned} \tag{11.44}$$

by our choices of κ and ω. Combining (11.43) and (11.44) gives

$$\dot{V} \le \frac{\tau\underline{\alpha}}{2}|x|^2$$

everywhere. Also, Lemma 10.2, our choice of κ, and the fact that $|\bar{m}(t)| \equiv 1$ combine to give $\kappa|x|^2 \le V(t,\tau,x) \le\ (\kappa+\tilde{c}^2)|x|^2$ and

$$\left|\frac{\partial V}{\partial \tau}(t,\tau,x)\right| \ \le\ \frac{\tilde{c}^2}{2}|x|^2 \le V(x,t,\tau).$$

It follows that $\dot{V}(t,\tau,x) \le -q(\tau)V(t,\tau,x)$, everywhere, where

$$q(\tau) \doteq -\frac{\tau\underline{\alpha}}{2(\kappa + \tilde{c}^2)},$$

so Assumption 11.2 is satisfied. This proves the lemma. □

Combining Lemma 11.2 and Theorem 11.1, we immediately obtain:

Theorem 11.4. *Let (11.38) admit constants $T > 0$, $\tilde{c} > 0$, $\underline{\alpha} > 0$, and $\bar{\alpha} > 1$ such that Assumption 11.4 is satisfied. Then for any constant*

$$\alpha > \frac{4T(\kappa + \tilde{c}^2)}{\underline{\alpha}^2} \sup\{|h'(r)| : r \in \mathbb{R}\}, \tag{11.45}$$

the function

$$V_{\alpha}^{\sharp}(t,x) \doteq \exp\left(-\frac{\alpha\underline{\alpha}}{2T(\kappa + \tilde{c}^2)} \int_{\frac{t}{\alpha}-T}^{\frac{t}{\alpha}} \int_{s}^{\frac{t}{\alpha}} h(l)\mathrm{d}l\,\mathrm{d}s\right) V\big(t, h(t/\alpha), x\big) \tag{11.46}$$

with the function $V(t,\tau,x) = x^\top P(t,\tau)x$ as defined in Lemma 11.2 is a global strict Lyapunov function for (11.38). In particular, (11.38) is UGAS for all constants α that satisfy (11.45).

11.7 Input-to-State Stability

As we noted in previous chapters, one important advantage of having explicit strict Lyapunov functions is that they make it possible to quantify the effects of uncertainty. We illustrate this by extending our results to slowly time-varying control affine systems

$$\dot{x} = f\big(t, p(t/\alpha), x\big) + g\big(t, p(t/\alpha), x\big)u \tag{11.47}$$

evolving on $\mathbb{R}^n$ with control values $u \in \mathbb{R}^m$, assuming $f : [0,\infty) \times \mathbb{R}^d \times \mathbb{R}^n \to \mathbb{R}^n$ and $g : [0,\infty) \times \mathbb{R}^d \times \mathbb{R}^n \to \mathbb{R}^{n\times m}$ are locally Lipschitz functions for which there exists $\alpha_4 \in \mathcal{K}_\infty$ satisfying

$$\big|f\big(t, p(t/\alpha), x\big)\big| + \big|g\big(t, p(t/\alpha), x\big)\big| \le \alpha_4(|x|)$$

everywhere, and where $p : \mathbb{R} \to \mathbb{R}^d$ for some d is globally bounded with a globally bounded first derivative. As before, the inputs for (11.47) comprise the set $\mathcal{M}(\mathbb{R}^m)$ of all measurable essentially bounded functions $\mathbf{u} : [0,\infty) \to \mathbb{R}^m$ with the essential supremum norm $|\cdot|_\infty$. Our robustness result will be based on the ISS paradigm; see Chap. 1 for the ISS related definitions.

We assume throughout this subsection that Assumption 11.2 holds for some choice of $V \in C^1$, and that:

Assumption 11.5 *The function $\alpha_1 \in \mathcal{K}_\infty$ and the constant $c_a > 0$ from Assumption 11.2 also satisfy:*

1. $|V_x(t, p(t/\alpha), x)| \le c_a \sqrt{\alpha_1(|x|)}$; *and*
2. $|g(t, p(t/\alpha), x)| \le c_a \left(1 + \sqrt[4]{\alpha_1(|x|)}\right)$

for all $t \ge 0$, $\alpha > 0$, and $x \in \mathbb{R}^n$.

In the particular case where $\alpha_1(x) = |x|^2$, Assumption 11.5 stipulates linear growth on V_x, which will be the case when V has the classical form $x^\top P(t)x$ for a suitable bounded everywhere positive definite matrix P. The following was shown in [103]:

Theorem 11.5. *Let (11.5) satisfy Assumptions 11.2 and 11.5 for some choices of c_a, V, and α_1. Then for each constant*

$$\alpha > \frac{4Tc_a\bar{p}}{c_b}, \tag{11.48}$$

the dynamics (11.47) are ISS and

$$V_\alpha^\sharp(t,x) \doteq \exp\left(\frac{\alpha}{T}\int_{\frac{t}{\alpha}-T}^{\frac{t}{\alpha}}\int_s^{\frac{t}{\alpha}} q(p(l))\mathrm{d}l\,\mathrm{d}s\right) V\big(t, p(t/\alpha), x\big) \tag{11.49}$$

is an ISS Lyapunov function for (11.47).

Proof. Sketch. The proof is analogous to the proof of Theorem 11.1; we indicate the necessary changes in the earlier proof. Define the function

$$\chi(s) \doteq \frac{c_b\sqrt{\alpha_1(s)}}{2Tc_a^2\left(1 + \sqrt[4]{\alpha_1(s)}\right)},$$

where α_1 and c_a are as in Assumption 11.2. Then $\chi \in \mathcal{K}_\infty$. Our assumptions imply that

$$\big|V_x\big(t, p(t/\alpha), x\big)g\big(t, p(t/\alpha), x\big)u\big| \le \frac{c_b\alpha_1(|x|)}{2T} \le \frac{c_b}{2T}V\big(t, p(t/\alpha), x\big)$$

if $|u| \le \chi(|x|)$. Choose $\hat{V}$ as in (11.10) and $E(t,\alpha)$ as in (11.12). Then, along any trajectory $x(t)$ of (11.47) with inputs $\mathbf{u}$ satisfying $|\mathbf{u}|_\infty \le \chi(|x(t)|)$ for all $t \ge 0$, we have

$$\dot{\hat{V}} = -q\big(p(t/\alpha)\big)\hat{V}(t,x) + \left[\frac{c_a\bar{p}}{\alpha} + \frac{c_b}{2T}\right]\hat{V}(t,x)$$

everywhere and therefore also

$$
\begin{aligned}
\dot{V}_\alpha^\sharp &\leq E(t,\alpha)\left[\frac{c_a\bar{p}}{\alpha}+\frac{c_b}{2T}-\frac{1}{T}\int_{t/\alpha-T}^{t/\alpha} q(p(l))\mathrm{d}l\right]\hat{V}(t,x) \\
&\leq E(t,\alpha)\left[\frac{c_a\bar{p}}{\alpha}-\frac{c_b}{2T}\right]\hat{V}(t,x) \\
&\leq -\frac{c_b E(t,\alpha)}{4T}\hat{V}(t,x)
\end{aligned}
\tag{11.50}
$$

when (11.48) is satisfied, by reasoning exactly as in the earlier proof. This gives the necessary ISS Lyapunov function decay condition. Since systems admitting ISS Lyapunov functions are ISS, the result follows. □

Theorem 11.5 readily applies to our friction example from Sect. 11.6.3, allowing us to conclude that (11.37) is an ISS Lyapunov function for the slowly time-varying controlled friction dynamic

$$
\begin{cases}
\dot{x}_1 = x_2 \\
\dot{x}_2 = -\sigma_1(t/\alpha)x_2 - k(t)x_1 + g(t,t/\alpha,x)u \\
\qquad - \left\{\sigma_2(t/\alpha) + \sigma_3(t/\alpha)e^{-\beta_1\mu(x_2)}\right\}\operatorname{sat}(x_2)
\end{cases}
\tag{11.51}
$$

for any g satisfying the requirements of Assumption 11.5 with $\alpha_1(s) = s^2/2$, assuming the constants $c_a > 0$ and $\alpha > 0$ are sufficiently large.

To illustrate the ISS property, we simulated (11.51) with the choices

$$
\begin{aligned}
&\alpha = 100,\ \ \sigma_1(t) = \sigma_2(t) = \sigma_3(t) = k(t) = 1 + e^{-t}, \\
&\beta_1 = 1,\ \ \mu(x_2) \doteq \arctan^2(x_2),\ \ \operatorname{sat}(x_2) = \tanh(10x_2), \\
&g(t,t/\alpha,t) \equiv 1,\ \ \text{and}\ \ u \equiv 0.05.
\end{aligned}
\tag{11.52}
$$

We report the results in Fig. 11.1. Consistent with the ISS estimate, $x_1(t)$ converges to 0 with an overflow depending on the magnitude of the disturbance. Moreover, just as in the rapidly time-varying case, we can use our strict Lyapunov function to explicitly construct the functions $\beta \in \mathcal{KL}$ and $\gamma \in \mathcal{K}_\infty$ in the ISS estimate, which makes it possible to precisely quantify the effects of the uncertainty. Similar ISS results can be obtained for the identification and pendulum examples we considered above.

11.8 Comments

The stability of slowly time-varying systems is a classical topic. A first result seems to have been [143], which shows that

$$
\dot{x}(t) = A(t)x(t) \tag{11.53}
$$

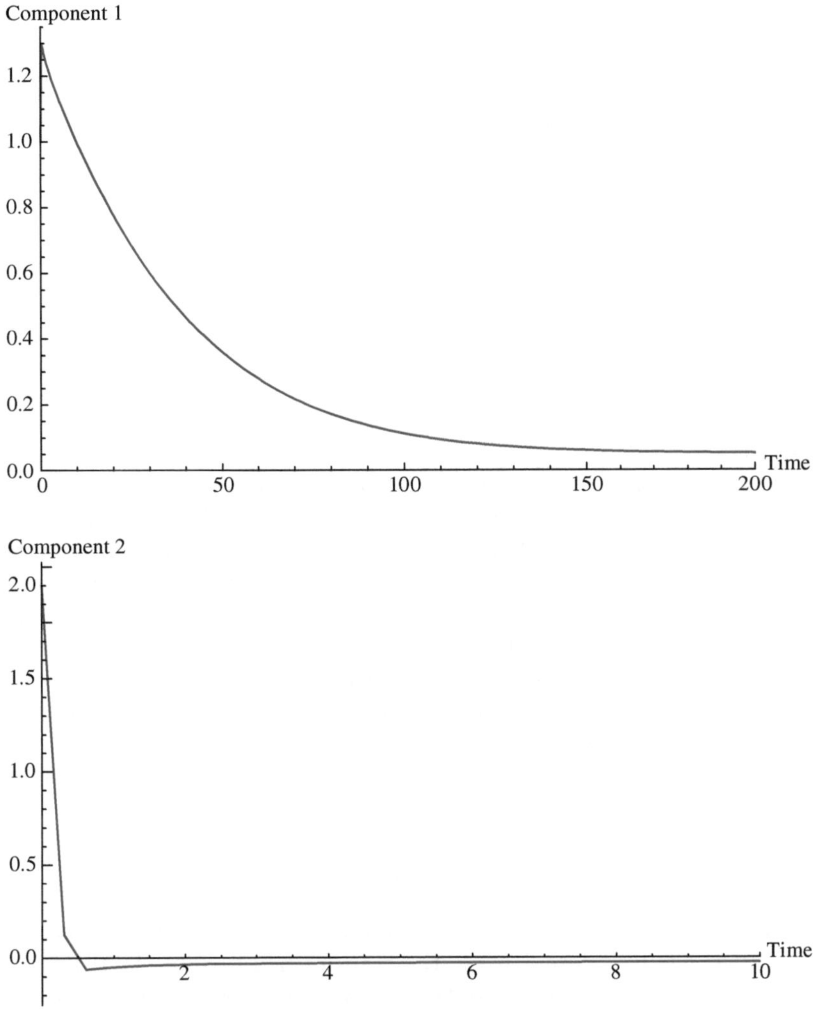

Fig. 11.1 State of (11.51) with choices (11.52)

is UGAS provided:

Assumption 11.6 *The following conditions hold:*

1. *the bounded matrix A admits a constant $\sigma_0 > 0$ such that* $\mathrm{Re}\lambda_i[A(t)] \leq -2\sigma_0$ *for all i and t; and*
2. $a_M \doteq \sup_{t\geq 0} ||\dot{A}(t)||$ *is sufficiently small.*

Later, [37] noted that Assumption 11.6 implies that (11.53) admits the global Lyapunov function

$$V(t,x) = x^\top [\varepsilon_1 I + P(t)]x$$

when $\varepsilon_1 > 0$ is a small enough constant and P is computed from the Riccati equation

$$A^\top(t)P(t) + P(t)A(t) = -3I.$$

In [26], Coppel showed that (11.53) is exponentially stable provided Condition 1. from Assumption 11.6 holds and

$$||\dot{A}|| < \frac{\sigma_0^2}{(4M\ln(M))^2},$$

where M satisfies an estimate of the form

$$\big|\big|\exp\big(\tau A(t)\big)\big|\big| \leq Me^{-\sigma_0 \tau}$$

for all $t \geq 0$ and $\tau \geq 0$.

The earliest result for systems that violate Condition 1. from Assumption 11.6 seems to be [154], which requires that the eigenvalues of A stay in the left half plane on average but not necessarily for all times. The frozen dynamics approach to establishing stability of slowly time-varying nonlinear systems appeared in the 1996 version of Khalil's book [70]. The slowly time-varying pendulum model (11.1) was studied in [137], where it was shown that (11.1) is UGAS when b_2 is periodic for some period $T > 0$, valued in $[-1.5, -0.5]$, and right continuous with only finitely many discontinuities on $[0, T]$; see [70] for related results, and [154] for results that are restricted to the linear case. Our Assumption 11.2 is a variant of those of [137, Theorem 2]. See also [71] for singular perturbation methods for building strict Lyapunov functions for systems of the form

$$\begin{cases} \dot{x} = f(x, z) \\ \varepsilon\dot{z} = g(x, z) \end{cases} \tag{11.54}$$

for small values of the constant parameter $\varepsilon > 0$. The system (11.54) is related to, but conceptually different from, the transformed system (11.16).

Compared with the known results [70, 137, 154], one novel feature of our Theorem 11.1 is that we allow general nonlinear systems including cases where the function q can take both positive and negative values (which corresponds to the allowance in [154] of eigenvalues that wander into the right half plane while remaining in the strict left half plane on average). Moreover, none of these earlier works gave explicit constructions for strict Lyapunov functions for general slowly time-varying systems. Also, we provided new methods for constructing explicit closed form strict ISS Lyapunov functions for slowly time-varying control systems, in terms of a suitable family of generalized Lyapunov-like functions for the frozen dynamics. This is significant because Lyapunov functions play essential roles in robustness analysis and controller design.

It is reasonable to expect that the results of this chapter can be extended to systems with measurement errors, or which are components of larger controlled hybrid dynamical systems. However, to the best of our knowledge, no such extensions have been carried out. It would also be of interest to cover slowly time-varying systems with outputs, and to construct corresponding IOS Lyapunov functions; see [170, 171] for further background on systems with outputs and Chap. 6 for some first results on explicitly constructing IOS Lyapunov functions for non-autonomous systems with a single continuous time scale, in terms of given non-strict Lyapunov functions.

Chapter 12
Hybrid Time-Varying Systems

Abstract In previous chapters, we saw how to explicitly construct global strict Lyapunov functions for continuous and discrete time systems, in terms of oftentimes readily available non-strict Lyapunov functions. This led to more explicit formulas for stabilizing feedbacks, as well as explicit quantizations of the effects of uncertainties, in the context of ISS. However, there are many cases where continuous and discrete time systems in and of themselves are inadequate to describe and prescribe the motion of dynamical systems. Instead, the system is presented in a hybrid way, with continuous and discrete subsystems and ways to switch between the subsystems. Moreover, standard constructive nonlinear control methods for continuous and discrete systems are not applicable to hybrid systems. In this chapter, we present some first results on constructive nonlinear control for hybrid systems that provide explicit ISS Lyapunov functions, in terms of given non-strict Lyapunov functions for the continuous and discrete subsystems, as well as a hybrid version of Matrosov's Theorem. We illustrate our results using a hybrid version of the identification dynamics we saw in previous chapters.

12.1 Motivation

Consider the continuous time dynamics

$$\dot{x} = -\bar{m}(t)\bar{m}^\top(t)x \tag{12.1}$$

where $\bar{m} : [0, \infty) \to \mathbb{R}^n$ is continuous and admits positive constants a, b, and c for which

$$\begin{aligned} &aI_n \leq \int_t^{t+c} \bar{m}(\tau)\bar{m}^\top(\tau)\,\mathrm{d}\tau \leq bI_n \ \text{ and} \\ &|\bar{m}(t)| = 1 \ \forall t \geq 0. \end{aligned}$$

This is a special case of the more general identification type dynamics that we saw in previous chapters, and it is well-known that (12.1) is globally exponentially stable [136]. Set

$$\lambda \doteq \frac{c}{2} + \frac{c^4}{4a}.$$

By reducing $a \in (0, 1)$ as needed, we can assume $\lambda > 1 + a$.

Next let $p : \mathbb{Z} \to [0, 1]$ be any bounded function that admits constants $l \in \mathbb{Z}_{\geq 0}$ and $\delta \in (0, l)$ such that[1]

$$\sum_{i=k-l}^{k} p(i) \geq \delta \quad \forall k \in \mathbb{Z}_{\geq 0} \tag{12.2}$$

and pick any discrete time dynamics

$$x^+ = [1 - p(k + 1)]x + p(k + 1)h(k, x, u), \tag{12.3}$$

where h admits $\chi \in \mathcal{K}_\infty$ such that the following holds for all $x \in \mathbb{R}^n$ and $u \in \mathbb{R}^m$:

$$\{|x| \geq \chi(|u|)\} \;\Rightarrow\; |h(k, x, u)| \leq \frac{|x|}{6\sqrt{\lambda}} \;\forall k. \tag{12.4}$$

The dynamics (12.3) freezes x whenever $p(k + 1) = 0$. Notice that we are allowing cases such as $h(k, x, u) = A(k, x) + B(k, x)u$ where B is bounded and $|A(k, x)| \leq |x|/(6\lambda)$ everywhere as well as more general systems where the system might not be affine in the control u; the choice of the bound on $|h(t, k, u)|$ in (12.4) will become clear in what follows.

Finally, let us consider the combined hybrid system

$$\begin{cases} x^+ = [1 - p(k + 1)]x + p(k + 1)h(k, x, u), & x \in D \\ \dot{x} = -\bar{m}(t)\bar{m}^\top(t)x, & x \in C \end{cases} \tag{12.5}$$

for $k \in \mathbb{Z}_{\geq 0}$, $t \geq 0$, and controls $u \in \mathbb{R}^d$, where $C, D \subseteq \mathbb{R}^n$ are given and $\bar{m}$, p, and h are as above. The system (12.5) should be understood in terms of hybrid trajectories on hybrid time domains; see [20] or Sect. 12.2 for precise definitions. In Chap. 10, we saw how to construct a global strict Lyapunov function for the continuous part (12.1) of the dynamics, which leads to an explicit ISS estimate that quantifies the effects of uncertainty in the dynamics. It is natural to extend ISS to the hybrid system (12.5); see Sect. 12.2 for the definitions of hybrid ISS and hybrid ISS Lyapunov functions. However, it is by no means clear how to construct an explicit, closed form ISS Lyapunov function for the entire hybrid dynamics (12.5).

---

[1] Condition (12.2) is the discrete time analog of the PE condition from preceding chapters. Therefore, we call a bounded non-negative function satisfying an estimate of the form (12.2) a (discrete time) PE function.

Just as in the continuous and discrete time cases, it is important to know whether hybrid dynamics such as (12.7) are ISS, and then to have explicit ISS Lyapunov functions for robustness analysis. One consequence of our work in this chapter will be that (12.5) has the ISS-CLF

$$\begin{aligned} V^\sharp(t,k,x) \doteq & \left\{ x^\top \left(\lambda I + \int_{t-c}^{t} \int_{s}^{t} \bar{m}(l)\bar{m}^\top(l)\,\mathrm{d}l\,\mathrm{d}s \right) x \right\} \times \\ & \left[2 + \frac{a}{2(\lambda + c^2/2)^2} \left\{ \frac{1}{8} + \frac{1}{4(l+1)} \sum_{s=k-l}^{k} \sum_{j=s}^{k} p(j) \right\} \right] \end{aligned} \tag{12.6}$$

when λ is large enough, and so is ISS. We prove this in Sect. 12.5 as a special case of our general CLF constructions.

The dynamics (12.5) is a special case of the hybrid dynamics

$$\begin{cases} x^+ = F(k,x,u),\ x \in D,\ k \in \mathbb{Z}_{\geq 0} \\ \dot{x} \ = G(t,x,u),\ x \in C,\ t \in [0,\infty) \end{cases} \tag{12.7}$$

for given sets $C, D \subseteq \mathbb{R}^n$, discrete dynamics F, continuous dynamics G, and control values $u \in \mathbb{R}^d$. The map F describes the jumps of the state x which occur when $x \in D$, while G describes the flow that occurs when $x \in C$; see Sect. 12.2.2 for the precise solution concept. There is a sizable literature on models of the form (12.7), including an abstract converse Lyapunov function theorem involving strict Lyapunov functions [20]. However, as for continuous and discrete time systems, the Lyapunov constructions in the hybrid converse theory are abstract and so may not lend themselves to applications where explicit closed form expressions for strict Lyapunov functions and stabilizing feedbacks are desirable. Moreover, it is important to be able to handle time-varying systems, because there are many applications where the time-invariant dynamics can be stabilized using time-varying controllers but cannot be stabilized by time-invariant feedback; see Chap. 1, or [94, 159]. The results of [20] are limited to time-invariant systems.

In this chapter, we address all of these issues by providing *general* methods for constructing *explicit closed form* global strict Lyapunov functions for time-varying discrete time and hybrid control systems. Our methods are based on suitable extensions of the strictification approach from Chap. 6. Roughly speaking, we strictify appropriate non-strict Lyapunov functions for the continuous and discrete time subsystems separately, and then we show how to merge the results into a strict Lyapunov function for the entire hybrid system. This requires an appropriate compatibility condition involving the continuous and discrete time non-strict Lyapunov functions, which is satisfied for (12.5) and other cases of interest. The results of this chapter were announced in [92].

12.2 Preliminaries

Throughout this chapter, all inequalities and equalities should be understood to hold globally unless otherwise indicated, and we leave out the arguments of our functions when no confusion would arise.

12.2.1 ISS and PE in Continuous and Discrete Time

Recall the standard classes of comparison functions $\mathcal{K}_\infty$ and $\mathcal{KL}$ from Chap. 1. We continue to set $\mathbb{Z}_{\geq 0} = \{0, 1, 2, \ldots\}$. To define stability for the more complex hybrid system (12.7), we need the following more general class of comparison functions that accommodates time pairs $(t, k) \in [0, \infty) \times \mathbb{Z}$. Let $\mathcal{KLL}$ denote the set of all functions $\beta : [0, \infty) \times [0, \infty) \times [0, \infty) \to [0, \infty)$ such that for each $\bar{t} \geq 0$, the functions $(s, t) \mapsto \beta(s, t, \bar{t})$ and $(s, t) \mapsto \beta(s, \bar{t}, t)$ are of class $\mathcal{KL}$. We say that a function $\Theta : [0, \infty) \times \mathbb{Z}_{\geq 0} \times \mathbb{R}^n \times \mathbb{R}^d \to \mathbb{R} : (t, k, x, u) \mapsto \Theta(t, k, x, u)$ (which may be independent of t, k, or u) is *uniformly state-bounded* and write $\Theta \in \mathcal{USB}$ provided there exists $\mu \in \mathcal{K}_\infty$ such that $|\Theta(t, k, x, u)| \leq \mu(|x|)$ for all $t \geq 0$, $k \in \mathbb{Z}_{\geq 0}$, $x \in \mathbb{R}^n$, and $u \in \mathbb{R}^d$. More generally, a vector-valued function $H : [0, \infty) \times \mathbb{Z}_{\geq 0} \times \mathbb{R}^n \times \mathbb{R}^d \to \mathbb{R}^j : (t, k, x, u) \mapsto H(t, k, x, u)$ is of *class* $\mathcal{USB}$, written $H \in \mathcal{USB}$, provided $(t, k, x, u) \mapsto |H(t, k, x, u)|$ is of class $\mathcal{USB}$.

We also say that Θ is *uniformly proper and positive definite (UPPD)* and write $\Theta \in \mathcal{UPPD}$ provided there are $\alpha_1, \alpha_2 \in \mathcal{K}_\infty$ such that $\alpha_1(|x|) \leq \Theta(t, k, x, u) \leq \alpha_2(|x|)$ for all $t \in [0, \infty)$, $k \in \mathbb{Z}_{\geq 0}$, $x \in \mathbb{R}^n$, and $u \in \mathbb{R}^d$. We say that Θ is (ω_1, ω_2)*-periodic* provided $\omega_1 \in [0, \infty)$ and $\omega_2 \in \mathbb{Z}_{\geq 0}$ satisfy $\Theta(t - \omega_1, k - \omega_2, x, u) = \Theta(t, k, x, u)$ for all $(t, k, x, u) \in \times[0, \infty) \times \mathbb{Z}_{\geq 0} \times \mathbb{R}^n \times \mathbb{R}^d$. When Θ is independent of t (resp., k), we define ω_2-periodicity (resp., ω_1-periodicity) analogously. A continuous function α mapping a subset of Euclidean space containing the origin into $[0, \infty)$ is *positive definite* (written $\alpha \in \mathcal{PD}$) provided α is zero at the origin and positive at all other points in its domain. We also recall the convention that for a C^1 function ρ defined on $[0, \infty)$, we interpret $\rho'(0)$ as a one-sided derivative, and continuity of ρ' at 0 as one-sided continuity.

To study hybrid systems (12.7), we first deal separately with their continuous and discrete subsystems, which we assume to have state space $\mathcal{X} = \mathbb{R}^n$ and control set $U = \mathbb{R}^m$. We study time-varying discrete time systems

$$x_{k+1} = F(k, x_k, u_k) \tag{12.8}$$

where $F \in \mathcal{USB}$ and u_k is the control value at time k. The system (12.8) is also denoted by $x^+ = F(k, x, u)$. Our continuous time systems take the form

$$\dot{x} = G(t, x, u) \tag{12.9}$$

where $G \in \mathcal{USB}$ is locally Lipschitz. The control functions for (12.8) and (12.9) comprise the set $\mathcal{M}_p$ of all bounded piecewise continuous functions $\mathbf{u} : [0, \infty) \to \mathbb{R}^d$.[2] We always assume that (12.9) is *forward complete*, i.e., for each $x_0 \in \mathbb{R}^n$, $\mathbf{u} \in \mathcal{M}_p$, and $t_0 \geq 0$, there is a unique solution $t \mapsto \phi(t, t_0, x_0, \mathbf{u})$ for (12.9) defined on $[t_0, \infty)$ that satisfies $\phi(t_0, t_0, x_0, \mathbf{u}) = x_0$. We interpret the solutions of (12.9) in the Lebesgue sense of $\dot{x}(t) = G(t, x(t), \mathbf{u}(t))$ for almost all (a.a.) $t \geq 0$. We also use $k \mapsto \phi(k, k_0, x_0, \mathbf{u})$ to denote the discrete time solution of (12.8) satisfying $\phi(k_0, k_0, x_0, \mathbf{u}) = x_0$ whenever no confusion would result; we always assume that (12.8) is forward complete as well. Given any function $V : [0, \infty) \times \mathbb{Z}_{\geq 0} \times \mathbb{R}^n \to \mathbb{R} : (t, k, x) \mapsto V(t, k, x)$, we define

$$\begin{aligned} \Delta_k V(t, k, x, u) &\doteq V\big(t, k+1, F(k, x, u)\big) - V(t, k, x) \\ \mathcal{D}V(t, k, x, u) &\doteq \tfrac{\partial V}{\partial t}(t, k, x) + \tfrac{\partial V}{\partial x}(t, k, x) G(t, x, u) \ , \end{aligned}$$

where we also assume that $(t, x) \mapsto V(t, x, k)$ is C^1 for each $k \in \mathbb{Z}_{\geq 0}$ when defining $\mathcal{D}V$. We write $\Delta_k V(k, x, u)$ instead of $\Delta_k V(t, k, x, u)$ when V is independent of t (and similarly for $\mathcal{D}V$). In Chapters 1 and 2, we defined ISS and ISS Lyapunov functions for continuous time systems. We use the following analogs for (12.8):

Definition 12.1. We define ISS-CLF and ISS as follows:

1. Let $V \in \mathcal{UPPD}$. We call V a *(strict) ISS CLF* for (12.8) provided there exist $\chi \in \mathcal{K}_\infty$ and $\alpha_3 \in \mathcal{PD}$ such that for all $x \in \mathbb{R}^n$, $u \in \mathbb{R}^d$, and $k \in \mathbb{Z}_{\geq 0}$,
$$|x| \geq \chi(|u|) \ \Rightarrow \ \Delta_k V(k, x, u) \ \leq \ -\alpha_3(|x|). \tag{12.10}$$
2. We call (12.8) *ISS* provided there exist $\beta \in \mathcal{KL}$ and $\gamma \in \mathcal{K}_\infty$ such that for all $x_0 \in \mathbb{R}^n$, $k_0 \in \mathbb{Z}_{\geq 0}$, and $\mathbf{u} \in \mathcal{M}_p$, we have
$$|\phi(k, k_0, x_0, \mathbf{u})| \ \leq \ \beta(|x_0|, k - k_0) + \gamma(|\mathbf{u}|_\infty) \tag{12.11}$$
for all $k \geq k_0$.

For the case of no controls, ISS reduces to UGAS; then we refer to an ISS-CLF simply as a *(strict) Lyapunov function.* The corresponding CLF, GAS, and ISS definitions for (12.9) are obtained from Definition 12.1 by replacing k with t, $\mathbb{Z}_{\geq 0}$ with $[0, \infty)$, and $\Delta_k V$ with $\mathcal{D}V$. When (12.8) has no controls, we also write $\Delta_k V(k, x)$ instead of $\Delta_k V(k, x, u)$ (and similarly for $\mathcal{D}V$).

It is worth noting that we do not require α_3 in (12.10) to be of class $\mathcal{K}_\infty$. However, when $\alpha_3 \in \mathcal{K}_\infty$ and F has no controls, the existence of a strict Lyapunov function V for $x^+ = F(k, x)$ is known to imply that this system is GAS [123, Theorem 8]. In fact, in that case we have $\Delta_k V(k, x) \leq -\alpha(V(k, x))$

[2] The piecewise continuity is essential because we need to evaluate $\mathbf{u}$ at specific times $\mathbf{u}(k) =: \mathbf{u}_k$ which would not be possible if we allowed a more general class of *measurable* essentially bounded inputs.

everywhere, where $\alpha \doteq \alpha_3 \circ \alpha_2^{-1} \in \mathcal{K}_\infty$ and α_2 as in the $\mathcal{UPPD}$ condition on V. On the other hand, by replacing $\alpha_3(|x|)$ in (12.10) by

$$\Theta(V(k,x)) \doteq \min\left\{\alpha_3(s) : \alpha_2^{-1}(V(k,x)) \le s \le \alpha_1^{-1}(V(k,x))\right\},$$

and minorizing Θ as needed so that it is increasing on $[0,1]$ and non-increasing on $[1,\infty)$ (as in Lemma A.7) and then C^1, we can use Lemma A.6 to get a strict a Lyapunov function V satisfying $\Delta_k V(k,x) \le -\alpha_3(|x|)$ everywhere with a new $\alpha_3 \in \mathcal{K}_\infty$. Combining this with the stability result from [123] and the invariance argument from [67] to get γ in the ISS estimate gives:

Lemma 12.1. *If (12.8) admits a strict ISS-CLF, then it is ISS.*

See Lemma 12.3 for a generalization to hybrid systems. We also use the following PE definitions:

Definition 12.2. We use PE in the following ways:

1. A bounded function $p : \mathbb{Z} \to [0,\infty)$ is of *discrete PE type with parameters l and δ* (written $p \in \mathcal{P}_{\rm dis}(l,\delta)$) provided $l \in \mathbb{N}$ and $\delta \in (0,l)$ are constants such that
$$\sum_{i=k-l}^{k} p(i) \ \ge \ \delta \quad \forall k \in \mathbb{Z}. \tag{12.12}$$
2. We say that a bounded continuous function $q : \mathbb{R} \to [0,\infty)$ is of *continuous PE type with parameters τ and ε* and write $q \in \mathcal{P}_{\rm cts}(\tau,\varepsilon)$ provided the constants $\tau \ge 0$ and $\varepsilon > 0$ satisfy
$$\int_{t-\tau}^{t} q(r)\,\mathrm{d}r \ \ge \ \varepsilon \quad \forall t \in \mathbb{R}. \tag{12.13}$$
3. We set $\mathcal{P}_{\rm dis} = \cup\{\mathcal{P}_{\rm dis}(l,\delta) : l \in \mathbb{Z}_{\ge 0}, \delta > 0\}$ and $\mathcal{P}_{\rm cts} = \cup\{\mathcal{P}_{\rm cts}(\tau,\varepsilon) : \tau > 0, \varepsilon > 0\}$.

Elements of $\mathcal{P}_{\rm dis}$ and $\mathcal{P}_{\rm cts}$ are called *PE parameters* and arise in a variety of applications [91, 98]. We use the following simple observations:

Lemma 12.2. *Let $l \in \mathbb{N}$ and $\tau, \varepsilon, \delta > 0$ be constants, and let $p \in \mathcal{P}_{\rm dis}(l,\delta)$ and $q \in \mathcal{P}_{\rm cts}(\tau,\varepsilon)$ be bounded above by $\bar{p}$ and $\bar{q}$, respectively. Define the functions $S : \mathbb{Z} \to [0,\infty)$ and $R : \mathbb{R} \to [0,\infty)$ by*

$$S(k) \doteq \sum_{s=k-l}^{k} \sum_{j=s}^{k} p(j) \ \ \textit{and} \ \ R(t) \doteq \int_{t-\tau}^{t} \int_{z}^{t} q(y)\,\mathrm{d}y\,\mathrm{d}z. \tag{12.14}$$

Then

$$S(k) \le \bar{p}(l+1)^2 \ \forall k \in \mathbb{Z} \ \ \textit{and} \ \ R(t) \le \tau^2 \frac{\bar{q}}{2} \ \forall t \in \mathbb{R}. \tag{12.15}$$

If p is l-periodic, then so is S. If q is τ-periodic, then so is R.

The preceding lemma follows from switching the order of the summations, a simple change of variables, and the formula $1 + 2 + \ldots + m = m(m+1)/2$.

12.2.2 Hybrid Systems

Following [20, 24], we interpret the hybrid system (12.7) in terms of hybrid trajectories on hybrid time domains. The relevant definitions are as follows. A *compact hybrid time domain* is a subset $E \subseteq [0, \infty) \times \mathbb{Z}_{\geq 0}$ of the form

$$\cup_{k=0}^{K-1}\big([t_k, t_{k+1}] \times \{k\}\big)$$

for some finite sequence $0 = t_0 \leq t_1 \leq \ldots \leq t_K$. A *hybrid time domain* is a subset $E \subseteq [0, \infty) \times \mathbb{Z}_{\geq 0}$ with the property that for all $(T, J) \in E$,

$$E \cap \big([0, T] \times \{0, 1, \ldots, J\}\big)$$

is a compact hybrid time domain. Therefore, E is a hybrid time domain provided it is a finite or infinite union of sets of the form $[t_k, t_{k+1}] \times \{k\}$ with $\{t_k\}$ non-decreasing in $[0, \infty)$, with a possible additional "last" set having the form $[t_k, T) \times \{k\}$ with T finite or infinite. To simplify the notation, we use

$$\cup_{k \in J}\big([t_k, t_{k+1}] \times \{k\}\big)$$

to denote a generic hybrid time domain with the understanding that (i) either $J = \mathbb{Z}_{\geq 0}$ or J is a finite set of the form $\{0, 1, 2, \ldots, j_{\max}\}$ and (ii) $[t_k, t_{k+1}]$ may mean $[t_k, t_{k+1})$ if J is finite and $k = j_{\max}$.

A *hybrid arc* is a function $x(t, k)$ defined on a hybrid time domain $\mathrm{dom}(x)$ such that $t \mapsto x(t, k)$ is locally absolutely continuous for each k. Given sets $C, D \subseteq \mathbb{R}^n$ and F and G satisfying the assumptions above, the solutions of the corresponding *hybrid control system*

$$\mathcal{H} \doteq \begin{cases} \dot{x} &= G(t, x, u), \quad\;\; x \in C \\ x_{k+1} &= F(k, x_k, u_k),\; x_k \in D \end{cases} \tag{12.16}$$

are defined as follows.

Definition 12.3. Given $\mathbf{u} \in \mathcal{M}_p$, a *hybrid trajectory for (12.16) (for the input* $\mathbf{u}$*)* is a hybrid arc $x(t, k)$ that satisfies:

1. for all $k \in \mathbb{Z}_{\geq 0}$ and a.a. t such that $(t, k) \in \mathrm{dom}(x)$, we have $x(t, k) \in C$ and $\frac{\partial}{\partial t} x(t, k) = G(t, x(t, k), \mathbf{u}(t))$; and
2. if $(t, k) \in \mathrm{dom}(x)$ and $(t, k+1) \in \mathrm{dom}(x)$, then $x(t, k) \in D$ and $x(t, k+1) = F(k, x(t, k), \mathbf{u}(k))$.

Notice that continuous time solutions of (12.9) in C, and discrete time solutions of (12.8) in D starting with $k_0 = 0$, are special cases of hybrid trajectories of (12.16), with no switchings between the continuous and discrete subsystems. In the same way, the next definition reduces to Definition 12.1 for discrete time ISS and discrete time ISS-CLFs, in the special case where $C = \emptyset$ and $D = \mathbb{R}^n$:

Definition 12.4. We use ISS in the following ways:

1. Let $V \in \mathcal{UPPD}$ be C^1 in x and t. We call V a *(strict) ISS-CLF* for $\mathcal{H}$ provided there exist $\chi \in \mathcal{K}_\infty$ and a positive definite function α_3 such that for all $(t, k, x, u) \in [0, \infty) \times \mathbb{Z}_{\geq 0} \times \mathbb{R}^n \times \mathbb{R}^d$, we have

$$|x| \geq \chi(|u|) \Rightarrow \begin{cases} \Delta_k V(t,k,x,u) \leq -\alpha_3(|x|) \text{ if } x \in D; \\ \text{and } \mathcal{D}V(t,k,x,u) \leq -\alpha_3(|x|) \text{ if } x \in C. \end{cases} \tag{12.17}$$

If, in addition, there is a constant $r > 0$ such that

$$|x| \geq \chi(|u|) \Rightarrow \begin{cases} V(t,k+1,F(k,x,u)) \leq e^{-r} V(t,k,x) \text{ if } x \in D; \\ \text{and } \mathcal{D}V(t,k,x,u) \leq -rV(t,k,x) \text{ if } x \in C, \end{cases} \tag{12.18}$$

for all $t \geq 0$, $k \in \mathbb{Z}_{\geq 0}$, $x \in \mathbb{R}^n$, and $u \in \mathbb{R}^d$, then we call V an *exponential decay ISS-CLF* for $\mathcal{H}$.

2. We call $\mathcal{H}$ ISS provided there are $\beta \in \mathcal{KLL}$ and $\gamma \in \mathcal{K}_\infty$ such that: For each $\mathbf{u} \in \mathcal{M}_p$ and each trajectory $x(t,k)$ of $\mathcal{H}$ for $\mathbf{u}$ defined on each hybrid time domain $\cup_{k \in J}([t_k, t_{k+1}] \times \{k\})$, we have

$$|x(t,k)| \;\leq\; \beta\big(|x(t_0,0)|, k, t - t_k\big) + \gamma(|\mathbf{u}|_\infty) \tag{12.19}$$

for all $k \in J$ and all $t \in [t_k, t_{k+1}]$.

We close this section with the following hybrid analog of Lemma 12.1:

Lemma 12.3. *If $\mathcal{H}$ admits an ISS-CLF V, then it is ISS.*

Proof. We first prove the result for the UGAS case where the dynamics do not depend on u; the result will then follow from a variant of a standard invariance argument, e.g., the one used in [157]. First note that since α_3 in the decay estimate is independent of k, standard arguments (e.g., those in [157] applied with $a(x) \doteq \alpha_3(|x|)$) provide $\beta_1 \in \mathcal{KL}$ such that

$$|x(t,k)| \;\leq\; \beta_1\big(|x(t_k,k)|, t - t_k\big) \quad \forall k \in J, \;\; t \in [t_k, t_{k+1}] \tag{12.20}$$

for any hybrid trajectory $x(t,j)$ satisfying any initial condition $x(t_0,0) = x_0$ on any hybrid time domain $\cup_{k \in J}([t_k, t_{k+1}] \times \{k\})$.

Since α_3 is also independent of t, we can also find a $\beta_2 \in \mathcal{KL}$ such that

$$|x(t_k,k)| \le \beta_2(|x_0|,k) \quad \forall k \in J, \tag{12.21}$$

by the following construction. Applying Lemma A.6 to the discrete time decay condition and replacing V with $\kappa \circ V$ for a suitable choice of $\kappa \in \mathcal{K}_\infty$ without relabeling, we can construct a function $\gamma \in \mathcal{K}_\infty$ such that $\Delta_k V(t,k,x) \le -\gamma(V(t,k,x))$ for all $x \in D$ and all (t,k). Since the mapping $t \mapsto V(t,k,x(t,k))$ decays on (t_k, t_{k+1}) for each k, we get

$$\begin{aligned} V\big(t_{k+2},k+1,x(t_{k+2},k+1)\big) &\le V\big(t_{k+1},k+1,x\big(t_{k+1},k+1\big)\big) \\ &= V\big(t_{k+1},k+1,F\big(k,x(t_{k+1},k)\big)\big), \end{aligned}$$

and therefore

$$\begin{aligned} &V(t_{k+2},k+1,x(t_{k+2},k+1)) - V(t_{k+1},k,x(t_{k+1},k)) \\ &\le \Delta_k V\big(t_{k+1},k,x(t_{k+1},k)\big) \\ &\le -\gamma\big(V\big(t_{k+1},k,x(t_{k+1},k)\big)\big) \end{aligned}$$

everywhere, assuming t_k, t_{k+1}, and t_{k+2} are all interval endpoints in $\mathrm{dom}(x)$. If we now apply [123, Theorem 8] to $k \mapsto V(t_{k+1},k,x(t_{k+1},k))$, the fact that V is uniformly proper and positive definite provides functions $\alpha_1, \alpha \in \mathcal{K}_\infty$ and $\tilde{\beta}_2 \in \mathcal{KL}$ such that

$$\begin{aligned} &\alpha_1(|x|) \le V(t,k,x) \le \alpha_2(|x|) \;\; \forall (t,k,x) \;\text{ and} \\ &|x(t_{k+1},k)| \le \tilde{\beta}_2\big(|x(t_1,0)|,k\big) \;\; \forall k; \end{aligned} \tag{12.22}$$

the function $\tilde{\beta}_2$ is independent of the choice of the trajectory.

The discrete time decay condition therefore gives

$$\begin{aligned} |x(t_{k+1},k)| &\ge \alpha_2^{-1} \circ V\big(t_{k+1},k,x(t_{k+1},k)\big) \\ &\ge \alpha_2^{-1} \circ V\big(t_{k+1},k+1,x(t_{k+1},k+1)\big) \\ &\ge \alpha_2^{-1} \circ \alpha_1\big(|x(t_{k+1},k+1)|\big) \;\; \forall k \in J. \end{aligned}$$

By analogous reasoning, the continuous time decay condition gives

$$\begin{aligned} |x(t_1,0)| &\le \alpha_1^{-1} \circ V\big(t_1,0,x(t_1,0)\big) \\ &\le \alpha_1^{-1} \circ V\big(t_0,0,x(t_0,0)\big) \le \alpha_1^{-1} \circ \alpha_2(|x_0|). \end{aligned}$$

Combining the last three estimates, it follows that (12.21) is satisfied with

$$\beta_2(s,k) \doteq \alpha_1^{-1} \circ \alpha_2 \circ \tilde{\beta}_2\big(\alpha_1^{-1} \circ \alpha_2(s),k\big) + \frac{s}{k+1},$$

where the extra term $s/(k+1)$ is used to cover the case where $k = 0$. Combining (12.20) and (12.21) shows we can satisfy the requirements of Lemma 12.3 using $\beta(s,t,k) = \beta_1(\beta_2(s,k),t)$ when there are no controls $\mathbf{u}$.

To extend to the case where there are controls $\mathbf{u}$, notice that the preceding inequalities remain true when controls are present as long as the current state x is such that $|x| \ge \chi(|\mathbf{u}|_\infty)$. We can now reason as in the usual proof of the continuous time ISS estimate (e.g., as in [157]) to satisfy the ISS requirements with the same choice $\beta \in \mathcal{KLL}$. It remains to construct the function $\gamma \in \mathcal{K}_\infty$ in the ISS estimate. The argument from the first part of the proof provides $\kappa, \alpha_4, \sigma \in \mathcal{K}_\infty$ such that $\tilde{V} \doteq \kappa(V)$ satisfies

$$\Delta_k \tilde{V}(t,k,x,u) \leq -\alpha_4\big(\tilde{V}(t,k,x)\big) + \sigma(|u|)$$

for all $t \geq 0$, $k \in \mathbb{Z}_{\geq 0}$, $x \in D$, and $u \in \mathbb{R}^d$. By reducing $\alpha_4 \in \mathcal{K}_\infty$, we can assume that $s \mapsto s - \alpha_4(s)$ is increasing [67, Lemma 2.4]. Choose $\tilde{\alpha}_1, \tilde{\alpha}_2 \in \mathcal{K}_\infty$ such that

$$\tilde{\alpha}_1(|x|) \leq \tilde{V}(t,k,x) \leq \tilde{\alpha}_2(|x|)$$

for all (t,k,x).

Choose any $\mathbf{u} \in \mathcal{M}_p$ and any hybrid trajectory $x(t,k)$ for $\mathcal{H}$ and $\mathbf{u}$. Define the set $\mathcal{S} \doteq \{(t,k) \in \mathrm{dom}(x) : \tilde{V}(t,k,x(t,k)) \leq \bar{b}\}$, where $\bar{b} = \alpha_4^{-1}(2\sigma(|\mathbf{u}|_\infty))$. Set $\bar{c} = \alpha_2 \circ \chi(|\mathbf{u}|_\infty)$ where χ is from our ISS Lyapunov function assumption and $\alpha_2 \in \mathcal{K}_\infty$ satisfies (12.22). By enlarging σ from our ISS condition, we can assume that

$$V(t,k,x) \leq \bar{c} \Rightarrow \tilde{V}(t,k,x) \leq \bar{b}.$$

Claim: If there is a pair $(\bar{t},\bar{k}) \in \mathcal{S}$, then $(t,k) \in \mathcal{S}$ whenever $\mathrm{dom}(x) \ni (t,k) \succeq (\bar{t},\bar{k})$ where $\succeq$ is the lexicographical ordering on $\mathrm{dom}(x)$.[3] To verify this claim, suppose it were not true. We could then find a pair $(t,k) \in \mathrm{dom}(x)$ and a constant $\varepsilon > 0$ such that:

1. $(t,k) \succeq (\bar{t},\bar{k})$; and
2. $\tilde{V}\big(t,k,x(t,k)\big) \geq \bar{b} + \varepsilon$.

Choose (t',k') to be the smallest such pair $(t,k) \in \mathrm{dom}(x)$ in the ordering $\succeq$ for this choice of ε. If $t' = t_{k'}$ and $k' \geq 1$, then $(t',k'-1) \in \mathrm{dom}(x)$ and $x(t',k'-1) \in D$, by Condition 2. in Definition 12.3. Also, $(t',k'-1) \in \mathcal{S}$, because otherwise, we would have $V(t_{k'},k'-1,x(t_{k'},k'-1)) > \bar{c}$, hence $|x(t_{k'},k'-1)| > \chi(|\mathbf{u}|_\infty)$, so our decay condition on $\tilde{V}$ would imply that

$$\bar{b} + \varepsilon \leq \tilde{V}\big(t_{k'},k',x(t_{k'},k')\big) \leq \tilde{V}\big(t_{k'},k'-1,x(t_{k'},k'-1)\big),$$

contrary to the minimality of (t',k'). Since $(t',k'-1) \in \mathcal{S}$, the argument from [67, Sect. 2] applied to the dynamics $x_{r+1} = F(r,x_r,u(r))$ implies that $(t',k') \in \mathcal{S}$ as well, contradicting our choice of (t',k'). On the other hand, if $t' \neq t_{k'}$ or $k' = 0$, then $t' \in (t_{k'}, t_{k'+1}]$. Since $\tilde{V}(t',k',x(t',k')) > \bar{b}$, we then have $V(t',k',x(t',k')) > \bar{c}$, so $|x(t',k')| > \chi(|\mathbf{u}|_\infty)$. Therefore, property 1. from Definition 12.3 and (12.17) give a pair $(t^\star,k') \in \mathrm{dom}(x)$, with $t^\star < t'$ but $t^\star$ near t', for which $V(t^\star,k',x(t^\star,k')) \geq V(t',k',x(t',k'))$, which again contradicts the minimality of (t',k'). (We are using the fact that $\tilde{V}$ is increasing as a function of V.) This proves the claim.

---

[3] The ordering is defined as follows: $(t_2,k_2) \succeq (t_1,k_1)$ if and only if either (a) $t_2 > t_1$ or (b) $t_2 = t_1$ and $k_2 \geq k_1$.

By separately considering the cases where the hybrid time set $\mathcal{S} = \emptyset$ and $\mathcal{S} \neq \emptyset$, we can then choose

$$\gamma(s) = \tilde{\alpha}_1^{-1} \circ \alpha_4^{-1}(2\sigma(s))$$

to satisfy the ISS estimate. This proves Lemma 12.3. □

12.3 Strictification for Time-Varying Systems

We next extend our basic strictification results from Chap. 6 to discrete and hybrid time-varying systems. As in the continuous time case, we express the non-strict Lyapunov decay conditions in terms of PE parameters, and the non-strict decay conditions can be used to prove UGAS directly without constructing strict Lyapunov functions. However, it is well appreciated that explicit strict Lyapunov functions are useful in many applications, e.g., for quantifying the effects of uncertainty, which motivates our closed form expressions for our global strict Lyapunov functions.

12.3.1 Discrete Time Strictification

The following was shown in [93]:

Theorem 12.1. *Let $l \in \mathbb{Z}_{\geq 0}$, $\delta > 0$, $p \in \mathcal{P}_{\rm dis}(l,\delta)$, $V \in \mathcal{UPPD}$, $\chi \in \mathcal{K}_\infty$, and $\Theta \in \mathcal{PD}$ satisfy the following for all $k \in \mathbb{Z}_{\geq 0}$, $x \in \mathbb{R}^n$, and $u \in \mathbb{R}^d$:*

$$\begin{aligned} &|x| \geq \chi(|u|) \;\; \Rightarrow \\ &\Delta_k V(k,x,u) \;\leq\; -p(k+1)\Theta\big(V(k,x)\big). \end{aligned} \tag{12.23}$$

Then one can construct functions $\kappa, \Gamma \in \mathcal{K}_\infty$ such that

$$U(k,x) \;\doteq\; \kappa\big(V(k,x)\big) + \frac{\Gamma\big(V(k,x)\big)}{4(l+1)} \sum_{s=k-l}^{k} \sum_{j=s}^{k} p(j) \tag{12.24}$$

is a strict ISS-CLF for (12.8). In particular, (12.8) is ISS. If p and V are both l-periodic, then so is U.

Proof. Sketch. We only sketch the proof; see [93] for details. By minorizing $\Theta \in \mathcal{PD}$ as necessary without relabeling, we can assume that Θ is non-decreasing on $[0,1]$ and non-increasing on $[1,\infty)$; see Lemma A.7 for details. The following is shown in exactly the same way as Lemma A.6:

Lemma 12.4. *Let $\Theta \in \mathcal{PD}$ be as above and $p \in \mathcal{P}_{\text{dis}}$. Define $\mu : [0,\infty) \to [1,\infty)$, κ, and Ψ by*

$$\kappa(r) \doteq 2\int_0^r \mu(z)\,\mathrm{d}z, \quad \Psi(r) \doteq \Theta(2r)\mu(r), \quad \text{and}$$
$$\mu(r) = \begin{cases} 1 + 4r^2, \; 0 \le r \le 1/2 \\ \dfrac{4\Theta(1)r}{\Theta(2r)}, \; 1/2 \le r < \infty \end{cases}. \tag{12.25}$$

Let $\nu \in \mathcal{UPPD}$ and $\chi \in \mathcal{K}_\infty$ be such that for all $x \in \mathbb{R}^n$, $k \in \mathbb{Z}_{\ge 0}$, and $u \in \mathbb{R}^d$, we have:

$$|x| \ge \chi(|u|) \;\Rightarrow\; \Delta_k \nu(k,x,u) \le -p(k+1)\Theta(\nu(k,x)).$$

Then $\kappa \in \mathcal{K}_\infty$, $\Psi \in \mathcal{K}_\infty$, and $V \doteq \kappa(\nu) \in \mathcal{UPPD}$ satisfies the following for all $x \in \mathbb{R}^n$, $u \in \mathbb{R}^d$, and $k \in \mathbb{Z}_{\ge 0}$:

$$|x| \ge \chi(|u|) \;\Rightarrow\; \Delta_k V(k,x,u) \le -p(k+1)\Gamma\big(V(k,x)\big) \tag{12.26}$$

where $\Gamma \in \mathcal{K}_\infty$ is defined by $\Gamma(s) \doteq \Psi(\kappa^{-1}(s)/2)$.

Returning to the proof of the theorem, we can therefore assume that V satisfies (12.26) with $\Gamma \in \mathcal{K}_\infty$, possibly by replacing V with $\kappa(V)$ with $\kappa \in \mathcal{K}_\infty$ defined in Lemma 12.4 without relabeling. Next choose $S(k)$ as in Lemma 12.2 and U as in (12.24) with $\kappa(s) \equiv s$. Since Γ is increasing, we readily get

$$\begin{aligned}
\Delta_k U(k,x,u) &= V\big(k+1, F(k,x,u)\big) \\
&\quad + \frac{S(k+1)}{4(l+1)}\Gamma\big(V\big(k+1,F(k,x,u)\big)\big) \\
&\quad - V(k,x) - \frac{S(k)}{4(l+1)}\Gamma\big(V(k,x)\big) \\
&= \Delta_k V(k,x,u) + \frac{1}{4(l+1)} S(k+1)\Delta_k(\Gamma \circ V)(k,x) \\
&\quad + \tfrac{1}{4(l+1)}\Gamma(V(k,x))\,[S(k+1) - S(k)] \\
&\le \Delta_k V(k,x,u) \\
&\quad + \frac{1}{4(l+1)}\Gamma(V(k,x))\,[S(k+1) - S(k)],
\end{aligned} \tag{12.27}$$

for all $k \in \mathbb{Z}_{\ge 0}$, $x \in \mathbb{R}^n$, and $u \in \mathbb{R}^d$ satisfying $|x| \ge \chi(|u|)$. Using the definition of $S(k)$ and canceling terms gives

$$
\begin{aligned}
S(k+1) - S(k) &= \sum_{s=k-l}^{k} \sum_{j=s}^{k} p(j) - \sum_{j=k-l}^{k} p(j) + (l+1)p(k+1) \\
&\quad - \sum_{s=k-l}^{k} \sum_{j=s}^{k} p(j) \\
&= - \sum_{j=k-l}^{k} p(j) + (l+1)p(k+1) \,.
\end{aligned}
$$

Substituting into (12.27) and using (12.12) and (12.26) gives

$$
\begin{aligned}
\Delta_k U(k,x,u) \le \Delta_k V(k,x,u) &+ \frac{1}{4(l+1)} \Gamma\big(V(k,x)\big) \\
&\times \left((l+1)p(k+1) - \sum_{j=k-l}^{k} p(j) \right) ,
\end{aligned}
\tag{12.28}
$$

hence also

$$
\begin{aligned}
\Delta_k U(k,x,u) \le \Delta_k V(k,x,u) &+ \frac{p(k+1)\Gamma(V(k,x))}{4} \\
&- \frac{\Gamma\big(V(k,x)\big)}{4(l+1)} \sum_{j=k-l}^{k} p(j) \\
\le - \frac{\delta}{4(l+1)} \Gamma\big(V(k,x)\big),
\end{aligned}
\tag{12.29}
$$

for all $k \in \mathbb{Z}_{\ge 0}$, all $x \in \mathbb{R}^n$, and all $u \in \mathbb{R}^d$ satisfying $|x| \ge \chi(|u|)$. Combined with the fact that $V \in \mathcal{UPPD}$ and using the global boundedness of $S(k)$ from Lemma 12.2, we conclude that U is an ISS-CLF for (12.8). Hence, (12.8) is ISS, by Lemma 12.1. The assertion on periodicity follows from Lemma 12.2. This proves the theorem. □

Remark 12.1. A key point in the non-strict decay condition (12.23) is that the PE condition allows $p(k+1) = 0$ and so also $\Delta_k V(k,x,u) = 0$ for some values of k. Also, we do not require Θ in (12.23) to be of class $\mathcal{K}_\infty$. However, our proof of Theorem 12.1 shows that we can take $\kappa(s) \equiv s$ and $\Theta = \Gamma$ in the special case where $\Theta \in \mathcal{K}_\infty$.

12.3.2 Hybrid Strictification

We next extend Theorem 12.1 to the hybrid system (12.16), assuming for simplicity that Θ in (12.23) and its continuous time analog are $\Theta(s) = s$.

Theorem 12.2. *Let $V \in \mathcal{UPPD}$ be C^1 in x and t. Let $\delta, \varepsilon, \tau > 0$ and $l \in \mathbb{N}$ be given constants. Assume that there exist $\chi \in \mathcal{K}_\infty$, $r \in \mathcal{P}_{\mathrm{dis}}(l, \delta)$ and $q \in \mathcal{P}_{\mathrm{cts}}(\tau, \varepsilon)$ such that*

$$|x| \geq \chi(|u|) \;\Rightarrow\; \begin{cases} V(t, k+1, F(k, x, u)) \leq e^{-r(k+1)} V(t, x, k) \text{ if } x \in D; \\ \text{and } \mathcal{D}V(t, k, x, u) \leq -q(t) V(t, x, k) \text{ if } x \in C \end{cases} \tag{12.30}$$

for all $t \geq 0$, $k \in \mathbb{Z}_{\geq 0}$, $x \in \mathbb{R}^n$, and $u \in \mathbb{R}^d$. Then

$$\begin{aligned} V^\sharp(t, x, k) = \Bigg[& 2 + \frac{1}{4(l+1)} \sum_{s=k-l}^{k} \sum_{j=s}^{k} \left(1 - e^{-r(j)}\right) \\ & + \frac{1}{\tau} \int_{t-\tau}^{t} \int_{z}^{t} q(y)\, \mathrm{d}y\, \mathrm{d}z \Bigg] V(t, k, x) \end{aligned} \tag{12.31}$$

is an exponential decay ISS-CLF for (12.16) so (12.16) is ISS. If in addition V is (τ, l)-periodic and r and q are l-periodic and τ-periodic respectively, then $V^\sharp$ is also (τ, l)-periodic.

Proof. For each $k \in \mathbb{Z}_{\geq 0}$, consider the function

$$V_{\mathrm{cts}}(t, k, x) \doteq \left[1 + \frac{1}{\tau} \int_{t-\tau}^{t} \int_{z}^{t} q(y)\, \mathrm{d}y\, \mathrm{d}z\right] V(t, k, x).$$

For all $t \geq 0$ and $k \in \mathbb{Z}_{\geq 0}$, and for all $x \in C$ and $u \in \mathbb{R}^d$ satisfying $|x| \geq \chi(|u|)$, we have

$$\mathcal{D}V_{\mathrm{cts}}(t, k, x, u) \;\leq\; -\frac{\varepsilon}{\tau} V(t, k, x) \;\leq\; -\frac{\varepsilon}{\tau(1 + \tau\bar{q})} V_{\mathrm{cts}}(t, k, x),$$

where $\bar{q}$ is a global bound on q and we used Lemma 12.2. We next rewrite the first decay condition in (12.30) as

$$\Delta_k V(t, x, u) \leq -p(k+1) V(t, k, x)$$

for all $t \in [0, \infty)$, $k \in \mathbb{Z}_{\geq 0}$, $x \in D$, and $u \in \mathbb{R}^d$ satisfying $|x| \geq \chi(|u|)$, where

$$k \mapsto p(k) \doteq 1 - e^{-r(k)} \tag{12.32}$$

is again of PE type.

For each $t \geq 0$, consider the function

$$V_{\mathrm{dis}}(t, k, x) \doteq \left[1 + \frac{S(k)}{4(l+1)}\right] V(t, k, x),$$

where $S(k)$ is defined by (12.14) with the choice (12.32) of p.

Arguing as in the proof of Theorem 12.1 shows that

$$\Delta_k V_{\text{dis}}(t,k,x,u) \;\leq\; -\frac{\delta V(t,k,x)}{4(l+1)} \;\leq\; -\frac{\delta V_{\text{dis}}(t,k,x)}{4\mathcal{M}(l+1)}$$

for all $t \in [0,\infty)$, $k \in \mathbb{Z}_{\geq 0}$, $x \in D$, and $u \in \mathbb{R}^d$ satisfying $|x| \geq \chi(|u|)$, where $\mathcal{M} = 1 + 0.25\bar{p}(l+1)$ and $\bar{p}$ is a global bound on p. We can assume that $\delta < \mathcal{M}l$, in which case the discrete decay condition in (12.30) is satisfied with

$$r(k) \equiv \ln\big(4\mathcal{M}(l+1)\big) - \ln\big(4\mathcal{M}(l+1) - \delta\big) > 0$$

and with V replaced by V_{dis}. When $|x| \geq \chi(|u|)$, $t \geq 0$, and $k \in \mathbb{Z}_{\geq 0}$, we have the following:

1. $\mathcal{D}V(t,k,x,u) \leq 0$ when $x \in C$; and
2. $\Delta_k V(t,k,x,u) \leq 0$ when $x \in D$.

Hence, $\mathcal{D}V_{\text{dis}} \leq 0$ on C and $\Delta_k V_{\text{cts}} \leq 0$ on D when $|x| \geq \chi(|u|)$. We can find positive constants r_c and r_d such that

$$V_{\text{cts}}(t,k,x) \;\leq\; r_c V_{\text{dis}}(t,k,x) \;\leq\; r_d V_{\text{cts}}(t,k,x)$$

everywhere. It follows that

$$V^\sharp(t,k,x) \doteq V_{\text{cts}}(t,k,x) + V_{\text{dis}}(t,k,x),$$

i.e., (12.31), is an exponential decay ISS-CLF for $\mathcal{H}$. The periodicity and ISS assertions follow from Lemmas 12.2 and 12.3. □

Remark 12.2. By taking $D = \emptyset$ and $C = \mathbb{R}^n$, Theorem 12.2 includes continuous time dynamics (with the understanding that the term involving the double sum in (12.31) is not present). Similarly, by taking $C = \emptyset$ and $D = \mathbb{R}^n$, it includes discrete time dynamics (in which case there is no double integral term in $V^\sharp$).

12.4 Matrosov Constructions for Time-Varying Systems

In Chapters 3 and 8, we saw how to explicitly construct strict Lyapunov functions under suitable variants of Matrosov's Conditions. We next extend these results to hybrid time-varying systems, including cases where the decay condition on the non-strict Lyapunov function involves PE parameters. For simplicity, we only consider the case of one auxiliary function.

12.4.1 Discrete Time Construction

Consider the discrete time system

$$x_{k+1} = F(k, x_k) \tag{12.33}$$

with $F \in \mathcal{USB}$. We assume the following discrete time analog of the Matrosov Theorem conditions from the previous chapters:

Assumption 12.1 *There exist $V_1 \in \mathcal{UPPD}$; $V_2 \in \mathcal{USB}$, a function $\phi_2 \in \mathcal{K}_\infty$; everywhere non-negative functions $N_1, N_2 \in \mathcal{USB}$; a function $\chi : [0,\infty) \times \mathbb{Z}_{\geq 0} \times \mathbb{R}^n \to \mathbb{R}$; an everywhere positive increasing function ϕ_1; $W \in \mathcal{PD}$; and $p \in \mathcal{P}_{\text{dis}}$ such that*

$$\begin{aligned}
\Delta_k V_1(k,x) &\leq -N_1(k,x); \\
\Delta_k V_2(k,x) &\leq -N_2(k,x) + \chi\big(N_1(k,x),k,x\big); \\
\big|\chi\big(N_1(k,x),k,x\big)\big| &\leq \phi_1(|x|)\phi_2(N_1(k,x)); \text{ and} \\
N_1(k,x) + N_2(k,x) &\geq p(k+1)W(x)
\end{aligned}$$

hold for all $x \in \mathbb{R}^n$ and $k \in \mathbb{Z}_{\geq 0}$.

As in our earlier Matrosov results, we allow V_2 to take both positive and negative values. Under Assumption 12.1, it is not obvious how to explicitly construct corresponding global strict Lyapunov functions. Therefore, we prove the following result, which was announced in [92, 93]:

Theorem 12.3. *If (12.33) satisfies Assumption 12.1, then one can construct an explicit closed form strict Lyapunov function for (12.33). In particular, (12.33) is UGAS.*

Proof. Let $V_3 = V_1 + V_2$, and $\alpha_1, \alpha_2 \in \mathcal{K}_\infty$ satisfy the UPPD requirements for V_1. Consider the positive definite function

$$\lambda(s) = \min\{W(x) : x \in \mathbb{R}^n, \alpha_1(|x|) \leq s \leq \alpha_2(|x|)\},$$

which is positive definite because W is positive definite. It readily follows from Assumption 12.1 that

$$\begin{aligned}
\Delta_k V_3(k,x) &\leq -p(k+1)W(x) + \phi_1(|x|)\phi_2\big(N_1(k,x)\big) \\
&\leq -p(k+1)\lambda\big(V_1(k,x)\big) + \phi_1(|x|)\phi_2\big(N_1(k,x)\big).
\end{aligned} \tag{12.34}$$

By minorizing λ as needed without relabeling, we assume that it is non-decreasing on $[0,1/2]$, non-increasing on $[1/2,\infty)$ and C^1. The argument used to prove Lemma A.6 with the choice $\Theta(r) \doteq \lambda(r/2)$ gives an increasing continuous function $k_1 : [0,\infty) \to [1,\infty)$ such that $\Lambda_1(s) \doteq k_1(s)\lambda(s)$ is $\mathcal{K}_\infty$. Let

$$V_4 = k_1(V_1)V_3.$$

By (12.34),

$$\begin{aligned}
\Delta_k V_4(k,x) = & \big[k_1\big(V_1\big(k+1,F(k,x)\big)\big) - k_1\big(V_1(k,x)\big)\big] \\
& \times V_3(k+1,F(k,x)) \\
& + k_1(V_1(k,x))\big[V_3(k+1,F(k,x)) - V_3(k,x)\big] \\
\leq & \big[k_1(V_1(k+1,F(k,x))) - k_1(V_1(k,x))\big] \\
& \times V_3(k+1,F(k,x)) \\
& - k_1(V_1(k,x))p(k+1)\lambda(V_1(k,x)) \\
& + k_1(V_1(k,x))\phi_1(|x|)\phi_2(N_1(k,x))\,.
\end{aligned} \tag{12.35}$$

Using the facts that F and V_3 are of class $\mathcal{USB}$ and $V_1 \in \mathcal{UPPD}$, we can easily construct continuous increasing everywhere positive functions Γ and Λ_2 satisfying

$$\begin{aligned}
\Delta_k V_4(k,x) \leq & \big[-\Delta_k V_1(k,x)\big]\Gamma\big(V_1(k,x)\big) - p(k+1)\Lambda_1\big(V_1(k,x)\big) \\
& + \Lambda_2\big(V_1(k,x)\big)\phi_2\big(N_1(k,x)\big)\ .
\end{aligned} \tag{12.36}$$

This can be done by first constructing an increasing everywhere positive function $\tilde{\alpha}$ such that $|k_1'(r)| \leq \tilde{\alpha}(r)$.

Next consider the function $k_2(s) = s\Gamma(s)$. Then $k_2 \in \mathcal{K}_\infty$ and $k_2(s)/s$ is increasing. Hence,

$$\frac{k_2(b) - k_2(a)}{b-a} \geq \frac{k_2(b)}{b}$$

when $b > a \geq 0$, which gives $k_2(b) - k_2(a) \geq (b-a)\Gamma(b)$ if $b \geq a \geq 0$. Specializing to the cases $a = V_1(k+1,F(k,x))$ and $b = V_1(k,x)$ gives

$$\Delta_k(k_2 \circ V_1)(k,x) \ \leq\ \Delta_k V_1(k,x)\Gamma(V_1(k,x))$$

everywhere. Therefore, the function $V_5 \doteq V_4 + k_2(V_1)$ satisfies

$$\Delta_k V_5(k,x) \leq -p(k+1)\Lambda_1\big(V_1(k,x)\big) + \Lambda_2\big(V_1(k,x)\big)\phi_2\big(N_1(k,x)\big) \tag{12.37}$$

everywhere.

Arguing as in the proof of Theorem 12.1 except with Γ replaced by Λ_1 provides V_6 such that

$$\begin{aligned}
\Delta_k V_6(k,x) \leq & -\frac{\delta}{4(l+1)}\Lambda_1\big(V_1(k,x)\big) \\
& + \Lambda_2\big(V_1(k,x)\big)\phi_2\big(N_1(k,x)\big).
\end{aligned} \tag{12.38}$$

A special case of Lemma A.3 provides a function $k_3 \in C^1 \cap \mathcal{PD}$ so that

$$k_3(r) \;\leq\; \phi_2^{-1}\left(\frac{\delta}{8(l+1)}\frac{\Lambda_1(r)}{1+\Lambda_2(r)}\right)\frac{1}{1+\Lambda_2(r)} \quad \text{and} \quad |k_3'(r)| \;\leq\; 1 \tag{12.39}$$

for all $r \geq 0$, and therefore also

$$\phi_2\big(k_3(V_1)\Lambda_2(V_1)\big)\Lambda_2(V_1) \leq \frac{\delta}{8(l+1)}\Lambda_1(V_1) \tag{12.40}$$

everywhere. Choose $\mu_F, \alpha_6 \in \mathcal{K}_\infty$ such that

$$\big|V_6(k,x)\big| \;\leq\; \alpha_6(|x|) \;\text{ and }\; |F(k,x)| \leq \mu_F(|x|)$$

everywhere. Replacing Γ with $\alpha_6 \circ \mu_F \circ \alpha_1^{-1}$ in the argument we used to build k_2 and recalling that $\Delta_k V_1(k,x) \leq 0$ allows us to build a function $k_4 \in \mathcal{K}_\infty$ satisfying

$$\begin{aligned}\Delta_k(k_4 \circ V_1)(k,x) &\leq \big[\alpha_6 \circ \mu_F \circ \alpha_1^{-1} \circ V_1(k,x)\big]\Delta_k V_1(k,x)\\ &\leq \alpha_6(\mu_F(|x|))\Delta_k V_1(k,x)\end{aligned}$$

everywhere.

Recalling from (12.39) that $|k_3'| \leq 1$ everywhere now gives

$$\begin{aligned}\big|\Delta_k(k_3 \circ V_1)(k,x)V_6\big(k+1,F(k,x)\big)\big| &\leq [-\Delta_k V_1(k,x)]\big|V_6\big(k+1,F(k,x)\big)\big|\\ &\leq -\alpha_6(\mu_F(|x|))\Delta_k V_1(k,x)\end{aligned}$$

everywhere. Hence, the function $V_7 \doteq k_3(V_1)V_6 + k_4(V_1)$ is such that

$$\begin{aligned}\Delta_k V_7(k,x) &\leq \big|\Delta_k(k_3 \circ V_1)(k,x)V_6\big(k+1,F(k,x)\big)\big|\\ &\quad + k_3\big(V_1(k,x)\big)\Delta_k V_6(k,x) + \Delta_k(k_4 \circ V_1)(k,x)\\ &\leq -\tfrac{\delta}{4(l+1)}k_3(V_1)\Lambda_1(V_1) + k_3(V_1)\Lambda_2(V_1)\phi_2\big(N_1(k,x)\big)\,.\end{aligned} \tag{12.41}$$

Next note that for all functions $\mu \in \mathcal{K}_\infty$, we have $ab \leq \mu(a)a + \mu^{-1}(b)b$ for all $a,b \geq 0$, by separately considering the cases where $\mu(a) \geq b$ and $\mu^{-1}(b) \geq a$. Hence, for all $\mu \in \mathcal{K}_\infty$, (12.41) gives

$$\begin{aligned}\Delta_k V_7 \leq &-\tfrac{\delta}{4(l+1)}k_3(V_1)\Lambda_1(V_1) + \mu\big(k_3(V_1)\Lambda_2(V_1)\big)k_3(V_1)\Lambda_2(V_1)\\ &+\mu^{-1}\big(\phi_2(N_1(k,x))\big)\phi_2(N_1(k,x)).\end{aligned} \tag{12.42}$$

Specializing to the case where $\mu = \phi_2$, we get

$$\begin{aligned} \Delta_k V_7 &\le -\tfrac{\delta}{4(l+1)} k_3(V_1)\Lambda_1(V_1) + \phi_2\big(k_3(V_1)\Lambda_2(V_1)\big)k_3(V_1)\Lambda_2(V_1) \\ &\quad + N_1(k,x)\phi_2(N_1) \\ &\le -\tfrac{\delta}{8(l+1)} k_3(V_1)\Lambda_1(V_1) + N_1(k,x)\phi_2(N_1) \quad \text{(by (12.40))}. \end{aligned} \tag{12.43}$$

Recalling that $N_1 \in \mathcal{USB}$ and $\phi_2 \in \mathcal{K}_\infty$, we can readily construct a function $\phi_3 \in \mathcal{K}_\infty$ such that

$$\Delta_k V_7 \;\le\; -\frac{\delta}{8(l+1)} k_3(V_1)\Lambda_1(V_1) + N_1(k,x)\phi_3(V_1). \tag{12.44}$$

By reasoning as in our construction of k_2 above (with Γ replaced by ϕ_3), we can build a function $k_5 \in \mathcal{K}_\infty$ satisfying

$$\begin{aligned} \Delta_k(k_5 \circ V_1)(k,x) &\le \Delta_k V_1(k,x)\phi_3\big(V_1(k,x)\big) \\ &\le -N_1(k,x)\phi_3\big(V_1(k,x)\big). \end{aligned}$$

Choosing $V_8 \doteq V_7 + k_5(V_1)$ and the positive definite function

$$\alpha_3(s) \;\doteq\; \frac{\delta}{8(l+1)} \min\big\{k_3(u)\Lambda_1(u) : \alpha_1(s) \le u \le \alpha_2(s)\big\},$$

it follows that

$$\Delta_k V_8(k,x) \le -\frac{\delta}{8(l+1)} k_3(V_1(k,x))\Lambda_1\big(V_1(k,x)\big) \;\le\; -\alpha_3(|x|). \tag{12.45}$$

This is the desired Lyapunov decay condition, but the function V_8 is not guaranteed to be of class $\mathcal{UPPD}$. However, we can transform V_8 into the desired strict Lyapunov function by arguing as we did at the end of the proof of Theorem 5.1, as follows. Choose an everywhere positive increasing C^1 function k_* such that $k_*(V_1(k,x)) \ge |\sup_r V_8(r,x)| + 1$ everywhere; this can be done because V_1 is of class $\mathcal{UPPD}$ and V_8 is of class $\mathcal{USB}$. One then shows that $V_9 \doteq V_1 V_8 + k_*(V_1)V_1$ is of class $\mathcal{UPPD}$ and satisfies

$$\begin{aligned} \Delta_k V_9(k,x) &= \big\{\Delta_k V_1(k,x)V_8(k+1,x) + k_*(V_1(k,x))\Delta_k(V_1(k,x))\big\} \\ &\quad + V_1(k,x)\Delta_k V_8(k,x) + \Delta_k k_*\big(V_1(k,x)\big)V_1(k+1,x) \\ &\le -V_1(k,x)\alpha_3(|x|) \end{aligned}$$

where the term in braces is non-positive by our choice of k_* and we also used the nonpositivity of $\Delta_k k_*(V_1(k,x))V_1(k+1,x)$. Therefore, V_9 satisfies the requirements of the theorem. In conjunction with Lemma 12.1 for the case of no controls, this proves the theorem. □

12.4.2 Hybrid Version

We now merge our continuous and discrete time Matrosov constructions to cover hybrid systems

$$\mathcal{H}_{\rm nc} \doteq \begin{cases} \dot{x} &= G(t,x), \quad x \in C \\ x_{k+1} &= F(k,x_k),\ x_k \in D \end{cases} \tag{12.46}$$

assuming $F, G \in \mathcal{USB}$ and the following:

Assumption 12.2 *There exist $V_1 \in \mathcal{UPPD}$ and $V_2 \in \mathcal{USB}$ that are C^1 in (t,x); everywhere non-negative $N_1, N_2 \in \mathcal{USB}$; a function $\chi : [0,\infty)^2 \times \mathbb{Z}_{\geq 0} \times \mathbb{R}^n \to \mathbb{R}$; an everywhere positive increasing ϕ_1; and $W \in \mathcal{PD}$, $p \in \mathcal{P}_{\rm dis}$, $\phi_2 \in \mathcal{K}_\infty$, and $q \in \mathcal{P}_{\rm cts}$ such that the following hold for all $t \geq 0$ and $k \in \mathbb{Z}_{\geq 0}$:*

1. *For all $x \in D$, we have:*
   $\Delta_k V_1(t,k,x) \ \leq\ -N_1(t,k,x)$;
   $\Delta_k V_2(t,k,x) \ \leq\ -N_2(t,k,x) + \chi(N_1(t,k,x),t,k,x)$; *and*
   $N_1(t,k,x) + N_2(t,k,x) \ \geq\ p(k+1)W(x)$.
2. *For all $x \in C$, we have:*
   $\mathcal{D}V_1(t,k,x) \ \leq\ -N_1(t,k,x)$;
   $\mathcal{D}V_2(t,k,x) \ \leq\ -N_2(t,k,x) + \chi(N_1(t,k,x),t,k,x)$; *and*
   $N_1(t,k,x) + N_2(t,k,x) \ \geq\ q(t)W(x)$.
3. *For all $x \in \mathbb{R}^n$, we have:*
   $|\chi(N_1(t,k,x),t,k,x)| \ \leq\ \phi_1(|x|)\phi_2(N_1(t,k,x))$.

The meaning of Assumption 12.2 is that the continuous and discrete time subsystems of $\mathcal{H}_{\rm nc}$ satisfy appropriately compatible discrete and continuous Matrosov Conditions. It applies to discrete time systems by choosing $C = \emptyset$ and $D = \mathbb{R}^n$; in that case, its Condition 2. is true vacuously. The following was announced in [92, 93]:

Theorem 12.4. *If $\mathcal{H}_{\rm nc}$ satisfies Assumption 12.2, then one can construct an explicit closed form strict Lyapunov function for $\mathcal{H}_{\rm nc}$. In particular, $\mathcal{H}_{\rm nc}$ is UGAS.*

Proof. Take $p \in \mathcal{P}_{\rm dis}(l,\delta)$ and $q \in \mathcal{P}_{\rm cts}(\tau,\varepsilon)$. We apply the first part of the proof of Theorem 12.3 for each $t \geq 0$, using the functions $(k,x) \mapsto V_1(t,k,x)$ and $(k,x) \mapsto V_2(t,k,x)$. This produces functions V_5 and $\Lambda_1, \Lambda_2 \in C^1$, with $\Lambda_1 \in \mathcal{K}_\infty$, such that

$$\begin{aligned} \Delta_k V_5(t,k,x) \leq &-p(k+1)\Lambda_1\left(V_1(t,k,x)\right) \\ &+\Lambda_2\left(V_1(t,k,x)\right)\phi_2(N_1(t,k,x)) \end{aligned} \tag{12.47}$$

for all $t \geq 0$, $k \in \mathbb{Z}_{\geq 0}$, and $x \in D$. We now use the continuous time analog of the preceding argument from Lemma A.9 to construct a continuous version

V_5^{cts} of V_5 for which

$$\begin{aligned} \mathcal{D}V_5^{\text{cts}}(t,k,x) \leq & -q(t)\Lambda_1\left(V_1(t,k,x)\right) \\ & +\Lambda_2\left(V_1(t,k,x)\right)\phi_2(N_1(t,k,x)) \end{aligned} \tag{12.48}$$

for all $t \geq 0$, $k \in \mathbb{Z}_{\geq 0}$, and $x \in C$. In fact, by enlarging Γ, we can enlarge k_2 in such a way that V_5^{cts} and V_5 have the same formula.

Applying the strictification method of Theorem 12.1 to V_5 and recalling Remark 12.1, we get

$$V_6^{\text{dis}}(t,k,x) \doteq V_5(t,k,x) + \frac{1}{4(l+1)}S(k)\Lambda_1\big(V_1(t,k,x)\big).$$

This satisfies (12.38) with V_6 replaced by V_6^{dis} and V_1 and N_1 also depending on t. In the same way, we can apply the continuous time strictification approach from Chap. 6 as we did in the proof of Theorem 12.2 to the function V_5^{cts}. This gives a function

$$V_6^{\text{cts}}(t,k,x) \doteq V_5^{\text{cts}}(t,k,x) + \frac{1}{\tau}\left[\int_{t-\tau}^{t}\int_{z}^{t} q(\nu)\,\mathrm{d}\nu\,\mathrm{d}z\right]\Lambda_1\big(V_1(t,k,x)\big)$$

for which

$$\mathcal{D}V_6^{\text{cts}}(t,k,x) \;\leq\; -\frac{\varepsilon}{\tau}\Lambda_1\big(V_1(t,k,x)\big) + \Lambda_2\big(V_1(t,k,x)\big)\phi_2\big(N_1(t,k,x)\big)$$

when $x \in C$. We can assume that

$$\frac{\delta}{4(l+1)} < 1$$

without relabeling, by enlarging l as needed. Setting $V_6 = V_6^{\text{cts}} + V_6^{\text{dis}}$ and recalling that V_5^{cts} and V_5 have the same formula, we can enlarge the function Λ_2 sufficiently in such a way that

$$\begin{aligned} \Delta_k V_6(t,k,x) \leq & -\tfrac{\delta}{4(l+1)}\Lambda_1\big(V_1(t,k,x)\big) \\ & +\Lambda_2\big(V_1(t,k,x)\big)\phi_2\big(N_1(t,k,x)\big) \quad \forall x \in D \\ \mathcal{D}V_6(t,k,x) \leq & -\tfrac{\delta}{4(l+1)}\Lambda_1\big(V_1(t,k,x)\big) \\ & +\Lambda_2\big(V_1(t,k,x)\big)\phi_2\big(N_1(t,k,x)\big) \quad \forall x \in C \end{aligned} \tag{12.49}$$

hold for all $t \geq 0$ and $k \in \mathbb{Z}_{\geq 0}$.

Next, we follow the rest of the proof of Theorem 12.3, applied to V_6 for each choice of $t \geq 0$. This produces a function $V_9^{\text{dis}}(t,k,x)$ that satisfies the conclusion of the proof when $x \in D$. Similarly, we apply the continuous time analog of that part of the proof to V_6 for each k to get a function V_9^{cts}

satisfying

$$\mathcal{D}V_9^{\text{cts}}(t,k,x) \;\le\; -\tilde{\alpha}(|x|) \tag{12.50}$$

for all $t \ge 0$, $k \in \mathbb{Z}_{\ge 0}$, and $x \in C$ for a suitable function $\tilde{\alpha} \in \mathcal{PD}$. This continuous time argument is as in the discrete time case, except with $\Delta_k V_i$ replaced by $\mathcal{D}V_i$ for all i. It is similar to the proof we give for Lemma A.9. By enlarging the functions $k_4, k_5 \in \mathcal{K}_\infty$ and reducing $k_3 \in \mathcal{PD}$ in the discrete and continuous versions of the proof, we can assume that they are the same, which means that V_9^{cts} and V_9^{dis} have the same expression (as functions of V_6). Therefore, we can meet the requirements of the theorem with their common expression. In conjunction with Lemma 12.3, this proves the theorem. □

12.5 Illustrations

Assume that the continuous time system

$$\dot{x} = G(t,x,u) \tag{12.51}$$

admits $V \in C^1$, $q \in \mathcal{P}_{\text{cts}}$, and $\Gamma, \chi, \alpha_1, \alpha_2 \in \mathcal{K}_\infty$ satisfying:

1. $\alpha_1(|x|) \le V(t,x) \le \alpha_2(|x|)$ for all $x \in \mathbb{R}^n$ and $t \ge 0$;
2. $\mathcal{D}V(t,x,u) \le -q(t)\Gamma(V(t,x))$ for all $t \ge 0$, $x \in \mathbb{R}^n$ and $u \in \mathbb{R}^d$ satisfying $|x| \ge \chi(|u|)$; and
3. $x \mapsto V(t,x)$ is convex for each $t \in [0,\infty)$.

The preceding requirements all hold if for example

$$\dot{x} = G(t,x,u) \equiv A(t)x \tag{12.52}$$

is GAS and $A(t)$ is continuous and bounded, in which case we take

$$V(t,x) \;\doteq\; x^\top P(t)x \tag{12.53}$$

for a suitable matrix $P(t)$, e.g, by arguing as in [70, Sect. 4.6]. Given $C, D \subseteq \mathbb{R}^n$ and $p \in \mathcal{P}_{\text{dis}}$ valued in $[0,1]$, we give conditions on $h \in \mathcal{USB}$ guaranteeing that we can construct an ISS-CLF for

$$\mathcal{H}_p \doteq \begin{cases} \dot{x} &= G(t,x,u)\ , \quad x \in C \\ x_{k+1} &= F(k,x_k,u)\ , \ x_k \in D \end{cases}, \tag{12.54}$$

where

$$F(k,x,u) \doteq [1 - p(k+1)]x + p(k+1)h(k,x,u).$$

The construction to follow also works if instead of assuming that V is convex in x and p is valued in $[0,1]$, we just assume that $p(k) \in \{0,1\}$ for all $k \in \mathbb{Z}_{\ge 0}$.

To find our conditions, first note that by reducing $\Gamma \in \mathcal{K}_\infty$ as needed without relabeling, we can assume

$$\Gamma \in C^1 \cap \mathcal{K}_\infty \text{ and } \Gamma(s) \leq \frac{1}{2}\alpha_1\big(\alpha_2^{-1}(s)\big)$$

for all $s \geq 0$. Assume that

$$|h(k,x,u)| \leq \alpha_2^{-1}\big(0.5\alpha_1(|x|)\big) \tag{12.55}$$

for all $x \in D$, $k \in \mathbb{Z}_{\geq 0}$, and $u \in \mathbb{R}^d$ satisfying $|x| \geq \chi(|u|)$, which reduces to linear growth if V has the form (12.53) and P has globally bounded positive eigenvalues. Using the facts that V is convex in x and $p(k) \in [0,1]$ everywhere, we have

$$\begin{aligned} V\big(t, F(k,x,u)\big) - V(t,x) &\leq [1-p(k+1)]V(t,x) \\ &\quad + p(k+1)V\big(t, h(k,x,u)\big) - V(t,x) \\ &\leq p(k+1)\alpha_2\big(|h(k,x,u)|\big) \\ &\quad - p(k+1)\alpha_1(|x|) \end{aligned} \tag{12.56}$$

and therefore

$$\begin{aligned} V\big(t, F(k,x,u)\big) - V(t,x) &\leq -\tfrac{1}{2}p(k+1)\alpha_1(|x|) \\ &\leq -p(k+1)\Gamma\big(\alpha_2(|x|)\big) \\ &\leq -p(k+1)\Gamma\big(V(t,x)\big) \end{aligned} \tag{12.57}$$

whenever $|x| \geq \chi(|u|)$, $t \geq 0$, and $k \in \mathbb{Z}_{\geq 0}$. Arguing as in the proof of Theorem 12.2, we readily obtain an explicit global strict ISS-CLF

$$\begin{aligned} V^\sharp(t,k,x) &\doteq 2V(t,x) + \left[\frac{1}{4\tau}\int_{t-\tau}^{t}\int_{s}^{t} q(r)\,\mathrm{d}r\,\mathrm{d}s\right]\Gamma\big(V(t,x)\big) \\ &\quad + \left[\frac{1}{4(l+1)}\sum_{s=k-l}^{k}\sum_{j=s}^{k} p(j)\right]\Gamma\big(V(t,x)\big) \end{aligned}$$

for $\mathcal{H}_p$, where l and τ are as in the requirements $p \in \mathcal{P}_{\text{dis}}$ and $q \in \mathcal{P}_{\text{cts}}$, so $\mathcal{H}_p$ is ISS.

We now specialize the preceding construction to our example (12.5). As we saw in Sect. 10.6.2, its subsystem

$$\dot{x} = G(t,x,u) \doteq -\bar{m}(t)\bar{m}^\top(t)x$$

has the strict Lyapunov function

$$V(t,x) = x^\top \left(\lambda I + \int_{t-c}^{t} \int_{s}^{t} \bar{m}(l) m^\top(l)\, \mathrm{d}l\, \mathrm{d}s \right) x, \tag{12.58}$$

where λ and c are defined in Sect. 12.1. In fact,

$$\mathcal{D}V(t,x) \ \le \ -\frac{a}{2}|x|^2 \ \le \ -\Gamma(V(t,x)) \tag{12.59}$$

everywhere, in terms of the function

$$\Gamma(s) = \frac{as}{2\{\lambda + c^2/2\}^2}$$

and the constant a from Sect. 12.1. The choice of Γ easily follows from Lemma 12.2.

Taking

$$\alpha_1(s) = \lambda s^2 \ \text{ and } \ \alpha_2(s) = \left(\lambda + \frac{c^2}{2} \right) s^2,$$

we can easily check that $\Gamma(s) \le 0.5\alpha_1(\alpha_2^{-1}(s))$ everywhere. Moreover, we can use (12.4) to check that condition (12.55) on h is satisfied for large enough constants $\lambda > 1$. Recalling that $p \in \mathcal{P}_{\text{dis}}$ is valued in $[0,1]$ and V is convex in the state, the preceding construction applies with $q(t) \equiv 1$ and $\tau = 1$. Also, the strict Lyapunov function $V^\sharp$ becomes (12.6). It follows that (12.5) has the ISS-CLF (12.6) and so is ISS by Lemma 12.3, as we claimed in Sect. 12.1.

12.6 Comments

The hybrid systems framework we used in this chapter was systematically developed by [20, 24] and has been used extensively by several authors; see [176] for a recent survey. It is well appreciated that hybrid controllers are useful for stabilizing nonlinear systems that are not stabilizable by continuous time state feedbacks; see for example [140] for robust quasi-time optimal hybrid stabilization for Brockett's Integrator. Moreover, using the notions of graphical convergence and set convergence [142], one can characterize when a collection of arcs converges to a hybrid arc [48].

The hybrid framework of this chapter can incorporate hybrid automata, as well as switching systems $\dot{z} = f_q(z)$ with average dwell-time conditions, by using appropriate choices of the discrete sub-dynamics [176]. There is also a hybrid version of LaSalle Invariance for cases where the hybrid system (12.7) is time-invariant with no controls. It is expressed in terms of the functions $u_c(x) = \langle \nabla V(x), G(x) \rangle$ and $u_d(x) = V(F(x)) - V(x)$, where $V \in C^1$ is chosen so that $u_c(x)$ is non-positive on $\bar{C}$ and $u_d(x)$ is non-positive on $\bar{D}$ [176]. One can also develop hybrid converse Lyapunov function theory [21].

However, there has been little systematic work on constructive nonlinear control for hybrid systems. Our treatment here is based on [93] (which covers systems with no controls) and [92] (which announced the extensions to control systems). For an alternative construction of strict Lyapunov functions for discrete time systems with no controls, involving *infinite sums* of PE parameter values, see [120].

Part V
Appendices

Appendix A
Some Lemmas

A.1 Useful Families of Functions

We prove the lemmas needed in several chapters. We maintain our convention that all functions encountered should be understood to be sufficiently smooth, and that all (in)equalities should be understood to hold globally unless otherwise indicated. The first four lemmas we prove are from [111].

While not used explicitly in the main text, we use the following lemma to prove Lemma A.2:

Lemma A.1. *Let $F : [0, \infty) \times [0, \infty) \to [0, \infty)$ be a continuous function that admits a non-decreasing continuous function $\Theta : [0, \infty) \to (0, \infty)$ such that*

$$F(a, 0) = 0 \tag{A.1}$$

and

$$F(a, b) \leq \Theta(a)\Theta(b) \tag{A.2}$$

for all $(a, b) \in [0, \infty) \times [0, \infty)$. Then the function $Z : [0, \infty) \to \mathbb{R}$ defined by

$$Z(b) = \sup_{\alpha \geq 0} \frac{F(\alpha, b)}{(\alpha^2 + 1)\Theta(\alpha)} \tag{A.3}$$

is everywhere non-negative and continuous. Moreover, $Z(0) = 0$ and

$$F(a, b) \leq (a^2 + 1)\Theta(a)Z(b) \tag{A.4}$$

for all $(a, b) \in [0, \infty)^2$.

Proof. Let us prove that Z is well defined on $[0, \infty)$. To simplify the notation, we use the function

$$\overline{F}(\alpha, b) = \frac{F(\alpha, b)}{(\alpha^2 + 1)\Theta(\alpha)} \; . \tag{A.5}$$

Since (A.2) is satisfied, we have

$$\overline{F}(\alpha, b) \leq \frac{\Theta(\alpha)}{(\alpha^2+1)\Theta(\alpha)}\Theta(b) \leq \Theta(b) \,. \tag{A.6}$$

It follows that $Z(b) = \sup\{\overline{F}(\alpha, b) : \alpha \in [0, \infty)\}$ is a finite non-negative real number for all $b \geq 0$. From the definition of Z, (A.1), and (A.2), we deduce easily that Z is everywhere non-negative, that (A.4) is satisfied, and that $Z(0) = 0$. We now prove that this function is continuous. Let $b_c \geq 0$ and $\varepsilon > 0$ be given constants. For all $b \geq 0$,

$$Z(b) = \max\left\{ \sup_{\alpha\in[0,\alpha^*]} \overline{F}(\alpha, b), \sup_{\alpha\geq\alpha^*} \overline{F}(\alpha, b) \right\} \tag{A.7}$$

with $\alpha^* = \sqrt{\frac{2}{\varepsilon}\Theta(b_c+1)}$. From (A.6), it follows that

$$\begin{aligned} \sup_{\alpha\geq\alpha^*} \overline{F}(\alpha, b) &\leq \sup_{\alpha\geq\alpha^*} \frac{\Theta(\alpha)\Theta(b)}{(\alpha^2+1)\Theta(\alpha)} \\ &= \sup_{\alpha\geq\alpha^*} \frac{\Theta(b)}{\alpha^2+1} \leq \frac{\Theta(b)}{\frac{2}{\varepsilon}\Theta(b_c+1)+1} \end{aligned} \tag{A.8}$$

for all $b \geq 0$. Since Θ is non-decreasing, it follows that

$$\sup_{\alpha\geq\alpha^*} \overline{F}(\alpha, b) \leq \tfrac{\varepsilon}{2} \tag{A.9}$$

for all $b \in [0, b_c+1]$.

We deduce easily that

$$\sup_{\alpha\in[0,\alpha^*]} \overline{F}(\alpha, b) \leq Z(b) \leq \sup_{\alpha\in[0,\alpha^*]} \overline{F}(\alpha, b) + \frac{\varepsilon}{2} \tag{A.10}$$

for all $b \in [0, b_c+1]$. In particular,

$$\sup_{\alpha\in[0,\alpha^*]} \overline{F}(\alpha, b_c) \leq Z(b_c) \leq \sup_{\alpha\in[0,\alpha^*]} \overline{F}(\alpha, b_c) + \frac{\varepsilon}{2} \,. \tag{A.11}$$

From (A.10) and (A.11), we deduce that

$$|Z(b) - Z(b_c)| \leq \left| \sup_{\alpha\in[0,\alpha^*]} \overline{F}(\alpha, b) - \sup_{\alpha\in[0,\alpha^*]} \overline{F}(\alpha, b_c) \right| + \frac{\varepsilon}{2} \tag{A.12}$$

for all $b \in [0, b_c+1]$. The function

$$b \mapsto \max_{\alpha\in[0,\alpha^*]} \overline{F}(\alpha, b)$$

is continuous because $[0, \alpha^*]$ is a compact set. It follows that there exists a constant $\delta \in (0, 1]$ such that, for all $b \in [\max\{0, b_c - \delta\}, b_c + \delta]$,

$$\left| \sup_{\alpha \in [0,\alpha^*]} \overline{F}(\alpha, b) - \sup_{\alpha \in [0,\alpha^*]} \overline{F}(\alpha, b_c) \right| \leq \frac{\varepsilon}{2} \, . \tag{A.13}$$

From (A.12) and (A.13), we conclude that $|Z(b) - Z(b_c)| \leq \varepsilon$ for all $b \in [\max\{0, b_c - \delta\}, b_c + \delta]$. Hence, Z is continuous on $[0, \infty)$. This finishes the proof. □

We used the following lemma in Chap. 8:

Lemma A.2. *Let $n \geq 1$ and $q \geq 2$ be integers and $\chi_* : \mathbb{R}^{n+q-1} \to \mathbb{R}$ be an everywhere non-negative continuous function such that*

$$\chi_*(x, 0, ..., 0) = 0 \;\; \forall x \in \mathbb{R}^n . \tag{A.14}$$

Then, one can determine a continuous, everywhere positive, non-decreasing function ρ_* and a function $\phi_* \in \mathcal{K}_\infty$ such that

$$\chi_*(x, r_1, ..., r_{q-1}) \leq \phi_* \left(\sum_{k=1}^{q-1} r_k \right) \rho_*(|x|) \tag{A.15}$$

for all $x \in \mathbb{R}^n$ and all non-negative values $r_1, ..., r_{q-1}$.

Proof. Define $F_* : [0, \infty)^2 \to \mathbb{R}$ by

$$F_*(s, R) = \max_{(z, l_1, \ldots, l_{q-1}) \in E(s,R)} \chi_*(z, l_1, \ldots, l_{q-1}) \, , \tag{A.16}$$

where

$$E(s, R) = \left\{ (z, l_1, ..., l_{q-1}) \in \mathbb{R}^n \times [0, \infty)^{q-1} : |z| \leq s, l_k \in [0, R] \, , k = 1, ..., q-1 \right\} .$$

Then

$$\chi_*(x, r_1, \ldots, r_{q-1}) \leq F_* \left(|x|, \sum_{k=1}^{q-1} r_k \right) \tag{A.17}$$

for all $x \in \mathbb{R}^n$ and non-negative r_i's. Also, (A.14) gives

$$\begin{aligned} F_*(s, 0) &= \max_{(z, l_1, \ldots, l_{q-1}) \in E(s,0)} \chi_*(z, l_1, \ldots, l_{q-1}) \\ &= \max_{\{z \in \mathbb{R}^n : |z| \leq s\}} \chi_*(z, 0, \ldots, 0) \;=\; 0. \end{aligned} \tag{A.18}$$

Moreover, F_* is everywhere non-negative and non-decreasing with respect to each of its arguments. This implies that

$$F_*(s, R) \leq [F_*(s, s) + 1][F_*(R, R) + 1] \tag{A.19}$$

for all $s \in [0, \infty)$ and $R \in [0, \infty)$.

Therefore, Lemma A.1 applies to the function F_* and provides a continuous everywhere non-negative function Z that is zero at zero and such that

$$F_*(s, R) \leq (s^2 + 1)[F_*(s, s) + 1]Z(R) \tag{A.20}$$

for all $s \geq 0$ and $R \geq 0$. From (A.17), it follows that

$$|\chi_*(x, r_1, \ldots, r_{q-1})| \leq Z\left(\sum_{k=1}^{q-1} r_k\right) \rho_*(|x|) \tag{A.21}$$

for all $x \in \mathbb{R}^n$, $r_1 \geq 0$,...,$r_{q-1} \geq 0$, where

$$\rho_*(s) = (s^2 + 1)[F_*(s, s) + 1] \ . \tag{A.22}$$

This function is everywhere positive and non-decreasing on $[0, \infty)$ and

$$|\chi_*(x, r_1, \ldots, r_{q-1})| \leq \phi_*\left(\sum_{k=1}^{q-1} r_k\right) \rho_*(|x|) \tag{A.23}$$

for all $x \in \mathbb{R}^n$, $r_1 \in [0, \infty)$,...,$r_{q-1} \in [0, \infty)$, where

$$\phi_*(s) = s + \sup_{l \in [0,s]} Z(l) \ \ \forall s \geq 0 \ . \tag{A.24}$$

One can prove easily that ϕ_* is of class $\mathcal{K}_\infty$. This completes the proof of the lemma. □

Lemma A.3. *Let $w_1, w_2 : \mathbb{R}^n \to \mathbb{R}$ be any continuous positive definite functions, and let $V : [0, \infty) \times \mathbb{R}^n \to \mathbb{R}$ be any storage function. Let $N \in \mathbb{N}$ be arbitrary. Then one can construct a real-valued function $L \in C^N$ such that $L(0) = 0$, $L(s) > 0$ for all $s > 0$, and*

$$L\big(V(t, x)\big) \leq w_1(x) \tag{A.25}$$

and

$$\big|L'\big(V(t, x)\big)\big| \leq w_2(x) \tag{A.26}$$

hold for all $(t, x) \in [0, \infty) \times \mathbb{R}^n$.

Proof. We will presently construct an everywhere positive increasing C^N function ρ and a function $\alpha \in \mathcal{K}_\infty \cap C^N$ such that

$$\alpha\big(V(t, x)\big) \leq w_1(x)\rho\big(V(t, x)\big) \tag{A.27}$$

and

$$\alpha\big(V(t, x)\big) \leq w_2(x)\rho\big(V(t, x)\big) \ . \tag{A.28}$$

For the time being, we assume that these functions are known and introduce the function

$$L(s) \doteq \int_{\frac{s}{2}}^{s} \frac{\alpha(l)}{2(1+l^2)(1+\rho^2(2l))} \mathrm{d}l \,. \tag{A.29}$$

Then $L(0)=0$, $L(s)>0$ for all $s>0$, and L is of class C^N. Also, since both α and ρ are increasing, we get

$$L(s) \leq \int_{\frac{s}{2}}^{s} \frac{\alpha(s)}{2\left(1+\left(\frac{s}{2}\right)^2\right)(1+\rho^2(s))} \mathrm{d}l \leq \frac{\alpha(s)}{1+\rho^2(s)} \leq \frac{\alpha(s)}{\rho(s)} \tag{A.30}$$

for all $s \geq 0$. It follows that

$$L(V(t,x)) \leq \frac{\alpha(V(t,x))}{\rho(V(t,x))} \leq w_1(x) \,. \tag{A.31}$$

Therefore (A.25) is satisfied. On the other hand,

$$L'(s) = \frac{\alpha(s)}{2(1+s^2)(1+\rho^2(2s))} - \frac{\alpha\left(\frac{s}{2}\right)}{4\left(1+\left(\frac{s}{2}\right)^2\right)(1+\rho^2(s))} \quad \forall s \geq 0\,. \tag{A.32}$$

Since both α and ρ are increasing, it follows that

$$\begin{aligned} |L'(s)| &\leq \frac{\alpha(s)}{2(1+s^2)(1+\rho^2(2s))} + \frac{\alpha\left(\frac{s}{2}\right)}{4\left(1+\left(\frac{s}{2}\right)^2\right)(1+\rho^2(s))} \\ &\leq \frac{\alpha(s)}{2(1+\rho^2(s))} + \frac{\alpha(s)}{4(1+\rho^2(s))} \\ &\leq \frac{\alpha(s)}{\rho(s)} \quad \forall s \geq 0\,. \end{aligned} \tag{A.33}$$

Consequently, the inequalities

$$\left|L'(V(t,x))\right| \leq \frac{\alpha(V(t,x))}{\rho(V(t,x))} \leq w_2(x) \tag{A.34}$$

are satisfied, and therefore (A.26) is satisfied.

We end the proof by constructing an everywhere positive, increasing C^N function ρ and a function $\alpha \in \mathcal{K}_\infty \cap C^N$ such that (A.27) and (A.28) are satisfied. We introduce the four functions

$$w(x) = \min\{w_1(x), w_2(x)\}\,, \tag{A.35}$$

$$\delta_l(r) = \begin{cases} \min\limits_{\{z:|z|\in[1,r]\}} w(z), & \text{if } \; r \geq 1 \\ W_f, & \text{if } \; r \in [0,1]\,, \end{cases} \tag{A.36}$$

$$\delta_s(r) = \begin{cases} \min\limits_{\{z:|z|\in[r,1]\}} w(z), & \text{if} \quad r \in [0,1] \\ W_f, & \text{if} \quad r \geq 1 \ , \end{cases} \tag{A.37}$$

$$\text{and} \ \delta(r) = \frac{1}{W_f}\delta_s(r)\delta_l(r) \ , \tag{A.38}$$

where we use the constant

$$W_f = \min_{\{z:|z|=1\}} w(z) \ . \tag{A.39}$$

We have the following two cases:

1. If $|x| \leq 1$, then

$$\delta(|x|) \ = \ \frac{1}{W_f}\delta_s(|x|)\delta_l(|x|) \ = \ \delta_s(|x|) \ = \ \min_{\{z:|z|\in[|x|,1]\}} w(z) \leq w(x).$$

2. If $|x| \geq 1$, then

$$\delta(|x|) \ = \ \frac{1}{W_f}\delta_s(|x|)\delta_l(|x|) \ = \ \delta_l(|x|) \ = \ \min_{\{z:|z|\in[1,|x|]\}} w(z) \leq w(x).$$

It follows that for all $x \in \mathbb{R}^n$,

$$w(x) \geq \delta(|x|) = \frac{1}{W_f}\delta_s(|x|)\delta_l(|x|) \ . \tag{A.40}$$

Since w is a positive definite function, δ_l is an everywhere positive function on $[0,\infty)$. Therefore, (A.40) gives

$$\delta_s(|x|) \ \leq \ w(x)\frac{W_f}{\delta_l(|x|)} \ \ \forall x \in \mathbb{R}^n \ . \tag{A.41}$$

We introduce the two functions

$$\alpha_a(r) = r\delta_s(r) \ \text{ and } \ \rho_a(r) = \frac{W_f(1+r)}{\delta_l(r)} \ . \tag{A.42}$$

By (A.41),

$$\alpha_a(|x|) \ \leq \ w(x)\rho_a(|x|) \ \ \forall x \in \mathbb{R}^n \ . \tag{A.43}$$

Since w is positive definite and at least continuous, one can prove easily that $\delta_s(0) = 0$, $\delta_s(r) > 0$ if $r > 0$ and δ_s is non-decreasing and continuous. It follows that $\alpha_a \in \mathcal{K}_\infty$. For similar reasons, δ_l is continuous, everywhere positive and non-increasing. It follows that ρ_a is everywhere positive and increasing. Since V is a storage function, we can find functions $\gamma_1, \gamma_2 \in \mathcal{K}_\infty$ such that

$$\gamma_1(|x|) \leq V(t,x) \leq \gamma_2(|x|) \tag{A.44}$$

for all $t \in [0, \infty)$ and $x \in \mathbb{R}^n$. Using the properties of α_a and ρ_a and (A.44), we deduce that

$$\alpha_a(\gamma_2^{-1}(V(t,x)) \leq w(x)\rho_a(\gamma_1^{-1}(V(t,x))) \tag{A.45}$$

for all $(t,x) \in [0,\infty) \times \mathbb{R}^n$.

As an immediate consequence, we have

$$V^N(t,x)\alpha_b(V(t,x)) \ \leq \ w(x)[V(t,x)+1]^N \rho_b(V(t,x)) \ , \tag{A.46}$$

where

$$\alpha_b(r) = \alpha_a(\gamma_2^{-1}(r)) \ \text{ and } \ \rho_b(r) = \rho_a(\gamma_1^{-1}(r)). \tag{A.47}$$

Next consider the functions

$$\alpha(r) = \textstyle\int_0^r \int_0^{s_1} \ldots \int_0^{s_{N-1}} \alpha_b(s_N) \mathrm{d}s_N \ldots \mathrm{d}s_1 \tag{A.48}$$

and

$$\rho(r) = \textstyle\int_0^{r+1} \int_0^{s_1+1} \ldots \int_0^{s_{N-1}+1} (s_N+1)^N \rho_b(s_N) \mathrm{d}s_N \ldots \mathrm{d}s_1 \ . \tag{A.49}$$

For all $r \geq 0$, we then have

$$\alpha(r) \leq r^N \alpha_b(r) \ \text{ and } \ \rho(r) \geq (r+1)^N \rho_b(r) \ , \tag{A.50}$$

by replacing the lower bounds in the integrations in (A.49) with r. These inequalities and (A.46) yield

$$\alpha\big(V(t,x)\big) \ \leq \ w(x)\rho\big(V(t,x)\big) \ . \tag{A.51}$$

Since $0 \leq w(x) \leq w_1(x)$ and $0 \leq w(x) \leq w_2(x)$, we deduce that (A.27) and (A.28) are satisfied. One can check readily that ρ is everywhere positive, increasing and C^N, and that $\alpha \in \mathcal{K}_\infty \cap C^N$. This concludes the proof. □

Lemma A.4. *Let $\Omega : \mathbb{R}^n \to \mathbb{R}$ be a continuous function. Then, the function $\zeta : [0, \infty) \to \mathbb{R}$ defined by*

$$\zeta(r) = 1 + \textstyle\int_0^{r+1} \int_0^{s_1+1} \ldots \int_0^{s_{N-1}+1} \left[\sup_{\{z \in \mathbb{R}^n : |z| \leq s_N\}} |\Omega(z)|\right] \mathrm{d}s_N \ldots \mathrm{d}s_1$$

is everywhere positive, of class C^N, and non-decreasing and $|\Omega(x)| \leq \zeta(|x|)$ for all $x \in \mathbb{R}^n$.

Proof. From the definition of ζ, it follows immediately that ζ is everywhere positive, non-decreasing and of class C^N. To simplify the notation, we define the function

$$\Omega_s(r) = \sup_{\{z \in \mathbb{R}^n : |z| \leq r\}} |\Omega(z)| \ . \tag{A.52}$$

The function Ω_s is non-decreasing on $[0, \infty)$. Therefore, for all $s_{N-1} \geq 0$,

$$\int_0^{s_{N-1}+1} \Omega_s(s_N)\mathrm{d}s_N \ \geq\ \int_{s_{N-1}}^{s_{N-1}+1} \Omega_s(s_N)\mathrm{d}s_N \ \geq\ \Omega_s(s_{N-1})\ . \qquad \text{(A.53)}$$

By integrating both sides of (A.53) over $[s_{N-2}, s_{N-2}+1]$ and arguing inductively, we deduce that

$$\int_0^{r+1}\int_0^{s_1+1}\ldots\int_0^{s_{N-1}+1} \Omega_s(s_N)\mathrm{d}s_N\ldots\mathrm{d}s_1 \geq \Omega_s(r)\ \forall r\geq 0\ , \qquad \text{(A.54)}$$

It follows that

$$\zeta(|x|)\geq \Omega_s(|x|) = \sup_{\{z\in\mathbb{R}^n:|z|\leq|x|\}} |\Omega(z)| \geq |\Omega(x)| \qquad \text{(A.55)}$$

for all $x\in\mathbb{R}^n$, proving the lemma. □

We used the following lemma from [108], as Lemma 10.3 in Sect. 10.6.2:

Lemma A.5. *Let Assumption 10.2 hold and set*

$$P(t) = \left\{\frac{\tilde{c}}{2|f^\star|} + \frac{1}{4\alpha'}\tilde{c}^4|f^\star|\right\} I_n + \int_{t-\tilde{c}}^{t}\int_s^t \bar{m}(l)\bar{m}^\top(l)\,\mathrm{d}l\,\mathrm{d}s. \qquad \text{(A.56)}$$

Then $V(t,x)=x^\top P(t)x$ is a strict Lyapunov function for (10.38) for which $2V/\alpha'$ satisfies the conclusions of Lemma 2.2.

Proof. Sketch. Let κ denote the constant in braces in (10.39). Applying (10.13) and (10.14), we easily check that the time derivative of V along trajectories of (10.38) is

$$\begin{aligned}\dot{V} &= (2f^\star\kappa+\tilde{c})x^\top\bar{m}(t)\bar{m}^\top(t)x\\ &\quad +2f^\star x^\top\left[\int_{t-\tilde{c}}^t\int_s^t \bar{m}(l)\bar{m}^\top(l)\,\mathrm{d}l\,\mathrm{d}s\right]\bar{m}(t)\bar{m}^\top(t)x\\ &\quad -x^\top\left[\int_{t-\tilde{c}}^t \bar{m}(l)\bar{m}^\top(l)\,\mathrm{d}l\right]x\\ &\leq (2f^\star\kappa+\tilde{c})|\bar{m}^\top(t)x|^2+\tilde{c}^2|f^\star|\,|x||\bar{m}^\top(t)x| - \alpha'|x|^2\end{aligned}$$

by Condition 2.i. from Assumption 10.2. Since

$$\tilde{c}^2|f^\star||x||\bar{m}^\top(t)x| \ \leq\ \frac{1}{2}\alpha'|x|^2 + \frac{1}{2\alpha'}\tilde{c}^4|f^\star|^2|\bar{m}^\top(t)x|^2\ ,$$

we get

$$\dot{V}\ \leq\ \left(2f^\star\kappa+\tilde{c}+\frac{1}{2\alpha'}\tilde{c}^4|f^\star|^2\right)|\bar{m}^\top(t)x|^2 - \frac{1}{2}\alpha'|x|^2$$

everywhere. The result is now immediate from the choice of κ and the fact that $|\bar{m}(t)|=1$ everywhere. □

We used the following lemma from [93] in Chap. 12, where the notation is from Definition 12.2:

Lemma A.6. *Let* $\Theta \in \mathcal{PD} \cap C^1$ *be non-decreasing on* $[0,1]$ *and non-increasing on* $[1,\infty)$*, and let* $p \in \mathcal{P}_{\mathrm{dis}}$*. Choose the following functions* $\mu : [0,\infty) \to [1,\infty)$*,* κ*,* χ*, and* γ*:*

$$\kappa(r) \doteq 2\int_0^r \mu(z)\,dz, \quad \chi(r) \doteq \Theta(2r)\mu(r),$$
$$\gamma(s) \doteq \chi\big(\kappa^{-1}(s)/2\big), \text{ and } \mu(r) = \begin{cases} 1+4r^2, \ 0 \le r \le 1/2 \\ \dfrac{4\Theta(1)r}{\Theta(2r)}, \ 1/2 \le r < \infty \end{cases}. \tag{A.57}$$

Assume that $\nu \in \mathcal{UPPD}$ *is such that*

$$\Delta_k \nu(k,x) \le -p(k+1)\Theta\big(\nu(k,x)\big)$$

for all $x \in \mathbb{R}^n$ *and* $k \in \mathbb{Z}_{\ge 0}$*, along the trajectories of* $x^+ = F(k,x)$*, where* $F \in \mathcal{USB}$*. Then* $\kappa \in \mathcal{K}_\infty \cap C^1$*,* $\gamma \in \mathcal{K}_\infty$*,* $\chi \in \mathcal{K}_\infty$*, and* $V \doteq \kappa(\nu) \in \mathcal{UPPD}$ *satisfies*

$$\Delta_k V(k,x) \le -p(k+1)\gamma\big(V(k,x)\big) \tag{A.58}$$

for all $x \in \mathbb{R}^n$ *and* $k \in \mathbb{Z}_{\ge 0}$*.*

Proof. Sketch. Fixing $x \in \mathbb{R}^n$ and $k \in \mathbb{Z}_{\ge 0}$ and applying the Fundamental Theorem of Calculus to the function

$$s \mapsto \mathcal{F}(s) \doteq \kappa\big(s\nu(k+1, F(k,x)) + (1-s)\nu(k,x)\big)$$

gives $\Delta_k V(k,x) = \mathcal{F}(1) - \mathcal{F}(0) = \int_0^1 \mathcal{F}'(s)\,\mathrm{d}s$, i.e.,

$$\begin{aligned}
\Delta_k V(k,x) &= \left[\int_0^1 \kappa'\big(s\nu(k+1,F(k,x)) + (1-s)\nu(k,x)\big)\mathrm{d}s\right] \\
&\quad \times \big[\nu(k+1,F(k,x)) - \nu(k,x)\big] \\
&\le -p(k+1)\left[\int_0^1 \kappa'\big((1-s)\nu(k,x)\big)\mathrm{d}s\right]\Theta\big(\nu(k,x)\big) \\
&\le -p(k+1)\left[\int_0^{1/2} \kappa'\left(\frac{1}{2}\nu(k,x)\right)\mathrm{d}s\right]\Theta\big(\nu(k,x)\big) \\
&= -p(k+1)\mu\left(\tfrac{1}{2}\nu(k,x)\right)\Theta\big(\nu(k,x)\big)\,,
\end{aligned}$$

where we used the facts that κ and κ' are non-decreasing. The lemma now follows from our choices of γ and χ. □

A.2 Some Useful Inequalities

We used the following simple lemma from [105] on p.120:

Lemma A.7. *For each continuous positive definite function $\rho : [0,\infty) \to [0,\infty)$, we can find a function $\omega \in \mathcal{K}_\infty \cap C^1$ and an increasing everywhere positive function $K \in C^1$ such that*

$$\rho(r) \geq \frac{\omega(r)}{K(r)} \tag{A.59}$$

for all $r \geq 0$.

Proof. We can assume that ρ is increasing on $[0,1]$ and non-increasing on $[1,\infty)$; otherwise, replace it with the minorizing function

$$\rho_{\text{new}}(r) = \begin{cases} r\min\{\rho(q) : r \leq q \leq 1\}, \; 0 \leq r \leq 1 \\ \min\{\rho(q) : 1 \leq q \leq r\}, \;\; r \geq 1 \end{cases} \tag{A.60}$$

without relabeling. Notice that

$$\rho(r) = \frac{\omega_0(r)}{K_0(r)} \tag{A.61}$$

for all $r \geq 0$, where

$$\omega_0(r) = \begin{cases} \dfrac{\rho(r)}{\rho(1)}, \; 0 \leq r \leq 1 \\ r, \quad\;\; r \geq 1 \end{cases} \quad \text{and} \;\; K_0(r) = \begin{cases} \dfrac{1}{\rho(1)}, \; 0 \leq r \leq 1, \\ \dfrac{r}{\rho(r)}, \; r \geq 1 \,. \end{cases} \tag{A.62}$$

We can then satisfy (A.59) by picking any function $\omega \in \mathcal{K}_\infty \cap C^1$ such that $\omega(r) \leq \omega_0(r)$ for all $r \geq 0$ and any increasing C^1 function K such that $K(r) \geq K_0(r)$ for all $r \geq 0$. This proves the result. □

We used the following simple lemma in Chap. 5:

Lemma A.8. *For all $A \in (-1,\infty)$, the inequalities*

$$\frac{e^A}{1+A} \geq \frac{1}{6}(1+A^2)\,, \tag{A.63}$$

$$A - \ln(1+A) \geq \frac{A^2}{2(1+|A|)}\,, \text{ and} \tag{A.64}$$

$$|A| \leq 2\sqrt{[A-\ln(1+A)] + [A-\ln(1+A)]^2} \tag{A.65}$$

are all satisfied.

Proof. To prove (A.63), first assume that $A \in (-1, 0)$. Then

$$\frac{e^A}{1+A} \geq e^{-1} \geq \frac{1}{6}(1+A^2). \tag{A.66}$$

If, on the other hand $A \geq 0$, then

$$\begin{aligned}\frac{e^A}{1+A} &\geq \frac{1+A+\frac{1}{2}A^2+\frac{1}{6}A^3}{1+A} \\ &\geq 1+\frac{\frac{1}{2}A^2+\frac{1}{6}A^3}{1+A} \\ &= 1+A^2\frac{\frac{1}{2}+\frac{1}{6}A}{1+A} \\ &\geq 1+A^2\frac{\frac{1}{6}+\frac{1}{6}A}{1+A} = 1+\frac{1}{6}A^2.\end{aligned} \tag{A.67}$$

This proves (A.63).

To prove (A.64), assume first that $A \in (-1, 0)$. Then

$$A - \ln(1+A) = \int_0^A \frac{m}{1+m} \mathrm{d}m \geq \frac{A^2}{2}.$$

If on the other hand $A \geq 0$, then

$$A - \ln(1+A) = \int_0^A \frac{m}{1+m} \mathrm{d}m \geq \int_0^A \frac{m}{1+A} \mathrm{d}m = \frac{A^2}{2(1+A)},$$

which gives (A.64).

To prove (A.65), notice that (A.64) implies that for all $A > -1$, we have

$$2[A - \ln(1+A)] + 2|A|[A - \ln(1+A)] \geq A^2. \tag{A.68}$$

Combining (A.68) with the inequality

$$2|A|[A - \ln(1+A)] \leq \frac{1}{2}A^2 + 2[A - \ln(1+A)]^2,$$

we deduce that

$$2\sqrt{[A - \ln(1+A)] + [A - \ln(1+A)]^2} \geq |A| \tag{A.69}$$

which proves (A.65). This completes the proof. □

A.3 A Lower Bound for the Lotka-Volterra Model

We sketch the proof that the Lotka-Volterra error dynamics (5.48) satisfies (5.52) for some constant $\underline{d} > 0$, hence also (5.35) from Assumption 5.1 for some positive function $\rho \in C^{\infty}$; see [105] for more details. We continue to use the notation of Sect. 5.5. Consider the function

$$E(p,q) \;=\; p - q\ln\left(1+\frac{p}{q}\right),$$

which is defined for $p > -q$ when $q > 0$. Then $V_1(\tilde{x},\tilde{y}) = E(\tilde{x},x_*) + E(\tilde{y},y_*)$. We claim that we can find a constant

$$\delta \in \left(0, \tfrac{1}{2}\min\{x_*,y_*\}\right] \tag{A.70}$$

so that

$$\begin{aligned}\sum_{i=1}^{2}\mathcal{N}_i(\tilde{x},\tilde{y}) &= \tfrac{1}{2}\tilde{x}^2 + \left[(\tilde{x}+\alpha\tilde{y})(\tilde{x}+x_*)\right]^2\\ &\geq \frac{\delta^3 V_1(\tilde{x},\tilde{y})}{1+V_1^2(\tilde{x},\tilde{y})}\end{aligned} \tag{A.71}$$

for all $(\tilde{x},\tilde{y})$ in the set $\mathcal{D} = \{(\tilde{x},\tilde{y}) \in \mathcal{X} : \tilde{x} \leq -x_* + \delta \text{ or } \tilde{y} \leq -y_* + \delta\}$.

To check this claim, first note that for any δ satisfying (A.70),

$$\sum_{i=1}^{2}\mathcal{N}_i(\tilde{x},\tilde{y})$$

is bounded from below on $\mathcal{D}$ by a positive constant $\underline{m}(\delta)$ depending on δ. (Indeed, if $\tilde{x} \leq -x_* + \delta$, then

$$\sum_{i=1}^{2}\mathcal{N}_i(\tilde{x},\tilde{y}) \;\geq\; \frac{1}{8}x_*^2.$$

If on the other hand $\tilde{x} \geq -x_* + \delta$, then

$$\sum_{i=1}^{2}\mathcal{N}_i(\tilde{x},\tilde{y}) \;\geq\; \frac{1}{2}\delta^2\tilde{x}^2 + \delta^2(\tilde{x}+\alpha\tilde{y})^2.$$

We can also find a constant $c_* \in (0,1)$ so that

$$\frac{1}{2}\tilde{x}^2 + (\tilde{x}+\alpha\tilde{y})^2 \geq c_*(\tilde{x}^2+\tilde{y}^2),$$

which is bounded below by $c_*y_*^2/4$ when $\tilde{y} \leq -y_* + \delta$. Hence,

$$\sum_{i=1}^{2} \mathcal{N}_i(\tilde{x}, \tilde{y}) \geq \delta^2 \frac{c_*}{8} \min\{x_*^2, y_*^2\} \doteq \underline{m}(\delta)$$

on $\mathcal{D}$.) Reducing $\delta > 0$ guarantees that

$$\frac{\delta^3 V_1(\tilde{x}, \tilde{y})}{1 + V_1^2(\tilde{x}, \tilde{y})} \leq \underline{m}(\delta)$$

on $\mathcal{D}$, as claimed. Fix $\delta > 0$ satisfying the preceding requirements.

We next consider points in $\mathcal{X} \setminus \mathcal{D}$. First notice that for each constant $q > 0$, we can find a constant $c(q) > 1$ such that

$$E(p, q) \leq c(q) p^2 \ \forall p \geq -q + \delta.$$

Therefore,

$$\begin{aligned} V_1(\tilde{x}, \tilde{y}) &\leq \tilde{x}^2 \left(\frac{E(\tilde{x}, x_*)}{\tilde{x}^2} \right) + \tilde{y}^2 \left(\frac{E(\tilde{y}, y_*)}{\tilde{y}^2} \right) \\ &\leq [c(x_*) + c(y_*)](\tilde{x}^2 + \tilde{y}^2) \end{aligned}$$

on $\mathcal{X} \setminus \mathcal{D}$ when neither $\tilde{x}$ nor $\tilde{y}$ is zero. Similar reasoning gives

$$V_1(\tilde{x}, \tilde{y}) \leq [c(x_*) + c(y_*)](\tilde{x}^2 + \tilde{y}^2)$$

on all of $\mathcal{X} \setminus \mathcal{D}$ (by separately considering points where $\tilde{x} = 0$ and $\tilde{x} \neq 0$). Moreover, we can find a constant $\underline{c} > 0$ so that

$$\sum_{i=1}^{2} \mathcal{N}_i(\tilde{x}, \tilde{y}) \geq \underline{c}(\tilde{x}^2 + \tilde{y}^2)$$

on $\mathcal{X} \setminus \mathcal{D}$. Hence,

$$\begin{aligned} \textstyle\sum_{i=1}^{2} \mathcal{N}_i(\tilde{x}, \tilde{y}) &\geq \left(\frac{\underline{c}}{c(x_*) + c(y_*)} \right) [c(x_*) + c(y_*)](\tilde{x}^2 + \tilde{y}^2) \\ &\geq \left(\frac{\underline{c}}{c(x_*) + c(y_*)} \right) V_1(\tilde{x}, \tilde{y}) \end{aligned}$$

on $\mathcal{X} \setminus \mathcal{D}$, so we can take

$$\rho(r) = \min \left\{ \frac{\underline{c}}{c(x_*) + c(y_*)}, \delta^3 \right\} \frac{1}{1 + r^2}.$$

We deduce that the Lotka-Volterra model satisfies Assumption 5.1, as claimed.

A.4 ISS and iISS for the Lotka-Volterra Model

For completeness, we summarize the robustness arguments from [105] that are needed to prove Theorem 5.3 on the ISS and iISS of the Lotka-Volterra error dynamics; see Section 5.5.2 for the notation we employ in the sequel. The proof involves showing that

$$\mathcal{U}_K(\tilde{x},\tilde{y}) = V_2(\tilde{x},\tilde{y}) + KV_1(\tilde{x},\tilde{y}) \tag{A.72}$$

is an iISS Lyapunov function for (5.62) when the disturbance $\mathbf{u}$ is valued in $\frac{d}{2}\mathcal{B}_1$, and that

$$\overline{\mathcal{U}_K}(\tilde{x},\tilde{y}) = \mathcal{U}_K(\tilde{x},\tilde{y})e^{\mathcal{U}_K(\tilde{x},\tilde{y})} \tag{A.73}$$

is an ISS Lyapunov function for (5.62) when $\mathbf{u}$ is valued in $\bar{U}\mathcal{B}_1$, where V_1 and V_2 are as defined in Sect. 5.5. The argument proceeds as follows.

Along the trajectories of (5.62) in $\mathcal{X}^b$, our choice $\mathcal{B} = 1 + \frac{2}{d} + y_*$ readily gives

$$\dot{V}_1 \le -\tilde{x}^2 + \mathcal{B}|\mathbf{u}|\,. \tag{A.74}$$

Also, since $V_2(\tilde{x},\tilde{y}) = x\tilde{x}[\tilde{x}+\alpha\tilde{y}]$, we get

$$\begin{aligned}\dot{V}_2 &= -(\tilde{x}+\alpha\tilde{y})^2x^2 + \big\{-\tilde{x}(2\tilde{x}+x_*) - \alpha\tilde{y}\tilde{x}\big\}(\tilde{x}+\alpha\tilde{y})x\\ &\quad +\tilde{x}[\alpha^2\tilde{x} - \alpha\mathbf{u}]xy.\end{aligned}$$

From the triangular inequality, we get

$$\begin{aligned}\dot{V}_2 &\le -\tfrac{1}{2}\big(\tilde{x}+\alpha\tilde{y}\big)^2x^2 + \tfrac{1}{2}\big\{\tilde{x}(2\tilde{x}+x_*) + \alpha\tilde{y}\tilde{x}\big\}^2\\ &\quad +\alpha^2\tilde{x}^2xy - \alpha\tilde{x}\mathbf{u}xy\,.\end{aligned} \tag{A.75}$$

Since $(x,y)\in\mathcal{S}$, we deduce that

$$\begin{aligned}\dot{V}_2 &\le -\tfrac{1}{2}(\tilde{x}+\alpha\tilde{y})^2x^2 + \tfrac{(3+\alpha)^2\mathcal{B}^2}{2}\tilde{x}^2 + \big[\alpha^2\tilde{x}^2 + \alpha\mathcal{B}|\mathbf{u}|\big]\mathcal{B}^2\\ &= -\tfrac{1}{2}(\tilde{x}+\alpha\tilde{y})^2x^2 + \left[\tfrac{(3+\alpha)^2}{2} + \alpha^2\right]\mathcal{B}^2\tilde{x}^2 + \alpha\mathcal{B}^3|\mathbf{u}|\,.\end{aligned} \tag{A.76}$$

On the other hand, according to (A.64) with the choices $A = \tilde{x}/x_*$ and then $A = \tilde{y}/y_*$, we have

$$\begin{aligned}\tilde{x} - x_*\ln\left(1+\tfrac{\tilde{x}}{x_*}\right) &\ge \tfrac{\tilde{x}^2}{2(2x_*+x)}\,,\\ \tilde{y} - y_*\ln\left(1+\tfrac{\tilde{y}}{y_*}\right) &\ge \tfrac{\tilde{y}^2}{2(2y_*+y)}\,,\end{aligned} \tag{A.77}$$

and therefore

$$V_1(\tilde{x},\tilde{y}) \ge \tfrac{\tilde{x}^2}{2(2x_*+x)} + \tfrac{\tilde{y}^2}{2(2y_*+y)} \;\ge\; \tfrac{\tilde{x}^2+\tilde{y}^2}{6\mathcal{B}} \tag{A.78}$$

for all $(x, y) \in \mathcal{S}$. Moreover, for all $(x, y) \in \mathcal{S}$, we have

$$\begin{aligned} |V_2(\tilde{x}, \tilde{y})| &\leq (\tilde{x}^2 + \alpha|\tilde{x}\tilde{y}|)\mathcal{B} \\ &\leq \mathcal{B}\left(\tfrac{3}{2}\tilde{x}^2 + \tfrac{\alpha^2}{2}\tilde{y}^2\right) \leq \mathcal{B}\max\{\tfrac{3,\alpha^2}{2}\}\left(\tilde{x}^2 + \tilde{y}^2\right), \end{aligned} \tag{A.79}$$

so for all $(x, y) \in \mathcal{S}$,

$$|V_2(\tilde{x}, \tilde{y})| \leq \mathcal{B}^2 \max\{9, 3\alpha^2\} V_1(\tilde{x}, \tilde{y}) \,. \tag{A.80}$$

Also,

$$\mathcal{U}_K(\tilde{x}, \tilde{y}) \geq \left[-\mathcal{B}^2 \max\{9, 3\alpha^2\} + K\right] V_1(\tilde{x}, \tilde{y}) \,, \tag{A.81}$$

and our choice of $K \geq K_0$ gives

$$\mathcal{U}_K(\tilde{x}, \tilde{y}) \geq \mathcal{B}^2 \max\left\{9, 3\alpha^2\right\} V_1(\tilde{x}, \tilde{y}) \tag{A.82}$$

and

$$\dot{\mathcal{U}}_K \leq -\mathcal{Q}(\tilde{x}, \tilde{y}) + \overline{\mathcal{B}}|\mathbf{u}|, \tag{A.83}$$

where $\overline{\mathcal{B}} = \alpha\mathcal{B}^3 + K\mathcal{B}$ and

$$\mathcal{Q}(\tilde{x}, \tilde{y}) = \frac{1}{2}(\tilde{x} + \alpha\tilde{y})^2 x^2 + \frac{K_0}{2}\tilde{x}^2 \,. \tag{A.84}$$

We consider two cases:

Case 1. $\mathcal{Q}(\tilde{x}, \tilde{y}) \geq \theta$. Then

$$\dot{\mathcal{U}}_K \leq -\theta \frac{\mathcal{U}_K(\tilde{x}, \tilde{y})}{1 + \mathcal{U}_K(\tilde{x}, \tilde{y})} + \overline{\mathcal{B}}|\mathbf{u}| \,. \tag{A.85}$$

Case 2. $\mathcal{Q}(\tilde{x}, \tilde{y}) \leq \theta$. Then

$$|\tilde{x}| \leq \sqrt{\frac{2}{K_0}\theta} \tag{A.86}$$

and therefore our choice of θ implies that

$$\frac{x_*}{2} \leq x \,. \tag{A.87}$$

Moreover,

$$|\tilde{x} + \alpha\tilde{y}|x \leq \sqrt{2\theta} \,. \tag{A.88}$$

One can also use (A.87) to show that

$$|\tilde{y}| \leq \frac{2}{x_*\alpha}\sqrt{2\theta} + \frac{1}{\alpha}|\tilde{x}| \,. \tag{A.89}$$

We deduce from (A.86) that

$$|\tilde{y}| \leq \frac{2}{x_*\alpha}\sqrt{2\theta} + \frac{1}{\alpha}\sqrt{\frac{2}{K_0}\theta} \leq \frac{y_*}{2}, \tag{A.90}$$

by our choice of θ.

Next, one can easily prove that for all $A \in \left[-\frac{1}{2}, \frac{1}{2}\right]$,

$$A - \ln(1 + A) \leq A^2 \tag{A.91}$$

and therefore, when $|\tilde{x}| \leq \frac{x_*}{2}$ and $|\tilde{y}| \leq \frac{y_*}{2}$, we have

$$V_1(\tilde{x}, \tilde{y}) \leq \frac{\tilde{x}^2}{x_*} + \frac{\tilde{y}^2}{y_*}. \tag{A.92}$$

Since the definition of $\mathcal{U}_K$ and (A.80) imply that

$$\mathcal{U}_K(\tilde{x}, \tilde{y}) \leq \left(K + \mathcal{B}^2 \max\left\{9, 3\alpha^2\right\}\right) V_1(\tilde{x}, \tilde{y}),$$

we get

$$\mathcal{U}_K(\tilde{x}, \tilde{y}) \leq \overline{K}\left[\frac{\tilde{x}^2}{x_*} + \frac{\tilde{y}^2}{y_*}\right], \tag{A.93}$$

where $\bar{K} = K + \mathcal{B}^2 \max\left\{9, 3\alpha^2\right\}$. Also, (A.87) implies that

$$\mathcal{Q}(\tilde{x}, \tilde{y}) \geq \frac{x_*^2}{8}(\tilde{x} + \alpha\tilde{y})^2 + \frac{K_0}{2}\tilde{x}^2. \tag{A.94}$$

By separately considering the possibilities $|\tilde{x}| \geq \frac{1}{4}\alpha|\tilde{y}|$ and $|\tilde{x}| \leq \frac{1}{4}\alpha|\tilde{y}|$ and noting that $K_0 \geq 9\mathcal{B}^2 \geq 9$, it follows that

$$\begin{aligned} \mathcal{Q}(\tilde{x}, \tilde{y}) &\geq \frac{x_*^2}{16}\alpha^2\tilde{y}^2 + 2\tilde{x}^2 \\ &\geq \min\left\{\frac{x_*^2}{16}\alpha^2 y_*, 2x_*\right\}\left[\frac{\tilde{x}^2}{x_*} + \frac{\tilde{y}^2}{y_*}\right]. \end{aligned} \tag{A.95}$$

Combining (A.93) and (A.94) yields

$$\mathcal{U}_K(\tilde{x}, \tilde{y}) \leq \overline{K}\frac{\mathcal{Q}(\tilde{x}, \tilde{y})}{\min\left\{2x_*, \frac{x_*^2}{16}\alpha^2 y_*\right\}}. \tag{A.96}$$

From (A.83), we deduce that $\dot{\mathcal{U}}_K \leq -\widehat{K}\mathcal{U}_K(\tilde{x}, \tilde{y}) + \overline{\mathcal{B}}|\mathbf{u}|$.

Hence, in both cases,

$$\dot{\mathcal{U}}_K \leq -\mho\frac{\mathcal{U}_K(\tilde{x}, \tilde{y})}{1 + \mathcal{U}_K(\tilde{x}, \tilde{y})} + \overline{\mathcal{B}}|\mathbf{u}|, \tag{A.97}$$

where $\mho = \min\{\widehat{K}, \theta\}$. This shows that $\mathcal{U}_K$ is an iISS Lyapunov function for the Lotka-Volterra errors dynamics (5.62) when the disturbance $\mathbf{u}$ satisfies the less stringent bound $|\mathbf{u}|_\infty \leq \frac{d}{2}$. To prove the ISS assertion, notice that

$$\begin{aligned}\dot{\overline{\mathcal{U}_K}} &= e^{\mathcal{U}_K(\tilde{x},\tilde{y})}\left[1+\mathcal{U}_K(\tilde{x},\tilde{y})\right]\dot{\mathcal{U}}_K \\ &\leq e^{\mathcal{U}_K(\tilde{x},\tilde{y})}\left[\{-\mho+\overline{\mathcal{B}}|\mathbf{u}|\}\,\mathcal{U}_K(\tilde{x},\tilde{y})+\overline{\mathcal{B}}|\mathbf{u}|\right] .\end{aligned} \tag{A.98}$$

Therefore, when $|\mathbf{u}|_\infty \leq \frac{\mho}{2\overline{\mathcal{B}}}$, we have

$$\begin{aligned}\dot{\overline{\mathcal{U}_K}} &\leq e^{\mathcal{U}_K(\tilde{x},\tilde{y})}\left[-\tfrac{\mho}{2}\mathcal{U}_K(\tilde{x},\tilde{y})+\overline{\mathcal{B}}|\mathbf{u}|\right] \\ &\leq -\tfrac{\mho}{2}\overline{\mathcal{U}_K}(\tilde{x},\tilde{y})+\overline{\mathcal{B}}|\mathbf{u}|\left[e^{\mathcal{U}_K(\tilde{x},\tilde{y})}-1\right]+\overline{\mathcal{B}}|\mathbf{u}| .\end{aligned} \tag{A.99}$$

Using the inequalities $e^a - 1 \leq ae^a$ and $\overline{\mathcal{B}}|\mathbf{u}| \leq \frac{\mho}{4}$, we therefore obtain

$$\dot{\overline{\mathcal{U}_K}} \leq -\tfrac{\mho}{4}\overline{\mathcal{U}_K}(\tilde{x},\tilde{y})+\overline{\mathcal{B}}|\mathbf{u}| . \tag{A.100}$$

The desired ISS inequality now follows from standard arguments. □

A.5 Useful Integral

For a given constant $\delta > 0$, let

$$I = \int_0^{\frac{\pi}{2}} \frac{1}{(\cos^2(a)+\delta)^2}\mathrm{d}a . \tag{A.101}$$

Then the double angle formula gives

$$I = 4\int_0^{\frac{\pi}{2}} \frac{1}{(\cos(2a)+1+2\delta)^2}\mathrm{d}a . \tag{A.102}$$

Set $\beta = 1+2\delta$. Then

$$\begin{aligned}I = 4\int_0^{\frac{\pi}{2}} \frac{1}{(\cos(2a)+\beta)^2}\mathrm{d}a &= 2\int_0^{\pi} \frac{1}{(\cos(r)+\beta)^2}\mathrm{d}r \\ = 2\int_0^{\frac{\pi}{2}} \frac{1}{(\cos(r)+\beta)^2}\mathrm{d}r &+ 2\int_{\frac{\pi}{2}}^{\pi} \frac{1}{(\cos(r)+\beta)^2}\mathrm{d}r \\ = 2\int_0^{\frac{\pi}{2}} \frac{1}{(\cos(r)+\beta)^2}\mathrm{d}r &+ 2\int_0^{\frac{\pi}{2}} \frac{1}{(-\cos(r)+\beta)^2}\mathrm{d}r ,\end{aligned} \tag{A.103}$$

where the last integral is from the relation $\cos(r) = -\cos(\pi - r)$ and the substitution $y = \pi - r$. Let

$$t = \tan\left(\frac{r}{2}\right).$$

Then $\cos(r) = \frac{1-t^2}{1+t^2}$, so (A.103) gives

$$\begin{aligned} I &\leq 2\int_0^1 \frac{1}{\left(\frac{1-t^2}{1+t^2}+\beta\right)^2}\frac{2}{1+t^2}\mathrm{d}t + 2\int_0^1 \frac{1}{\left(-\frac{1-t^2}{1+t^2}+\beta\right)^2}\frac{2}{1+t^2}\mathrm{d}t \\ &= 4\left[\int_0^1 \frac{1+t^2}{(1-t^2+(1+2\delta)(1+t^2))^2}\mathrm{d}t \right. \\ &\quad \left. + \int_0^1 \frac{1+t^2}{(-1+t^2+(1+2\delta)(1+t^2))^2}\mathrm{d}t\right] \end{aligned} \tag{A.104}$$

and therefore

$$I \leq \int_0^1 \frac{1+t^2}{(1+\delta+\delta t^2)^2}\mathrm{d}t + \int_0^1 \frac{1+t^2}{(\delta+(1+\delta)t^2)^2}\mathrm{d}t\ . \tag{A.105}$$

Let

$$s = \sqrt{\frac{\delta}{1+\delta}}t \quad \text{and} \quad m = \sqrt{\frac{1+\delta}{\delta}}t.$$

Then

$$\begin{aligned} I &\leq \int_0^{\sqrt{\frac{\delta}{1+\delta}}} \frac{1+\frac{1+\delta}{\delta}s^2}{(1+\delta+(1+\delta)s^2)^2}\sqrt{\frac{1+\delta}{\delta}}\mathrm{d}s \\ &\quad + \int_0^{\sqrt{\frac{1+\delta}{\delta}}} \frac{1+\frac{\delta}{1+\delta}m^2}{(\delta+\delta m^2)^2}\sqrt{\frac{\delta}{1+\delta}}\mathrm{d}m \\ &\leq \tfrac{1}{\delta^{3/2}(1+\delta)^{3/2}}\int_0^{\sqrt{\frac{\delta}{1+\delta}}} \frac{\delta+(1+\delta)s^2}{(1+s^2)^2}\mathrm{d}s \\ &\quad + \tfrac{1}{\delta^{3/2}(1+\delta)^{3/2}}\int_0^{\sqrt{\frac{1+\delta}{\delta}}} \frac{1+\delta+\delta m^2}{(1+m^2)^2}\mathrm{d}m \\ &= \tfrac{1}{\delta^{3/2}(1+\delta)^{3/2}}\int_0^{\sqrt{\frac{\delta}{1+\delta}}} \frac{\delta+\delta s^2}{(1+s^2)^2}\mathrm{d}s + \tfrac{1}{\delta^{3/2}(1+\delta)^{3/2}}\int_0^{\sqrt{\frac{\delta}{1+\delta}}} \frac{s^2}{(1+s^2)^2}\mathrm{d}s \\ &\quad + \tfrac{1}{\delta^{3/2}(1+\delta)^{3/2}}\int_0^{\sqrt{\frac{1+\delta}{\delta}}} \frac{1+\delta+(\delta+1)m^2}{(1+m^2)^2}dm \\ &\quad + \tfrac{1}{\delta^{3/2}(1+\delta)^{3/2}}\int_0^{\sqrt{\frac{1+\delta}{\delta}}} \frac{-m^2}{(1+m^2)^2}\mathrm{d}m\ . \end{aligned}$$

This implies that

$$I \leq \frac{1}{\delta^{1/2}(1+\delta)^{3/2}} \int_0^{\sqrt{\frac{\delta}{1+\delta}}} \frac{1}{1+s^2} \mathrm{d}s + \frac{1}{\delta^{3/2}(1+\delta)^{3/2}} \int_0^{\sqrt{\frac{\delta}{1+\delta}}} \frac{s^2}{(1+s^2)^2} \mathrm{d}s$$
$$+ \frac{1}{\delta^{3/2}(1+\delta)^{1/2}} \int_0^{\sqrt{\frac{1+\delta}{\delta}}} \frac{1}{1+m^2} \mathrm{d}m - \frac{1}{\delta^{3/2}(1+\delta)^{3/2}} \int_0^{\sqrt{\frac{1+\delta}{\delta}}} \frac{m^2}{(1+m^2)^2} \mathrm{d}m$$

and therefore also

$$I \leq \frac{1}{\delta^{1/2}(1+\delta)^{3/2}} \arctan\left(\sqrt{\frac{\delta}{1+\delta}}\right) + \frac{1}{\delta^{3/2}(1+\delta)^{1/2}} \arctan\left(\sqrt{\frac{1+\delta}{\delta}}\right)$$
$$+ \frac{1}{\delta^{3/2}(1+\delta)^{3/2}} \int_0^{\sqrt{\frac{\delta}{1+\delta}}} \frac{s^2}{(1+s^2)^2} \mathrm{d}s$$
$$- \frac{1}{\delta^{3/2}(1+\delta)^{3/2}} \int_0^{\sqrt{\frac{1+\delta}{\delta}}} \frac{m^2}{(1+m^2)^2} \mathrm{d}m\ .$$

One can easily prove that

$$\int_0^A \frac{s^2}{(1+s^2)^2} \mathrm{d}s = -\frac{A}{2(1+A^2)} + \tfrac{1}{2} \arctan(A) \quad \forall A \geq 0\ . \tag{A.106}$$

We deduce that

$$I \leq \frac{1}{\delta^{1/2}(1+\delta)^{3/2}} \arctan\left(\sqrt{\frac{\delta}{1+\delta}}\right) + \frac{1}{\delta^{3/2}(1+\delta)^{1/2}} \arctan\left(\sqrt{\frac{1+\delta}{\delta}}\right)$$
$$+ \frac{1}{\delta^{3/2}(1+\delta)^{3/2}} \left[-\frac{\sqrt{\frac{\delta}{1+\delta}}}{2\left(1+\frac{\delta}{1+\delta}\right)} + \tfrac{1}{2} \arctan\left(\sqrt{\frac{\delta}{1+\delta}}\right) \right]$$
$$- \frac{1}{\delta^{3/2}(1+\delta)^{3/2}} \left[-\frac{\sqrt{\frac{1+\delta}{\delta}}}{2\left(1+\frac{1+\delta}{\delta}\right)} + \frac{1}{2} \arctan\left(\sqrt{\frac{1+\delta}{\delta}}\right) \right]$$
$$= \frac{\arctan\left(\sqrt{\frac{\delta}{1+\delta}}\right)}{\delta^{1/2}(1+\delta)^{3/2}} + \frac{\arctan\left(\sqrt{\frac{1+\delta}{\delta}}\right)}{\delta^{3/2}(1+\delta)^{1/2}}$$
$$+ \frac{\arctan\left(\sqrt{\frac{\delta}{1+\delta}}\right)}{2\delta^{3/2}(1+\delta)^{3/2}} - \frac{\arctan\left(\sqrt{\frac{1+\delta}{\delta}}\right)}{2\delta^{3/2}(1+\delta)^{3/2}}$$
$$- \frac{1}{2\delta(1+\delta)^2} \frac{1}{\left(1+\frac{\delta}{1+\delta}\right)} + \frac{1}{2\delta^2(1+\delta)} \frac{1}{\left(1+\frac{1+\delta}{\delta}\right)}$$

and so also

$$I \le \arctan\left(\sqrt{\frac{\delta}{1+\delta}}\right)\left[\frac{1}{\delta^{1/2}(1+\delta)^{3/2}} + \frac{1}{2\delta^{3/2}(1+\delta)^{3/2}}\right]$$
$$+ \arctan\left(\sqrt{\frac{1+\delta}{\delta}}\right)\left[\frac{1}{\delta^{3/2}(1+\delta)^{1/2}} - \frac{1}{2\delta^{3/2}(1+\delta)^{3/2}}\right].$$

The preceding inequalities and the fact that $\arctan(q) + \arctan(1/q) \equiv \frac{\pi}{2}$ on $(0,\infty)$ now give

$$\begin{aligned} I &\le \tfrac{1}{\delta^{1/2}(1+\delta)^{1/2}}\Big\{\arctan\Big(\sqrt{\tfrac{\delta}{1+\delta}}\Big)\Big[\tfrac{1}{1+\delta} + \tfrac{1}{2\delta}\Big] \\ &\quad + \arctan\Big(\sqrt{\tfrac{1+\delta}{\delta}}\Big)\Big[\tfrac{1}{\delta} - \tfrac{1}{2\delta(1+\delta)}\Big]\Big\} \\ &= \tfrac{1}{\delta^{1/2}(1+\delta)^{1/2}}\Big[\arctan\Big(\sqrt{\tfrac{\delta}{1+\delta}}\Big)\tfrac{1}{1+\delta}\tfrac{1+3\delta}{2\delta} \\ &\quad + \arctan\Big(\sqrt{\tfrac{1+\delta}{\delta}}\Big)\tfrac{1}{\delta}\tfrac{1+2\delta}{2(1+\delta)}\Big] \\ &\le \frac{\pi(1+3\delta)}{4\delta^{3/2}(1+\delta)^{3/2}} \le \frac{3\pi}{4\delta^{3/2}}. \end{aligned} \tag{A.107}$$

A.6 Continuous Time Matrosov Result with PE

We give the continuous time Matrosov construction that we used in the proof of Theorem 12.4. We assume the following Matrosov Conditions:

Assumption A.1 *There exist C^1 functions $V_1 : [0,\infty)\times\mathbb{R}^n \to [0,\infty)$ of class $\mathcal{UPPD}$ and $V_2 : [0,\infty)\times\mathbb{R}^n \to \mathbb{R}$ of class $\mathcal{USB}$; $\phi_2 \in \mathcal{K}_\infty$; everywhere non-negative functions $N_1, N_2 \in \mathcal{USB}$; a function $\chi : [0,\infty)\times[0,\infty)\times\mathbb{R}^n \to \mathbb{R}$; an everywhere positive increasing function ϕ_1; $W \in \mathcal{PD}$; constants $\tau > 0$ and $\varepsilon > 0$; and $q \in \mathcal{P}_{\rm cts}(\tau,\varepsilon)$ such that*

$$\begin{aligned} &\mathcal{D}V_1(t,x) \le -N_1(t,x); \\ &\mathcal{D}V_2(t,x) \le -N_2(t,x) + \chi(N_1(t,x),t,x); \\ &|\chi(N_1(t,x),t,x)| \le \phi_1(|x|)\phi_2(N_1(t,x)); \quad \textit{and} \\ &N_1(t,x) + N_2(t,x) \ge q(t)W(x) \end{aligned}$$

hold for all $x \in \mathbb{R}^n$ and $t \in [0,\infty)$.

Notice that V_2 can take both positive and negative values. We show:

Lemma A.9. *If (12.9) satisfies Assumption A.1, then one can construct an explicit strict Lyapunov function for (12.9). In particular, (12.9) is UGAS.*

Proof. We indicate the changes needed in the proof of Theorem 12.3. We define V_3 and λ as in the earlier proof, giving

$$\begin{aligned}\mathcal{D}V_3(t,x) &\leq -q(t)W(x) + \phi_1(|x|)\phi_2\big(N_1(t,x)\big)\\ &\leq -q(t)\lambda\big(V_1(t,x)\big) + \phi_1(|x|)\phi_2\big(N_1(t,x)\big)\end{aligned}$$

everywhere. We define k_1, $\Lambda_1 \in \mathcal{K}_\infty$, and

$$V_4 \doteq k_1(V_1)V_3$$

as before and therefore can find everywhere positive increasing functions Γ and Λ_2 satisfying

$$\begin{aligned}\mathcal{D}V_4(t,x) \leq [-\mathcal{D}V_1(t,x)]\Gamma\big(V_1(t,x)\big) - q(t)\Lambda_1\big(V_1(t,x)\big)\\ +\Lambda_2\big(V_1(t,x)\big)\phi_2\big(N_1(t,x)\big),\end{aligned} \tag{A.108}$$

again by the previous argument. Choosing a function $k_2 \in \mathcal{K}_\infty$ such that

$$k_2' \geq \Gamma$$

everywhere, we get

$$\mathcal{D}(k_2 \circ V_1) = k_2'(V_1)\mathcal{D}V_1 \leq \Gamma(V_1)\mathcal{D}V_1,$$

since $\mathcal{D}V_1 \leq 0$ everywhere. It follows that

$$V_5 \doteq V_4 + k_2(V_1)$$

satisfies

$$\mathcal{D}V_5(t,x) \leq -q(t)\Lambda_1\big(V_1(t,x)\big) + \Lambda_2\big(V_1(t,x)\big)\phi_2\big(N_1(t,x)\big)\,. \tag{A.109}$$

Using the continuous time strictification approach from Chap. 6, the fact that $\Lambda_1 \in C^1 \cap \mathcal{K}_\infty$ implies that

$$V_6(t,x) \doteq V_5(t,x) + \frac{1}{\tau}\left[\int_{t-\tau}^{t}\int_{s}^{t} q(r)\,\mathrm{d}r\,\mathrm{d}s\right]\Lambda_1(V_1(t,x)) \tag{A.110}$$

satisfies

$$\mathcal{D}V_6(t,x) \leq -\frac{\varepsilon}{\tau}\Lambda_1\big(V_1(t,x)\big) + \Lambda_2\big(V_1(t,x)\big)\phi_2\big(N_1(t,x)\big).$$

The argument from the proof of Theorem 12.3 now provides a function $k_3 \in \mathcal{PD} \cap C^1$ such that

$$k_3(r) \le \phi_2^{-1}\left(\frac{\frac{\varepsilon}{T}\Lambda_1(r)}{1+\Lambda_2(r)}\right)\frac{1}{1+\Lambda_2(r)},$$

and therefore

$$\phi_2\big(k_3(V_1)\Lambda_2(V_1)\big)\Lambda_2(V_1) \le \frac{\varepsilon}{T}\Lambda_1(V_1)$$

everywhere. Let α_1 be the lower bound function in the $\mathcal{UPPD}$ requirement on V_1, and pick $\alpha_6 \in \mathcal{K}_\infty$ such that

$$\alpha_6(|x|) \ge |V_6(t,x)|$$

for all $x \in \mathbb{R}^n$ and $t \ge 0$. Choose $k_4 \in \mathcal{K}_\infty \cap C^1$ such that

$$k_4'(s) \ge |k_3'(s)|(\alpha_6 \circ \alpha_1^{-1})(s)$$

everywhere. Then

$$k_4'(V_1) \ge |k_3'(V_1)V_6|$$

everywhere. We conclude that the function

$$V_7 \doteq k_3(V_1)V_6 + k_4(V_1)$$

satisfies

$$\begin{aligned} \mathcal{D}V_7 &\le -\tfrac{\varepsilon}{T}k_3(V_1)\Lambda_1(V_1) + k_3(V_1)\Lambda_2(V_1)\phi_2(N_1) \\ &\quad -|k_3'(V_1)V_6|\mathcal{D}V_1 + k_4'(V_1)\mathcal{D}V_1 \\ &\le -\tfrac{\varepsilon}{T}k_3(V_1)\Lambda_1(V_1) + k_3(V_1)\Lambda_2(V_1)\phi_2(N_1) \end{aligned}$$

everywhere, since $\mathcal{D}V_1 \le 0$ everywhere. The conclusion of the argument is similar to the corresponding part of the proof of Theorem 12.3, except with Δ_k replaced by $\mathcal{D}$ and $\frac{\delta}{8(l+1)}$ replaced by $\frac{\varepsilon}{2T}$. □

Appendix B
Converse Theory

B.1 Converse Lyapunov Function Theorem

For completeness, we provide a sketch of the proof of Theorem 2.1, which builds strict Lyapunov functions for time-varying UGAS systems in terms of the flow map. We follow the argument from [70, Appendix C.7]. The proof relies on the following result which is known as Massera's Lemma:

Lemma B.1. *Assume that* $g : [0,\infty) \to (0,\infty)$ *is a continuous non-increasing function satisfying* $g(t) \to 0$ *as* $t \to +\infty$, *and* $h : [0,\infty) \to (0,\infty)$ *is continuous and non-decreasing. Then there exists a function* G *satisfying the following two conditions:*

1. $G, G' \in \mathcal{K}$; *and*
2. *there is a constant* $\bar{k} > 0$ *such that*

$$\max\left\{\int_0^\infty G(u(t))\mathrm{d}t, \int_0^\infty G'(u(t))h(t)\mathrm{d}t\right\} \le \bar{k} \tag{B.1}$$

for all continuous functions u *satisfying* $0 < u(t) \le g(t)$ *for all* $t \ge 0$.

Proof. Fix a sequence $\{t_n\}$ in $[1,\infty)$ such that

$$g(t_n) \le \frac{1}{n+1} \quad \forall n \in \mathbb{N}.$$

We show that the requirements of the lemma are met with

$$G(r) = \int_0^r H(s)\mathrm{d}s, \text{ where } H(s) = \begin{cases} \dfrac{\exp(-\eta^{-1}(s))}{h(\eta^{-1}(s))}, & s > 0 \\ 0, & s = 0 \end{cases}$$

and η is any decreasing function that satisfies the following conditions:

1. η is affine on (t_n, t_{n+1}) and $\eta(t_n) = 1/n$ for each n; and

2. $\eta(s) = (t_1/s)^p$ on $(0, t_1]$ where $p \in \mathbb{N}$ is large enough so that the one sided derivatives at t_1 satisfy $\eta'(t_1^-) < \eta'(t_1^+)$.

Notice that $g(r) < \eta(r)$ for all $r \in [t_1, \infty)$.

To simplify the notation, let U_g denote the set of all continuous functions u satisfying

$$0 \ < \ u(t) \ \le \ g(t) \ \ \forall t \ge 0.$$

Since η^{-1} is also decreasing, we have

$$\eta^{-1}(u(t)) \ge \eta^{-1}(g(t)) > \eta^{-1}(\eta(t)) = t \tag{B.2}$$

for all $t \ge t_1$ and $u \in U_g$. Moreover, $H \in \mathcal{K}$, and (B.2) in conjunction with the fact that h is non-decreasing implies

$$\int_{t_1}^{\infty} G'(u(t))h(t)\mathrm{d}t \ \le \ \int_{t_1}^{\infty} e^{-t}\mathrm{d}t \ \le \ 1.$$

Since we can also bound $G'(u(t)) \le G'(g(t))$ on $(0, t_1]$, we get a bound on the second integral in (B.1) that is uniform in $u \in U_g$. Similarly, since $-\eta^{-1}(s) \le -t$ when $0 < s \le \eta(t)$ and h is non-decreasing, we get

$$\begin{aligned}\int_{t_1}^{\infty} G(u(t))\mathrm{d}t &\le \int_{t_1}^{\infty}\int_0^{\eta(t)} \frac{\exp(-\eta^{-1}(s))}{h(0)}\,\mathrm{d}s\,\mathrm{d}t \\ &\le \int_{t_1}^{\infty} \frac{e^{-t}}{h(0)}\eta(t)\mathrm{d}t \ < \ \infty\end{aligned}$$

and therefore also a uniform bound on the first integral in (B.1). This proves the lemma. □

Returning to the proof of the theorem, we apply the preceding lemma with the choices

$$\begin{aligned} g(s) &= \beta(r_0, s) \ \text{ and} \\ h(s) &= \exp(Ls) + \sup\{|f(r,x)| : r \ge 0, |x| \le s\}, \end{aligned} \tag{B.3}$$

where the constant L is chosen so that

$$\sup\left\{\left|\frac{\partial f}{\partial x}(t,x)\right| : x \in r\mathcal{B}_n, t \ge 0\right\} \ \le \ L. \tag{B.4}$$

Letting G be the function that results from the lemma, we now show that

$$V(t,x) = \int_t^{\infty} G(|\phi(\tau,t,x)|)\ \mathrm{d}\tau \tag{B.5}$$

satisfies the estimates of the theorem, where ϕ is the flow map of the system. The choice (B.4) of L implies that

$$\left|\frac{\partial \phi}{\partial x}(\tau, t, x)\right| \leq \exp(L[\tau - t]) \quad \text{and} \quad |\phi(\tau, t, x)| \geq |x|\exp(-L[\tau - t]) \tag{B.6}$$

on $[0, \infty)^2 \times (r\mathcal{B}_n)$; see [70, Appendix C.7]. This gives the bound

$$\begin{aligned} \left|\tfrac{\partial V}{\partial x}(t, x)\right| &= \left|\int_t^\infty G'(|\phi(\tau, t, x)|)\frac{\phi^\top}{|\phi|}\phi_x \mathrm{d}\tau\right| \\ &\leq \int_t^\infty G'(\beta(|x|, \tau - t))\exp(L[\tau - t])\mathrm{d}\tau \\ &\leq \int_0^\infty G'(\beta(|x|, s))e^{Ls}\mathrm{d}s \doteq \alpha_4(|x|) \end{aligned} \tag{B.7}$$

when $0 < |x| \leq r_0\mathcal{B}_n$, as long as $x \neq 0$. This gives the growth requirement on V_x from the statement of the theorem.

Next notice that our choice of β gives

$$V(t, x) \leq \int_t^\infty G(\beta(|x|, \tau - t))\,\mathrm{d}\tau = \int_0^\infty G(\beta(|x|, s))\,\mathrm{d}s \doteq \alpha_2(|x|).$$

Also, the second inequality in (B.6) gives

$$\begin{aligned} V(t, x) &\geq \int_t^\infty G(|x|\exp(-L[\tau - t])\,\mathrm{d}\tau \\ &= \int_0^\infty G(|x|\exp(-Ls)\,\mathrm{d}s \\ &\geq \int_0^{(\ln 2)/L} G(0.5|x|)\mathrm{d}s \doteq \alpha_1(|x|). \end{aligned}$$

This gives $\alpha_1(|x|) \leq V(t, x) \leq \alpha_2(|x|)$ for all $(t, x) \in [0, \infty) \times (r_0\mathcal{B}_n)$, and $\alpha_1, \alpha_2 \in \mathcal{K}$, so the first requirement of the theorem is met. Finally, the variational equality gives $\phi_t(\tau, t, x) + \phi_x(\tau, t, x)f(t, x) \equiv 0$ for all $\tau \geq t$ and therefore

$$\frac{\partial V}{\partial t} + \frac{\partial V}{\partial x}f(t, x) = -G(|x|)$$

when $|x| \leq r_0$. The preceding construction can be done globally in the special case where $\partial f/\partial x$ is globally bounded. Since $\phi(\tau, t, x)$ agrees with the autonomous flow $\phi(\tau - t, x)$ when f is independent of t, the last assertion of the theorem follows as well. □

Remark B.1. We can choose V to be as smooth as desired. This follows from the regularization arguments from [12, Sect. 4.1.5]. However, C^1 Lyapunov functions normally suffice for feedback design.

B.2 Time-Varying Converse ISS Result

A well-known converse Lyapunov function result says that if a time-invariant system (satisfying appropriate assumptions) is ISS, then it admits an ISS Lyapunov function; the proof uses the abstract strict Lyapunov function construction from [82]. The fact that this is true for time periodic *time-varying* systems as well was announced in [39] and can be deduced from the following special case of [12, Theorem 4.5]:[1]

Lemma B.2. *Let* $\mathcal{F} : [0,\infty) \times \mathbb{R}^n \rightrightarrows \mathbb{R}^n$ *be continuous and compact and convex-valued. Assume that the differential inclusion*

$$\dot{x} \in \mathcal{F}(t,x) \tag{B.8}$$

is UGAS. Then for any constant $\lambda > 0$, *there exists a* C^∞ *storage function* $V : [0,\infty) \times \mathbb{R}^n \to [0,\infty)$ *such that*

$$\frac{\partial V}{\partial t}(t,x) + \frac{\partial V}{\partial x}(t,x)v \le -\lambda V(t,x) \tag{B.9}$$

for all $t \ge 0$, $x \in \mathbb{R}^n$, *and* $v \in \mathcal{F}(t,x)$. *If, in addition,* $\mathcal{F}$ *is periodic in* t *with some period* T, *then we can choose* V *in such a way that it also has period* T *in* t.

Let us show how the converse ISS result follows from Lemma B.2. For simplicity, we assume that f is Lipschitz. If f is ISS, then [169] provides $\chi \in \mathcal{K}_\infty$ such that the constrained input system

$$\dot{x} = f_\chi(t,x,d) \doteq f(t,x,d\chi^{-1}(|x|)), \quad |d| \le 1 \tag{B.10}$$

is UGAS; i.e., there exists $\beta \in \mathcal{KL}$ such that for each $t_0 \ge 0$ and $x_0 \in \mathbb{R}^n$ and each trajectory y of (B.10) satisfying $y(t_0) = x_0$, we have $|y(t_0+h)| \le \beta(|x_0|, h)$ for all $h \ge 0$. By minorizing χ^{-1}, we can assume it is C^1. This means the locally Lipschitz set-valued dynamics

$$F(t,x) = \{f(t,x,u) : \chi(|u|) \le |x|\}$$

is UGAS, as is its convexification $\overline{\text{co}}(F)$, namely $(t,x) \mapsto \overline{\text{co}}\{F(t,x)\}$ where $\overline{\text{co}}$ denotes the closed convex hull [12]. Since $\mathcal{F} \doteq \overline{\text{co}}(F)$ is continuous and compact and convex-valued, and since we are assuming that f is periodic in t, Lemma B.2 provides a time periodic $V \in$ UBPPD such that (B.9) holds

[1] By a solution of a differential inclusion (B.8), we mean an absolutely continuous function $\phi : \mathcal{I} \to \mathbb{R}^n$ defined on some nonempty interval $\mathcal{I} = [t_0, t_{\max})$ with the property that $\dot{\phi}(t) \in \mathcal{F}(t,\phi(t))$ for almost all $t \in \mathcal{I}$. We say that (B.8) is UGAS (to the origin) provided there is a function $\beta \in \mathcal{KL}$ such that for all initial times $t_0 \ge 0$, all nonempty intervals $[t_0, t_{\max})$, and all solutions $\phi : [t_0, t_{\max}) \to \mathbb{R}^n$ of (B.8), we have $|\phi(t)| \le \beta(|\phi(t_0)|, t - t_0)$ for all $t \in [t_0, t_{\max})$. See [12] for the definition of continuity of a set-valued map.

for all $x \in \mathbb{R}^n$, $t \geq 0$, and $v \in F(t,x)$ with $\lambda = 1$. In particular, we can find a function $\alpha_1 \in \mathcal{K}_\infty$ such that $V(t,x) \geq \alpha_1(|x|)$ everywhere. Recalling the definition of F, we therefore have

$$
\begin{aligned}
&|x| \geq \chi(|u|) \;\Rightarrow\; f(t,x,u) \in F(t,x)\\
&\Rightarrow\; \dot{V}(t,x,u) \leq -V(t,x) \leq -\alpha_1(|x|)
\end{aligned}
$$

for all $t \geq 0$, so V is the desired strict ISS Lyapunov function for f. This proves the result.

References

1. D. Aeyels and J. Peuteman. New criteria for exponential and uniform asymptotic stability of nonlinear time-variant differential equations. In *Proceedings of the IEEE Conference on Decision and Control*, pages 3614–3619. San Diego, CA, December 1997.
2. D. Aeyels and J. Peuteman. A new asymptotic stability criterion for nonlinear time-variant differential equations. *IEEE Transactions on Automatic Control*, 43(7):968–971, 1998.
3. D. Aeyels and J. Peuteman. On exponential stability of nonlinear time-varying differential equations. *Automatica*, 35(6):1091–1100, 1999.
4. B. Anderson. Exponential stability of linear equations arising in adaptive identification. *IEEE Transactions on Automatic Control*, AC-22(1):84–88, February 1977.
5. D. Angeli. Input-to-state stability of PD-controlled robotic systems. *Automatica*, 35(7):1285–1290, 1999.
6. D. Angeli, B. Ingalls, E.D. Sontag, and Y. Wang. Separation principles for input-output and integral-input to state stability. *SIAM Journal on Control and Optimization*, 43(1):256–276, 2004.
7. D. Angeli and E.D. Sontag. Forward completeness, unboundedness observability, and their Lyapunov characterizations. *Systems and Control Letters*, 38(4-5):209–217, 1999.
8. D. Angeli, E.D. Sontag, and Y. Wang. A characterization of integral input-to-state stability. *IEEE Transactions on Automatic Control*, 45(6):1082–1097, 2000.
9. M. Arcak, D. Angeli, and E.D. Sontag. A unifying integral ISS framework for stability of nonlinear cascades. *SIAM Journal on Control and Optimization*, 40(6):1888–1904, 2002.
10. Z. Artstein. Stabilization with relaxed controls. *Nonlinear Analysis: Theory Methods and Applications*, 7(11):1163–1173, 1983.
11. A. Bacciotti. *Local Stabilizability of Nonlinear Control Systems*, volume 8 of *Advances in Mathematics for Applied Sciences.* World Scientific, London, UK, 1992.
12. A. Bacciotti and L. Rosier. *Liapunov Functions and Stability in Control Theory, Second Edition.* Communications and Control Engineering Series. Springer-Verlag, Berlin, 2005.
13. E. Barbashin and N. Krasovski. On the stability of motion in the large. *Dokl. Akad. Nauk, SSSR*, 86:453–456, 1952.
14. G. Bastin and D. Dochain. *On-Line Estimation and Adaptive Control of Bioreactors.* Elsevier, Amsterdam, 1990.
15. M.S. Berger and M. Berger. *Perspectives in Nonlinearity: An Introduction to Nonlinear Analysis.* W.A. Benjamin, New York, 1968.

16. O. Bernard, Z. Hadj-Sadok, D. Dochain, A. Genovesi, and J.-P. Steyer. Dynamical model development and parameter identification for an anaerobic wastewater treatment process. *Biotechnology and Bioengineering*, 75(4):424–438, 2001.
17. N.P. Bhatia and G.P. Szegö. *Stability Theory of Dynamical Systems.* Springer-Verlag, New York, 1970.
18. R. Brockett. Asymptotic stability and feedback stabilization. In R.W. Brockett, R.S. Millman, and H.J. Sussmann, editors, *Differential Geometric Control Theory: Proceedings of the Conference Held at Michigan Technological University, June 28-July 2, 1982*, volume 27 of *Progress in Mathematics*, pages 181–191. Birkhäuser Boston, Boston, MA, 1983.
19. C. Byrnes and A. Isidori. New results and examples in nonlinear feedback stabilization. *Systems and Control Letters*, 12(5):437–442, 1989.
20. C. Cai, A. Teel, and R. Goebel. Converse Lyapunov theorems and robust asymptotic stability for hybrid systems. In *Proceedings of the American Control Conference*, pages 12–17. Portland, OR, June 2005.
21. C. Cai, A. Teel, and R. Goebel. Smooth Lyapunov functions for hybrid systems. Part 1: Existence is equivalent to robustness. *IEEE Transactions on Automatic Control*, 52(7):1264–1277, 2007.
22. F. Clarke, Y. Ledyaev, R. Stern, and P. Wolenski. *Nonsmooth Analysis and Control Theory*, volume 178 of *Graduate Texts in Mathematics.* Springer-Verlag, New York, 1998.
23. E. Coddington and N. Levinson. *Theory of Ordinary Differential Equations.* Krieger Publishing Company, Malabor, FL, 1984.
24. P. Collins. A trajectory-space approach to hybrid systems. In *Proceedings of the International Symposium on the Mathematical Theory of Networks and Systems.* Katholiek Univ. Leuven, Belgium, August 2004.
25. F. Colonius and W. Kliemann. *The Dynamics of Control.* Birkhäuser, Boston, 2000.
26. W. Coppel. *Dichotomies in Stability Theory*, volume 629 of *Lecture Notes in Mathematics.* Springer-Verlag, Berlin-New York, 1978.
27. J-M. Coron. Global asymptotic stabilization for controllable systems without drift. *Mathematics of Control, Signals, and Systems*, 5(3):295–312, 1992.
28. J-M. Coron. *Control and Nonlinearity*, volume 136 of *Mathematical Surveys and Monographs.* American Mathematical Society, Providence, RI, 2007.
29. J-M. Coron, B. d'Andréa Novel, and G. Bastin. A strict Lyapunov function for boundary control of hyperbolic systems of conservation laws. *IEEE Transactions on Automatic Control*, 52(1):2–11, 2007.
30. J-M. Coron and E-Y. Kerai. Explicit feedbacks stabilizing the attitude of a rigid spacecraft with two control torques. *Automatica*, 32(5):669–677, 1996.
31. J-M. Coron and L. Praly. Adding an integrator for the stabilization problem. *Systems and Control Letters*, 17(2):89–104, 1991.
32. B.T. Costic, S.P. Nagarkatti, D.M. Dawson, and M.S. de Queiroz. Autobalancing dcal controller for a rotating unbalanced disk. *Mechatronics - An International Journal*, 12(5):685–712, 2002.
33. P. Crouch. Spacecraft attitude control and stabilization: applications of geometric control theory to rigid body models. *IEEE Transactions on Automatic Control*, 29(4):321–331, 1984.
34. P. De Leenheer and D. Aeyels. Stabilization of positive linear systems. *Systems and Control Letters*, 44(4):259–271, 2001.
35. C. De Persis, R. De Santis, and A.S. Morse. Supervisory control with state-dependent dwell-time logic and constraints. *Automatica*, 40(2):269–275, 2004.
36. M. de Queiroz, D. Dawson, S. Nagarkatti, and F. Zhang. *Lyapunov-Based Control of Mechanical Systems.* Control Engineering. Birkhäuser Boston Inc., Boston, MA, 2000.
37. C. Desoer. Slowly varying system $\dot{x} = A(t)x$. *IEEE Transactions on Automatic Control*, AC-14:780–781, 1969.

38. D.K. Do, Z-P. Jiang, and J. Pan. On global tracking control of a VTOL aircraft without velocity measurement. *IEEE Transactions on Automatic Control*, 48(12):2212–2217, 2003.
39. H. Edwards, Y. Lin, and Y. Wang. On input-to-state stability for time-varying nonlinear systems. In *Proceedings of the IEEE Conference on Decision and Control*, pages 3501–3506. Sydney, Australia, December 2000.
40. L. Faubourg and J-B. Pomet. Strict control-Lyapunov functions for homogeneous Jurdjevic-Quinn type systems. In H. Huijberts, H. Nijmeijer, A. van der Schaft, and J. Scherpen, editors, *Nonlinear Control Systems Design Symposium (NOLCOS'98)*, pages 823–829. IFAC, 1998.
41. L. Faubourg and J-B. Pomet. Control Lyapunov functions for homogeneous "Jurdjevic-Quinn" systems. *ESAIM: Control, Optimisation and Calculus of Variations*, 5:293–311, 2000.
42. T.I. Fossen. *Guidance and Control of Ocean Vehicles*. John Wiley & Sons, Chichester, 1994.
43. R. Francisco, F. Mazenc, and S. Mondié. Global asymptotic stabilization of a PVTOL aircraft model with delay in the input. In J. Chiasson and J-J. Loiseau, editors, *Applications of Time Delay Systems*, volume 352 of *Lecture Notes in Control and Information Sciences*, pages 343–356. Springer Berlin, Heidelberg, Germany, 2007.
44. R. Freeman and L. Praly. Integrator backstepping for bounded controls and control rates. *IEEE Transactions on Automatic Control*, 43(2):258–262, 1998.
45. J-P. Gauthier. *Structure de Systèmes Non-Linéaires*. Éditions du CNRS, Paris, France, 1984.
46. J.M. Godhavn. Nonlinear tracking of underactuated surface vessels. In *Proceedings of the IEEE Conference on Decision and Control*, pages 987–991. Kobe, Japan, December 1996.
47. R. Goebel, J. Hespanha, A. Teel, C. Cai, and R. Sanfelice. Hybrid systems: generalized solutions and robust stability. In *Proceedings of the IFAC Symposium on Nonlinear Control Systems*. Stuttgart, Germany, September 2004.
48. R. Goebel and A. Teel. Solutions to hybrid inclusions via set and graphical convergence with stability theory applications. *Automatica*, 42(4):573–587, 2006.
49. J-L. Gouzé, A. Rapaport, and M.Z. Hadj-Sadok. Interval observers for uncertain biological systems. *Ecological Modeling*, 133(1):45–56, 2000.
50. L. Grüne, E.D. Sontag, and F. Wirth. Asymptotic stability equals exponential stability, and ISS equals finite energy gain—if you twist your eyes. *Systems and Control Letters*, 38(2):127–134, 1999.
51. K. Gu, V.L. Kharitonov, and J. Chen. *Stability of Time-Delay Systems*. Birkhäuser, Boston, 2002.
52. K. Gu and S-I. Niculescu. Stability analysis of time-delay systems: A Lyapunov approach. In A. Loría and F. Lamnabhi-Lagarrigue, editors, *Advanced Topics in Control Systems Theory. Lecture Notes from FAP 2005*, volume 328 of *Lecture Notes in Control and Information Sciences*, pages 139-170. Springer Berlin, Heidelberg, 2006.
53. J. Guckenheimer and P. Holmes. *Nonlinear Oscillations, Dynamical Systems, and Bifurcations of Vector Fields*. Springer-Verlag, New York, 1983.
54. W. Hahn. *Stability of Motion*. Springer-Verlag New York, Inc., New York, 1967.
55. J. Hale. *Ordinary Differential Equations, Second Edition*. Robert E. Krieger Publishing Co. Inc., Huntington, N.Y., 1980.
56. J. Hauser, S. Sastry, and G. Meyer. Nonlinear control design for slightly nonminimum phase systems: Application to V/STOL aircraft. *Automatica*, 28(4):665–679, 1992.
57. E. Heinzle, I.J. Dunn, and G.B. Ryhiner. Modeling and control for anaerobic wastewater treatment. In *Bioprocess Design and Control*, volume 48 of *Advances in Biochemical Engineering/Biotechnology*, pages 79–114. Springer Berlin, Heidelberg, Germany, 1993.
58. S. Hsu. Limiting behavior for competing species. *SIAM Journal on Applied Mathematics*, 34(4):760–763, 1978.

59. P. Ioannou and J. Sun. *Robust Adaptive Control.* Prentice Hall, Englewood Cliffs, NJ, 1996.
60. A. Isidori. *Nonlinear Control Systems. II.* Communications and Control Engineering Series. Springer-Verlag London Ltd., London, UK, 1999.
61. H. Ito. Scaling supply rates of iss systems for stability of feedback interconnected nonlinear systems. In *Proceedings of the IEEE Conference on Decision and Control*, pages 5074–5079. Maui, HI, December 2003.
62. M. Jankovic. Control Lyapunov-Razumikhin functions and robust stabilization of time delay systems. *IEEE Transactions on Automatic Control*, 46(7):1048–1060, 2001.
63. Z-P. Jiang. Global tracking control of underactuated ships by Lyapunov's direct method. *Automatica*, 38(2):301–309, 2002.
64. Z-P. Jiang, Y. Lin, and Y. Wang. Stabilization of time-varying nonlinear systems: A control Lyapunov function approach. In *Proceedings of the IEEE International Conference on Control and Automation*, pages 404–409. Guangzhou, China, 2007.
65. Z-P. Jiang and H. Nijmeijer. Tracking control of mobile robots: a case study in backstepping. *Automatica*, 33(7):1393–1399, 1997.
66. Z-P. Jiang and H. Nijmeijer. A recursive technique for tracking control of nonholonomic systems in chained form. *IEEE Transactions on Automatic Control*, 42(2):265–279, 1999.
67. Z-P. Jiang, E. Sontag, and Y. Wang. Input-to-state stability for discrete-time nonlinear systems. In *Proceedings of the 14th IFAC World Congress, Volume E*, pages 277–282. Beijing, China, 1999.
68. V. Jurdjevic and J.P. Quinn. Controllability and stability. *Journal of Differential Equations*, 28(3):381–389, 1978.
69. G. Kaliora and A. Astolfi. Nonlinear control of feedforward systems with bounded signals. *IEEE Transactions on Automatic Control*, 49(11):1975–1990, 2004.
70. H. Khalil. *Nonlinear Systems, Third Edition.* Prentice Hall, Upper Saddle River, NJ, 2002.
71. P. Kokotović, H. Khalil, and J. O'Reilly. *Singular Perturbation Methods in Control. Analysis and Design*, volume 25 of *Classics in Applied Mathematics.* Society for Industrial and Applied Mathematics, Philadelphia PA, 1999.
72. V.B. Kolmanovskii and A. Myshkis. *Introduction to the Theory and Applications of Functional Differential Equations.* Kluwer Academic Publishers, Dordrecht, The Netherlands, 1999.
73. M. Krichman, E.D. Sontag, and Y. Wang. Input-output-to-state stability. *SIAM Journal on Control and Optimization*, 39(6):1874–1928, 2001.
74. M. Krstić. Invariant manifolds and asymptotic properties of adaptive nonlinear stabilizers. *IEEE Transactions on Automatic Control*, 41(6):817–829, 1996.
75. M. Krstić, I. Kanellakopoulos, and P. Kokotovic. *Nonlinear and Adaptive Control Design.* Wiley, New York, 1995.
76. M. Krstić and A. Smyshlyaev. Backstepping boundary control for first-order hyperbolic PDEs and application to systems with actuator and senor delays. *Systems and Control Letters*, 57(9):750–758, 2008.
77. J. LaSalle. Stability theory for ordinary differential equations. *Journal of Differential Equations*, 4:57–65, 1968.
78. E. Lefeber. *Tracking Control of Nonlinear Mechanical Systems.* PhD Thesis. University of Twente, Enschede, The Netherlands, 2000.
79. A. Leung. Limiting behavior for several interacting populations. *Mathematical Biosciences*, 29(1-2):85–98, 1976.
80. F. Lin, W. Zhang, and R.D. Brandt. Robust hovering control of a PVTOL aircraft. *IEEE Transactions on Control Systems Technology*, 7(3):343–351, 1999.
81. Y. Lin and E. Sontag. Control-Lyapunov universal formulae for restricted inputs. *Control: Theory and Advanced Technology*, 10:1981–2004, 1995.

82. Y. Lin, E.D. Sontag, and Y. Wang. A smooth converse Lyapunov theorem for robust stability. *SIAM Journal on Control and Optimization*, 34(1):124–160, 1996.
83. E.N. Lorenz. Deterministic nonperiodic flow. *Journal of the Atmospheric Sciences*, 20(2):130–141, 1963.
84. A. Loría, R. Kelly, and A. Teel. Uniform parametric convergence in the adaptive control of mechanical systems. *European Journal of Control*, 11(2):87–100, 2005.
85. A. Loría and E. Panteley. Uniform exponential stability of linear time-varying systems: revisited. *Systems and Control Letters*, 47(1):13–24, 2002.
86. A. Loría, E. Panteley, D. Popović, and A. Teel. A nested Matrosov theorem and persistency of excitation for uniform convergence in stable nonautonomous systems. *IEEE Transactions on Automatic Control*, 50(2):183–198, 2005.
87. A. Loría, E. Panteley, and A. Teel. Integral characterizations of uniform asymptotic and exponential stability with applications. *Mathematics of Control, Signals, and Systems*, 15(3):177–201, 2002.
88. A. Loría and A. Zavala-Río. Adaptive tracking control of chaotic systems with applications to synchronisation. *IEEE Transactions on Circuits and Systems*, 49(9):2019–2029, 2007.
89. L. Mailleret, O. Bernard, and J-P. Steyer. Nonlinear adaptive control for bioreactors with unknown kinetics. *Automatica*, 40(8):1379–1385, August 2004.
90. M. Malisoff and F. Mazenc. Further constructions of strict Lyapunov functions for time-varying systems. In *Proceedings of the American Control Conference*, pages 1889-1894. Portland, OR, June 2005.
91. M. Malisoff and F. Mazenc. Further remarks on strict input-to-state stable Lyapunov functions for time-varying systems. *Automatica*, 41(11):1973–1978, 2005.
92. M. Malisoff and F. Mazenc. On control-Lyapunov functions for hybrid time-varying systems. In *Proceedings of the IEEE Conference on Decision and Control*, pages 3265–3270. San Diego, CA, December 2006.
93. M. Malisoff and F. Mazenc. Constructions of strict Lyapunov functions for discrete time and hybrid time-varying systems. *Nonlinear Analysis: Hybrid Systems*, 2(2):394–407, 2008.
94. M. Malisoff, L. Rifford, and E.D. Sontag. Global asymptotic controllability implies input-to-state stabilization. *SIAM Journal on Control and Optimization*, 42(6):2221–2238, 2004.
95. L. Marconi, A. Isidori, and A. Serrani. Autonomous vertical landing on an oscillating platform: an internal-model based approach. *Automatica*, 38(1):21–32, January 2002.
96. P. Martin, S. Devasia, and B. Paden. A different look at output tracking: control of a VTOL aircraft. *Automatica*, 32(1):101–107, 1996.
97. V. M. Matrosov. On the stability of motion. *Journal of Applied Mathematics and Mechanics*, 26:1337–1353, 1963.
98. F. Mazenc. Strict Lyapunov functions for time-varying systems. *Automatica*, 39(2):349–353, 2003.
99. F. Mazenc and S. Bowong. Backstepping with bounded feedbacks for time-varying systems. *SIAM Journal on Control and Optimization*, 43(3):856–871, 2004.
100. F. Mazenc, M. de Queiroz, and M. Malisoff. Uniform global asymptotic stability of adaptively controlled nonlinear systems via strict Lyapunov functions. In *Proceedings of the American Society of Mechanical Engineers (ASME) Dynamic Systems and Control Conference*. Ann Arbor, MI, October 2008.
101. F. Mazenc, M.S. de Queiroz, and M. Malisoff. Uniform global asymptotic stability of a class of adaptively controlled nonlinear systems. *IEEE Transactions on Automatic Control*, 54(5), 2009, to appear.
102. F. Mazenc and M. Malisoff. Further constructions of control-Lyapunov functions and stabilizing feedbacks for systems satisfying the Jurdjevic-Quinn conditions. *IEEE Transactions on Automatic Control*, 51(2):360–365, 2006.
103. F. Mazenc and M. Malisoff. Further results on Lyapunov functions for slowly time-varying systems. *Mathematics of Control, Signals, and Systems*, 19(1):1–21, 2007.

104. F. Mazenc and M. Malisoff. Lyapunov functions under LaSalle conditions with an application to Lotka-Volterra systems. In *Proceedings of the 2009 American Control Conference*. St. Louis, MO, June 2009, to appear.
105. F. Mazenc and M. Malisoff. Strict Lyapunov function constructions under LaSalle conditions with an application to Lotka-Volterra systems. Preprint.
106. F. Mazenc, M. Malisoff, and O. Bernard. Lyapunov functions and robustness analysis under Matrosov conditions with an application to biological systems. In *Proceedings of the American Control Conference*, pages 2933–2938. Seattle, WA, June 2008.
107. F. Mazenc, M. Malisoff, and P. De Leenheer. On the stability of periodic solutions in the perturbed chemostat. *Mathematical Biosciences and Engineering*, 4(2):319–338, 2007.
108. F. Mazenc, M. Malisoff, and M. de Queiroz. Further results on strict Lyapunov functions for rapidly time-varying nonlinear systems. *Automatica*, 42(10):1663–1671, 2006.
109. F. Mazenc, M. Malisoff, and Z. Lin. Further results on input-to-state stability for nonlinear systems with delayed feedbacks. *Automatica*, 44(9):2415–2421, 2008.
110. F. Mazenc and D. Nešić. Strong Lyapunov functions for systems satisfying the conditions of La Salle. *IEEE Transactions on Automatic Control*, 49(6):1026–1030, 2004.
111. F. Mazenc and D. Nešić. Lyapunov functions for time-varying systems satisfying generalized conditions of Matrosov theorem. *Mathematics of Control, Signals, and Systems*, 19(2):151–182, 2007.
112. F. Mazenc, K. Pettersen, and H. Nijmeijer. Global uniform asymptotic stabilization of an underactuated surface vessel. *IEEE Transactions on Automatic Control*, 47(10):1759–1762, 2002.
113. F. Mazenc and L. Praly. Adding integrations, saturated controls, and stabilization for feedforward systems. *IEEE Transactions on Automatic Control*, 41(11):1559–1578, 1996.
114. W. Michiels and S-I. Niculescu. *Stability and Stabilization of Time-delay Systems: An Eigenvalue-Based Approach*. Society for Industrial and Applied Mathematics, Philadelphia, PA, 2007.
115. J. Milnor. Differential topology. In *1964 Lectures on Modern Mathematics Volume II*, pages 165–183. Wiley, New York, 1964.
116. A. Morgan and K. Narendra. On the uniform asymptotic stability of certain linear nonautonomous differential equations. *SIAM Journal on Control and Optimization*, 15(1):5–24, 1977.
117. P. Morin and C. Samson. Time-varying exponential stabilization of a rigid spacecraft with two controls. *IEEE Transactions on Automatic Control*, 42(4):528–534, 1997.
118. P. Morin, C. Samson, J-B. Pomet, and Z-P. Jiang. Time-varying feedback stabilization of the attitude of a rigid spacecraft with two controls. *Systems and Control Letters*, 25(5):375–385, 1995.
119. K. Narendra and A. Annaswamy. Persistent excitation in adaptive systems. *International Journal of Control*, 45(1):127–160, 1987.
120. D. Nešić and A. Loría. On uniform asymptotic stability of time-varying parameterized discrete-time-cascades. *IEEE Transactions on Automatic Control*, 49(6):875–887, 2004.
121. D. Nešić and A. Teel. Changing supply functions in input-to-state stable systems: The discrete-time case. *IEEE Transactions on Automatic Control*, 46(6):960–962, 2001.
122. D. Nešić and A. Teel. Input-to-state stability for nonlinear time-varying systems via averaging. *Mathematics of Control Signals and Systems*, 14(3):257–280, 2001.
123. D. Nešić, A. Teel, and E.D. Sontag. Formulas relating $\mathcal{KL}$ stability estimates of discrete-time and sampled-data nonlinear systems. *Systems and Control Letters*, 38(1):49–60, 1999.

124. S-I. Niculescu. *Delay Effects on Stability: A Robust Control Approach*, volume 269 of *Lecture Notes in Control and Information Sciences*. Springer Berlin, Heidelberg, Germany, 2001.
125. S-I. Niculescu, J-M. Dion, L. Dugard, and H. Li. Stability of linear systems with delayed state: an LMI approach. *European Journal of Automation*, 31(Special Issue on Analysis and Control of Time-Delay Systems):955–970, 1997.
126. R. Outbib and G. Sallet. Stabilizability of the angular velocity of a rigid body revisited. *Systems and Control Letters*, 18(2):93–98, 1992.
127. R. Outbib and J. Vivalda. On the stabilization of smooth nonlinear systems. *IEEE Transactions on Automatic Control*, 44(1):200–202, 1999.
128. B. Paden and R. Panja. A globally asymptotically stable 'PD+' controller for robot manipulators. *International Journal of Control*, 47(6):1697-1712, 1988.
129. A. Palomino, P. Castillo, I. Fantoni, R. Lozano, and C. Pégard. Control strategy using vision for the stabilization of an experimental pvtol aircraft setup. In *Proceedings of the IEEE Conference on Decision and Control*, pages 4487–4492. Maui, HI, December 2003.
130. P. Pepe, I. Karafyllis, and Z-P. Jiang. A Lyapunov-Krasovskii methodology for ISS and iISS of time-delay systems. *Systems and Control Letters*, 55(12):1006–1014, 2006.
131. L. Perko. *Differential Equations and Dynamical Systems, Third Edition*, volume 7 of *Texts in Applied Mathematics*. Springer-Verlag, New York, 2001.
132. K.Y. Pettersen and O. Egeland. Exponential stabilization of an underactuated surface vessel. In *Proceedings of the IEEE Conference on Decision and Control*, pages 967–971. Kobe, Japan, December 1996.
133. K.Y. Pettersen and H. Nijmeijer. Global practical stabilization and tracking for an underactuated ship - a combined averaging and backstepping approach. In *Proceedings of the IFAC Conference on Systems Structure Control*, pages 59–64. Nantes, France, July 1998.
134. K.Y. Pettersen and H. Nijmeijer. Tracking control of an underactuated surface vessel. In *Proceedings of the IEEE Conference on Decision and Control*, pages 4561–4566. Tampa, FL, December 1998.
135. J. Peuteman and D. Aeyels. Averaging results and the study of uniform asymptotic stability of homogeneous differential equations that are not fast time-varying. *SIAM Journal on Control and Optimization*, 37(4):997–1010, 1999.
136. J. Peuteman and D. Aeyels. Exponential stability of nonlinear time-varying differential equaitons and partial averaging. *Mathematics of Control, Signals, and Systems*, 15(1):42–70, 2002.
137. J. Peuteman and D. Aeyels. Exponential stability of slowly time-varying nonlinear systems. *Mathematics of Control, Signals, and Systems*, 15(3):202–228, 2002.
138. L. Praly, B. d'Andréa Novel, and J-M. Coron. Lyapunov design of stabilizing controllers for cascaded systems. *IEEE Transactions on Automatic Control*, 36(10):1177–1181, 1991.
139. L. Praly and Y. Wang. Stabilization in spite of matched unmodeled dynamics and an equivalent definition of input-to-state stability. *Mathematics of Control, Signals, and Systems*, 9(1):1–33, 1996.
140. C. Prieur and E. Trélat. Robust optimal stabilization of the Brockett integrator via a hybrid feedback. *Mathematics of Control, Signals, and Systems*, 17(3):201–216, 2005.
141. L. Rifford. Semiconcave control-Lyapunov functions and stabilizing feedbacks. *SIAM Journal on Control and Optimization*, 41(3):659–681, 2002.
142. R.T. Rockafellar and R.J.B. Wets. *Variational Analysis*, volume 317 of *Grundlehren der Mathematischen Wissenschaften*. Springer-Verlag, Berlin, Germany, 1998.
143. H. Rosenbrock. The stability of time-dependent control systems. *Journal of Electronics and Control*, 15:73–80, July 1963.
144. O.E. Rössler. An equation for continuous chaos. *Physics Letters A*, 57(5):397–398, 1976.

145. N. Rouche and J. Mawhin. *Ordinary Differential Equations II: Stability and Periodical Solutions.* Pitman Publishing Ltd., London, UK, 1980.
146. C. Samson. Control of chained systems application to path following and time-varying point-stabilization of mobile robots. *IEEE Transactions on Automatic Control*, 40(1):64–77, January 1995.
147. S. Saperstone and J. Yorke. Controllability of linear oscillatory systems using positive controls. *SIAM Journal on Control*, 9(2):253–262, 1971.
148. S. Sastry. *Nonlinear Systems*, volume 10 of *Interdisciplinary Applied Mathematics.* Springer-Verlag, New York, 1999.
149. R. Sepulchre, M. Janković, and P. V. Kokotović. *Constructive Nonlinear Control.* Communications and Control Engineering Series. Springer-Verlag, Berlin, 1997.
150. J-J. Slotine and W. Li. Theoretical issues in adaptive manipulator control. In *Proceedings of the 5th Yale Workshop on Applied Adaptive Systems Theory*, pages 252–258. New Haven, CT, 1987.
151. J-J. Slotine and W. Li. Adaptive manipulator control: A case study. *IEEE Transactions on Automatic Control*, AC-33:995–1003, 1988.
152. J-J. Slotine and W. Li. *Applied Nonlinear Control.* Prentice Hall, Englewood Cliffs, NJ, 1991.
153. H. Smith and P. Waltman. *The Theory of the Chemostat.* Cambridge University Press, Cambridge, UK, 1995.
154. V. Solo. On the stability of slowly time-varying linear systems. *Mathematics of Control, Signals, and Systems*, 7(4):331–350, 1994.
155. E.D. Sontag. A Lyapunov-like characterization of asymptotic controllability. *SIAM Journal on Control and Optimization*, 21(3):462–471, 1983.
156. E.D. Sontag. Stabilizability, i/o stability and coprime factorizations. In *Proceedings of the IEEE Conference on Decision and Control*, pages 457–458. Austin, TX, December 1988.
157. E.D. Sontag. Smooth stabilization implies coprime factorization. *IEEE Transactions on Automatic Control*, 34(4):435–443, April 1989.
158. E.D. Sontag. A "universal" construction of Artstein's theorem on nonlinear stabilization. *Systems and Control Letters*, 13(2):117–123, 1989.
159. E.D. Sontag. Feedback stabilization of nonlinear systems. In M.A. Kaashoek, J.H. van Schuppen, and A.C.M. Ran, editors, *Robust Control of Linear Systems and Nonlinear Control. Proceedings of the International Symposium MTNS-89*, volume 4 of *Progress in Systems Control Theory*, pages 61–81. Birkhäuser Boston, Boston, MA, 1990.
160. E.D. Sontag. Comments on integral variants of ISS. *Systems and Control Letters*, 34(1-2):93–100, 1998.
161. E.D. Sontag. *Mathematical Control Theory, Second Edition*, volume 6 of *Texts in Applied Mathematics.* Springer-Verlag, New York, 1998.
162. E.D. Sontag. Clocks and insensitivity to small measurement errors. *ESAIM: Control, Optimisation, and Calculus of Variations*, 4:537–557, 1999.
163. E.D. Sontag. Stability and stabilization: discontinuities and the effect of disturbances. In F.H. Clarke and R.J. Stern, editors, *Nonlinear Analysis, Differential Equations and Control*, volume 528 of *NATO Science Series: Mathematical and Physical Sciences*, pages 551–598. Kluwer Academic Publishers, Dordrecht, 1999.
164. E.D. Sontag. The ISS philosophy as a unifying framework for stability-like behavior. In A. Isidori, F. Lamnabhi-Lagarrigue, and W. Respondek, editors, *Nonlinear Control in the Year 2000, Volume 2*, volume 259 of *Lecture Notes in Control and Information Sciences*, pages 443–467. Springer, London, UK, 2001.
165. E.D. Sontag. Input-to-state stability: Basic concepts and results. In P. Nistri and G. Stefani, editors, *Nonlinear and Optimal Control Theory. Lectures Given at the C.I.M.E. Summer School Held in Cetraro, Italy June 19–29, 2004*, volume 1932 of *Lecture Notes in Mathematics*, pages 163–220. Springer, Berlin, Germany, 2008.

166. E.D. Sontag and M. Krichman. An example of a GAS system which can be destabilized by an integrable perturbation. *IEEE Transactions on Automatic Control*, 48(6):1046–1049, 2003.
167. E.D. Sontag and A. Teel. Changing supply functions in input/state stable systems. *IEEE Transactions on Automatic Control*, 40(8):1476–1478, 1995.
168. E.D. Sontag and Y. Wang. New characterization of input-to-state stability. *IEEE Transactions on Automatic Control*, 41(9):1283–1294, 1995.
169. E.D. Sontag and Y. Wang. On characterizations of the input-to-state stability property. *Systems and Control Letters*, 24(5):351–359, 1995.
170. E.D. Sontag and Y. Wang. Notions of input to output stability. *Systems and Control Letters*, 38(4-5):235–248, 1999.
171. E.D. Sontag and Y. Wang. Lyapunov characterizations of input to output stability. *SIAM Journal on Control and Optimization*, 39(1):226–249, 2000.
172. J-P. Steyer, J-C. Bouvier, T. Conte, P. Gras, and P. Sousbie. Evaluation of a four year experience with a fully instrumented anaerobic digestion process. *Water Science and Technology*, 45(4-5):495–502, 2002.
173. H. Sussmann. Subanalytic sets and feedback control. *Journal of Differential Equations*, 31(1):31–52, 1979.
174. A. Teel. A nonlinear small gain theorem for the analysis of control systems with saturation. *IEEE Transactions on Automatic Control*, 41(9):1256–1270, 1996.
175. A. Teel. Connections between Razumikhin-type theorems and the ISS nonlinear small gain theorems. *IEEE Transactions on Automatic Control*, 43(7):960–964, 1998.
176. A. Teel. Robust hybrid control systems: An overview of some recent results. In C. Bonivento, A. Isidori, L. Marconi, and C. Rossi, editors, *Advances in Control Theory and Applications*, volume 353 of *Lecture Notes in Control and Information Sciences*, pages 279–302. Springer Berlin, Heidelberg, Germany, 2007.
177. A. Teel and J. Hespanha. Examples of GES systems that can be driven to infinity by arbitrarily small additive decaying exponentials. *IEEE Transactions on Automatic Control*, 49(8):1407–1410, 2004.
178. A. Teel, R.M. Murray, and G. Walsh. Nonholonomic control systems: from steering to stabilization with sinusoids. In *Proceedings of the IEEE Conference on Decision and Control*, pages 1603–1609. Tucson, AZ, December 1992.
179. J. Tsinias. Sufficient Lyapunov-like conditions for stabilization. *Mathematics of Control, Signals, and Systems*, 2(4):343–357, 1989.
180. J. Tsinias. Input to state stability properties of nonlinear systems and applications to bounded feedback stabilization using saturation. *ESAIM: Control, Optimisation and Calculus of Variations*, 2:57–85, 1997.
181. J. Tsinias. Backstepping design for time-varying systems and application to partial-static feedback and asymptotic tracking. *Systems and Control Letters*, 39(4):219–227, 2000.
182. J. Tsinias and I. Karafyllis. ISS property for time-varying systems and application to partial-static feedback stabilization and asymptotic tracking. *IEEE Transactions on Automatic Control*, 44(1):2179–2184, 1999.
183. A. van der Schaft. *L_2-Gain and Passivity Techniques in Nonlinear Control, Second Edition.* Communications and Control Engineering Series. Springer-Verlag London Ltd., London, UK, 2000.
184. V.I. Zubov. *Methods of A.M. Lyapunov and Their Application.* P. Noordhoff Ltd., Groningen, The Netherlands, 1964.

Index

MIX
Papier aus verantwortungsvollen Quellen
Paper from responsible sources
FSC® C105338

If you have any concerns about our products, you can contact us on
ProductSafety@springernature.com

In case Publisher is established outside the EU, the EU authorized representative is:
Springer Nature Customer Service Center GmbH
Europaplatz 3, 69115 Heidelberg, Germany

Printed by Libri Plureos GmbH
in Hamburg, Germany